THE ROUTLEDGE HANDBOOK OF MEMORY AND PLACE

This Handbook explores the latest cross-disciplinary research on the inter-relationship between memory studies, place, and identity.

In the works of dynamic memory, there is room for multiple stories, versions of the past and place understandings, and often resistance to mainstream narratives. Places may live on long after their physical destruction. This collection provides insights into the significant and diverse role memory plays in our understanding of the world around us, in a variety of spaces and temporalities, and through a variety of disciplinary and professional lenses. Many of the chapters in this Handbook explore place-making, its significance in everyday lives, and its loss. Processes of displacement, where people's place attachments are violently torn asunder, are also considered. Ranging from oral history to forensic anthropology, from folklore studies to cultural geographies and beyond, the chapters in this Handbook reveal multiple and often unexpected facets of the fascinating relationship between place and memory, from the individual to the collective.

This is a multi- and intra-disciplinary collection of the latest, most influential approaches to the interwoven and dynamic issues of place and memory. It will be of great use to researchers and academics working across Geography, Tourism, Heritage, Anthropology, Memory Studies, and Archaeology.

Sarah De Nardi is a Lecturer in Heritage and Tourism at Western Sydney University, Australia.

Hilary Orange is an Honorary Research Associate at the UCL Institute of Archaeology, specialising in the contemporary past, particularly on deindustrialisation and industrial heritage.

Steven High is a Professor of History at Concordia University's Centre for Oral History and Digital Storytelling, Canada.

Eerika Koskinen-Koivisto is an ethnologist and Emil Aaltonen research fellow at the Department of History and Ethnology, the University of Jyväskylä, Finland.

THE ROUTLEDGE HANDBOOK OF MEMORY AND PLACE

*Edited by Sarah De Nardi, Hilary Orange,
Steven High, and Eerika Koskinen-Koivisto*

LONDON AND NEW YORK

First published 2020
by Routledge
2 Park Square, Milton Park, Abingdon, Oxon OX14 4RN

and by Routledge
52 Vanderbilt Avenue, New York, NY 10017

Routledge is an imprint of the Taylor & Francis Group, an informa business

© 2020 selection and editorial matter, Sarah De Nardi, Hilary Orange, Steven High, and Eerika Koskinen-Koivisto; individual chapters, the contributors

The right of Sarah De Nardi, Hilary Orange, Steven High, and Eerika Koskinen-Koivisto to be identified as the authors of the editorial material, and of the authors for their individual chapters, has been asserted in accordance with sections 77 and 78 of the Copyright, Designs and Patents Act 1988.

All rights reserved. No part of this book may be reprinted or reproduced or utilised in any form or by any electronic, mechanical, or other means, now known or hereafter invented, including photocopying and recording, or in any information storage or retrieval system, without permission in writing from the publishers.

Trademark notice: Product or corporate names may be trademarks or registered trademarks, and are used only for identification and explanation without intent to infringe.

British Library Cataloguing-in-Publication Data
A catalogue record for this book is available from the British Library

Library of Congress Cataloging-in-Publication Data
A catalog record for this book has been requested

ISBN: 978-0-8153-8630-8 (hbk)
ISBN: 978-0-8153-5426-0 (ebk)

Typeset in Bembo
by Apex CoVantage, LLC

Printed and bound in Great Britain by
TJ International Ltd, Padstow, Cornwall

CONTENTS

List of illustrations x
About the editors xiii
List of contributors xiv
Acknowledgements xxiii

    Introduction 1
    Sarah De Nardi, Hilary Orange, Steven High, and Eerika Koskinen-Koivisto

PART I
Mobility **9**

    Introduction 9
    Sarah De Nardi

1 The restorative museum: understanding the work of memory at the Museum
   of Refugee Memory in Skala Loutron, Lesvos, Greece 13
   Andrea Witcomb and Alexandra Bounia

2 Urban heritage between silenced memories and 'rootless' inhabitants: the case
   of the Adriatic coast in Slovenia 22
   Katja Hrobat Virloget

3 Uncanny District Six: removals, remains, and deferred regeneration 31
   Sean Field

4 Colonial Complexity in the British Landscape: an African-centric autoethnography 42
   Shawn Sobers

5 Mapping memories of exile   52
   Sébastien Caquard, Emory Shaw, José Alavez, and Stefanie Dimitrovas

PART II
Difficult memories   **67**

   Introduction   67
   Sarah De Nardi and Eerika Koskinen-Koivisto

6 Memory and space: (Re)reading Halbwachs   69
   Sarah Gensburger

7 Remembering Belene Island: commemorating a site of violence   77
   Lilia Topouzova

8 The landscapes of death among the Selk'nams: place, mobility, memory,
   and forgetting   89
   Melisa A. Salerno

9 Forensic archaeology and the production of memorial sites: situating the mass
   grave in a wider memory landscape   99
   Layla Renshaw

10 Urban bombsites   109
   Gabriel Moshenska

PART III
Memoryscapes   **117**

   Introduction   117
   Sarah De Nardi and Steven High

11 When memoryscapes move: 'Comfort Women' memorials as transnational   119
   Jihwan Yoon and Derek H. Alderman

12 The spatiality of memoryscapes, public memory, and commemoration   129
   Anett Árvay and Kenneth Foote

13 Stó:lō memoryscapes as Indigenous ways of knowing: Stó:lō history from
   stone and fire   138
   Keith Thor Carlson with Naxaxalhts'i (Albert 'Sonny' McHalsie)

14 Pots, tunnels, and mountains: myth, memory, and landscape at Great Zimbabwe, Zimbabwe   148
    Ashton Sinamai

15 Learning by doing: memoryscape as an educational tool   158
    Toby Butler

PART IV
Industry   171

Introduction   171
Steven High and Hilary Orange

16 Post-industrial memoryscapes: combatting working-class erasure in North America and Europe   175
    Lachlan MacKinnon

17 Remembering spaces of work   185
    Emma Pleasant and Tim Strangleman

18 Memory and post-industrial landscapes in Govan (Scotland)   193
    Martin Conlon

19 'Hidden in plain sight': uncovering the gendered heritage of an industrial landscape   203
    Lucy Taksa

20 Remembered into place   214
    Jeff Benjamin

21 Thinking volumetrically about urban memory: the buried memories and networked remembrances of underground railways   223
    Samuel Merrill

PART V
The body   233

Introduction   233
Sarah De Nardi and Hilary Orange

22 Memorialising war: rethinking heritage and affect in the context of Pearl Harbor   237
    Emma Waterton

23 *Lieux de mémoire* through the senses: memory, state-sponsored history, and sensory experience 249
Shanti Sumartojo

24 Memory and the photological landscape 254
Dan Hicks

25 Walking, writing, reading place and memory 261
Ceri Morgan

26 Mnemonic mapping practices 268
Patrick Laviolette, Anu Printsmann, and Hannes Palang

27 Facilitating voicing and listening in the context of post-conflict performances of memory: The Colombian scenario 277
Luis C. Sotelo

PART VI
Shared traditions 287

Introduction 287
Sarah De Nardi and Hilary Orange

28 Folklore, politics, and place-making in Northern Ireland 291
Ray Cashman

29 Rewilding as heritage-making: new natural heritage and renewed memories in Portugal 305
Nadia Bartolini and Caitlin DeSilvey

30 Taste and memory in action: translating academic knowledge to public knowledge 315
C. Nadia Seremetakis

31 Foodshed as memoryscape: legacies of innovation and ambivalence in New England's agricultural economy 325
Cathy Stanton

32 Historicising historical re-enactment and urban heritagescapes: engaging with past and place through historical pageantry, c. 1900–1950s 334
Tanja Vahtikari

PART VII
Ritual   343

    Introduction   343
    Eerika Koskinen-Koivisto

33 'My death waits there among the flowers': popular music shrines in London as memory and remembrance   345
    Hilary Orange and Paul Graves-Brown

34 An ethnography of memory in the secret valleys of the Himalayas: sacred topographies of mind in two *Beyul* pilgrimages   357
    Hayley Saul

35 Cremation and contemporary churchyards   367
    Howard Williams and Elizabeth Williams

36 Ritual, place, and memory in ancient Rome   384
    Ana Mayorgas

37 Ritually recycling the landscape   392
    Ceri Houlbrook

38 Contested memory in the holy springs of Western Siberia   400
    Jeanmarie Rouhier-Willoughby

Index   *408*

ILLUSTRATIONS

Figures

1.1	Monument to the 1922 Greek refugees from Asia Minor, Epano Skala, Mytilene.	14
1.2	The chess set created by students and staff of the University of the Aegean, Mytilene.	15
3.1	A city view from the prior location of our Lemmington Terrace home.	32
3.2	District Six Museum floor map. See Constitution Street, below De Waal Drive.	34
3.3	On the steps of 63 Lemmington Terrace.	36
4.1	Front pages of *The Daily Record*, 1936 and 2017.	50
5.1	A screenshot of a map designed with Atlascine that combines the five Rwandan stories (each with its own colour) and differentiating places that were associated with violence (opaque) from places that were not (translucent) in the Lake Kivu region of East Africa.	54
5.2a	Screenshot of the ESRI Story Map associated to her life story. The map can be viewed at arcg.is/0S1HP1.	57
5.2b	Alexandra Philoctète (left) and Stefanie Dimitrovas (right) designing Alexandra's life story map.	58
5.3a	A screenshot of the final map at the global scale. A video recording of the map as it evolves over time with the recorded interview can be viewed online at vimeo.com/246040198.	58
5.3b	C.K. (left) and Emory Shaw (right) designing C.K.'s story map.	59
5.4a	Two screenshots of the 'Tree Map' viewable online at prezi.com/0pg3auer9jwc.	60
5.4b	Lilia Bitar (left) and Nasim Abaeian (right) working on their tree map.	61
5.5a	Three still frames of the Meghri Bakarian's life story made in collaboration with Khadija Baker. The video can be viewed at vimeo.com/244583982.	62
5.5b	Khadija Baker (left) and Meghri Bakarian (right) developing their project.	64
7.1	A secret police map of Belene Island, the site of the Belene camp and prison, from 1956. Site II, the location of the camp, is at the far left. The image is perseverved in the archive of Bulgaria's former Ministry of the Interior. AMVR, f.23, op.1, a.u.149.	79
7.2	The camp watchtower and the memorial cross and crescent at Site II of the former Belene camp, Belene, 2006. Photo stills from the documentary film *The Mosquito Problem and Other Stories* (2007).	82
7.3	Site II of the Belene camp and prison (1954–1955), rendered from memory in 1977.	85

7.4	Site II of the Belene camp and prison (1952–1960; 3,500 to 5,000 people) rendered from memory in 1977.	86
7.5	Description of the camp compound of Site II, the Belene camp and prison complex.	86
10.1	Boys digging an allotment on a bombsite in London, 1942.	112
11.1	Statue of Peace in Seoul, Korea is an important focal point for public memorial services in support of 'Comfort Women' and Wednesday Demonstrations against the Japanese government.	120
11.2	Comfort Women memorial in Bergen County, New Jersey quilted, discursively and spatially, with and alongside memorials to other human rights struggles.	125
12.1	Examples of site-specific memorials where the location of memorials is tied to particular places. These include (clockwise from top) a reconstructed barracks at the Recsk prison camp honouring victims of political repression during the communist period; grave poles in Budapest's Parcel 301 cemetery honouring victims of the 1956 Hungarian Uprising; and a memorial at Pákozd honouring one of the greatest victories of the Hungarian revolutionary army during the 1848–1849 War of Independence.	131
12.2	A view of central Budapest from Parliament to Freedom Square [*Szabadságtér*] indicating a few of the many important sites and memorials in this area. The spatial positioning and alignment of these memorials illustrates the process of symbolic accretion in which the positioning of one memorial leads to the siting of others to create assemblages of public art and architecture in public spaces.	133
14.1	The Great Enclosure at Great Zimbabwe.	150
14.2	The Allan Wilson crypt (1902) is still in its original place just outside the Site Museum at Great Zimbabwe.	151
14.3	*Pfuko yaKuvanji* one of the two vessels that are said to have 'walked' in the Great Zimbabwe landscape.	153
15.1	The Ports of Call public workshop programme.	159
15.2	A poster advertising the trails featuring Brenda Ridge outside the Asta Centre, Silvertown in East London.	162
15.3	The *Channel Heritage Way RIX Wiki*.	165
15.4	The *Brixton Munch* food-related memoryscape.	167
15.5	Occupy LSX Audio Trail web page.	168
18.1	The Fairfield cranes in 1999.	194
21.1	Attendees of Meier's annual memorial vigil lay offerings beneath his plaque on the twentieth anniversary of his death.	228
21.2	Visitors during an Aldwych open day in 2011.	230
22.1	Layout of the Pearl Harbor Memorial Complex.	238
22.2	The USS *Arizona* Memorial.	240
22.3	Looking back at the Pearl Harbor memorial complex from the USS *Bowfin*.	245
24.1	Evaluation trench for the A435 Alcester-Evesham Bypass, Warwickshire, May 1993. The excavator nearest to the camera is the author	256
24.2	Evaluation trench for the A435 Alcester-Evesham Bypass, Warwickshire, May 1993. The blurred excavator to the right is the author	257
28.1	Murals in the Bogside, Derry, as seen from the city walls.	294
28.2	*The Petrol Bomber*, Battle of the Bogside. Mural by the Bogside Artists.	294
28.3	Civil Rights Leader, Bernadette Devlin. Mural by the Bogside Artists.	294
28.4	British soldier. Mural by the Bogside Artists.	296
28.5	*The Rioter*. Mural by the Bogside Artists.	297
28.6	Mural of Bobby Sands.	298

28.7	Carrickanaltar mass rock.	300
28.8	Nationalist poster depicting an open-air mass.	300
28.9	Replica of the Drumawark cross in Co. Donegal.	301
28.10	The Free Derry gable, Bogside.	302
29.1	Map of the Côa Valley in Portugal.	306
32.1	Working-class history was represented by several units in the historical pageant staged as part of Helsinki's quadricentennial celebrations, 11 June 1950. The Civil War of 1918 between the Whites and the Reds, still a divisive experience within the post-war Finnish society, was not included amongst the pageant themes.	338
32.2	One of the displays depicting the Finlayson factory passing by the Finlayson factory gate at the 1954 historical pageant in Tampere.	340
33.1	Mural of Aladdin Sane/Bowie in Tunstall Road, Broxton. Mural painted by Jimmy C. in 2013.	347
33.2	Bronze bust of Marc Bolan, Barnes Common.	350
33.3	The tribute wall to Freddie Mercury, Earl's Court. The tributes were removed in November 2017.	351
33.4	The statue of Amy Winehouse by sculptor Scott Eaton, Stables Market, Camden.	352
33.5	The shrine to George Michael, Highgate.	353
34.1	Looking up the Dagam Namgo valley to the horseshoe of white peaks, the physical centre of the Beyul.	360
34.2	One of the footprints of Guru Rinpoche that occupies the Dagam Namgo Beyul, sitting atop a boulder that is intricately carved with 'om mani padme hum' prayers.	362
34.3	'Snake rock,' the engraved impression of a *naga* being, often known as 'rainbow bridges' between realms in the Beyul.	364
35.1	Cremation memorials beside the medieval church of St Peter's, Delamere (Cheshire)	371
35.2	Cremation memorials at the east side of the medieval church of St Mary's, Chirk, Wrexham.	372
35.3	Cremation memorials on the south side of the churchyard extension of St Mary's, Chirk, Wrexham.	373
35.4	Cremation memorials in defined plots at St James' Christleton (Cheshire).	374
35.5a and b	Cremation memorials beside paths at (a) St Dunawd's, Bangor on Dee (Wrexham), (b) St Chad's, Farndon (Cheshire)	375
35.6	Cremation graves outside the original churchyard wall in a graveyard extension, St Marcella's, Marchwiel (Wrexham).	376
35.7a and b	Cremation memorials beside churchyard walls at (a) St Boniface, Bunbury (Cheshire) and (b) St Mary's, Eccleston (Cheshire).	377
35.8	Cremation memorials in relation to churchyard extensions at St Peter's, Waverton (Cheshire).	378
35.9a and b	Cremation memorials at St Andrew's, Tarvin (Cheshire), (a) a garden of remembrance as a focal point of inhumation graves, and (b) cremation graves against the outside of the old churchyard wall.	379
37.1	The remains of the Isle Maree tree: coin-encrusted spars, propped up against each other.	397

Table

22.1	How does it make you feel?	242

ABOUT THE EDITORS

Sarah De Nardi is a Lecturer in Heritage and Tourism at Western Sydney University (Australia). Her research practice explores the ways people enact place through memory, the imagination and storytelling practices. She conducts participatory, community-led mapping and visualisation practices that channel sense of place from the perspective of local stakeholders, activists, and avocational scholars. She does community-centred fieldwork in Italy, the United Kingdom, Pakistan, and Australia. Her recent monograph *The Poetics of Conflict Experience: Materiality and Embodiment in Second World War Italy* (Routledge) traces the experiences and memories of individuals and communities 'caught up' in the Italian conflict from a multitude of interconnected emotional and sensory perspectives.

Steven High is a Professor of History at Concordia University's Centre for Oral History and Digital Storytelling. He is the author of *Oral History at the Crossroads: Sharing Stories of Displacement and Survival* (2015), *Going Public: The Art of Participatory Practice* (2017, with Liz Miller and Ted Little), and *The Deindustrialized World: Confronting Ruination in Postindustrial Places* (2017, with Lachlan MacKinnon and Andrew Perchard).

Eerika Koskinen-Koivisto is an ethnologist and Emil Aaltonen research fellow at the Department of History and Ethnology, the University of Jyväskylä. In her current research, she examines transnational destinies and family histories of the Second World War, analysing the ways in which displacement are transmitted and communicated within families through stories and objects of memory. Her other research interests include dark and difficult heritage, life stories, ethnographic methods, and nostalgia. Her main publications include the monography *Her Own Worth – Negotiations of Subjectivity in the Life Narrative of a Female Labourer* (2014) and the co-edited volume *Transnational Death* (2019).

Hilary Orange is an archaeologist specialising in the contemporary past, particularly on deindustrialisation and industrial heritage. She edited the book *Reanimating Industrial Spaces: Conducting Memory Work in Post-Industrial Spaces* (Routledge, 2015) and co-edited the volume *The Good, the Bad and the Unbuilt: Handling the Heritage of the Recent Past* (BAR, 2012). She is an Honorary Research Associate at UCL Institute of Archaeology and was, from 2016–2018, an Alexander von Humboldt Research Fellow based at the *Institut für soziale Bewegungen* (Ruhr-Universität Bochum). She has written on deindustrialisation, artificial light, Cornish heritage, and more recently on popular music (with Paul Graves-Brown).

CONTRIBUTORS

José Alavez is a PhD student at Concordia University's Geography, Urban, and Environmental Department. Drawing on new cartographic theories and practices such as Deep Mapping, his research focuses on studying the stories of individuals who experienced death in the context of migration. Additionally, Jose holds a BA in Human Geography from The Metropolitan Autonomous University in Mexico City and a master's degree in Geomatics from CONACYT's Research Centre of Geospatial Information. His previous academic work includes the study of everyday geographies and domestic spaces of the homeless in Mexico City and new approaches to link Arts and Humanities with Cartography.

Derek H. Alderman is Professor of Geography at the University of Tennessee and Past President of the American Association of Geographers. He is a cultural and historical geographer specialising in race, public memory, commemorative landscape formation, critical place name study, and heritage tourism. Alderman is the author of over 120 articles and book chapters, the co-author (with Owen Dwyer) of the award-winning book *Civil Rights Memorials and the Geography of Memory* (2008), and co-editor (with Reuben Rose-Redwood and Maoz Azaryahu) of *The Political Life of Urban Streetscapes: Naming, Politics, and Place* (2017).

Anett Árvay is a senior assistant professor and the head of the Hungarian Studies programs at the University of Szeged, Hungary. She gained her MA in Hungarian, English linguistics and literature, and Hungarian studies at the University of Szeged in 1996. She completed her doctoral studies in applied linguistics at the Eötvös Loránd University, Budapest, in 2007. She has published articles in the field of pragmatics, methodology, and cultural geography of Hungary. Currently she and Kenneth Foote are at work on a book project on Hungarian collective memory entitled *Contested Places, Contested Pasts: Public Memory and Commemoration in Contemporary Hungary*.

Nadia Bartolini is an Associate Research Fellow at the University of Exeter. Her research interests are memory and the built environment, cultural and natural heritage, spirituality, creative research methods, and public engagement. Her publications include 'Critical Urban Heritage: From Palimpsest to Brecciation' (*International Journal of Heritage Studies*, 2014); 'The Politics of Vibrant Matter: Consistency, Containment and the Concrete of Mussolini's Bunker' (*Journal of Material Culture*, 2015); 'The Place of Spirit: Modernity and the Geographies of Spirituality' (with Sara MacKian and Steve Pile, *Progress in Human Geography*, 2017); and *Spaces of Spirituality* (with Sara MacKian and Steve Pile, Routledge, 2018).

Contributors

Jeff Benjamin lives and respires in an eroded plateau formed by millions of years of deposition in an inland sea, near the southern portion of an ancient meteor strike whose centre is what is now called Panther Mountain near Boiceville, New York. He reads and writes about the industrial past – aided and assisted by the company of flickers, bluebirds, finches, milkweed, ermine, fox, mugwort, wood sorrel, and coyotes. Occasionally he also makes art and likes to write letters.

Alexandra Bounia currently directs the MA in Museum and Gallery Practice at UCL in Qatar. She is a Professor of Museology at the University of the Aegean in Greece. Her research interests are on the history, theory, and management of collections and museums; the interpretation of material culture; and contemporary collecting. Her most recent book is entitled *The Political Museum: Power, Conflict and Identity in Cyprus* (co-authored with Theopisti Stylianou-Lambert) (2016, Routledge).

Toby Butler is a heritage consultant, oral history trainer, and research fellow at Birkbeck, University of London. He has directed and worked on international oral history projects and is known for his work exploring how history and memory can be used to interpret places and their pasts. Toby has an interest in using place-based multi-media and has created oral history trails along the River Thames and in Victoria Park. He is the project director for the *Ports of Call* project, which has been working with community groups and artists around the docks of East London to map and historically interpret the area and more recently the *Bethnal Green Disaster Memorial Project*.

Sébastien Caquard is an associate professor in the department of Geography, Planning and Environment at Concordia University (Montréal). His research lies at the intersection between mapping, technologies, and the humanities. In his current research, he seeks to explore how technological maps can help to better understand the complex relationships that exist between places and narratives. Caquard has also participated in the production of multiple online mapping platforms and leads the development of Atlascine, an award-winning online mapping application dedicated to the mapping of stories for research purposes. Caquard is the founder and director of the Geomedia lab (http://geomedialab.org/), which was established in 2011 at Concordia University.

Keith Thor Carlson is Professor of History at the University of Saskatchewan where he holds the Research Chair in Indigenous and Community-Engaged History. He has authored five books, including the multi-award-winning *The Power of Place, The Problem of Time: Aboriginal identity and Historical Consciousness in the Cauldron of Colonialism* (University of Toronto Press, 2010), and edited or co-edited four additional books including *The Stó:lō-Coast Salish Historical Atlas* (University of Washington Press, 2001) and *Orality and Literacy: Reflections Across Disciplines* (University of Toronto Press, 2011).

Ray Cashman is a Professor of Folklore at Indiana University, director of the IU Folklore Institute, and editor of *The Journal of Folklore Research*. He is author of *Storytelling on the Northern Irish Border: Characters and Community* (2008) and *Packy Jim: Folklore and Worldview on the Irish Border* (2016). With Tom Mould and Pravina Shukla, he edited *The Individual and Tradition: Folkloristic Perspectives* (2011). He is co-editor of two Indiana University Press book series, *Encounters: Explorations in Folklore and Ethnomusicology* and *Irish Culture, Memory, and Place*. He is currently researching aspects of folklore and emplaced memory concerning the Irish Famine (1845–1852).

Martin Conlon is a PhD student at the University of Strathclyde, based in the Scottish Oral History Centre. His research focuses on the monumentality and materiality of post-industrial sites – particularly those in Glasgow and the surrounding areas – using memory practices to interrogate and better understand deindustrialisation processes. He has worked broadly across the heritage sector in Scotland with a range of organisations, including Historic Environment Scotland and Archaeology Scotland. He currently teaches History at the University of Strathclyde.

Caitlin DeSilvey is an Associate Professor of Cultural Geography at the University of Exeter, where she has been employed since 2007. Her research explores the cultural significance of material and environmental change, with a focus on heritage contexts. She has worked on a range of interdisciplinary projects, supported by funding from UK research councils (AHRC, EPSRC, NERC), the Royal Geographical Society, the Norwegian Research Council and the European Social Fund. Recent publications include *Anticipatory History* (2011, with Simon Naylor and Colin Sackett), *Visible Mending* (2013, with Steven Bond and James R. Ryan), and *Curated Decay: Heritage Beyond Saving* (2017).

Stefanie Dimitrovas graduated from Concordia University in 2015 with a BSc in Environmental Science. She was a research assistant for the FRQSC-funded project entitled *Pour une cartographie émotionelle de récits de vie*, a partnership with the Centre for Oral History and Digital Storytelling. For her honors thesis, Stefanie conducted a critical analysis of online story mapping applications using a life story recorded as part of the Montreal Life Stories Project as a case study. The critical analysis is also included in the Mappemonde publication *Story Maps & Co./État de l'art de la cartographie des récits sur Internet* and is meant as a guide for storytellers who want to map their stories in selecting which application to use.

Sean Field currently works at the Historical Studies Department of the University of Cape Town in South Africa. He has published in various international journals and anthologies, and his monograph: *Oral History, Community and Displacement: Imagining Memories in Post-Apartheid South Africa* (Palgrave Macmillan, 2012) received the US Oral History Association (OHA) annual book award. His current writing is a critical response to debates around trauma theories within oral historiography, and how such theories have a troubling public life in post-apartheid South Africa.

Kenneth Foote is a Professor of Geography at the University of Connecticut. He has served as president of the Association of American Geographers in 2010–2011 as well as President of the U.S. National Council for Geographic Education in 2006 and has previously taught at the University of Texas in Austin and the University of Colorado Boulder. Much of his work focuses on the social and geographical dynamics of public memory and commemoration, especially the response to violence as expressed in the landscapes of the United States and Europe. His major work in this area is *Shadowed Ground: America's Landscapes of Violence and Tragedy* (2003).

Sarah Gensburger is a sociologist of memory. She obtained her PhD from the Ecole des Hautes Etudes en Sciences Sociales (EHESS, 2006). Her dissertation on the process of remembrance, entitled 'Righteous among the Nations,' won the French Political Science Association prize for the best dissertation (Paris, 2007) as well as a special award from the Auschwitz Foundation (Brussels, 2007). She is currently a tenured researcher in social sciences at the French National Center for Scientific Research (CNRS). She published several books, among which are these two in English, *National Policy, Global Memory* (Berghahn Books, 2016) and *Witnessing the Robbing of the Jews* (Indiana University Press, 2015).

Paul Graves-Brown is an archaeologist specialising in the contemporary past. He edited the book *Matter, Materiality and Modern Culture* (Routledge, 2000) and co-edited *The Oxford Handbook of the Archaeology of the Contemporary World* (2013). He has written on car culture, the Kalashnikov AK47, shopping malls, popular music, containerisation of shipping, and also the privatisation of public spaces. He is a musician, a composer (of sorts), and (very occasionally) a producer.

Contributors

Dan Hicks is a Professor of Contemporary Archaeology at the University of Oxford and Curator of World Archaeology at the Pitt Rivers Museum. His recent books include *LANDE: the Calais 'Jungle' and Beyond* (with Sarah Mallet, Bristol University Press, 2019) and *Archaeology and Photography* (edited with Lesley Mcfadyen, Bloomsbury, 2019). Follow him on Twitter: @ProfDanHicks

Ceri Houlbrook is an Early Career Research Fellow in Folklore and History at the University of Hertfordshire. Her primary research interests are the ritual practices and popular beliefs of post-medieval Britain and contemporary folklore. She has published papers on such modern-day customs as coin-trees and love-locks and has recently published a monograph on the former: *The Magic of Coin-Trees from Religion to Recreation: The roots of a ritual (2018)*, with Palgrave Macmillan. Co-edited works include *The Materiality of Magic (2015)* and a special issue on 'Cataloguing Magic' for the *Journal of Material Religion*. She is currently developing a Masters in Folklore Studies at the University of Hertfordshire.

Katja Hrobat Virloget is a research fellow and assistant professor at the Faculty of Humanities University of Primorska, where she teaches in the field of anthropology/ethnology, heritage, heritage tourism. Her research fields are anthropology of memory, heritage, nationalism, identity, folklore, ethnogeography, mythical landscape, and the connection between archaeology and ethnology. She is the head of the Institute for Intercultural studies UP FHS. As partner she has lead the European project *Heroes we love – Ideology, Identity and Socialist Art in New Europe* with the Art Gallery Maribor (Creative Europe) and currently the Slovenian-Croatian project of Mythical park (Interreg).

Patrick Laviolette (PhD UCL) from 2015 to 2019 has been the co-editor of EASA's journal *Social Anthropology/Anthropologie Sociale* and is the incoming editor of the *Anthropological Journal of European Cultures* from 2019 to 2023. He is the author of *The Landscaping of Metaphor and Cultural Identity* (Peter Lang, 2011) and *Extreme Landscapes of Leisure: Not a Hap-Hazardous Sport* (Ashgate, 2011; re-issued by Routledge in 2016). In 2013 he co-edited the volume *Things in Culture – Culture in Things* (Tartu University Press). Currently he is completing a monograph provisionally entitled *The Hitchhiking Diaries*.

Lachlan MacKinnon is an Assistant Professor of History at Cape Breton University. He is a co-editor of *The Deindustrialized World: Confronting Ruination in Post-industrial Places* (2017) and has published extensively on issues relating to the economic, social, and cultural ramifications of deindustrialisation. His research has appeared in journals such as *Labour/Le Travail* and *Labor: Studies in Working Class History*. His current research interrogates the transnational intersections of regional economic planning and deindustrialisation in Atlantic Canada and the Scottish Highlands.

Ana Mayorgas is Associate Professor of Ancient History at the Complutense University of Madrid. Previously she was a Fulbright postdoctoral fellow at the University of California, Berkeley (2006–2008), and a postdoctoral researcher at the University Carlos III of Madrid (2009–2011). She specialises in Roman history and her research interests include memory and historiography, culture and identity, and education, orality, and literacy in ancient societies. She is the author of *La memoria de Roma: oralidad, escritura y memoria en la República romana* (2007) and *Arqueología de la palabra. Oralidad y escritura en el mundo antiguo* (2010).

Samuel Merrill is an interdisciplinary researcher currently specialising in digital sociology. His research interests centre on social movements, collective memory, cultural heritage, and digital media with respect to a broadly conceived underground (spatial, political, creative, and technological). Based at Umeå University's Digital Social Research Unit in Northern Sweden, he has a doctorate in Cultural Geography from University College London, United Kingdom. His PhD thesis won first prize in the Peter Lang 2014 Young

Scholar in Memory Studies Competition and was published as *Networked Remembrance: Excavating Buried Memories in the Railways Beneath London and Berlin* in 2017.

Ceri Morgan is a senior lecturer at Keele University, United Kingdom. She works on literary geographies in Québec fiction, place writing, geopoetics, walking studies, and GeoHumanities. Recent projects include a digital map of literary Montreal entitled, 'Fictional Montreal/Montréal fictif' (British Academy small research grant 2016–2017), produced with media artist, Philip Lichti; and *Seams* (2018) – a show about coal mining developed and performed with participatory performance company, Restoke, a walking-writing group at Keele University called the Dawdlers, and community participants. Morgan is currently working on a project on walking and chronic pain with screendance artist, Anna Macdonald.

Gabriel Moshenska is an Associate Professor of Public Archaeology at UCL Institute of Archaeology. He teaches conflict archaeology, public archaeology, and the archaeology of the modern world. His many and varied research interests include the intellectual history of nineteenth-century archaeology, material cultures of childhood in conflict zones, antiquarian themes in supernatural fiction, the archaeology and heritage of the Mau Mau Emergency, and the archaeology of climate change in medieval Asturias, Spain.

Naxaxalhts'i (Dr. Albert 'Sonny' McHalsie) is a nationally acclaimed historical researcher and cultural interpreter based at the Stó:lō Research and Resource Management Centre in Chilliwack B.C. where he additionally serves as the Cultural Advisor to the Stó:lō Xwexwilmexw Treaty Association. Sonny has a strong publication record, including most recently co-editing *Towards a New Ethnohistory* (University of Manitoba Press, 2018). The guiding principal of his research is the ancient Stó:lō tradition of 'being of good mind.' His areas of expertise include Stó:lō place names, legendary narratives, fishing, and oral history. He is a member of the Shxw'ōwhamel First Nation, and is a proud father and grandfather.

Hannes Palang is a geographer and former president of the Permanent European Conference for the Study of the Rural Landscape (2006–2014). He is the co-editor of *Landscape Interfaces* (Kluwer, 2003), *European Rural Landscapes: Persistence and Change in a Globalising Environment* (Kluwer, 2004), and *Landscape and Seasonality* (Kluwer, 2007). He has published extensively on landscape change and perception.

Emma Pleasant is studying for a PhD in Sociology at the School of Social Policy, Sociology and Social Research (SSPSSR) at the University of Kent, Canterbury. Her current research focuses on the impact of deindustrialisation on occupational cultures and working-class communities. Emma's research interests include working-class studies, deindustrialisation, the sociology of work, memory studies, heritage, and oral history. She is a Co-Chair of the Memory Studies Association working group on 'Work, Class and Memory.' Emma is also an active memory of the Working-Class Studies Association as Co-Manager of the social media accounts.

Anu Printsmann (*MSc*) is a cartographer, a human and landscape geographer working currently as a researcher for the Centre for Landscape and Culture in School of Humanities at Tallinn University. She has been the board member of Estonian Geographical Society (2010–2015). She has co-edited a book *Landscape and Seasonality* (Kluwer, 2007) and special issues for *European Countryside* (2010 and 2012) and *Norsk Geografisk Tidsskrift – Norwegian Journal of Geography* (2017). Her main research interests include industrial landscape, heritage, perception, planning, and life stories.

Layla Renshaw is an Associate Professor in the Department of Applied and Human Sciences, Kingston University, London, where she teaches forensic archaeology and anthropology. She is the author of

Exhuming Loss: Memory, Materiality and Mass Graves of the Spanish Civil War. Her research interests focus on the relationship between human remains and traumatic memory, and the public perception of forensics. Her current work concerns the recovery of war dead from post-colonial contexts, the representation of contemporary migrant deaths, and the identification of First World War soldiers at Fromelles, exploring the link between genetic testing and memory.

Jeanmarie Rouhier-Willoughby is a Professor of Russian Studies and Folklore at the University of Kentucky. Her research focuses on legends and the religious revival in post-socialist Russia. She is editor of the journal *Folklorica* and is currently president of the International Society for Contemporary Legend Research.

Melisa A. Salerno is a researcher at the Multidisciplinary Institute of History and Human Sciences at the National Council for Scientific and Technical Research (CONICET, Argentina). She completed her PhD in archaeology at the University of Buenos Aires. She is interested in historical archaeology, silenced groups, bodily experience, and identities. She has edited several books, including *Memories from Darkness: Archaeology of Repression and Resistance in Latin America*, with Pedro Funari and Andrés Zarankin (2009); and *Coming to Senses: Topics in Sensory Archaeology*, with Andrés Zarankin and Roberto Pellini (2015).

Hayley Saul is a Senior Lecturer at Western Sydney University and Director of the Himalayan Exploration and Archaeological Research Team (H.E.A.R.T). Following her PhD at the University of York she undertook AHRC funded postdoctoral research with the Early Pottery in East Asia Project and was a Japan Society for the Promotion of Science fellow on the Japanese Archaeo-Ceramic Residue Research Strategy project (JARRS), to investigate cuisine in prehistoric hunter-gatherer groups across Eurasia. Her current research project, 'The Chaturale Museum of Cuisine' with a community on the outskirts of Kathmandu, draws together Hayley's interests in culinary heritage and mountain archaeology.

C. Nadia Seremetakis is Professor of Cultural Anthropology at the University of the Peloponnese, Greece. She has been actively engaged in public anthropology via public lectures, media presentations, and the design and organisation of public cultural programs and events in two continents. She has been the subject of a television documentary film herself. An invited member of the National Committee for the Intangible Cultural Heritage, Unesco-Hellenic Ministry of Culture since 2012, she has also served as temporary advisor to W.H.O. (Europe), advisor to the Hellenic Minister of Health, and publishing director of the series *Everyday Life and Culture* in Athens.

Emory Shaw completed a BA in Urban Systems at McGill University and an MSc in Geography at Concordia University. His Master's thesis explored the many ways of studying and representing urban spaces on social media. Since 2015, Emory also remained an active member of the Geomedia Lab at Concordia, facilitating mapping workshops and teaching courses on GIS and the Geoweb, as well as researching on methods for mapping life stories and contributing to the development of an online story mapping research tool.

Ashton Sinamai is a Zimbabwean archaeologist with a PhD in Cultural Heritage and Museum Studies from Deakin University, Australia. He worked as an archaeologist in Zimbabwe and Chief Curator (Archaeology) in Namibia and has worked as an ICCROM resource person in Zimbabwe, Botswana, Kenya, Sudan, and Lesotho. He has also taught Archaeology and Cultural Heritage Studies at the Midlands State University in Zimbabwe. He is currently a Marie Curie Incoming International Fellow at the University of York, United Kingdom, and an Adjunct Research Fellow at Flinders University, Australia. His research focus is on memory studies, landscape narratives, and how these can be used to map forgotten and abbreviated cultural landscapes.

Contributors

Shawn Sobers is an Associate Professor of Lens Based Media at University of the West of England (UWE), Bristol. He is director of the Bristol Photography Research Group, and the UWE Equity Research Group. He trained as a Social Anthropologist at School of Oriental and African Studies (SOAS), London, and has carried out a wide range of research projects spanning diverse topics, ranging from legacies of the slavery, African presence in Georgian and Victorian Britain, disability and walking, Rastafari language and culture, creative citizenship in social media, Ethiopian connections with the city of Bath, and how artists have responded to a statue of a slave trader in the city of Bristol.

Luis C. Sotelo is the Canada Research Chair in Oral History Performance and Associate Professor in the Department of Theatre at Concordia University (Montreal, Quebec, Canada). With support from the Canada Foundation for Innovation, he is establishing an *Acts of Listening Lab* at Concordia University's Centre for Oral History and Digital Storytelling. The aim of the Lab is to support the production, standardised collection, storage, classification, and analysis of data on listening in the context of participatory Oral History Performance events.

Cathy Stanton is a Senior Lecturer in Anthropology at Tufts University and an active public humanist. She is the author of *The Lowell Experiment: Public History in a Postindustrial City* (2006) and co-author, with Michelle Moon, of *Public History and the Food System: Adding the Missing Ingredient* (2018). Her research has encompassed historical reenactment, culture-led redevelopment in postindustrial places, and histories of food and farming. She has produced ethnographic studies of park-affiliated groups for the US National Park Service, including an award-winning study of farming in Columbia County, New York.

Tim Strangleman is a Professor of Sociology in the School of Social Policy, Sociology and Social Research (SSPSSR) at the University of Kent, Canterbury. He has researched and written widely on work, deindustrialisation, working-class communities, memory, and nostalgia using oral history, visual methods, and ethnographic approaches. He has been an active member of the Working Class Studies Association and has served as its President. He is the author of *Work and Society: Sociological Approaches, Themes and Methods* (Routledge, 2008) with Tracey Warren; and *Voices of Guinness: An Oral History of the Park Royal Brewery* (Oxford, 2019). He is one of the co-editors of the forthcoming *Routledge International Handbook of Working Class Studies* (Routledge, 2019).

Shanti Sumartojo is Associate Professor of Design Research in the Department of Design and a member of the Emerging Technologies Lab at Monash University (Melbourne). Her research explores how people experience their spatial surroundings, including both material and immaterial aspects, with a focus on the built environment, using ethnographic and creative practice methodologies. This includes ongoing work on memorials and commemorative sites at the end of the First World War centenary. Her most recent books are *Atmospheres and the Experiential World: Theory and Methods* (with Sarah Pink) and *Commemorating Race and Empire in the Great War Centenary* (with Ben Wellings).

Lucy Taksa is a Professor of Management and the Director of the Centre for Workforce Futures at Macquarie University in Sydney. Previously, she was President of the Australian Society for the Study of Labour History, 2006–2009, and a Chair of the NSW State Archives Board, 2007–2012. She has published on scientific management; gendered workplace cultures in transport, nursing, and finance occupations; gender identities and (mis)representations of women leaders; migrant workers; memory; and industrial heritage. Her work on the history and heritage of the Eveleigh Railway Workshops has been funded by the Australian Research Council and by the NSW Government's Migration Heritage Centre.

Contributors

Lilia Topouzova is a Lecturer in the Department of Historical and Cultural Studies at the University of Toronto (Scarborough). Previously, she was a Post-Doctoral Fellow at Brown University and Social Sciences and Humanities and Research Council (SSHRC) and a Post-Doctoral Fellow at the Centre for Oral History and Digital Storytelling at Concordia University. She is completing a book manuscript titled *The Bulgarian Gulag: History & Legacy*. Her documentary films include *The Mosquito Problem & Other Stories* (writer, 2007), *Saturnia* (co-writer, co-director, co-producer, 2012), and she is currently in production of her third film, *Anaanaga: My Mother* (co-writer, co-director, co-producer).

Tanja Vahtikari is a Senior Lecturer in international history at Tampere University, Finland. She is also a member of the Centre of Excellence 'History of Experience,' funded by the Academy of Finland. Her academic interests focus on history of heritage, how people in the first half of the twentieth century engaged with the past in multiple ways, ranging from intellectual to bodily and emotional. Further interests include public history, urban memory, and post-war community reconstruction. She is the author of *Valuing World Heritage Cities* (Routledge, 2017) and (with Gabor Sonkoly) the European Commission Policy Review: *Innovation in Cultural Heritage Research. For an integrated European Research Policy* (2018).

Emma Waterton is a Professor in the School of Social Sciences at Western Sydney University, and an Institute Fellow in the Institute for Culture and Society. Her research explores the interface between heritage, identity, memory, and affect. She is the author of *Politics, Policy and the Discourses of Heritage in Britain* (Palgrave Macmillan, 2010), and she has co-authored other volumes: *Heritage, Communities and Archaeology* (with Laurajane Smith, Duckworth, 2009) and *The Semiotics of Heritage Tourism* (with Steve Watson, Channel View Publications, 2004). She co-edits the book series *Critical Studies in Heritage, Emotion and Affect* (Routledge).

Elizabeth Williams graduated with a single honours degree in Archaeology from Trinity College Carmarthen before gaining a Postgraduate Diploma in Archaeology, Heritage and Museum Interpretation from the University of Leicester. She is co-author (with Howard Williams) of 'Digging for the dead: archaeological practice as mortuary commemoration', *Public Archaeology* 6(1): 45–61.

Howard Williams is a Professor of Archaeology at the University of Chester. His research interests focus on medieval, post-medieval, and contemporary mortuary archaeology, archaeologies of memory, and the history of archaeology. His edited books include *Archaeologists and the Dead* (with Melanie Giles, OUP, 2016), *Cremation and the Archaeology of Death* (with Jessica Cerezo-Román and Anna Wessman, OUP, 2017) and *The Public Archaeology of Death* (with Ben Wills-Eve and Jennifer Osborne, Equinox, 2018). His monograph is titled *Death and Memory in Early Medieval Britain* (CUP, 2006) and he recently served as Honorary Editor for the Royal Archaeological Institute's *Archaeological Journal* (2013–2017).

Andrea Witcomb is a Professor of Cultural Heritage and Museum Studies at Deakin University, Australia. Her books include *From the Barracks to the Burrup: The National Trust in Western Australia* with Kate Gregory (UNSW Press, 2010), co-editor with Chris Healy of *South Pacific Museums: Experiments in Culture* (Monash epress, 2006, 2012), and, with Kylie Message, *Museum Theory* (Wiley Blackwell, 2015). Her work teases out the ways in which objects and accompanying interpretation strategies can be used to build affective modes of interpretation aimed at supporting revisionist interpretations of the past, focusing on how museums have engaged with the history of migration, represented cultural diversity, and difficult histories.

Jihwan Yoon is a Visiting Assistant Professor of Geography at Konkuk University, Seoul, South Korea. His research focuses on the post-colonial space of East Asia, memory politics, struggles for public space, and the cultural space of marginalised subjects. Based on interdisciplinary research activities at the University of Tennessee and knowledge of public memory, he has recently finished his dissertation, 'The Korean Comfort Women Commemorative Campaign: Role of Intersectionality, Symbolic Space, and Transnational Circulation in Politics of Memory and Human Rights.' Much of his work is now being extended to the transnational conflicts over history and landscapes of traumatic memories in East Asia to raise a comprehensive understanding and awareness of wartime atrocities.

ACKNOWLEDGEMENTS

The editors would like to thank Dr Danielle Drozdzewski of Stockholm University for starting up this Handbook project with them and contributing priceless ideas and contacts among the contributors, although she sadly had to withdraw as an editor at a later stage. Thanks also go to the anonymous proposal reviewers for strengthening and sharpening the focus of the volume.

INTRODUCTION

Sarah De Nardi, Hilary Orange, Steven High, and Eerika Koskinen-Koivisto

Why memory and place matters

The Routledge Handbook of Memory and Place explores the latest research on the interrelationship between important notions of memory, place, and identity across disciplines and scholarly traditions. Memory is relevant to everyone, always, anywhere. Memories and remembrance fuel, shape, and give life to our positioning as individuals, communities and nations. Remembering is also a fundamental act of being human, whether acts of remembrance are palpable or ephemeral, individual or collective, present or absent, painful or celebratory.

Across the globe, desires for independence and robust national identifications are tied to cultural differences and memories of ancestral sovereignty. In European countries like Italy, Poland, and Germany, moreover, the memory of the Holocaust and Fascism haunts everyday politics to date. In Australia and New Zealand, Anzac Day is a powerfully contested hotbed of contention, with echoes of colonial violence mingling with wartime pride and nationalistic values. In these countries, and many more, remembering the nation's past collectively, and according to a state-oriented politics of memory, is central to the nation-building project. On the Asian subcontinent, (post)colonialism figures prominently in remembrances of territory and the nation state. Here, the separation of Pakistan from India in 1947 and memories of the partition offer a compelling social and cultural framework to understand relations between the two countries. The disputed territory of Kashmir is a bone of contention that harks back to the partition days: the area's wish to stay independent after the withdrawal of the British Empire still marks the landscape in myriad emotional, cultural, military, and socio-spatial ways.

The global distribution and multidisciplinary span of authors in this *Handbook* provides not only a fresh perspective on emplaced memory/situated memory debates, but importantly presents the work of some the most innovative thinkers in a single volume. The spatio-temporal reflections making up *The Routledge Handbook of Memory and Place*, then, arrive at a crucial time in human history when the stories and remnants of the past are becoming critically important to understanding the present-day geopolitics of the fast-moving future-orientated lifestyles of the Global North, as well as parts of Asia and Australia.

Therefore, as the necessarily concise review of some key works has suggested, scholars engaged in memory work are enlisting a rich array of new approaches to studying the interplay of place, memory, and identity (De Nardi 2018). The inception and genesis of the present *Handbook* is per se an indication that times are changing, and that we can only benefit from bringing along new ontological possibilities and methodological thrusts in the study of how memory works. Indeed, we have sought to build on this research by collecting a sample of the exciting and innovative work currently undertaken today across the disciplinary spectrum.

Emplacing memory

Such a pervasive, haunting energy is as unpredictable as it is powerful – memory permeates places and attaches itself onto things, big and small. As oral historian Alessandro Portelli (1997: 32) suggests, 'In memory, time becomes "place."' Mnemonic traces and layers can be found everywhere as places (and things) gather 'stories, attitudes, opinions, and practices in a way that cannot be measured by instruments' (Turkel 2007: 227). Layers of history are thus sedimented over time (Cubitt 2007). Thus, memory is 'living history, the remembered past that exists in the present' (Frisch 1990: xxiii).

Albeit alive, remembering does not happen in a political vacuum, but occurs within wider structures of power and inequality. Increasingly, social justice movements are reclaiming place as an arena of debate and vindication of rights. The social and political valence of memory may, then, come to the fore through interventions and reimaginings of geopolitics that vary in scale from the street and neighbourhood to the nation state and continent (MacDonald 2013). The creation of brand-new memory narratives blends with shared traditions to shape presents and futures in contemporary politics.

Wider processes of physical and social ruination can therefore make public remembering more difficult, even at times dangerous. Forced forgetting is thus an integral part of ethnic cleansing, war, genocide, and, one might even say, nationalism itself. Even the more mundane structural violence of capitalism and deindustrialisation (see MacKinnon [Chapters 16], Taksa [Chapter 19], Pleasant and Strangleman [Chapter 17], and Conlon [Chapter 18] in this volume) can leave working-class memory in ruins. Displacement, the tearing asunder of past associations with place, is thus a fundamental part of this volume. At a bigger scale, the predominance of human agency in the making and sharing of memory is increasingly being challenged, however. Memory is hard to pin down, entangled as it is in more-than-human assemblages of interaction, understanding, and perception. Memory is embodied in voices and silences, caught in the contradiction of life itself. Indeed, tacit knowledge is embodied in life experiences (Cruikshank 2006: 9).

Memory and sense of place have to do with more than 'physical' monuments and structures. In other words, the idea and imagined reality of place coexists with established and factual understandings of a topographical, social, and demographic nature (Tolia-Kelly 2004). Yet places play only a part in what is experienced as a holistic sense of identity, self-positioning, and what material feminist Nancy Tuana has called 'viscous porosity' (2008). For Tuana, 'viscosity' places emphasis on

> resistance to changing form, . . . a more helpful image than 'fluidity,' which is too likely to promote a notion of open possibilities and to overlook sites of resistance and opposition or attention to the complex ways in which material agency is often involved in interactions, including, but not limited to, human agency.
>
> *(2008: 194)*

We may argue that memory in its many intersections with place plays a crucial role in such interactions. For Rudy Koshar (2000), memory applies to both the material and symbolic elements of memory and place. That is, memory is attached to objects and markers that we can perceive such as monuments, rituals, processions, or street names, but also coalesces around the sense of place closer to a structure of feeling and the sensory dimension of memory than to a concrete 'thing.' Imagination (an element within the creation of nostalgia) may be a 'collective practice that operates in ways similar to those suggested for collective memory' (Pink 2009: 45). Whether positive or negative, then, memories are powerful and complex forces linking experiences, emotions, places, and things. In terms of memory's temporalities, a focus on the link between place, experience, and memory challenges the assumption that encounter(s) 'allow a focus on the embodied nature of social distinctions and the unpredictable ways in which similarity and difference are negotiated *in the moment*' (Wilson 2016: 5, emphasis in original).

An increasingly scattered mapping: charting memory and place

Without intending to pigeon-hole thinkers and scholars within the bounds of specific areas of scholarship, we can broadly identify trends in subject areas. Thus, heritage specialists and archaeologists (Orange 2015; Moshenska 2015; McAtackney 2016) are turning their attention to the ways that people's personal or communal memories and recollections shape and negotiate 'heritage' and 'archaeology' as a way of dwelling in the world. Cultural geographers (Harvey 1996; Tolia-Kelly 2004; Drozdzewski 2014; De Nardi 2016, 2018; Crang & Tolia-Kelly 2010; Lorimer 2015) are ever-more aware that memory fuels our understanding of place and identity, in all their facets, including in the mundane places we frequent in our daily routines. Ethnologists and folklorists routinely work at the intersection of memory, folklore, storytelling, and place, and their perspective informs discourses on micro- and macrocultural interpretations of the linkages between people, place, and memory (Cashman 2008; Hrobat Virloget 2007; Koskinen-Koivisto 2011; Hrobat Virloget et al. 2016). Through a focus on popular memory and small-scale emplaced remembrance processes, such research extricates representations and emotional attachments to locales, monuments, objects, traditions, and processes. This is one of the ways that ethnologists, geographers, archaeologists, anthropologists, historians, and sociologists (among the rest) can single out place-specific patterns in environmental and material experiences of the world.

The visualisation of pasts and present as they intersect in memory and perception is also an increasingly powerful theme in research across disciplines. In consequence, geographer Doreen Massey has urged us to think of place as something more than a point on a map: 'Places as depicted on maps are places caught in a moment; they are slices through time' (1995: 188). And yet, experience of place is not a moment frozen in time, but an often socialised, somatic, political affectual encounter with the multiscale dimensions of time and place (High and Lewis 2007). Some of the contributors to this collection base their chapters upon the revolutionary ideas of maps as markers of time, identity, and experience as much as of place (Laviolette et al. [Chapter 26]; Caquard et al. [Chapter 5]). Maps, as with any other visualisation, can become conduits of social justice and catharsis if built around inclusion, openness, and experience. This conjunction is the stuff of memory itself, an assemblage of which maps and stories are but a part.

Therefore, by integrating memory as a fundamental 'piece' in the human and more-than-human puzzle of experience and place (Seremetakis 2018; Latimer & Miele 2013), we can glean deeper understandings of what makes people (and things, to an extent) feel in, or out of, place. And it is important to bear in mind that Indigenous memory paradigms and processes differ from Western ways of remembering in place (See chapters by Carlson and Naxaxalhts'i (McHalsie) [Chapter 13], as well as Salerno [Chapter 8]).

Diffractions of perspectives: about this volume

Why is this *Handbook* timely? Scholars working within the Humanities and Social Sciences have recently entered a constructive dialogue with diverse literatures that grapple with language, practice, artefacts, loss, absence, temporality, trust, materiality, ethics, self, and the larger body politic, 'nature,' but they have mostly found themselves settling into a shared framework aligned with the affective and emotional realms. As memory is no longer seen as the stuffy, stagnant remit of reminiscence and the mainstay of resistance to change, new methodologies are being sought and deployed; new horizons are reached and probed (Drozdzewski & Birdsall 2018). Moreover, although scholars across academic fields are increasingly working with the memory metaconcept, there is a perceived lack of cross-disciplinary engagement. Historian Alon Confino, for example, has suggested that oral history and ethnography have developed separately from memory studies. For him, memory studies have focused mainly on 'how the past was publicly represented' (2004: 409). National publics and the state thus loom large in memory studies. Oral history, by contrast, is 'built around people': individuals and marginalised communities, mainly (Thompson 2000: 23).

To celebrate the contribution of the current scholarship briefly delineated, and to explore exciting new potential directions in which we may take this work, *The Routledge Handbook of Memory and Place* brings together international scholars from diverse disciplinary and research frameworks, encompassing the fields of cultural geography, history, sociology, the arts, archaeology, anthropology, literature, performance, ethnology, and political science. In commissioning the various chapters contained within the volume, we asked authors to reflect on the links and relationships between memory, place, and identity that lead them to comprehending how a politics of memory operates in, on, and with (re)productions of places and identities in the present and in the past.

Together, these chapters operate a sort of diffraction (after Haraway 1997 and Barad 2007) that opens issues and complicates themes rather than seeking closure. In Donna Haraway's conceptualisation, diffraction differs from 'refraction,' as diffraction does not just reflect but scatters and intersects things. For Karen Barad, diffraction patterns are 'patterns of difference that make a difference . . . the fundamental constituents that make up the world' (2007: 72). Like memory and place, these diffractions work best when read and 'plugged' into one another, to create a dizzying, yet meaningful, assemblage of perspectives, approaches, and themes.

Part of this dynamic diffracted engagement is possible thanks to variations in voice and perspective, as well as subject matter. The contributors to this *Handbook* work in, and have addressed and engaged with, the ideas of memory and place in exciting, diverse and unusual ways. They range from early career researchers to tenured professors. We were keen to include new voices and fresh perspectives, and to engage them into a dynamic assemblage of perspectives.

All in all, most contributors to this volume have approached the intersections of memory and place through the lens of lived experience and the everyday landscapes we live, work, and move through as we remember. Each author fleshes out the myriad ways that remembering 'somewhere' comes to be and what it may mean, drawing on original case studies and/or by offering novel interpretations of theoretical canons of 'memory' and 'place,' sometimes through the diffractive prism of 'memoryscape,' a useful conceptual portmanteau to which we have dedicated a whole section of the *Handbook*. Others have set out to investigate the many facets of memory starting from uncannier and more haunting perspectives, working their way from the Abject or Other to the everyday and knowable (Salerno [Chapter 8], Renshaw [Chapter 9], Rouhier-Willoughby [Chapter 38]). Some have revisited familiar concepts through a deep engagement with theory and the epistemologies of remembrance in the present (Hicks [Chapter 24], Benjamin [Chapter 20]). For Haraway (1997), diffraction does not generate 'the same' displaced, as reflection and refraction do. Diffraction is a mapping of interference, not of replication, reflection, or reproduction. A 'diffraction pattern does not map where differences appear, but rather maps where the effects of differences appear' (Barad 2014: 172).

Oral history and reminiscence are increasingly being turned to for what they tell us about identity and place(s), either as one source among many or as our primary pathway into the past. The resurgence of an interest in memory's linkages with places has affected several disciplines and areas of research in the humanities, social sciences, media studies, communication, and political science and beyond. Studies on urban and environmental history (Laakkonen 2011; see also DeSilvey 2012 and Keul 2013) and conflict experiences (e.g. Moshenska 2010; De Nardi 2016; Seitsonen & Koskinen-Koivisto 2017), for example, have exemplified these different layers of memories and the ways in which they have shaped – and continue to shape – the ways in which we engage with our landscape and heritage. These are among the scholarly 'users' and disciplinarians who will benefit the most from this *Handbook*. As Luisa Passerini has argued, to take memory seriously we have to let it structure or frame our analysis. (1987: 8)

We now turn our attention to some of the main themes within memory studies and within this volume.

Introduction

With more people seeking refuge in another country in 2016 than in any year since the Second World War, we feel that this massive geographic (but also socio-economic) dis*place*ment is an important moment to consider people's relationship with time and place. In the section on memory of 'Mobility' (Part 1), authors reflect on the impact of movement and displacement and often enforced flight on the structures and power dynamics of memory-making and memory-enacting. Sometimes after the experience of physical loss, for example, place attachment can be 'activated retrospectively' (Low 1992: 167). According to Sean Field (see also this volume, Chapter 3), 'linguistically, "loss" suggests absence, but this loss of home and community has an ongoing emotional presence' (2008: 115).

Anxiety for a lost past, or a lost place or person in a lost past, can result in nostalgia, a form of place-based remembering, linked to (re)producing and maintaining a sense of identity. Nostalgia has been described as a 'historical emotion,' and even as a 'symptom of our age' (Boym 2001). According to Svetlana Boym, nostalgia is best understood as a 'longing for a lost time and lost home', but it could also be a workplace (see Pleasant and Strangleman, Chapter 17). In some cases, nostalgia can therefore be construed as a defence mechanism against historical upheavals, individual or societal trauma, and societal change. Nostalgia becomes a way of keeping the past in the present, and a force for change – a place for people and movements to gather strength and 'gain inspiration' (see Glazer 2005; Waterton 2005; Spitzer 1999; Sugiman 2004; Cashman 2006).

Nostalgia can take on a dizzying, uncanny effect when coupled with the fraught feeling of relief rather than longing after some place that is gone, rotting, or destroyed (Navaro-Yashin 2009). Not all memories are cherished and their markers, albeit visible, may cease to make sense if the frame of reference by which they are viewed changes. This is the case of places that suffered from war and conflict (Koskinen-Koivisto 2016; De Nardi 2016; Moshenska, this volume, Chapter 10) but especially the loci of colonialism and imperialism when viewed through the optics of the descendants of colonial subjects (Sobers, this volume, Chapter 4). Often, this kind of 'double vision' occurs when a social group or individual inhabit multiple personhoods.

Memory is also embodied as we experience the past, present, and sometimes the future through our senses. Scholars dealing with the non-verbal and the non-representational discover that speech does not 'express or represent thought, since thought is for the most part inchoate until it is spoken (or written)' (Csordas 1990: 25). Part 5 on 'The body' thus 'fleshes out' the tacit and sometimes impossible to articulate linkages between what is remembered, imagined, created, and what is physically felt and transmitted. Emplaced memories, then, have a unique materiality that contributors to the *Handbook* explore and challenge in their individual chapters. Authors analyse the way objects endure through time and encapsulate a 'silent' trajectory of human-object relations, sustaining the past in the present. The home, mementoes, childhood trinkets, murals and graffiti, and even burial places fix the presence of the past in everyday materiality and familial social relations. The everyday and mundane and the sacred can intersect in unexpected ways. In the final section on 'Ritual' (Part 7), Orange and Graves-Brown (Chapter 33) revisit celebrity shrines, reallocating meaning and affectual entwinements to the ephemeral materiality of the tribute and the enduring longing for a lost idol. In Rouhier-Willoughby's chapter (Chapter 38), sacred springs in Siberia clash and entangle themselves in the memory of the Gulag camps, testimony to the hard-to-define nature of memory as it coalesces in place, the speaker of many meanings and wearer of many 'faces'.

Contributors to this volume also position their analytical lens on the role played by the imagination in everyday place-making practices. Increasing attention is being paid to the role of everyday mythologies and the senses in our more-than-representational life worlds. In the section dedicated to 'Ritual' (Part 7), contributions do not only critique and deconstruct received understandings as 'sacred' places as timeless and almost organically growing out of their surrounding 'landscape'; instead, these chapters reveal genealogies of the sacred, contemporary para-religion, and the politics of time as major players in the discussion of what makes and maintains the holiness of a place and how its sacredness is reified and preserved through memorialisation, gestures, myths, storytelling, and other meaningful mnemonic stances and practices.

The imagined has, naturally, more to play in human existence than through ritual and belief. The imagined inhabits everyday life, caught up in the stuff of perception and identity. When tangible and intangible traces of history and memory become the luxury of the few, we need research that investigates and lays out the impossibility of divorcing place from memory, or coerced forgetting (Black 2013; Yoon and Alderman [Chapter 11], and Salerno [Chapter 8], in this volume). Storied memory has the potential to transform places, people, and events through the subjective production of new, revised, and counter narratives (de Certeau 1984). Edward Said reminds us that 'stories are at the heart of what explorers and novelists say about strange regions of the world; they also become the method colonized people use to assert their own identity and the existence of their own history' (1993, xii). Contributors to *The Routledge Handbook of Memory and Place* interrogate and integrate popular ecologies and imaginaries of the place/memory relationship, as well as addressing the grammars of decolonisation, inequality, and oppression.

We argue that emplacing memory is fundamental to our understanding of memory and place workings in synergy. We also believe in working with the entanglements of memory and place to better understand the contemporary world, to propel us forward into fairer academic and social practices and futures. We hope this volume speaks to some of these concerns, contextualising this powerful research within the zeitgeist of contemporary academic concerns for social justice, ethics, and scholarly openness.

References

Barad, K. (2007). *Meeting the universe half way: Quantum physics and the entanglement of matter and meaning*. Durham, NC: Duke University Press.
Barad, K. (2014). Diffracting diffraction: Cutting together-apart. *Parallax*, 20 (3): 168–187.
Black, M. (2013). Expellees tell tales: Partisan blood drinkers and the cultural history of violence after World War II. *History and Memory*, 25 (1): 77–110.
Boym, S. (2001). *The future of nostalgia*. New York: Basic Books.
Cashman, R. (2006). Critical nostalgia and material culture in Northern Ireland. *Journal of American Folklore*, 119: 137–160.
Cashman, R. (2008). *Storytelling on the Northern Irish border: Characters and community*. Bloomington: Indiana University Press.
Confino, A. (2004). Telling about Germany: Narrative of memory and culture. *The Journal of Modern History*, 76 (2): 389–416.
Crang, M., and Tolia-Kelly, D.P. (2010). Nation, race, and affect: Senses and sensibilities at national heritage sites. *Environment and Planning A*, 42: 2315–2331.
Cruikshank, J. (2006). *Do glaciers listen? Local knowledge, colonial encounters, and social imagination*. Vancouver: UBC Press.
Csordas, T. (1990). Embodiment as a paradigm for anthropology. *Ethos*, 18 (1): 5–47.
Cubitt, G. (2007). *History and memory*. Manchester: Manchester University Press.
de Certeau, M. (1984). *The practice of everyday life*. Berkeley, CA: University of California Press.
De Nardi, S. (2016). *The poetics of conflict experience: Materiality and embodiment in Second World War Italy*. London: Routledge.
De Nardi, S. (2018). Community memory mapping as a visual ethnography of post-war Northeast England. In: D. Drozdzewski and C. Birdsall, eds. *Doing memory research: New methods and approaches*. London: Palgrave Macmillan, pp. 191–209.
DeSilvey, C. (2012). Making sense of transience: An anticipatory history. *Cultural Geographies*, 19 (1): 31–54.
Drozdzewski, D. (2014). When the everyday and the sacred Collide: Positioning Płaszów in the Kraków landscape. *Landscape Research*, 39 (3): 255–266.
Drozdzewski, D., and Birdsall, C. eds. (2018). *Doing memory research: New methods and approaches*. London: Palgrave Macmillan.
Field, S. (2008). Imagining communities: Memory, loss, and resilience in post-Apartheid Cape Town. In: P. Hamilton and L. Shopes, eds. *Oral history and public memories*. Philadelphia: Temple University Press.
Frisch, M. (1990). *A shared authority: Essays on the craft and meaning of oral and public history*. Albany, NY: State University of New York Press.
Glazer, P. (2005). *Radical nostalgia: Spanish Civil War commemoration in America*. Rochester: University of Rochester Press.
Haraway, D. (1997). *Modest witness. Feminism and technoscience*. New York: Routledge.
Harvey, D. (1996). *Justice, nature and the geography of difference*. Oxford: Wiley.

Hrobat Virloget, K. (2007). Use of oral tradition in archaeology. *European Journal of Archaeology*, 10: 31–57.

Hrobat Virloget, K., Poljak Istenič, S., Čebron Lipovec, N., and Habinc, M. (2016). Abandoned spaces, mute memories: On marginalized inhabitants in the urban centres of Slovenia. *Proceedings of the SANU Ethnographic Institute Гласник Етнографског института САНУ*, 64 (1): 77–90.

Keul, A. (2013). Performing the swamp, producing the wetland: Social spatialization in the Atchafalaya Basin. *Geoforum*, 45: 315–324.

Koshar, R. (2000). *From monuments to traces: Artifacts of German memory 1870–1990*. Berkeley, CA: University of California Press.

Koskinen-Koivisto, E. (2011). Disappearing landscapes: Embodied experience and metaphoric space in the life story of a female factory worker. *Ethnologia Scandinavica*, 41: 25–39.

Koskinen-Koivisto, E. (2016). Reminder of dark heritage of human kind – Experiences of Finnish cemetery tourists of visiting the Norvajärvi German Cemetery. *Thanatos*, 5 (1): 23–41.

Laakkonen, S. (2011). Asphalt kids and the matrix city: Reminiscences of children's urban environmental history. *Urban History*, 38 (2): 301–323.

Latimer, J., and Miele, M. (2013). Naturecultures? Science, affect and the non-human. *Theory, Culture & Society*, 30 (7–8): 5–31.

Lorimer, H. (2015). *Wildlife in the anthropocene*. Minneapolis: University of Minnesota Press.

Low, S.M. (1992). Symbolic ties that bind: Place attachment in the plaza. *In*: I. Altman and S.M. Low, eds. *Place attachment*. New York: Plenum, pp. 165–185.

MacDonald, S. (2013). *Memorylands. heritage and identity in Europe today*. London: Routledge.

Massey, D. (1995). Places and their pasts. *History Workshop Journal*, 39: 182–192.

McAtackney, L. (2016). Re-remembering the Troubles: Community memorials, memory and identity in post-conflict Northern Ireland. *In*: E. Epinoux and F. Healy, eds. *Post-Celtic tiger Ireland: Exploring new cultural spaces*. Cambridge: Cambridge Scholars Publishing, pp. 42–64.

Moshenska, G. (2010). Working with memory in the archaeology of modern conflicts. *Cambridge Archaeological Journal*, 20 (1): 33–48.

Moshenska, G. (2015). Memory: Towards the reclamation of a vital concept. *In*: K. Lafrenz Samuel and T. Rico, eds. *Heritage keywords: Rhetoric and redescription in cultural heritage*. Boulder: University Press of Colorado, pp. 197–207.

Navaro-Yashin, Y. (2009). Affective spaces, melancholic objects: Ruination and the production of anthropological knowledge. *Journal of the Royal Anthropological Institute*, (N.S.), 15: 1–18.

Orange, H. ed. (2015). *Reanimating industrial spaces: Conducting memory work in post-industrial societies*. Walnut Creek, CA: Left Coast Press.

Passerini, L. (1987). *Fascism in popular memory: The cultural experience of the Turin working-class*. Cambridge: Cambridge University Press.

Pink, S. (2009). *Doing sensory ethnography*. London: Sage Publications.

Portelli, A. (1997). *The battle of Valle Giulia: Oral history and the art of dialogue*. Madison: University of Wisconsin Press.

Said, E.W. (1993). *Culture and imperialism*. New York: Vintage.

Seitsonen, O., and Koskinen-Koivisto, E. (2017). "Where the F . . . is Vuotso?" Heritage of Second World War forced movement and destruction in a Sámi reindeer herding community in Finnish Lapland. *International Journal of Heritage Studies*, 24 (4): 421–441.

Seremetakis, N. (2018). *Sensing the everyday*. London: Routledge.

Spitzer, L. (1999). Back through the future: Nostalgic memory and critical memory in a refuge from Nazism. *In*: M. Bal, J. Crew and L. Spitzer, eds. *Acts of memory: Cultural recall in the present*. Lebanon NH: Dartmouth College Press, pp. 87–104.

Sugiman, P.H. (2004). Memories of internment: Narrating Japanese Canadian women's life stories. *The Canadian Journal of Sociology*, 29 (3): 359–388.

Thompson, P. (2000). *The voice of the past: Oral history*. Third Edition. Oxford: Oxford University Press.

Tolia-Kelly, D. (2004). Locating processes of identification: Studying the precipitates of re-memory through artefacts in the British Asian home. *Transactions of the Institute of British Geographers*, 29 (3): 314–329.

Tuana, N. (2008). Viscous porosity: Witnessing Katrina. *In*: S. Alaimo and S. Hekman, eds. *Material feminisms*. Bloomington: Indiana University Press, pp. 188–213.

Turkel, W.J. (2007). *The archive of place: Unearthing the pasts of the Chilicotin plateau*. Vancouver: UBC Press.

Waterton, E. (2005). Whose sense of place? Reconciling archaeological perspectives with community values: Cultural landscapes in England. *International Journal of Heritage Studies*, 11 (4): 309–325.

Wilson, H. (2016). Witnessing and affect: Altering, imagining and making new spaces to remember the Great War in modern Britain. *In*: D. Drozdzewski, S. De Nardi and E. Waterton, eds. *Memory, place and identity. Commemoration and remembrance of war and conflict*. London: Routledge, pp. 221–235.

PART I

Mobility

Sarah De Nardi

Introduction

The first part of *The Routledge Handbook of Memory and Place* takes us to the shifting memory worlds of mobility, migration, changing ideas of place, diaspora, and displacement. Mobility is at the core of the more-than-representational frameworks within which many increasingly operate, as it reflects the unstable nature of place and identity.

Place practices do not just underline and nurture presences. They can also mourn absences and expose wounds (Navaro-Yashin 2009; De Nardi 2017). Communities that once defined themselves spatially do not necessarily need to be located in a place to nurture a strong sense of belonging. Imagination and inherited notions of place, long-distance longings and nostalgia for a homeland, a town, a street, can all fulfil the need for closeness to a thing, a place, or even an 'idea' of another place which may or may not correspond to an objective truth. At the same time the hauntings and the experiences of an 'elsewhere' remain strong.

Transformation and mobility are shaping the ways that we conceptualise place and the social. The ideas that affect and emotion are expressed through social activism and political resistance (see Ahmed 2004 and Askins 2009 among others), or that they emerge in transformations (Richard & Rudnyckyj 2009) are not new. Chapters in this part frame the mobility of memoryscapes as so many expressions of decolonising dissent in their exploration of identity and remembrance as forms of place-based resistance-performance.

In the workings of mobile, dynamic memory-enacting, there is room for multiple stories, multiple versions of the past, and resistance to mainstream narratives: de Certeau calls these acts of resistance 'tactics' or 'coping mechanisms' (de Certeau 1984: xxii); non-hegemonic, marginalised voices have thus the possibility of infiltrating their own innumerable differences, multiple identities and motives into the dominant text (de Certeau 1984: 41) even in mundane, everyday spaces and places (e.g. Moles 2009).

In this vein, we may conceptualise memory of colonialism as a challenging encounter. Pratt talked about the 'contact zone' as a space of colonial encounter where 'cultures met, clashed and grappled with each other' in circumstances of highly unequal power relations (1991: 34). For Pratt, the contact zone was characterised by 'rage, incomprehension, and pain,' but also by 'exhilarating moments of wonder and revelation, mutual understanding, and new wisdom' (1991: 39). Further, Parkin (1999) has written about the 'souvenirs' of wartime refugees who take with them emotionally valuable domestic objects or photographs that then become relics of their devastated world. As they merge their displaced identities within 'mementos in flight,' refugees waiting for resettlement are striving for a new stability that will allow them to reclaim their identities.

Andrea Witcomb and Alexandra Bounia's opening chapter interweaves narratives of twentieth-century Asia Minor and Greek diasporas following the Lausanne Convention with the current displacement plight of Syrian refugees at Skala Loutron on the island of Lesvos, Greece. Lesvos, together with the Museum of Refugee Memory on which the chapter centres, is an example of the workings of a shared memory which acts and operates a duty of care. Painful memories of the forced exodus of Asia Minor Greeks resonate in the present-day context, interwoven as they are with the agonising memories of the recent flux of people to Mytilene from Syria, via Turkey. This chapter channels the ways that the Museum foregrounds the processes of collection and the honouring of humble diasporic objects. These processes enable the creation of a community of care beyond direct memory and personal experience.

Katja Hrobat-Virloget brings us northwards in the Mediterranean to the Adriatic coast of Slovenia where 90% of mainly Italian-speaking dwellers emigrated, while the 'ghost' towns were settled by people from Slovenia and ex-Yugoslavia. This chapter represents a poignant foray into ghostly geographies of abandonment and melancholy cultural memory, in which the postmemory performances of former Yugoslavian and Italian ethnic populations intersect through the materialities and different emplacements in the region. The author makes a case for identifying processes of (non)heritagisation, appropriation, and silencing as the forces shaping the dominant remembrance and dwelling practices in a border area. She evokes the uncanniness of in-between places and things, haunted by present absences and loose ends of lives interrupted by the contingencies and emotional upheavals of borderland conflict.

Sean Field's chapter leads us to Cape Town in South Africa, where familial memory and place-making practices intertwine with the searing memory of racism and Apartheid in an autobiographical reflection on race, place, and identity. The local and global scales of historical consciousness and affectual remembrance interweave in diffracting understandings of what District Six was, is, and will become. The optics of family storytelling intersects the cultural consciousness of Apartheid in this powerful chapter on the uncanniness of memory and of 'memory in place' as they work against each other. Whose remembrance has the most worth? Here, imagined geographies of home, of whiteness, and of class overlap with actual topographies of racial segregation in a reflection on the reliability and fairness of memory processes.

Shawn Sobers also leads the reader on an auto-ethnographic pilgrimage of places of memory and postmemory. Starting from the unsettling question of 'How much of this was funded from the labour of my ancestors?', the author retraces the emplacement and flow of memories of colonial England and the remnants of transnational-African memoryscapes in the West Country and its stately homes. The overarching theme of this chapter is an extended reflection on nation and identity; Sobers' contribution serves as a poetic revisiting of the postmemory of enslaved Africans presented in a juxtaposition of experiences between the author himself and his teenage daughters coming to terms (or not) with a postmemory of sorts.

The final chapter by Sebastien Caquard, Emory Shaw, José Alavez, and Stefanie Dimitrovas traces the journeys of people and memories to Canada through the innovative medium of personal and migration story maps. In cultural geography especially, there is a reluctant hesitation to express or display cultural data cartographically: as Perkins argued, 'theoreticians of the new critical cartography usually employ *words* to extol the virtues of socially informed critiques of mapping, leaving to other people the messy and contingent process of creating maps as visualizations' (2003: 381). The authors of this chapter challenge this perceived shortcoming; their chapter reports on individual life stories rendered spatial and shared through non-Euclidean cartographic visualisations that chart emotion and experience as much as they pin down places. The map is a metaphor and an artefact, the materialisation of the dynamic nature of remembrance from place to place, accruing values and experiences in its motion.

Together, these chapters speak to the many dimensions of cultural mobility and displacement through diaspora, slavery, and war and conflict. The contributions to this part bear witness to the need to remember, respectfully, the plight of ourselves – as individuals, family members, and citizens – and others in ways that can move forward and even, when possible, heal.

References

Ahmed, S. (2004). Affective economies. *Social Text*, 22: 114–139.

Askins, K. (2009). That's just what I do. Placing emotion in academic activism. *Emotion Space and Society*, 2: 4–13.

de Certeau, M. (1984). *The practice of everyday life*. Berkeley, CA: University of California Press.

De Nardi, S. (2017). *The poetics of conflict: Experience. materiality and embodiment in Second World War Italy*. London: Routledge.

Moles, K. (2009). A landscape of memories: Layers of meaning in a Dublin park. *In:* M. Anico and E. Peralta, eds. *Heritage and identity: Engagement and demission in the contemporary world*. London: Routledge, pp. 129–140.

Navaro-Yashin, Y. (2009). Affective spaces, melancholic objects: Ruination and the production of anthropological knowledge. *Journal of the Royal Anthropological Institute* (N.S.), 15: 1–18.

Parkin, D.J. (1999). Mementoes as transitional objects in human displacement. *Journal of Material Culture*, 4 (3): 303–320.

Perkins, C. (2003). Cartography – cultures of mapping: Power in practice. *Progress in Human Geography*, 28 (3): 381–391.

Pratt, M.L. (1991). Arts of the contact zone. *Profession*, 91: 33–40.

Richard, A., and Rudnyckyj, D. (2009). Economies of affect. *Journal of the Royal Anthropological Institute* (N.S.), 15: 57–77.

1
THE RESTORATIVE MUSEUM

Understanding the work of memory at the Museum of Refugee Memory in Skala Loutron, Lesvos, Greece

Andrea Witcomb and Alexandra Bounia

Introduction

Lesvos is one of the Greek islands nearest to the coast of Asia Minor. Facing Ayvalik, on the Turkish coast, it stands as sentinel between Europe and Asia, as well as between past and present, receiving and sending people from across the seas. Traces of ancient Greece, Rome, and the Ottoman Empire, as well as the rise of the modern Greek state, can be seen and felt on its streets today — fenced off ruins, old Ottoman houses, now mostly in a dilapidated state but with their wooden window frames clearly announcing their origins, and empty mosques on the way to becoming a ruin. The clear lines of its nineteenth-century Greek houses, freshly painted in whites and blues, with their classical lines stand in clear contrast as does the orthodox Greek basilica in the centre of Mytilene, the island's capital city.

Down at the old port, the Epano Skala, there is a statue of a mother with a child standing with her back to Turkey, facing the old main street of Mytilene as a commemorative offering to the 1922 exodus of Asia Minor Greeks, who came to Greece in their hundred thousands (estimated more than 1,200,000), fleeing the burning of their houses and lands by the Turks in retaliation for the Greece's attempt to conquer what it saw as Greek Asia Minor (Figure 1.1).

A deluded attempt to recover the ancient Hellenic World, the result was a traumatic exchange of people, as Christians were sent back to Greece and Muslims were sent to Turkey after centuries of residence in lands that were now defined as on the other side of the newly established borders between Greece and Turkey (Clogg 2002). Mother and child stand quietly but powerfully, especially in the present-day context when memories of the most recent flux of people to Mytilene from Syria, via Turkey, is well and truly alive. At the local campus of the University of the Aegean, where students and staff engaged in a project to collect the detritus of the last wave of refugees in May 2016, there is a chess set made by the students out of the black rubber boats and orange life vests worn by those lucky enough to have them (Figure 1.2). Like the statue, the game stands sentinel to another episode in human history where political ambitions lead to vast human suffering.

Mytilene, like Lesvos itself, offers us a window into the way in which history not only repeats itself, but to the ways in which humans both commemorate, remember, forget, and re-remember past connections and violent breaks. This is nowhere more so than at a small local community museum, founded by third-generation descendants of the Asia Minor Greeks expelled from Turkey in 1922. The Museum of Refugee Memory, as it is called, can be found at another Skala, this time at Skala Loutron, a little village about an hour and a half's drive through picturesque countryside, south of Mytilene. It is to this museum that we now turn in an attempt to explore not only how time and space collapse into one another as a result of the memory

Figure 1.1 Monument to the 1922 Greek refugees from Asia Minor, Epano Skala, Mytilene.
Source: Photograph by Andrea Witcomb.

work undertaken by its founders, but also to explore the ways in which such memory work results in what we call the 'restorative museum' – a museum that seeks to restore to the souls of the dead their humanity and which does so by asking present-day visitors to empathise with the plight of these refugees and, by extension, the plight of the present wave of refugees. Our analysis is based on a joint visit in April 2017, including a detailed guided tour with Efthalia Tourli, one of the founders of this Museum as well as on previous visits and interviews with the guides/founders conducted by Alexandra Bounia in 2008 and 2013.[1]

Figure 1.2 The chess set created by students and staff of the University of the Aegean, Mytilene.
Source: Photograph by Andrea Witcomb.

The Museum of Refugee Memory

The Museum of Refugee Memory was established in 2003 and opened to the public in 2006. It was the initiative of a local cultural group called 'The Dolphin,' established in 1990 mostly by local fishermen and members of their families who are descendants of the original Greek Asia Minor refugees who settled in the area. The Museum is just one of the activities of this group and forms part of their wish to 'retrace their roots and to encourage/empower Asia Minor memories and identity' (Interview, Stratos Valachis 2008). It is housed in the old, now disused, school of the village, the actual space where all the people who established and currently run the museum were introduced into the official version of Greek history. This is, therefore, a place that they still associate with the *topos* of where historical truth should be delivered. In the context of this museum, however, that *topos* becomes a place to deliver their version of 'historical truth as they have learnt it,' rather than as the official history books represent it, as we will discuss further on. The skala on which Loutra is now located was once the uninhabited piece of land that provided the inhabitants of the small village of Loutra up the hill access to the sea up until 1931; it was only then that 25 two-room houses were built to provide shelter for 25 families that had arrived in Lesvos as refugees in 1922, mainly from the ancient town of Phokaia on the coast across the sea, as a result of what in Greek history books is usually called the 'Asia Minor catastrophe' – the expulsion of the Greek orthodox population who lived in Asia Minor by Neo-Turks and the subsequent exchange of populations in 1923 as a result of the Lausanne

Treaty between Greece and Turkey.[2] The refugee identity is still very strong, despite the fact that there is hardly anyone still alive who actually experienced the events.

While the descendants of these refugees do not have first-hand experience of these events, they do recall the narratives of their parents and grandparents, having internalised them in a way that makes these memories their own, as if they had actually experienced these events, in a classic example of what Marianne Hirsch (2001) calls a post-memory. Of interest here is Hirsh's argument that the labour of remembering, for the second generation, is an affective effort. Such an effort, as will become clear, makes this Museum an integral part of the founders' strategy for creating what Avishai Margalit (2002) calls a 'shared memory' – a collective enterprise that aims firstly to retain the bonds among the members of the community through a process of remembering what brought them together and, secondly, to expand this community of memory by promoting an ethics of caring (see also Creet 2011: 14). Thought about from this perspective, it is also possible to argue that what concerns the founders of this museum is not so much the need to testify to the veracity of historical events as to the impact of these events on their families and communities. The labour involved in this form of testimony, involving the work of post-memory, produces what Sundholm (2011: 122, 124), following the work of Alexander (2004), calls a 'cultural trauma.' The task is essentially a social one, oriented to present-day needs and involving a series of ethical relationships rather than a concern with moral values or establishing the truth or otherwise of a particular claim around the veracity of any given event. Thus, this small museum is used by members of the community not just to produce and maintain the memory of the collective trauma that has shaped their community, but also to share their testimony of this collective cultural trauma in the hope of making this relevant to other people today. This is why the guides to the Museum often express the – often considered as naïve – anticipation that knowing what has happened in the past to their families and therefore themselves will prevent the same from happening in the future. As Efthalia Tourli, one of three people who receive visitors in the museum argues: 'I am worrying for you, young people, because you have not been taught about the mistakes of the past,' stating that 'we created this museum in order to make our roots stronger . . . because no land survives without its roots' (Interview with E. Tourli 2013). Through her intervention as a guide, this Museum becomes a public practice of remembrance; it is at once inherently pedagogical (Simon 2014), a place for learning what has not been taught in schools, and a public practice of healing and caretaking for the future, a place where roots can grow stronger to support the next generation.

It is this ethics of caring which, we argue, makes the primary function of this Museum that of restoration, rather than preservation or even interpretation. Unlike many other community museums that speak to the need to prove the truth of a historical event by presenting testimony after testimony and cataloguing the practices of perpetrators, this Museum's function is not to prove what happened, but to make a group of people whole again, to heal what has become a collective 'cultural trauma.' As we shall see, they try to do so by expanding the 'we' so as to reach a level of 'thick relations' (Margalit 2002) that, in their turn, establish an ethics of memory in which we are drawn to recognise the cultural trauma of others. In what follows, we will look in some detail as to how this is done, both through the intervention of the guide but also through the particular assemblage of the exhibition itself. As we discuss later, the combination of these two elements of the Museum produce a number of strategies, all of which 'restore' or give back something that is regarded by the group as lost in the moment of trauma and its aftermath. These include the restoration of what the descendants of the original refugees consider 'their rightful place' to the refugees themselves. This is done, as we will argue, by providing a sacred keeping place for their souls, embodied in their *keimelia* or sacred treasures. The restitution of the community's 'rightful' social status is also achieved by their integration into the Greek national narrative as well as by the provision of a purpose beyond the group, a purpose that involves the recognition of the suffering of others in what becomes a shared responsibility to care for the trauma of others.

The power of keimelia *(κειμήλια)*

The exhibition consists of a collection of artefacts donated or offered as long-term loans by local people. The aesthetic value of the objects is secondary as is their ethnographic and historical interest. Their primary importance is, instead, attributed by the founders of the museum to their value as heirlooms and to their ability to trigger memory and storytelling. 'This is a museum of historical documentation,' claims Efthalia, who collected most of the objects herself by persuading their owners to offer them to the museum. She does not mean that the aim is to document history, if by that term we mean an attempt to explain the causes of a historical event or even to 'prove' that it happened. The veracity of the event itself is not in question. The objects are not, therefore, evidence of the event itself. Instead, they embody the personhood or 'soul' of those who were forced to flee. Efthalia thus often uses the term '*keimelia*' when she refers to the objects on display: the term derives from the ancient Greek verb '*keimai*' (to rest) and has designated durable and storable valuables since the time of Homer. In the Homeric epics, *keimelia* usually *keitai* (rest) in the *thalamos* of the palace. This is both a treasure room and a sacred space, for it shares the values of a tomb: it is a place of transition, a place that crosses over to the world beyond, and is, consequently, inviolable, sacrosanct. This is how Efthalia and her people view the Museum of Refugee Memory: a sacred space that can be used to safeguard what is important not just in terms of worldly values, but in the most preeminent, sacred, religious manner; they might not be important artefacts in terms of size, value, rarity or academic importance, but they are indeed *keimelia*, precious conveyors of individual and collective memory, while the Museum of Refugee Memory is a *thalamos* for them to be treasured, but also displayed for a sacred purpose – a purpose that has two sides to it. The first is to enable these objects to become horcruxes, magical objects used to house part of the Asia Minor refugees' souls, thus giving them the home that they lost back. Efthalia often refers to these objects as the 'souls' of the refugees, whereas the museum is often mentioned in her presentations as a place where these souls are kept, expressed, or simply continue to exist. The second purpose is to transform the visitors from outsiders to insiders, through the process of witnessing.

As a consequence of this, the organising principle of the exhibition is not a narrative – the narration is offered by the guided tours, by Efthalia or Stratis, as we discuss further on. Each display case – the handiwork of local people as well – houses artefacts organised in relational groups. The taxonomy is rather simple: clothes, personal documents, embroideries. The labels accompanying the object inform the visitor of the name of the object, its date expressed in rough approximates, and the name of the donor or the person who has loaned it to the museum. More often than not, the name of another person, a family member or a friend of the first generation of refugees, to whom this object belonged, accompanies the one of the donor. For instance, the following label accompanies what appears to be an unexceptional cigarette case which is given high status by its location in a glass display case of its own, right at the entrance to the Museum:

> This cigarette case was manufactured in a monastery in Izmir (end of [the] 19th century). It belonged to the prominent financier of Izmir, Stylianos Seferis (Seferiadis), father of the Nobel award winner poet Georgios Seferis (Nobel Award for Literature 1963). I dedicate this object to the memory of my good friend Dora Tsatsos, niece of the poet and daughter of the ex-President of the Republic, Konstantinos Tsatsos, who gave it to me in 1970. The donation of this object to the Museum has been approved by Georgios Symeonidis, son of Dora Tsatsos and godson of Georgios Seferis. Signed: Theodoros Toutountzidis, 2009. Director of the 1st Professional Lyceum of Mytilene.

The complicated level of relationships that structure this Museum and the values attributed to the artefacts as holders of the power to embody relationships are clearly illustrated in this example: religious undertones (the object is dedicated to the memory of an individual who is no longer alive, like all the other objects in the Museum, in a manner similar to that of a dedication to a Greek orthodox church) are combined with the associative value with a prominent family of Asia Minor descent who has done well for themselves, a value that the donor claims for himself as well, since he documents in detail his closeness to a socially and intellectually prestigious family. This provides a connecting thread between the village, the island and its inhabitants to Asia Minor, particularly to Izmir (the capital of the region they came from, which due to its history has acquired an almost mythical undertone in Greek narratives), as well as to an important family for modern Greek history, i.e. the country's president. A direct lineage is thus created for the families of Skala Loutron, re-instating them to a status that they feel has been taken away from them because of their refugee status and their need to flee and leave all their possessions, material and immaterial, behind. Efthalia claims that, in the Museum, 'we see that the refugees have managed, despite their status to introduce their level of culture – which was very developed in all arts in Asia Minor – to mainland Greece, which at the time was culturally underdeveloped' (Interview, 2017).

In other words, the artefacts in the homemade display cases of the Museum are more than heirlooms: they are the 'souls' of the dead, they are enablers towards the embodiment of sentiments. 'We did our duty to history by creating this sacred place, thereby housing the souls of our people to make it possible to understand, to experience their culture, their values' (Interview, 2013). The objects, in their materiality, provide anchors for what is understood as a timeless continuation (from antiquity to modern Greece) despite the relocation. Those who died and those who are still alive are connected through these artefacts that are meant to provide their testimony, being the evidence of what they once were and what they continue to be. They are a part of their roots, of what they had to leave behind, but they are also proofs of hope and success.

Unlike other museums in Greece and elsewhere, however, the objects in the Museum of Refugee Memory are not expected to speak for themselves. The role of articulating, enacting, and performing memory has been taken over by the museum founders who take visitors around, providing more than just a tour. Every visit presents the opportunity for offering testimony, for sharing the sacred duty of passing on the knowledge they have accumulated. It is 'pure history as I have learnt it,' claims Efthalia. And at another point of her presentation, she says: 'My sources of information are my own people' (both interviews 2013, 2017). Personal and family history become an equivalent of truth, authenticity, and purity. It is the knowledge that has been passed down from previous generations to this one through storytelling and through the entrustment of *keimelia*. It is the knowledge that the members of the community have acquired through their families, their neighbours, and the other members of their community. Based on their experiences, this knowledge is thus regarded as more 'true,' 'authentic,' and 'pure' than the knowledge one can get from history books which only deals with formal events, leaving out the impact of these on individual experiences. Because of this belief, the guides to this Museum feel that they have a responsibility to share their families' experiences, primarily with their children and grandchildren, and then with the visitors. The basic mechanism for such transference of knowledge is, we argue, a process that involves both the offering of a testimony and the making of a testament, creating an emotional encounter for museum visitors which positions them both as witnesses and as inheritors to something that must be passed on.

As Nanette Auerhahn & Dori Laub (1990) argued in their work on Holocaust testimonies, testimonies are in large part structured around the hope that the witnesses will recognise the testifier as another human being. They are, in essence, a device for structuring a form of relations between people that encourages not only active listening, but a sense of shared responsibility. The job of the witness is not only to listen but to engage in a form of ongoing witnessing. This is not unlike Roger Simon's discussion of a testament, which, as he puts it (2006: 193), embodies the ideas of a transferral between one party and another, both in relation to property as well as a religious covenant between God and his people. Significantly for our purposes, Simon also argues that it is possible to understand a testament as also relating to the process of putting together an assemblage of objects one intends to bequeath to someone else. When taken as a set of ideas within which to understand the work being done around the *Kemeilia* then, it is possible to see that the effort to gather and then display the collection is a form of sacred covenant with 'the souls of the dead' which, when set in motion through testimony, sets up what Simon argues is a 'terrible gift' that demands reciprocity – that is, it demands a form of witnessing that has a sacred character to it. Both testimony and testament become an intrinsic part of the collection, which is understood as consisting of 'inalienable objects'[3] that are meant to be inherited as a 'terrible' yet precious 'gift' which, as Simon argued, makes demands on those who receive it (Simon 2006: 189). Another way to put this is that to inherit something is to inherit a responsibility into the future. Efthalia is an excellent example of the way this works, making it clear to all who come to the Museum that their intention is to make every visitor an heir to their history, often by using religious metaphors. For example, in one visit with students from the University of the Aegean she proclaimed that 'I want you, young people, to hear this story unadulterated, just like it is.' In her animated discussion with those students, she often used the term 'transmigrate' – in Greek μεταλαμπαδεύω or *metalambadevo*, literally meaning to carry the light/candle further – to describe what her personal intention was as well as the museum's mission. The metaphor also clearly implicates the students into this light – for they become the bearers or inheritors of it, taking on, Efthalia hopes, their own sacred duty to continue to give hope to the souls of the dead.

As Simon (2006: 194) argues, however, 'to transmit a testament is not akin to simply passing a baton framed by the expectation of appreciation and preservation'. Time is an important parameter, since it is expected that the testament will remain as relevant to the present and the future as it was in the past. Making the message relevant and bringing it up to date has been very much within the efforts of the Museum. Since 2015, when the current refugee crisis from Syria reached unprecedented levels and the island of Lesvos found itself once again as a landing place for the flow of refugees leaving the shores of Turkey, the testimony on offer from the Museum was enriched to include those who are currently dislocated, forced to experience violent migrations similar to the one the families of the people of the Museum had to endure.

This was not, however, the first time that this Museum had chosen to embrace the misfortunes of others, recognising that the trauma of expulsion and of forced migration is not theirs alone. As we explained earlier, the Treaty of Lausanne set the guidelines for an exchange of populations between Greece and Turkey which also saw approximately 380,000 Muslims leave their homes and homeland in Greece to start life anew in Turkey under forced migration. A special display case in the Museum, which is dedicated to them, honours their trauma and thus makes a further argument regarding the need to recognise the humanity of others, extending to them the recognition of their suffering. It includes random artefacts, from documents to tablecloths and embroideries to small funerary stones, that have been 'rescued' from around the island. 'They [the Muslims] were refugees, too,' claims Stratis, another key person of the Museum, a special needs teacher himself, 'who lost their land and were forced to move away.' Efthalia explains to the students during their visit the reason they felt the need to have this case in the Museum: 'The two people [Christians and Muslims/Greeks and Turks] co-existed in harmony, there were friendships . . . my own people were advised to leave by Turks, my grandfather's friends . . . we respect them . . . we want to respect the balance and to tell you [the visitors] the real story of Asia Minor people; . . . of their friendships as well' (Interview 2013).

The connection with contemporary trauma and the collapse of time and space was thus present before 2015. As Efthalia claimed in the winter of 2013: 'The Asia Minor refugee has a pain in her soul; an economic migrant [like those arriving to the island before 2015] makes a decision to go elsewhere, to find a better future in another land. We respect these migrants completely, but . . . our story is much more painful.' After 2015, the connections with the present are enriched to include the new phenomenon: 'We received in the Museum newcomers [refugees who have just arrived], from Syria and elsewhere. We take them around, we talk to them about us. We understand them, because we have been in their shoes' (Interview 2017). This praxis embodies the claim, made by Efthalia (2013) that 'the most characteristic aspect/attitude of this village is solidarity. . . . There is pride, there is dignity, but also solidarity,' giving it an aura of truth.

Performance, re-enactment of memory, and participation

This invitation to connect, to become a witness, is also embedded in another aspect of a visit to this Museum – and that is the performative character of the visit which, we argue, makes the Museum and its inheritance more real, more existent. Once again, this has two aspects to it, one from the point of view of the guides and one from that of the visitor. While personal recollections bring the *keimelia* to life and give them back to the souls of the dead, visitors are invited to situate their personal memories and stories within the Museum's narrative. This is done through the provision of two murals depicting maps of Lesvos, Greece, the Asia Minor coast, and Anatolia which form central components of the exhibition. The museum people call them 'interactive maps' despite the fact that they do not provide a form of 'technological interactivity' (Witcomb 2003). There is nothing technological about them, nor an opportunity for the visitors to actually make something happen to them in a physical manner. They are, however, very powerful tools for reconstructing the homeland of the diaspora, becoming another way in which the souls of the dead can be returned 'home.' The maps are used as focal points for the discussion with visitors during the tour, to visualise where Greek villages once existed in Anatolia and the Turkish coast and where the 'new villages' (often having their names starting with the word 'new,' like 'New Smyrna,' a refugee suburb of Athens) were. Visitors are encouraged to look at the map, to be impressed by the sheer number of dots, and, if they come from a refugee background, to try to locate the village their families came from. If they cannot find it, they are invited to send information to the people of the Museum so that the village is added onto the map in the future. This can be a particularly emotional moment for some of the visitors. In 2013, a University of the Aegean student burst into tears right in front of one of these maps. It became clear that the map had reminded her of her grandparents, who had recently passed away. Sharing emotional memories is part of the experience of the Museum; it allows visitors whose ancestors came from Asia Minor to claim their connection to the diaspora as well as to the collective experience of uprooting (Kitzman 2011: 95). For those who established the Museum, the effect is to connect the emotional experiences of their families to other families, thus turning their personal and family trauma into a cultural trauma (cf. Sundholm 2011) that can be shared, ensuring at the same time that there is a legacy for their work. In engaging with the maps, visitors have an opportunity to enact their own relationship to the people and the events they have just been witnesses to, creating, in the process, a new community of memory that is deeply emotional and embodied, transcending both time and place. In connecting themselves to this place, those visitors reinstate the intricate social network that once supported the Asia Minor refugees, giving them back what Efthalia called their 'social status' while also promoting an ethics of caring and 'solidarity' as the primary value of the village.

In many ways, what the people who run the Museum of Refugee Memories are doing when they open the Museum's doors is to enable both their visitors and themselves to reach catharsis, recalling Aristotle's argument that catharsis is achieved by imitating actions that embody tragedy:

> Tragedy, then, is an imitation of an action that is serious and complete, one that has some greatness. It imitates in words with pleasant accompaniments, each type belonging separately to the different parts of the work. It imitates people performing actions and does not rely on narration. Through pity and fear it achieves purification (catharsis) from such feelings.
> *(Aristotle, Poetics, 1149b21–29 translated by Irwin & Fine 1995: 544)*

The Museum then, in offering catharsis, enables individuals, families, the local community, and all those who come into contact with it, by the act of witnessing their testimony, to restore their faith in themselves as being of value to others. The act of testifying, both through personal testimony and through bringing the *keimelia* to life, restores not only the souls of the dead in time and place but also gives to their descendants and those who listen to them their own role in space and time, restoring to them a social and, by extension, an ethical identity. The Museum both restores the humanity of the first wave of refugees and seeks to ensure that this recognition continues to others in the future.

Notes

1 The joint visit was made possible by an Erasmus+ Teaching and Training, Higher Education KA107 International Mobility Grant.
2 'Convention Concerning the Exchange of Greek and Turkish Population,' signed in Lausanne on 30 January 1923.
3 For the term, see Weiner 1992.

References

Alexander, J.C. (2004). Toward a theory of cultural trauma. *In:* J.C. Alexander, R. Eyerman, G. Bernhard, N. Smelser and P. Sztompka, eds. *Cultural trauma and collective identity*. Berkeley, CA: University of California Press, pp. 1–30.
Auerhahn, N.C., and Laub, D. (1990). Holocaust testimony. *Holocaust and Genocide Studies*, 5 (4): 447–462.
Clogg, R. (2002). *A concise history of Greece*. Cambridge: Cambridge University Press.
Creet, J. (2011). Introduction. *In:* J. Creet and A. Kitzmann, eds. *Memory and migration: Multidisciplinary approaches to memory studies*. Toronto: University of Toronto Press, pp. 3–26.
Hirsch, M. (2001). Surviving images: Holocaust photographs and the work of post-memory. *Yale Journal of Criticism*, 14 (1): 5–37.
Irwin, T., and Fine, G. (1995). *Aristotle selections*. Indianapolis and Cambridge: Hackett Publishing Company.
Kitzman, A. (2011). Frames of memory: WWII German expellees in Canada. *In:* J. Creet and A. Kitzmann, eds. *Memory and migration: Multidisciplinary approaches to memory studies*. Toronto: University of Toronto Press, pp. 93–119.
Margalit, A. (2002). *The ethics of memory*. Harvard: Harvard University Press.
Simon, R. (2006). The terrible gift: Museums and the possibility of hope without consolation. *Museum Management and Curatorship*, 21 (3): 187–204.
Simon, R. (2014). *A pedagogy of witnessing: Curatorial practice and the pursuit of social justice*. Albany, NY: State University of New York Press.
Sundholm, J. (2011). The cultural trauma process, or the ethics and mobility of memory. *In:* J. Creet and A. Kitzmann, eds. *Memory and migration: Multidisciplinary approaches to memory studies*. Toronto: University of Toronto Press, pp. 120–134.
Weiner, A. (1992). *Inalienable possessions: The paradox of keeping-while-giving*. Berkeley and Oxford: University of California Press.
Witcomb, A. (2003). *Reinventing the museum: Beyond the mausoleum*. London: Routledge.

Interviews

Efthalia Tourli, Interview 14 December 2013.
Efthalia Tourli, Interview 2 May 2017.
Stratos Valachis, Interview 12 November 2008.

2

URBAN HERITAGE BETWEEN SILENCED MEMORIES AND 'ROOTLESS' INHABITANTS

The case of the Adriatic coast in Slovenia

Katja Hrobat Virloget

Introduction. Population transfers in Istria, former Yugoslavia

This chapter analyses a nation's ignored past reflected in the form of 'wounds' or abandoned urban heritage. The main questions addressed will be how the processes of (non)heritagisation, appropriation, and silencing relate to the complex relations of power in a contested land where drastic change of urban population took place. Istria on the Adriatic coast in Slovenia affords an excellent opportunity to study questions related to relationships between the dominant and silenced memories, hegemonic and alternative heritages, identities, place attachment etc.[1]

Belonging to the Serenìssima Republic of Venice for five centuries, the Istrian region passed under Austrian rule in the nineteenth century and became part of the Italian Kingdom after the First World War. Repressive fascist anti-Slavic policies and forced Italianisation of the diverse, multi-ethnic population of Istria instigated the migration of 105,000 Slovenians and Croats from the border region of Venezia-Giulia (of which Istria represented roughly one-third) (Verginella 2015: 59–60). After the Second World War, the region was torn in two by the establishment of a temporary buffer state between the 'Democratic West' and the 'Communist East': the 'Free Territory of Trieste' (FTT). FTT was divided into two zones: Zone A containing the area around Trieste, which was held by the Allies and was integrated into Italy after FTT's dissolution in 1954. Zone B was held by the Yugoslavian army and integrated into the former Yugoslavian republics of Croatia and Slovenia (Pirjevec 2000), now independent states.

After the merging of ethnically mixed Istria with Yugoslavia, 90% of the predominantly Italian-speaking population emigrated, mainly from the urban areas (200,000 to 350,000 people left Istria as a whole [Ballinger 2003: 1, 275, n.1]) and, according to Slovenian authorities, 27,810 among them left from areas that fell under Slovenian jurisdiction between 1945 and 1958. They were mostly Italians (70%), but also Slovenians and Croats (Cunja 2004: 89; Troha 1997: 59). The (mostly) Italian emigration from Istria after the Second World War is called the 'Istrian exodus' by Italians while the Slovenians and Croats refer to those migrations as 'opting' due to the right to opt for Italian citizenship – with the consequent obligation to move to Italy – arising from two international treaties, the Paris Peace Treaty (1947) and London Memorandum (1954). 'Istrian exodus' has been stirring conflict in political discourses between Italy on one side and Slovenia and Croatia on the other for more than six decades (Hrobat Virloget 2015a: 159–162; Dota 2010; Verginella 2000; Ballinger 2003: 42–45).

The Yugoslavian authorities filled the void that remained after the Italians had left by stimulating the inflow of people from inland Slovenia and the rest of Yugoslavia. This eventually completely transformed the ethnic, social, and cultural face of Istria (Gombač 2005: 11). In 1960, a few years after the final phases of the 'exodus' took place, the proportion of native residents dropped to 49%, according to registry offices, reaching 65% in rural areas and 33% in urban settings. The difference between the rural and urban population accounts for the fact that the Italian population was concentrated in urban areas, while the adjacent rural population was largely Slovenian (Titl 1961; Kalc 2019: 149).

Italian and Slovenian historians agree that the consequence of these migrations was the ethnic homogenisation of contested lands in favour of the annexation to either Italy or Yugoslavia: Italian denationalisation of the Zone B of FTT and its 'Yugoslavisation' with the immigration of Yugoslavians, and 'Italianisation' of the Zone A of FTT with the immigration of the Istrian Italians and emigration of Slovenians (Volk 2003: 289–301; Pupo 2000: 203).

A number of scholars discuss Istrian migrations in a broader framework of massive population transfers in Central and Eastern European resulting from allies' policies in the post–Second World War period when the ethnic homogenisation of nation states was considered as the only possibility to prevent violence and assure peace and stability (e.g. Ther 2001; Corni 2015; Gousseff 2015; Pupo 2015). Recently Pamela Ballinger (2015) offered an alternative approach, reaching outside the classical frame of population transfers, by interpreting the 'Istrian exodus' as a (post)imperial process accompanying the defeat of fascism and Italy losing its newly acquired Balkan and African territories.

Disrupted tradition, heritage, and memories of 'rootless' inhabitants

If we adopt the Guillaume perception of heritage as the 'ideological apparatus of memory' (as cited in Candau 2005: 119), then the question arises: what does urban heritage represent to the currently prevailing population of Istrian towns, the immigrants? An interviewee, who came to Koper/Capodistria[2] with his family after the Second World War as a refugee escaping fascist oppression endured in the inter-war period from what is now Italian territory, admitted that new settlers have no attachment to place and lack any trans-generational memories linked to their new home environment:

> That's what we miss here where we stayed . . . the connections, the stories, knowing what happened here in this house, for example, who lived here. . . . These ties have been broken when the majority left. That's why we don't have any attitude, let's say, towards certain buildings. If it was about our own ancestors, it'd be different.
>
> *(Hrobat Virloget 2015a: 174)*

With departure, the primary ties of the population who left their homes with the place have been broken. After the Second World War, the immigrants from inland Slovenia and former Yugoslavia did not identify with Mediterranean space and Venetian heritage of Istrian towns. To the immigrants, some among who saw the sea for the first time, their new place of residence represented something foreign, something to which they only had a pragmatic relation. An interviewee, an immigrant to Koper from the former Yugoslavian republic of Serbia, explains that his and the life of his immigrant friends took place around factories, on football grounds, and around housing blocks, all of which are places lying outside the old town centre. The Venetian heritage of Istria does not hold any value for him as his affections, memories, and roots remain with his place of origin, where he is still considering returning to and building a house. As researchers of migrations argue, the return home as the 'natural' outcome of the migration process derives from the strong tie between a person and her or his land of origin (homeland) (Čapo 2015: 2013). However, staying outside the scope of migration studies, what seems clear in terms of attachment to place and memories is the lack of identification of the majority of population of Istrian towns with their living environment. This is probably one of the main reasons for the abandonment of urban centres in coastal towns. In particular, the urban centres, where urban tissue consists predominately of pre-war architecture of Venetian heritage, are neglected or inappropriately renovated with post-war architectural interventions reflecting the appropriation of space by the new, Yugoslavian colonisers (Čebron Lipovec 2012: 29). While tourists chose Piran as one of the 50 most romantic cities in Europe because they are only acquainted with its touristic (summer) aspect, the city remains empty throughout winters, with no urban life, empty apartments (many of them serve as second homes), and a high rate of alcoholism and drug addiction. Buildings in these cities are not being restored, there is no typical urban bustle on the streets, no tradition of a year-round life cycle in tune with the sea.

Besides the abandonment of urban heritage, a break with local tradition can be observed. Some local town festivals have been abandoned, like the festivity of Piran linked to the local salt production. It was recently reinvented, although the new urban population does not identify with it. An attempt has been made some years ago to relaunch a catholic festival, a maritime pilgrimage between Strunjan/Stugnano and Piran, but it failed due to the shortage of boat owners willing to participate. An interesting memory was shared by an Italian fishermen family who left but returned after a week of enduring horrific circumstances in refugee centres in Italy. Upon their return, they would teach Slovenian farmers, living in the surroundings of Koper, how to fish according to local traditions (Menih 2011: 131–137). Fishing tradition, namely, is one among traditions which have been interrupted by the 'exodus' of the majority of Italian-speaking fishermen. Fishing tradition associated with ownership of the sea and coast represents a part of contested heritage where each side, Slovenian and Italian, demonstrates its primacy by appropriation within nationalistic/ethnic discourses (Rogelja & Spreizer 2017: 50–60; Ballinger 2006).

The process of the marginalisation of urban space and the people observed in our case study are similar to what the geographer Stanko Pelc (2018: 35) calls the marginalisation of nature or pushing to the edge (margin) of human or, in our case, 'urban' priorities. The abandonment begins with the decline of the quality of life, resulting from the inability of inhabitants to properly maintain their living space. The reasons are usually economic, often in conjunction with policies in force. The reason for the abandonment of Istrian urban heritage dos not only lie in the lack of identification of immigrants with their new urban environment, but is also rooted in economic issues, as most inhabitants of the town centre are descendants of economically deprived Yugoslavian working-class immigrants who massively migrated to the coast during the 1960s and 1970s to answer local demand for workforce (Kalc 2019: 158). With the declaration of independence of Slovenia in the early 1990s, only one-half century later, they experienced deep social marginalisation and became 'second-class' citizens (without minority rights, some of them literally 'erased' from the register of Slovenian citizens) (Hrobat Virloget et al. 2016: 80, 85; Zorn & Lipovec Čebron 2008).

Related research on population exchange reveals that despite sharing the same nationality, the newcomers and native inhabitants are separated by a symbolic line (Čapo Žmegač 2002, 117–130; Čapo 2015; Hirschon 1989: 30–35). There are several types of boundaries which have emerged within the multilingual (Slovenian, Italian, Croatian, Bosnian, Serbian etc.) population of Istrian towns. A strong divide was formed between native Italian Istrians and newcomers while another boundary separates Italians and Slovenians from newcomers from former Yugoslavian republics (Hrobat Virloget 2015a: 175–177). The latter is based on orientalising discourse according to which the 'native' and 'new' Istrians of Slovenian origin perceive themselves differently from 'Non-Istrians' of the 'Balkan type' (Ballinger 2003: 266–273; Hrobat Virloget 2017: 147–148). A Slovenian whose family escaped from Trieste in Italy shared this comment on the arrival of the working class from the former republics of Yugoslavia:

> A different culture breaks in. . . . There were fewer differences, much fewer between us (Slovenians) and Italians, who actually lived in the same territory. . . . But then a completely different culture strikes. They were picking people there, in villages, you know, they were herdsmen. . . .

They were carrying bags and the like.... You brought people, shepherds more or less, and you put them in a highly developed urban environment.

Apart from ethnic/national diversification, the clash between different cultures is also evident in different cultural and social value systems of urban and rural population. A similar cultural conflict was noted before the Second World War in Istria, in the period of fascism, when Italian urban population had largely taken a distinctly superior attitude towards the Slavic-speaking population of the Istrian hinterland. In this case, the antagonism between the two communities was perhaps not only a consequence of fascist oppression, but also of conflict between two different cultural and social value systems, i.e. between the urban and rural populations (Brumen 2000: 124–133; Hrobat Virloget 2015a: 162–164). In contrast to the self-perception of 'Istrians' as highly urbanised, the perception of the Yugoslavian immigrants was quite distinct as demonstrated by an interviewee from Croatia:

The lowest standard! Hygienic! In former Yugoslavia, not only in Slovenia! These towns, Koper, Izola, Piran.... They did not have toilets, water and sewage systems. Nothing! ...At that time even towns in Kosovo and Macedonia had this arranged in a better way.

The answer to the question what urban heritage of Istrian towns means to the majority of their current population, the newcomers, is thus evident. Not a lot, since there are no transgenerational memories attached to it and there is no attachment to place. As everywhere else, newcomers have to engage with the 'foreign' that has to become 'home,' likewise they have to establish links with the new place and its people.

Contested heritage, place attachment, and nostalgia within silenced memories

Istrian urban environment is currently perceived as heritage only by the people who – despite different dominant national identity – have stayed here and became the officially recognised Italian minority. In socialist Yugoslavia, the Italian population was collectively held responsible for war crimes as well as for the preceding three decades of violent fascist politics of ethnic cleansing and imposing superiority of Italian culture (Ballinger 2003: 129–167, 207–244; Hrobat 2015a: 164–169; Hrobat Virloget et al. 2016: 81). Maurice Halbwachs (2001) suggested that in the social process of establishing a consensual collective memory, the memory that does not fit the dominant image of the past is rejected or stigmatised. This is the case with the memories of Istrian Italians who stayed behind after the 'exodus,' they remained silenced in the sphere of dominant Slovenian collective memory centred on national victimisation and heroic resistance to the fascist oppression as cornerstones of Slovenian national identity (Hrobat Virloget & Čebron Lipovec 2017; Fikfak 2009: 359; Hrobat Virloget 2015a: 161). A similar phenomenon is 'the memory gap' of Czechs regarding the expulsion of Sudeten Germans, who, during communist times, were associated with capitalist exploitation, international aggression, fascism, oppression, and also with being German. This perception has not changed dramatically to the present day (Spalová 2016: 16–19), which is also the case with Italians in Istria. The notion of collective national guilt for the expulsion of Germans or Italians is countered by arguments that this was merely a reaction to the horrors inflicted by these people during the war. Adopting this interpretation, the nation remains morally untouched while reclaiming the status of victim (Spalová 2016: 20–22; Hrobat Virloget 2015a, 2015b). Focusing the national memory on concepts such as being a victim and the guilt of another nation makes granting the status of the victim to 'the other' and dealing with one's own guilt and accountability virtually impossible (Assmann 2007: 17). Oblivion is integral to the process of remembering and it enables the memory to free itself of the heaviest traces of one's past, so as to deny them and, thence, construct a positive self-image. In fact, oblivion is not a lack of memory, but a censure (Candau 2005: 94). Memory constructs, however, do not necessarily involve falsifications of history or 'strategic *selection* of expedient recollections.' (Assmann 2007: 17)

Although different communities live together in Istria, they live as 'strangers either way' (Čapo Žmegač 2007) at least in relation to the contrasting collective memories concerning the post-war period (Hrobat Virloget 2015a, 2015b; Hrobat Virloget & Čebron Lipovec 2017). In Istrian towns, urban architecture from the Austrian and Venetian period is sometimes renovated; however, in Slovenian public discourse, there is (intentionally?) no link between Venetian heritage and Italian culture. This denial of a link is painful for the Italian-speaking inhabitants of Istria who, in accordance with the Italian identity construction, identify themselves with Venetian heritage. As an Italian interviewee said about the urban heritage of Koper 'every wall says it is Venetian, it is Italian' (Hrobat Virloget 2015a: 177–176). The memories of the remaining Istrian Italians are anchored in the material arrangement of urban space which provides them with (emotional) support and an anchor for their identity (Halbwachs 1971: 130, 2001: 151–152, 175–176; see Gensburger (2019), this volume, Chapter 6). Their nostalgic memories of 'good old times' have been materialised in an urban landscape, for them, it is a symbol of the Istrian Italian heritage and identity predating the total visual, ethnic, linguistic, economical etc. break after the Second World War. Another Italian interviewee commented:

> It's sad today to walk across Istria, everything is so abandoned. Once the forests resembled parks, everything was cleaned. Afterwards everything was abandoned because local products had no value /in socialism/. Olive oil was worth nothing ... 100-year-old olive trees were cut down.... My family has always taken care of the house, but some of them, they were destroying old houses in order to build new lodgings to settle in.

Researchers note that nostalgia is value loaded, often a consequence of change or of the fear of change, it treats the past as harmonious and idyllic in the contrast with the present, perceived in a more or less negative or pessimistic light (Koskinen-Koivisto 2017). The memories of Istrian Italians are embedded in the idyllic picture of past environment, destroyed by the inundation of newcomers of a different ethnic origin and by the 'exodus,' which broke their urban social networks:

> Capodistria was like an enlarged family, why? I was walking, visiting my aunt close by, a bit further was my grandmother, then there was the sister of my father.... Capodistria was this /town/ centre and there were all these families, because in Capodistria life was lived outside, life happened on the streets.

The nostalgic memories of the feelings of homeliness that materialised in the idyllic pictures of the town are in line with the interpretation of nostalgia by Bryan S. Turner (as quoted in Koskinen-Koivisto 2017: 196), which is based on the 'division between lost golden times and a sense of being at home against the coldness and foreignness of today's world.' The foreignness in this case derives from feelings of being a foreigner, although remaining at home – in the same, but drastically changed, hometown (Hrobat Virloget 2015a: 164–176). Describing her hometown Pula/Pola in Croatia, an Italian recounts:

> We had rights, all of them, but my town has nevertheless changed. All the names were in Croatian, nothing in Italian, of course.... I can't recall feeling a foreigner in my own country back then, but later surely.

While Slovenian and Croatian toponyms were strictly forbidden during the period of the fascist oppression before the Second World War, the previous Italian toponymy was replaced with Slovenian and Croatian names after the annexation of Istria to Yugoslavia. Language was one of the main causes of dispute between Italian Istrians, representing the majority in Istrian towns, and the first immigrants, who insisted on communicating in the language of the 'winners' (Hrobat Virloget 2015a: 165; Gombač 2005). As Vincent

Veschambre (2008: 10) notes, besides heritage, its destruction, new architecture production etc., changing toponymy is one of the most frequent interventions in space by colonisers expressing a strong symbolic marking of a new political force and appropriation of a space.

These 'spatialised' memories of Istrian Italians have been and still are wounded by many interventions in the historic built environment or simply by observing the decay. An Italian remarked on Venetian architecture:

> The palace... It has been in reconstruction for so long.... But it's always closed, abandoned, they don't take any care of it.... The same goes for many things in Capodistria, more respect should be paid to the environment.... From the trees which are so easily cut down.... They don't have this sentiment, they say we are Mediterranean.... Us?... I never felt Mediterranean.... This is the Adriatic. We were born on the sea. We love light.... Those who came, well, embrace this light! No! Let's cover everything! Heavy jutting roofs, everything covered! One does this in a different climate, not here. So they have no sensibility for the local and they bring things from other environments.... But this here is a different type of environment also in terms of culture, climate and all these aspects.

Besides feelings of being hurt by inappropriate attitudes towards 'their' heritage, this Italian narrative clearly shows a strong symbolic divide with the newcomers, this time in relation to the built environment. The process of '(non)heritagisation' (Harvey 2001) reflects the complex relations of power and appropriation of space. Demolition or any form of heritage is devalorisation, it implies a logic of making it invisible, denying the memories of those who identify with certain places and spaces. It can be interpreted as symbolic violence, an expression of power, symbolic dispossession, negation of a social group, institution, or space. To remove the inscription of groups onto a certain space, their signs have to be destroyed because the legitimacy of the population derives from their 'antiquity in a place,' thus heritage is the best site for appropriation of space (Veschambre 2008: 7–15, 115–117). In this context the recent attempts by the Italian communities of Istria to proclaim the town's old cemeteries (in Piran and Izola/Isola) as cultural heritage can be understood as a public call for the legitimisation of their group, now a minority, in this territory. As an Italian interviewee (and Italian names on old tombstones) suggests with respect to primacy in this territory: 'they /the newcomers/ were the ones who came from elsewhere, because we were already here!' In these competing discourses, including heritage appropriations by the hegemonic or authorised discourse, the memories and heritage of Italians living in Istria belong to what Laurajane Smith (2006) calls alternative, marginal, silent, silenced, or subaltern heritage discourse.

Memory and heritage entail a complex and continuing process of negotiating and fighting for what to remember and what to forget, what to preserve and what to discard (Harrison et al. 2008: 8). Through the cultural process of 'heritagisation,' a symbolic value is attached to material remains of the past. Ancient landmarks are used to gain authority for new traditions (Halbwachs 1971; Harvey 2001: 320, 331). Heritage is a selective tradition, a discourse practice, which gives cultural meanings to an 'imagined community' (Anderson 1983) which connects each individual into a wider national narrative. Those who cannot identify with these terms remain outside, they cannot 'properly belong.' Heritage is influenced by those who colonised the past, yet a change in circumstances, a twist in history, is enough for it to become a subject of revision and conflict (Hall 2008: 220–221).

A clear case of the contentiousness and competitiveness as prominent characteristics of contemporary heritage interpretations embraced in the term 'dissonant heritage' (Tunbridge & Ashworth 1996) occurred recently (in 2016) at the occasion of restoring the original colour of a Venetian building in the main Piran square called 'Benečanka' ('Venetian building/woman'). The majority of the inhabitants of Piran, newcomers after the Second World War, expressed strong resistance to the decision of local authorities (led by the Institute for the Protection of Cultural Heritage) to change the current red colour to the original white. The building had a red facade when newcomers first came and they were used to it as it stood for more than 70 years. The minority inhabitants, the Italians, however, were largely approving of the authoritarian change of colour to white, the colour used before the town experienced a drastic population change. As an Italian claimed in a public discussion, a violent appropriation of heritage can leave deep traumatic wounds:

> you feel it as violence, when the colour you are used to, something you were born with is changed. But the same thing happened when the colour was changed at the end of the 1950s. Many people suffered because the facades changed colours, because the original, Istrian-Venetian toponyms do not exist anymore etc.[3]

Conclusion. Towards inclusive heritage?

During the span of the twentieth century, urban communities in former Yugoslavia as well as in all Eastern and Central Europe have experienced an almost total change from heterogenous, multi-ethnic to more or less mono-ethnic and 'uniform' communities. The new urban population in communist states was influenced by the hegemonic discourse of new national states which changed urban histories ideologically, in line with their own national, uniform vision of the past (Ruble 2003; Sezneva 2003). The case discussed in this chapter is outstanding in terms of the almost total change of urban population resulting from population transfers associated with ethnic homogenisation of national states after the Second World War. The Istrian case speaks about silenced memories, alternative heritage discourses, appropriation, and interruptions of histories, but also represents an excellent case for studying questions of identity, memories, and heritage linked to the colonisation of a multi-ethnic territory by people of a single national identity, Yugoslavian, which, after the disintegration of Yugoslavia, transformed into new identities of 'others,' people from the former republics of Yugoslavia.

By analysing the link between memory and space, this chapter questions the value of urban heritage in Istrian coastal urban towns to its present-day inhabitants. It has been shown that the majority of urban population, composed of newcomers from various places in Slovenia and former Yugoslavia, do not feel attached to the space they dwell in, because they are not linked to it either by trans-generational memories or by tradition. The ones who do identify with urban heritage are in the minority, the Italians, silenced within the heritage discourse of the dominant 'other' – the Slovenian collective memory. Interrupted traditions, processes of appropriation of space and traditions, and alternative attempts to legitimise an ethnic group, silenced due to collective criminalisation, have been shown.

Let me conclude with an almost naïve question: after more than six decades, can a society heal the wounds embedded in the decaying urban heritage by finding a way to listen to 'the other' and compose an inclusive interpretation of heritage? As Stuart Hall argues, post-colonial nations in present-day multicultural societies need a revision of their dominant heritages, an inscription of the marginalised onto the centre and inclusion of 'their' histories into 'ours' (Hall 2008: 228), regardless of whether 'others' have been colonised or have colonised.

Notes

1 Research was performed in the scope of the ARRS project entitled 'Migration control in the Slovenian area from the times of Austria-Hungary to indipendent Slovenia,' project head Aleksej Kalc (J6–8250).
2 Bilingual toponyms are used only at first mention of a place, later only a single name is used: either Slovenian or Italian form according to the language of the author or interlocutor.
3 For this part of the recorded speech and the discussions on heritage conservations problems linked to it, I am grateful to art historian Neža Čebron Lipovec.

References

Anderson, B. (1983). *Imagined communities: Reflections on the origin and spread of nationalism*. London: Verso.
Assmann, A. (2007). Europe: A community of memory? *Bulletin of the German Historical Institute*, 40: 11–25.
Ballinger, P. (2003). *History in exile: Memory and identity at the borders of the Balkans*. Princeton, NJ and Oxford: Princeton University Press.
Ballinger, P. (2006). Lines in the water, peoples on the map: Maritime museums and the representation of cultural boundaries in the Upper Adriatic. *Narodna Umjetnost*, 43 (1): 15–39.
Ballinger, P. (2015). Remapping the Istrian exodus. New interpretive frameworks. In: K. Hrobat, C. Gousseff and G. Corni, eds. *At home but foreigners. populations' transfers in 20th century Istria*. Koper: Annales University Press, pp. 71–93.
Brumen, B. (2000). *Sv. Peter in njegovi časi: Socialni spomini, časi in identiteta v istrski vasi Sv. Peter*. Ljubljana: Založba.
Candau, J. (2005). *Anthropologie de la mémoire*. Paris: Armand Collin.
Čapo Žmegač, J. (2002). *Srijemski Hrvati. Etnološka studija migracije, identifikacije i interakcije*. Zagreb: Duriex.
Čapo, J. (2015). Population movements as instances of 'co-ethnic' encounters: A critique. In: K. Hrobat Virloget, C. Gousseff and G. Corni, eds. *At home but foreigners. Population transfers in 20th Century Istria*. Koper: Univerzitetna založba Annales, pp. 209–222.
Čapo Žmegač, J. (2007). *Strangers either way: The lives of Croatian refugees in their new home*. New York: Berghahn Books.
Čebron Lipovec, N. (2012). Arhitekturni pomniki izgradnje Kopra po drugi svetovni vojni. *Annales. Series Historia et Sociologia*, 22 (1): 211–232.
Corni, G. (2015). Commentary. In: K. Hrobat Virloget, C. Gousseff and G. Corni, eds. *At home but foreigners: Population transfers in 20th century Istria*. Koper: Univerzitetna založba Annales, pp. 15–23.
Cunja, L. (2004). *Škofije na Morganovi liniji*. Koper, Škofije: Lipa.
Dota, F. (2010). *Zaraćeno poraće: Konfliktni i konkurentski narativi o stradanju i ieljevanju Talijana iz Istre*. Zagreb: Srednja evropa.
Fikfak, J. (2009). Cultural and social representations on the border: From disagreement to coexistence. *Human Affairs*, 19 (4): 350–362.
Gensburger, S. (2019). Memory and space: (Re)reading Halbwachs. In: S. De Nardi, H. Orange, E. Koskinen-Koivisto and S. High, eds. *The Routledge handbook of memory and place*. London: Routledge, pp. xx–xx.
Gombač, J. (2005). *Ezuli ali optanti? Zgodovinski primer v luči sodobne teorije*. Ljubljana: Založba ZRC.
Gousseff, C. (2015). Presentation. In: K. Hrobat Virloget, C. Gousseff and G. Corni, eds. *At home but foreigners: Population transfers in 20th century Istria*. Koper: Univerzitetna založba Annales, pp. 11–13.
Halbwachs, M. (1971). *La topographie légendaire des évangelis en terre sainte: Etude de mémoire collective*. Paris: Presses Universitaires de France.
Halbwachs, M. (2001). *Kolektivni spomin*. Ljubljana: Studia Humanitatis.
Hall, S. (2008). Whose heritage? Un-settling 'The Heritage', reimagining the post-nation. In: R. Harrison, G. Fairclough, J.H. Jnr Jameson and J. Schofield, eds. *The heritage reader*. London: Routledge, pp. 219–228.
Harrison, R., Fairclough, G., Jameson, Jnr, J.H., and Schofield, J. (2008). Introduction. In: R. Harrison, G. Fairclough, J.H. Jnr Jameson and J. Schofield, eds. *The heritage reader*. London: Routledge, pp. 1–12.
Harvey, D.C. (2001). Heritage pasts and heritage presents: Temporality, meaning and the scope of heritage studies. *International Journal of Heritage Studies*, 7 (4): 319–338.
Hirschon, R. (1989). *Heirs of the Greek catastrophe: The social life of Asia Minor refugees in Pireus*. Oxford: Clarendon Press.
Hrobat Virloget, K. (2015a). The burden of the past. Silenced and divided memories of the post-war Istrian society. In: K. Hrobat Virloget, C. Gousseff and G. Corni, eds. *At home but foreigners: Population transfers in 20th century Istria*. Koper: Univerzitetna založba Annales, pp. 159–188.
Hrobat Virloget, K. (2015b). Breme preteklosti. Spomini na sobivanje in migracije v slovenski Istri po drugi svetovni vojni. *Acta Histriae*, 23 (3): 531–554.
Hrobat Virloget, K. (2017). Šop ključev: Spomini na eksodus in sobivanje prebivalstva v Istri po drugi svetovni vojni. In: *Istarske i kvarnerske teme: Radovi sa znanstvenog skupa Etnološke i folklorističke znanosti u Kvarnerskome primorju i Istri u 19. i 20. stoljeću: Rijeka, 11–12 listopada 2013. Zbornik za narodni život i običaje* 59, pp. 137–158.
Hrobat Virloget, K., and Čebron Lipovec, N. (2017). Heroes we love? Monuments to the National Liberation Movement in Istria between memories, care, and collective silence. *Studia Ethnologica Croatica*, 29: 47–71.
Hrobat Virloget, K., Poljak Istenič, S., Čebron Lipovec, N., and Habinc, M. (2016). Abandoned spaces, mute memories: On marginalized inhabitants in the urban centres of Slovenia. *Glasnik Etnografskog instituta*, 64: 77–90.
Kalc, A. (2019). The other side of the "Istrian exodus": Immigration and social restoration in Slovenian coastal towns in the 1950s. *Dve domovini/Two homelands*, 49: 145–162. doi:10.3986/dd.v0i49.7258.
Koskinen-Koivisto, E. (2017). Negotiating the past at the kitchen table: Nostalgia as a narrative strategy in an intergenerational context. *Journal of Finnish Studies*, 19: 7–23.

Menih, K. (2011). *Koprčani*. Piran: Mediteranum.

Pelc, S. (2018). Marginality and sustainability. *In:* S. Pelc and M. Koderman, eds. *Nature, tourism and ethnicity as drivers of (de)marginalization*. Cham: Springer Link, pp. 31–42.

Pirjevec, J. (2000). *Trst je naš, boj Slovencev za morje (1848–1954)*. Ljubljana: Nova revija.

Pupo, R. (2000). L'esodo degli Italiani da Zara, da Fiume e dall'Istria: Un quadro fattuale. *In:* C. Marina, M. Dogo and R. Pupo, eds. *Esodi: Trasferimenti forzati di popolazione nel Novecento europeo*. Neapelj: Edizioni scientifiche italiane (Quaderni di Clio), pp. 183–208.

Pupo, R. (2015). Italian historiography on the Istrian exodus: Topics and perspectives. *In:* K. Hrobat Virloget, C. Gousseff and G. Corni, eds. *At home but foreigners: Population transfers in 20th century Istria*. Koper: Univerzitetna založba Annales, pp. 25–47.

Rogelja, N., and Spreizer, J. (2017). *Spreizer fish on the move: Fishing between discourses and borders in the Northern Adriatic*. Cham: Springer.

Ruble, B.A. (2003). Living apart together: The city, contested identity, and democratic transitions. *In:* J. Czaplicka and B. Ruble, eds. *Composing urban history and the construction of civic identities*. Baltimore and Washington, DC: John Hopkins University Press with Woodrow Wilson Center Press, pp. 1–21.

Sezneva, O. (2003). The dual history: Politics of the past in Kaliningrad, former Koenigsberg. *In:* J. Czaplicka and B. Ruble, eds. *Composing urban history and the construction of civic identities*. Baltimore and Washington, DC: John Hopkins University Press with Woodrow Wilson Center Press, pp. 58–85.

Smith, L. (2006). *Uses of heritage*. London: Routledge.

Spalová, B. (2016). Remembering the German past in the Czech lands: A key moment between communicative and cultural memory. *History and Anthropology*, 28: 1, 84–109.

Ther, P. (2001). A century of forced migration: The origins and consequences of 'ethnic cleansing.' *In:* T. Philipp and A. Siljak, eds. *Redrawing nations: Ethnic cleansing in East-Central Europe, 1944–1948*. Oxford: Rowman and Littlefield Publishers, pp. 43–74.

Titl, J. (1961). *Populacijske spremembe v Koprskem primorju: Koprski okraj bivše cone B*. Koper: J. Titl.

Troha, N. (1997). STO – Svobodno tržaško ozemlje. *In:* S. Valentinčič, ed. *Zbornik Primorske – 50 let*. Koper: Primorske Novice, pp. 56–60.

Tunbridge, J.E., and Ashworth, G.J. (1996). *Dissonant heritage: The management of the past as a resource in conflict*. Chichester and New York: J. Wiley.

Verginella, M. (2000). L'esodo istriano nella storiografia slovena. *In:* C. Marina, M. Dogo and R. Pupo, eds. *Esodi: Trasferimenti forzati di popolazione nel Novecento europeo*. Neapelj: Edizioni scientifiche italiane (Quaderni di Clio), pp. 269–277.

Verginella, M. (2015). Writing historiography on migrations at the meeting point of nations in the Northern Adriatic. *In:* K. Hrobat Virloget, C. Gousseff and G. Corni eds. *At home but foreigners: Population transfers in 20th century Istria*. Koper: Univerzitetna založba Annales, pp. 49–70.

Veschambre, V. (2008). *Traces et mémoires urbaines. Enjeux sociaux de la patrimonilasation et de la démolition*. Rennes: PUR.

Volk, S. (2003). *Istra v Trstu. Naselitev istrskih in dalmatinskih ezulov in nacionalna bonifikacija na Tržaškem 1954–1966*. Koper: Univerza na Primorskem, Znanstveno-raziskovalno središče Koper, Zgodovinsko društvo za južno Primorsko.

Zorn, J., and Lipovec Čebron, U. eds. (2008). *Once upon an erasure: From citizens to illegal residents in Republic of Slovenia*. Ljublojana: Študentska založba.

3
UNCANNY DISTRICT SIX
Removals, remains, and deferred regeneration

Sean Field

Introduction

The city of Cape Town is stunningly beautiful. Its historic colonial centre nestles within the city bowl area surrounded by mountains and oceans. But this environmental beauty and its appealing surfaces conceal painful sites of memory. Paradoxically, hiding in plain sight, on the edge of the city centre is the perceived wasteland of rubble and weeds where the District Six community once lived (Figure 3.1). District Six was a pre-apartheid community erased through racist forced displacements. In the 1950s, under apartheid, all South Africans were racially classified coloured, black, Asian, and white. And, under the Group Areas Act, the apartheid state forcibly displaced and dispersed all District Six residents across the city and beyond. The architecture of District Six was bulldozed into smithereens, only places of worship were not demolished. The state planned a wealthy white inner-city suburb, but this did not happen as former residents resisted this apartheid development aim. To the present day, District Six looms large in the popular imagination of the city. Much has been written about its history and the post-apartheid memorial strategies steered by the District Six Museum and former residents.[1] I sketch the District Six history before and after removals, but my central focus is rather the personal meanings of this public space. This site contradicts both the seductive beauty of the landscape and the recurrent nostalgic memory patterns about community life before displacements. This chapter is also an excavation of photographic and oral remains obscured within my family archive by racism and overshadowed by the iconic District Six story.

The bleak space – where District Six thrived before erasure – is frequently referred to as 'empty,' 'empty land,' 'empty traumatic space,' and other variations. But this so-called empty space is neither materially nor symbolically empty. What remains is a proliferation of hauntings and other meanings (Jonker & Till 2009; O'Connell 2015). There are also many material traces: on the site, in people's homes, and those curated at the District Six Museum (Rassool & Prosalendis 2001). My aim is to move beyond the perceived emptiness of the space to look at District Six as a site of many mirrors, which reflect the mixed meanings of 'heterotopia' (Foucault 1986). Most typically, District Six is imaginatively remembered as the ideal home and community of the past and future. But such nostalgic reification psychically displaces 'homely' images onto a utopian plane, which conceals the 'unhomely' – the uncanny – that emerges from the familiar.[2] I argue that uncanny associations are evoked by spaces and objects in the present but the past sources for such affects are childhoods, especially unconscious early childhood. This means we need to discursively deploy both historical and psychoanalytic modes of thinking to explain how differing temporalities – conscious and unconscious – create uncanny senses of place.

Figure 3.1 A city view from the prior location of our Lemmington Terrace home.

Anxious ambivalences have dominated the writing of this chapter. There are three broad reasons. Firstly, several authors have noted that writing about the uncanny is necessarily shaped by its ambiguity and strangeness (Royle 2003). Secondly, it is evoked by working through my family dynamics linked to District Six. My family lived there from 1956 to 1962 and were not forcibly removed. The family archive I engage with here is in no way more important than the pain of the District Six community that still reverberates across generations.[3] Thirdly, writing about early childhood involves encounters with my parents' racism and other parental issues since my birth in 1961 in District Six. I argue that the uncanny might be perceived as out of place with public memories of District Six, but a psychoanalytic reading shows that *uncanny evocations* provide pathways to analyse a city's memoryscape as more than its nostalgic or beautiful surfaces. I conclude with reflections about the deferred regeneration of District Six.

Before and after removals

The figure of the forcibly removed forms a kind of absent presence at the centre of the contemporary discourses of the city. Just as s/he haunts the post-apartheid city, its lines of yearning and desire, are etched deep in living memory, form a supervening grid through which the city is experienced and erupt in the present.

(Murray, Shepherd & Hall 2007)

District Six in Cape Town, Sophia Town in Johannesburg, and Cato Manor in Durban are the most well-known urban examples of apartheid erasure. Before 1948, colonial and segregationist South African governments entrenched racial divisions through job and accommodation reservation across the country. While often exaggerated in liberal imaginaries, in Cape Town, residential pockets such as District Six, Woodstock, Mowbray, Rondebosch, Claremont, Retreat, and other spaces were, to varying degrees, culturally heterogeneous until the late 1950s (Field 2001).

District Six originated in the 1830s and many early residents were emancipated indigenous KhoiSan and slaves from Asia and Africa, who gradually adopted the label 'coloured' and, by the late nineteenth century, 'coloured identity had crystallized' (Adhikari 2005: 2). Moreover, from the 1860s with the emergence of Kimberly diamond and Johannesburg gold mining, Cape Town as a British imperial port grew and many sailors, migrants, and passing travelers from the North and East found residence in nearby District Six.

In the twentieth century, soldiers returning from the South African War (1899–1901) and both World Wars found a home there. More significantly, the area had a diverse popular cultural life, which included various music genres, dance bands, New Year's carnivals, youth gangs, and a thriving 'bioscope' scene (Jeppie & Soudien 1990; Mainguard 2017). It also had a network of traders and street corner shops, and many found employment in the nearby city centre, the harbour, and textile and garment factories. By the 1940s, District Six was deemed a 'slum' by city planners and its proximity to the city centre gave it economic potential and its cultural diversity was anathema to the onset of apartheid.

After District Six was zoned a 'white area' in 1966, removals occurred from 1968 to 1982. The ubiquitous 'knock on the door' was the dreaded moment when Group Areas inspectors came to inform families when they were to be removed. More than 60,000 people were forcibly removed from District Six and over 200,000 across the city in what was a massive apartheid program of social re-engineering.[4] District Sixer's classified as 'African' were relocated to the apartheid township of Guguletu (or endorsed out the region to the so-called rural 'homelands') and people classified as 'coloured' went to Bonteheuwel, Manenberg, Hanover Park, Mitchells Plein, and other new townships. The apartheid designed townships of the Cape Flats became known as the 'dumping grounds' of people displaced from the older inner-city suburbs (Field 2001). The widespread anguish and anger that was evoked by these racist removals gave impetus to the formation of an anti-apartheid coalition of organisations, for example, 'The Hands-off-District-Six' campaign. They were largely successful at blocking economic development in the 1980s (Jeppie & Soudien 1990). The one exception was the controversial construction of a university campus on the western side of District Six.

The Hands-Off campaign was a forerunner to the formation of the Museum. In 1995 a temporary exhibition about District Six was planned to run for only two weeks. But it was such a popular success that it developed into a permanent museum. The Museum placed considerable emphasis on visitor experiences and public dialogues. Former residents inscribe their stories onto memory cloths and are interviewed on site. The rekindling of place stimulates the imaginative 'holding together' of both self and community identity. The most profound activity is former residents writing their family homes onto a street map of District Six (Rassool & Prosalendis 2001). The map is located beneath the entrance floor of the Museum (Figure 3.2) and works as a mnemonic device, which recalls place and symbolically 'returns' former residents (Coombes 2004; McEachern 2001). The museum offers former residents with opportunities to re-connect with each other at exhibitions, book launches, music events, and public education programs, which creates a social framework to hold memories, emotions, and imaginings of District Six. These public projects have rebuilt community networks and are a form of social regeneration, but it remains partial and in-process (Field 2012). The kind of community people spatially return to will neither be the community before erasure nor the community imaginatively remembered thereafter.

Figure 3.2 District Six Museum floor map. See Constitution Street, below De Waal Drive.

Family remains

> the uncanny is that class of the frightening which leads back to what is known of old and long familiar. How is this possible, in what circumstances can the familiar become uncanny and frightening?
>
> *(Freud 1919: 220)*

It is strange that the initial three years of human life remain outside conscious memory. Yet we have fundamental experiences in those early years.[5] In later life stages, we might have flashbacks, auditory echoes, embodied memories, and uncanny sensations from early childhood, but such elusive phenomena require discursive psychoanalysis to make sense of. This acknowledged, I explore traces of my early childhood, which were mirrored to me through stories others told about my upbringing. I interpret three overlapping memory traces from the family archive: firstly, a conversational disclosure by my brother; secondly, a photograph of my mother with me; thirdly, an oral history interview with my mother.

The first trace. My family lived in a narrow semi-detached house on Lemmington Terrace from 1956 to 1962. I was born there in 1961 but have no memories of this home and the surrounding spaces. I have two siblings, Yvonne (born 1951) and Ronald (born 1953). I grew up with many stories about our Lemmington Terrace home, which my parents referred to as being in the suburb of Vredehoek. That suburb is *above* a highway, then called De Waal Drive, whereas District Six is directly *below* the highway Figure 3.2).

During a conversation with my brother – in the late 1980s – I referred to our time in Lemmington Terrace, Vredehoek. My brother's corrective response surprised me:

> That wasn't in Vredehoek, you know how racist dad and mom are, they said that because we had coloured neighbours. Lemmington Terrace is in District Six. They were ashamed we lived in District Six, so they told everyone we lived in Vredehoek because that is a white area.[6]

Lemmington Terrace was in fact on Upper Constitution Street (some residents refer to it as 'old Constitution Street'). Upper Constitution Street according to municipal maps is very much in District Six and from the stories of other residents it was unequivocally seen as part of the District Six community, albeit on the edge of the area. From 1953 to 1956 my family rented in Vredehoek where 'we had rooms,' but the 1956 move to Lemmington Terrace was significant as it was the first family house they rented. But in my mother's narrative, District Six was described as 'down there' or 'down where it was rough,' as if District Six only began several blocks below from where we lived. Lemmington Terrace had several white families but there were certainly coloured families too. Yet my mother portrays 'all' of Lemmington Terrace as white, and speaks of coloured children playing in the road as if they mysteriously came from elsewhere.

My official birth certificate – completed by my mother – does not indicate the suburb where we lived but indicates our home address as: '63 Lemmington Terrace, *off De Waal Drive.*' This documentary fudging together with the oral denial of our home being in District Six is indeed evidence of racism. This was partly fuelled by their shame over the family's positioning within a racialised class hierarchy. They had a deep desire to be 'better off' and throughout my childhood my father drummed into my head, 'you must study harder, so you can be better off than we are.' Their desire for upward class mobility involved attempts to distance the family from South Africans of colour, and efforts to improve their material position entailed the 'underestimated emotion of envy' (Steedman 1986: 6). This is not exceptional; when I was conducting PhD interviews with white working-class residents of other culturally diverse communities, such as Kensington, many engaged in the same racist denials or incorrect labelling of where they lived (Field 1996).

Before I began excavating my family archive for other articles (Field 2013), I used to believe that it was just a historical co-incidence that my family lived in District Six, but it was no co-incidence. District Six absorbed migrant former soldiers like my British army veteran father. And it absorbed poor coloured, African, and white rural inhabitants seeking a foothold in the city, such as the unskilled farm girl that was my mother. Her parents were tenant farmers in the rural hinterland of the Western Cape. My mother only had one year of high school when her father compelled her to leave school to work, and my father had three years of high school in Birmingham, before joining the British Army in 1940, coming to South Africa in 1948. In 1962, my family moved from District Six to Maitland, which was a white working-class neighbourhood. In 1969, they moved from Maitland to Bothazig in the northern suburbs. Bothazig was a housing project of the apartheid, Department of Community Development. The suburb was named after that department's minister, PW Botha, who later became an infamous apartheid Prime Minister. It is often forgotten, but the apartheid state was also a welfare state for the white working class, and my family and I were beneficiaries of those policies.

The second trace. With no irony, my physical conception was constantly described by my mother as, 'Sean was a mistake, but he was a lovely mistake.' The photograph is small (8 x 9 cm). It shows my mother smiling at me in her arms (Figure 3.3). It was taken on 11 December 1961, my christening day. My mother adopts a Madonna pose. Perhaps my Catholic father and photographer composed the scene. In her words, 'You have a christening robe on that was in the family – it was from my mother originally – I have a pretty dress on to, it was tan and white, this is the steps of Lemmington Terrace, sitting with you.' This is my only visual glimpse of that house, and a rare linking of oral and visual traces.

She looks happy in the photograph, but the image depicts more than that. My birth was a life or death moment: my parents had conflicting positive/negative blood groups, and hence the rhesus-factor came into

Figure 3.3 On the steps of 63 Lemmington Terrace.

conflict with me as the third child. In-utero, my mother's anti-bodies were attacking my blood as an alien presence. This was five years before the invention of in-utero blood transfusions. So, I was a rhesus baby, two-months premature, severely jaundiced, weighing only three pounds. I had a ten-day stay in an incubator and two months in the hospital. My birth was a 'shock to her system' (she had no prior knowledge of the rhesus factor), which was compounded by the mother-child separation. Repeatedly she said, 'But I went and saw you every day, even hitch-hiked on De Waal Drive' to get to Groote Schuur hospital. This photograph was roughly taken a week after my arrival in Lemmington Terrace after two months in hospital, it frames not just happiness but also her immense relief at my survival.

My near-death birth marked, what she admits, the beginning of her 'over-protectiveness.' She keeps me 'in the cot' next to the parental bed until age three. And I stay in the parental bedroom in Maitland until 29 May 1969, the day we moved to Bothazig. I was exactly seven years and ten months old. Others see this photograph as a perfect picture of motherly love, but to me it feels alien with uncanny echoes from within the familiarity of her anxieties.

The third trace. In 1999, I conducted three interview sessions with my mother. These interviews helped me contextualise my birth, family stories, and to interpret the photograph. However, these interviews were conducted eight months after she was diagnosed with frontal lobe dementia. At about 5.30 am, Sunday morning 31 July 2001, my then wife told me to feel her pregnant stomach so I could have my first touch of in-utero movements of our first child, Ella. Five minutes later, my father phoned to inform us that my mother had just died. The close timing of life as entry and death as exit evoked an uncanny mixture of feelings that morning.

While listening to the interviews of my deceased mother, her familiar voice evoked, on the one hand, an admiration for her considerable efforts to nurture me. On the other hand, my anger at her inability to let go. Her love knew no boundaries and was driven by her childhood traumas and unfulfilled needs in her adult life. My mother was a childlike parent, constantly craving positive mirroring and love, which my emotionally detached father could not provide. She was also threatened by her children's need to independently explore spaces beyond her control. The mother-child dynamic is:

> less significant and identifiable as an object than as a process that is identified with cumulative internal and external transformations . . . the uncanny: such moments feel familiar, sacred, reverential, but are fundamentally outside of cognitive experience. They are registered through an experience in being, rather than mind, because they express that part of us where the experience of rapport with the other was the essence of life before words existed.
>
> (Bollas 1987: 32)

In working through these traces in relation to my mother, my narcissistic impulse to be placed within District Six recedes. For many years I dismissed this narcissism as disingenuous. But by writing this chapter, I instead chose to empathically track that impulse to its infantile historical source, which allowed me to excavate fragments of my self buried beneath her engulfing behaviour that began in District Six and continued in other community spaces. Over the years, I have driven along the highway that passes Groote Schuur hospital, then District Six, and into the city. My body silently shudders as I pass these spatial reminders of her relentless anxiety. I now accept that her conditional mirroring – which left much unseen – was the best she was able to provide. This self-reflexive writing process has been productive and allows me to pursue a broader aim: to make sense of uncanny District Six and its many mirrors.

Uncanny District Six

The nineteenth-century umbrella category 'coloured' masks the clustering of creole identities, which cohere through popular cultural processes in colonial inner-city spaces such as District Six. These identities are not reducible to so-called miscegenation but are, in part, a colonial political construct and, in part, a product of creolisation (Constant-Martin 2013) and popular agencies (Adhikari 2005). While there is a historiography on the production of coloured identities and spaces (Erasmus 2001), a genealogy of the category 'coloured' from the late Dutch colonial period (1652–1806) and during the British imperial regime (1806–1910) is still to be written. That historical genealogy will make more explicable the imagined elevation of District Six to iconic status within the popular imaginaries of people who self-identify as coloured across the Cape and beyond. The apartheid erasure of District Six painfully sedimented that iconic image through the injuries of racist displacement. However, nostalgic constructions of pre-apartheid District Six as a culturally diverse haven obscures how the formation of this community occurred during post-emancipation British colonialism and liberal paternalism. A hankering that draws on such imperial discourses can still be found in family photograph collections, oral memories, and conservative political sensibilities of older coloured generations. With these discourses in mind, I focus specifically on the relationship between historical losses, the uncanny, and the happy family trope that dominates the nostalgia for a pre-erasure District Six.

Nostalgia is an imaginative process of framing evocative memories driven by psychic defenses and reinforced through shared memory patterns with others. This is a response to the historical loss of home and community which was especially devastating for those born and raised in District Six and who experienced removals as the severing of emotional attachments to their community spaces. For many it was also *as if* their inner-self was fragmented and imaginary mental places were reconstructed to sustain a degree of self-cohesion in the post-removals context marked by the social fragmentation of community and family networks (Field 2012: 87–100). Moreover, for the thousands who have for decades lived in poverty-stricken, violent ganglands across the Cape Flats, a major everyday concern is safety. These anxious safe/unsafe discourses are saturated with narratives of broken families damaged by domestic and gang violence, sexual abuse, drugs, and other vicissitudes of working-class life. Note also that poverty, inequality, and violence has dramatically increased under post-apartheid governance. Both common law criminal violence and the structural violence of poverty dominate precarious lives on the Cape Flats. For older generations, a reconstruction of the pre-erasure District Six as a safe and peaceful community where 'we had nothing, but we were all happy together' is a common response. In this context, it is little wonder that forms of nostalgia provide some solace while enduring the daily stresses of the unsafe present. And, yet, the victims of apartheid removals managed to forge 'new communities' on the Cape Flats (Salo 2005).

These interweaving explanations are crucial, but engaging family archives – as I have done in the prior section (see also O'Connell 2015) – suggests that more attention needs to be given to the uncanny evocations from childhood in understanding how nostalgia functions and how it induces the forgetting or silencing of uncomfortable aspects of intra-community contestations and histories of family dynamics before displacement. But nostalgia also involves desire or, more significantly, unfulfilled desires:

> A longing for a home that no longer exists or has never existed . . . modern nostalgia is a mourning for the impossibility of return, for the loss of an enchanted world with clear borders and values, a nostalgia for an absolute, a home that is both physical and spiritual, the Edenic unity of time and space before entry into history.
>
> *(Boym 2001: xiii)*

Paradoxically, nostalgia removes the subject from history, while the desire to be mirrored in history and to gain restitution continues. Temporally, the District Six nostalgic narrative is as follows: everything was fine before apartheid, during 'the good peaceful times,' but apartheid destroyed everything and 'bad violent times' commenced. This split-view of before apartheid as utopia versus the dystopia of present and future reinforces a crude binary: utopia as imagined perfection and dystopia as imagined negation. While bearing in mind these imagined extremities, we need to create holding frames for other possibilities to be articulated. The answer lies in heterotopia as the space of mixed mirrors.

> The space in which we live, which draws us out of ourselves, in which the erosion of our lives, our time and our history occurs, the space that claws and gnaws at us, is also in itself a heterogeneous space. . . . The mirror functions as a heterotopia in this respect: it makes this place that I occupy at the moment when I look at myself in the glass at once absolutely real, connected with all the space that surrounds it and absolutely unreal, since in order to be perceived it has to pass through this virtual point which is over there. As for heterotopias as such how can they be described?
>
> *(Foucault 1986: 23–24)*

Heterotopia holds utopian and dystopic views, at times in conflicting temporal sequence and with uncanny affects. Foucault's framing of 'heterotopia' could be usefully deployed in writing a spatial history of colonial and apartheid planning, regulation, and governance of inner-city spaces such as District Six. Moreover, heterotopia provides a conceptual frame to apprehend the public circulation of memories and meanings. The curatorial strategies of the District Six Museum have created such a heterotopic framework that holds a range of imagined memories and ways for identities to be empathically mirrored back to former residents and others who experienced apartheid displacements across the city. But mourning continues as legitimate desires for a spatial return are not being fulfilled. The *realisable hope* for a spatial return overlaps with an *unrealisable fantasy* to return to a pre-injurious emotional state, which is articulated through the trope of the happy family.

The trope of the happy family requires us to distinguish forms of loss experienced during childhood in contrast to those that occur at later life stages. As Steedman puts it, 'All children experience a first loss, a first exclusion; lives shape themselves around this sense of being cut off and denied' (1986: 6). However, such early childhood losses might be better termed an 'absence' (LaCapra 2001) or a 'constitutive lack' (Lacan 1977). Early childhood losses and exclusions produce problematic or faulty mirroring, which are the primary cause of narcissist wounding, a central psychoanalytic conception (Freud 1991 [1914]). These forms of faulty mirroring, to varying degrees, occur within all families, but the happy family trope of nostalgia silences or conceals these difficult or painful aspects of parent-child relationships. These early childhood scenes constitute unconscious frames through which people experience and comprehend (or are unable to register and comprehend) the historical losses inflicted in later family life or by institutions beyond the family such as the apartheid state's dismantling of the District Six community.

The District Six space of today has been called 'haunted' by many (Jonker & Till 2009). These hauntings are fueled by uncanny affects from pre-linguistic and early childhood and compounded by apartheid's severing of children's emotional attachments to homes, streets, and other communal spaces. The uncanny is not necessarily traumatic in origin but its pre-linguistic childhood source means that it is a weird companion to trauma. But there is an unresolved conceptual problem, psychoanalytic theory has not clearly worked out the relationship between trauma and the uncanny (Masschelen 2011), given that both are not directly registered in language. Moreover, I differ with 'event-centred' notions of trauma – such as post-traumatic stress syndrome (PTSD) – and prefer the Laplanchian view, which focuses on the temporal dialectic between early childhood scenes and later life events to produce traumatic traces (Laplanche 1976). In brief, neither the interiority of childhood family scenes nor the intrusions of exterior violence are sufficient explanations in of themselves to explain the afterwardsness of trauma or the uncanny. It is through the *intersection between differing temporalities* where potential answers lie. On the one hand, the present-to-the past reminders from the mirrors of heterotopia (which include nostalgia and dystopia), and on the other hand, the past-to-the-present unconscious transmission of enigmatic messages that emanate from childhood. As mnemonic objects are encountered, or people move through spaces, affects travel from these temporal directions to *converge and evoke* uncanny senses of place.

These are my brief conceptual contours for future research to historically trace the discursive webs between the materiality and affectivity of spaces. We also need to historicise the socio-political, economic, and psychic conditions of possibility that give rise to the mixture of nostalgic memories, fantasies, and myths that frame people's intersubjective attachments to spaces and what unfolds after forced displacement. In summary, the undecidability of the uncanny dominates the District Six memoryscape: where on-going nostalgia has become 'strangely familiar' and the site of erasure continues to be 'disquietly unfamiliar' (Kohon 2016: 13) and incongruent with the city landscape. The site remains as a spatial wound that never heals nor fulfils the desired return to a mythical pre-injurious emotional state.

Finally, the partial regeneration of the District Six community by former residents and the Museum is significant. The Museum named its new public education space 'the Homecoming Centre.' But the community return to the original site has been repeatedly stalled. At a 2004 ceremony, former President Mandela handed the keys to 24 families, but since then only another 115 houses have been built. A further 1,260 families were given the legal right to return and were promised housing, but they are still waiting.

Moreover, with the national re-opening of calls for restitution claims in 2014, a waiting list of a further 3,000 former District Six claimants have been reported (Rawoot 2016). However, the erratic process of return shames both city and national government departments. As the District Six Museum director, Bonita Bennett, put it:

> A pall of disappointment hangs over the District Six community. Despite the successes of this community's land claims, the way in which the Land Restitution Act has found expression in practice has been slow beyond comprehension, occasionally opaque in its processes and often painful.
>
> *(Bennett 2018)*

When the homecoming eventually happens for this community and cycles of social life are rekindled, degrees of restoration of self through community is conceivable. What might be the most meaningful for aging former residents will be to see their children, grandchildren, or great grandchildren living in the new space and community that District Six will become. Sadly, even such opportunities are fading as former residents pass on and both restitution and regeneration are repeatedly deferred.

Notes

1. For histories of District Six, see Jeppie and Soudien (1990). On the work of the District Six Museum, see Coombes (2004), Julius (2008), and Rassool and Prosalendis (2001).
2. For different definitions of the uncanny: the psychoanalytic (Freud 1919), deconstruction (Royle 2003), and phenomenological (Trigg 2012).
3. For popular accounts of family histories from District Six, see Fortune (1996), Ngcelwane (1998) and Ebrahim (1999). On photo albums and family life in District Six, see O' Connell (2015).
4. After the 1990s' first round of restitution applications, over 130,000 Cape Town residents had restitution claims verified. The second round of applications are ongoing since 2016.
5. Note four developments: the infant's recognition that it is a separate body/being from the primary caregiver ('mirror stage' between six to nine months); learning to walk (between nine and fifteen months); and the entry into the symbolic world of language. And, by age five, 'object relationships' between the child's sense of self and the world are unconsciously configured.
6. This was an unrecorded conversation with my brother. All other oral quotations are from recorded interview sessions conducted by the author with Mrs. Hermie Field at Hermanus (2, 4, and 12 April 1999).

References

Adhikari, M. (2005). *Not white enough, not black enough, racial identity in the South African coloured community*. Cape Town: Double Storey Books.
Bennett, B. (2018). Remembering District Six well is our finest weapon. *The Cape Times*, 9 February. Accessed 2 December 2018, www.iol.co.za/capetimes/opinion/remembering-district-six-well-is-our-finest-weapon-13193002
Bollas, C. (1987). *The shadow of the object, psychoanalysis of the unthought known*. New York: Columbia University Press.
Boym, S. (2001). *The future of nostalgia*. New York: Basic Books.
Constant-Martin, D. (2013). *Sounding the Cape: Music, identity and politics in South Africa*. Cape Town: African Minds Books.
Coombes, A. (2004). *History after apartheid: Visual culture and public memory in a democratic South Africa*. Johannesburg: Wits University Press.
Ebrahim, N. (1999). *Noor's story, my life in District Six*. Cape Town: District Six Museum.
Erasmus, Z. (2001). *Coloured by history, shaped by place, new perspectives on coloured identities in Cape Town*. Cape Town: Kwela books.
Field, S. (1996). *The power of exclusion: Moving memories from Windermere to the Cape Flats, 1920s to 1990s*. PhD, University of Essex, Colchester.
Field, S. (2001). *Lost communities, living memories: Remembering forced removals in Cape Town*. Cape Town: David Philip Publishers.

Field, S. (2012). *Oral history, community and displacement: Imagining memories in post-apartheid South Africa*. New York: Palgrave Macmillan.
Field, S. (2013). "Shooting at Shadows": Private John Field, war stories and why he would not be interviewed. *Oral History*, 41 (2): 75–86.
Fortune, L. (1996). *The house in Tyne Street, childhood memories of District Six*. Cape Town: Kwela Books.
Foucault, M. (1986). Of other spaces: Utopias and heterotopias, trans. J. Miskowiec. *Diacritics*, 16: 22–27.
Freud, S. (1919). *The uncanny*. London: Hogarth Press.
Freud, S. (1991 [1914]). *On narcissism*. New Haven, CT: Yale University Press.
Jeppie, S., and Soudien, C. (1990). *The struggle for District Six: Past and present*. Cape Town: Buchu Books.
Jonker, J., and Till, K. (2009). Mapping and excavating spectral traces in post-apartheid Cape Town. *Memory Studies*, 2 (3): 303–335.
Julius, C. (2008). "Digging deeper than the eye approves": Oral histories and their use in the "Digging deeper" exhibition of the District Six Museum. *Kronos. A Journal for Southern African histories*, 34: 106–138.
Kohon, G. (2016). *Reflections on the aesthetic experience, psychoanalysis and the uncanny*. London: Routledge.
Lacan, J. (1977). *Ecrits, a selection*. London: Routledge.
LaCapra, D. (2001). *Writing history, writing trauma*. Baltimore: Johns Hopkins University Press.
LaPlanche, J. (1976). *Life and death in psychoanalysis*. Baltimore: Johns Hopkins University Press.
Mainguard, J. (2017). Cinemagoing in District Six, Cape Town, 1920s to 1960s: History, politics, memory. *Memory Studies*, 10 (1): 17–34.
Masschelen, A. (2011). *The Freudian uncanny in late 20th century theory*. Albany, NY: State University of New York Press.
McEachern, C. (2001). Mapping memories: Politics, place and identity in the District Six Museum. *In*: A. Zegeye, ed. *Social identities in the new South Africa*. Cape Town: Kwela Books, pp. 243–248.
Murray, N., Shepherd, N., and Hall, M. eds. (2007). *Desire lines, space, memory and identity in the post-Apartheid city*. London: Routledge.
Ngcelwane, N. (1998). *Sala Kahle, District Six, an African woman's perspective*. Cape Town: Kwela Books.
O' Connell, S. (2015). Injury, illumination and freedom: Thinking about the afterlives of apartheid through family albums of District Six, Cape Town. *International Journal of Transitional Justice*, 1–19.
Rassool, C., and Prosalendis, S. (2001). *Recalling community in Cape Town*. Cape Town: District Six Museum.
Rawoot, I. (2016). District Six fails to rise from the ashes of apartheid. *Mail and Guardian*, 19 February. Accessed 2 December 2018, https://mg.co.za/article/2016-02-18-district-six-is-failing-to-rise-from-the-ashes-of-apartheid
Royle, N. (2003). *The uncanny*. Manchester: Manchester University Press.
Salo, E. (2005). Negotiating gender and personhood in the new South Africa, adolescent women and gangsters in Manenberg Township on the Cape Flats. *In*: S. Robbins, ed. *Limits to liberation after apartheid, citizenship, governance and culture*. Cape Town: David Philip, pp. 173–189.
Steedman, C. (1986). *Landscape for a good woman: A story of two lives*. London: Virago Press.
Trigg, D. (2012). *The memory of place, a phenomenology of the uncanny*. Athens: Ohio University Press.

4

COLONIAL COMPLEXITY IN THE BRITISH LANDSCAPE

An African-centric autoethnography

Shawn Sobers

Background to research

This chapter will explore the notion of what sites of collective significance in the United Kingdom can mean to people of African descent,[1] whose ancestral lands were formerly colonised by the British Empire. With such a premise, the methodological position of this study needs to be stated from the outset. Although I am of African descent, I am not attempting to make claims for a notional African diasporic community. An autoethnographic methodology (Chang 2008) has therefore been adopted to avoid any attempt at speaking for others, or to be mistaken as such. Autoethnography asks the researcher to draw on personal experience to make broader theoretical points, and to make him or herself as transparent and vulnerable in the process in order to highlight how research works and how knowledge is gathered (Behar 1997). This is a personal, physical, theoretical, and emotional journey exploring these ideas, drawing on primary experience, observations, and broader primary and secondary research. As stated by African American anthropologist Lanita Jacobs-Huey (2002: 791), 'The Natives are gazing and talking back.'

The significance of this study from a personal perspective stems from two specific moments I witnessed, which provides the schematic framework of this chapter, offering two distinct ways the notion of colonial complexity can be considered.

Part one – sites of association

In 2007 I was co-leading a research project which saw us taking groups of African heritage people to various National Trust properties. On different visits I saw both young and older people become upset when the tour guides failed to mention the historical African associations with the properties. This current study will not be looking back at those specific examples (published elsewhere, see Sobers & Mitchell 2013), but will look at a more recent example from 2017 of Newstead Abbey in Nottinghamshire, and how sites of association can be reconciled within the body politic.

Part two – sites of pilgrimage

The second key moment was in 2015, speaking with a friend, writer Judah Tafari, who lamented not feeling there was anywhere in Britain that he can visit with his family that has African heritage significance that was not related to slavery (which he was tired of seeing as the only African narrative). Judah said there was nowhere to visit other than Fairfield House, the former residence of Ethiopian Emperor Haile Selassie I during his exile from Mussolini's invasion from 1936–1940 (Tafari 2016). I have written elsewhere about how Fairfield House has become a site of pilgrimage for members of the Rastafari faith (Sobers 2017), and in this chapter I share the findings of an alternative Rastafari pilgrimage undertaken in 2017 of other sites the Emperor visited during exile. This saw me inventing my own three-day 899-mile solo pilgrimage across England and Scotland.

Embodied ownership of post-colonial landscapes

The two key moments described, for me, combine to present what Stuart Hall argued was the post-colonial 'fissured' notion of identity on the British Isles to which heritage industries need to respond (Hall 1999: 7):

> The emblems of Empire do, of course, fitfully appear in the Heritage. However, in general, 'Empire' is increasingly subject to a widespread selective amnesia and disavowal. And when it does appear, it is largely narrated from the viewpoint of the colonisers. Its master narrative is sustained in the scenes, images and the artefacts which testify to Britain's success in imposing its will, culture and institutions, and inscribing its civilising mission across the world.

Hall presents challenges to formal heritage and educational institutions to think about how they could re-imagine incorporating 'Black British' narratives into their work, and how, similar to Jacobs-Huey's provocation, they can chart how African descent people themselves create their own collections and heritage sites. The research in this chapter draws upon heritage industry discourses, though the muse for this study is not a formal set of recommendations for how they could do their work in a more inclusive manner, but rather the African heritage individuals themselves. As a creative practitioner, namely in photography and filmmaking, I also draw upon those discourses in relation to landscape and identity.

From my experience, as an African heritage person growing up in the United Kingdom, it takes motivation to engage with national heritage sites steeped in British history. Whilst I have enjoyed looking around stately homes and royal palaces, I still cannot shake the question: 'How much of this was funded from the labour of my ancestors?' Like many of my generation growing up in Caribbean families during 1970s and 1980s, we did not take annual summer holidays on the British Riviera or in Europe. We went on holiday about once a decade when we had enough money to summer on whichever island with familial ties (for me, Barbados). So the British landscape always felt alien, heightened when I educated myself in the thorns of history. As described by Gilroy (1993: 1) 'Striving to be both European and black requires some specific forms of double consciousness.' Thus to engage with the British landscape beyond the aesthetic requires a notional motivation for an emotional ownership of that landscape. In her personal exploration of her journey to reconnect with the United States, bell hooks describes her experience of returning to her hometown of Kentucky, which she had left 30 years previously, stifled by the impact of the state's racist history (hooks 2009: 18).

hooks goes onto describe how she read the memoirs of 'black Kentuckians,' of a state steeped in slavery and racial exploitation, yet seeing the 'inventive ways black folks deployed to survive and thrive in the midst of exploitation and oppression' (ibid., 20). After spending 30 years away from Kentucky, and seeing racism as the 'norm' throughout the States, on returning, hooks surprises herself and 'found there essential remnants of a culture of belonging, a sense of the meaning and vitality of geographical place' (ibid). From growing up in a place she began to despise because of its racist history and the 'dysfunctional' impact on the emotional state of people, including her family, she eventually made it her home.

The notion of home is an overarching metaphor in this conversation. Whether a home from childhood or more recent, permanent or temporary, an idealised notion of home provides the premise of a place the dweller should feel comfortable, safe, and rested (Bachelard 1994: 4).

This is the leap asked of children of the former British Empire such as myself: to extend the notion of home beyond immediate dwelling places, to reach into the civic and rural fabric of the United Kingdom. Not to forget the atrocities of slavery and Empire, or even be comfortable with traumatic histories, but to feel comfortable calling this place home, with acceptance of the ownership of its past, present, and future. The call for this leap is as romanticised as this notion of home itself. The key word in is the criteria of a home is *safe* – will children of the Empire feel safe to call this place their home, and will they be given access to full ownership of the United Kingdom's past, present, and future by the resident European heritage hosts? At the time of writing, the United Kingdom is negotiating to leave the European Union, thus complicating matters further.

The provocation behind my challenge is whether the descendants of my children's generation and younger should be influenced by the cultural politics of my generation and older? My heart says no, they need to look to the future and be their own people. My head urges that they need protection from the politics of history in case it resurfaces. Unsurprisingly Stuart Hall (in the same text quotedearlier), pre-empting my concerns, would be cautious of full adoption. He argues that the 'peculiarity' of African descendent British subjects is that, no matter how long we have lived in this country, assimilate, and become 'deeply familiar' within the landscape, due to the fact we remain (physically) black we remain 'culturally inexplicable' to a white population who have never had to give their heritage much thought (Hall 1999: 12). The diasporic and transatlantic bonds of this peculiarity concern the first theme under study: sites of association.

Part one – sites of association

As previously mentioned, when I visit stately homes or other splendid buildings, the thought of connections with slavery and Empire is present. If I am at the property for leisure, or there for work reasons (unrelated to heritage) such as a meeting or conference, the thought remains silent, but I might look for clues, dates, and familiar names. Living in Bristol, which has well-documented connections with the slave trade (Coules 2007), these thoughts are never absent and the same can be said for most cities in Britain, not just port cities. Not only do fine buildings suggest connections to slavery, but any luxurious surroundings concentrate the mind on such connections. However to remain a functioning 'sane' individual, unless connected to a particular project, these ideas are silenced, as to follow up on every association would lead to unviable life obsession.

This persistent awareness is the enactment of the *double-consciousness* to which Gilroy referred. Of looking at oneself from the perspective of those implicated in oppression, Du Bois states,

> It is a peculiar sensation, this double-consciousness, this sense of always looking at one's self through the eyes of others, of measuring one's soul by the tape of a world that looks on in amused contempt and pity. One ever feels his twoness – an American, a Negro; two souls, two thoughts, two unreconciled strivings; two warring ideals in one dark body, whose dogged strength alone keeps it from being torn asunder.
>
> *(Du Bois 1903: 168)*

With this is mind, it can be a relief to enter such a building with an overt agenda to discover a property's connections with slavery, rather than through furtive glances. When I first visited Newstead Abbey, Nottinghamshire (on 20 December 2017), it was as part of a participatory film project about the building's history, drawing upon original (forthcoming) research by Dr. Helen Bates and Dr. Susanne Seymour, Lisa Robinson and the Slave Trade Legacies[2] group. The house's most famous occupant, poet and political reformer Lord Byron, neglected the house to near dereliction, selling it to his former school-friend and slave-trader Thomas Wildman in 1818. Wildman renovated Newstead Abbey to its former glory, adding the extensions and modifications which made it the property it is today. As Simon Brown, the property's curator, said whilst showing me around, 'structurally, the Newstead we see today is Wildman's, not Byron's.'

Fans of Byron, visiting in their tens of thousands annually from around the world, have slave-trader Wildman to thank. An additional personal association arose, I noted that the Wildman family had bought their Jamaican plantation (Quebec) from slave-trader and politician William Beckford, whose son, the novelist and art collector also named William Beckford, is well known in my birth town of Bath; the entire city is overlooked by the prominent landmark Beckford's Tower. Beckford junior inherited his father's wealth in 1770 at the age of ten, making him one of the richest people in the country. Standing in Newstead Abbey, it was if I was transported back to the landscape of my birth town 163 miles away, following a financial trail through the Caribbean. As a child growing up in Bath, I was fascinated by Beckford's Tower, regularly looking out for it high in the landscape. It wasn't until I was in my 30s and lived in Bristol that I discovered the historical connection between the tower and transatlantic slavery I was disappointed that this indelible stain had polluted my previous appreciation.

This link between Newstead Abbey and the city of Bath is apt, as they both play a similar function in how the horrors of the Transatlantic slave trade can be viewed from the perspective of the British Isles, namely keeping up a polite facade in genteel society. Bath plays this role by geography, and Newstead Abbey by metaphor. As Bristol rose to prominence in the early eighteenth century, expanding its economy as a successful slaving port, much of the wealth flowed 30 miles east to Bath, with the merchants spending their plantation-gained money in the spas and health resorts of the city that grew wealthy from entertaining the genteel elite. According to Perry (2013: 108),

> Bath in the 18th Century was not just a larger version of a resort, such as Tunbridge Wells, or even a 'polite' provincial centre, such as York, but a town intimately linked with the fortunes of Britain's transatlantic trade.

Bristol thus shielded Bath as a site of association, the former containing the overt signs of ships, cranes, and warehouses, caught red-handed with blood-money, while Bath washed those hands clean like Pontius Pilate. The facade of the genteel elite is also present in the narrative of Newstead Abbey, with Thomas Wildman using money raised thousands of miles away at his father's slaving plantation to fund the purchase and renovation of the dilapidated Newstead Abbey. The equivalent of Bath's luxurious health spas and entertainment for the elite at Newstead became the figure of Byron himself. By the time Wildman purchased the property, Byron was so famous that Wildman could rebuild the property to become a museum celebrating the life and work of the great poet. Byron has no known connections with slavery (other than selling his house to Wildman), and the conceit of polite society was maintained, with the violence behind the money hidden across the Atlantic.

Hall asks, 'What would "England" mean without its cathedrals, churches, castles and country houses, its gardens, thatched cottages and hedgerowed landscapes?' (1999: 4). He argues that such entities in the landscape become English identity itself, presenting a social fabric as it wants to be, even if at the core of such fabrications it is a veneer folly, facial screens hiding the engine room. Hall suggests 'It is one of the ways in which the nation slowly constructs for itself a sort of collective social memory' (ibid). Both Newstead Abbey and Bath play that role for England in the eighteenth and nineteenth centuries; they present a romantic visual representation of how life could be, all the while funded by a shameful reality.

At Newstead Abbey, Byron left a poem which prophetically speaks to this notion, albeit in an unusual circumstance. Byron's dog Boatswain died of rabies in 1808, and the poem he wrote in tribute, 'Epitaph to a Dog,' is inscribed on the elaborate tomb in the grounds of Newstead Abbey. The poem sees Byron compare the virtuous and loyal nature of his animal companion with the corrupted conceit of human nature. The extract below is what Byron had to say about humanity:

> While man, vain insect! hopes to be forgiven,
> And claims himself a sole exclusive heaven.
> Oh man! thou feeble tenant of an hour,

Debased by slavery, or corrupt by power,
Who knows thee well, must quit thee with disgust,
Degraded mass of animated dust!

What to do with history like this today?

A community group from Nottingham called Slave Trade Legacies have been working together since 2015 to explore connections between Nottinghamshire and the transatlantic slave-trade. The group worked with Newstead Abbey and produced a body of work to begin telling this untold story. I worked with the group to produce a film based on a poem by Michelle 'Mother' Hubbard titled Blood Sugar, which was written through a participatory process with the group. The film is now screened on a monitor in a new room at the entrance to Newstead Abbey, designed to give an overview of the whole property at the start of a visit.

The physical inclusion of this narrative installed into the house is part of an overall refresh of the current interpretive material and includes new approaches to how the Byron and Wildman stories are told. Wildman's links to slavery plantations in Jamaica plays a small but significant part in this redesign.

The Slave Trade Legacies group is approximately 20 in number, a combination of Nottingham-born residents and those who have moved from elsewhere. The group consists of a range of ages (approximately 40–80), all self-defined as black. They have taken it upon themselves to take ownership of the Newstead landscape, not for their own benefit, but for the next visitors to that space. The embodied ownership of site-specific action sees the group become active members of civil society, custodians of the knowledge they have unearthed, presenting it back for public consumption, rather than their own private knowledge banks.

In this instance, the leap to ownership of the landscape comprises action beyond awareness. I would argue that it is possible to still feel ownership over a landscape without then having to take action to rectify previous injustices. As with hooks, working on our interior personalities and feeling comfortable and safe in a place is enough; it is already a big leap. There is a difference here between individual embodied knowledge, and activism of the collective body politic.

Autoethnographic postscript to section

As I was typing those last words about Beckford's Tower, I received a private Facebook message from someone I didn't know. She asked if I was related to any Sobers family in Barbados, Trinidad, or Grenada, as she was tracing her family tree. I replied saying all the Sobers family I knew were from Barbados, and we started trying to fit the pieces together. No direct link has been made yet, but it is still early days.

I mentioned it to my 16-year-old daughter, and she was immediately interested. Unbeknown to me she had been conducting her own research into the family name. She shared with me some of the things she discovered, and asked me about slavery, and where the plantation would have been in Barbados. We decided we would visit it one day. I told her about Beckford's Tower, which she also knew well from the Bath landscape. She guessed that most of the buildings of Bath and Bristol were connected with the slave trade. I replied saying yes, it is good to be aware but not consumed by that awareness.

I told her we would resume the conversation, and I returned to my laptop and wrote this Postscript. I shared with the woman the web page my daughter had found about the family name, and she replied saying she had previously seen it and that that is what inspired her to start the family research.

It seems the ghosts are calling home.

Part two – sites of pilgrimage

For the second part of this research, I invented my own pilgrimage and went on a three-day 899-mile journey alone. This pilgrimage took place between 21 and 23 December 2017, the day after I visited Newstead Abbey.

For many years I have volunteered at Fairfield House (Bath), the former residence of exiled Ethiopian Emperor Haile Selassie I and his wife, Empress Menen Asfaw. The royal household lived there from 1936–1943, with their children (Princes and Princesses), grandchildren, government advisors, Ethiopian Orthodox priests, nannies, cooks, and other members of the household (Bowers 2016: 91). Fairfield House has now become a pilgrimage site for the worldwide Rastafari community who view the Emperor as their deity (Tafari, 2016: 221–251). It is also a site of importance for many Ethiopians who live in, or visit, England, and members of the Orthodox Church. Visitors to the house have included members of the Ethiopian Royal Family themselves, some of whom lived there with the Emperor in exile. The Emperor is remembered fondly by the people of Bath, and in 1999 I directed a documentary for ITV-West (*Footsteps of the Emperor*) interviewing local people who either worked for the Emperor, met him, had family connections, or with other stories to tell. In 1958 the Emperor gifted the house to the City for use by the elderly. It is still used for that purpose today, also containing a museum, art gallery space, and community resources.

During his time in exile, the Emperor visited a number of sites around the United Kingdom. In the short space of time I had for my pilgrimage, I prioritised the furthest away, which happened to be Wemyss Bay, on the coast of the Firth of Clyde, Inverclyde, Scotland. The Bay was the site of the now demolished Wemyss Castle, where the Emperor visited in 1936, immortalised in a photograph. I then worked my way back south visiting Kelburn Castle, Abbey Hotel, and Holy Trinity Church in Malvern, where he was known to stay, visiting his granddaughters at a nearby boarding-school. Here, I discuss my experiences in Wemyss Bay.

For this pilgrimage I was interested in the impact this African presence had on the British landscape, and how the idea of this African presence could connect me with an unfamiliar British landscape. I was not looking for physical or anecdotal memories, though of course these would be welcomed. Neither was I looking for any spiritual reawakening or epiphany. The historical context provided me with a good excuse to travel, so I grasped it. I linked the pilgrimage to an Arts and Humanities Research Council (AHRC) funded project titled, 'Ethiopian Stories in a British Landscape.'

There were a number of rules I set myself in advance, as follows:

- **I would go alone.**
  I wanted this to be as much about self-discovery than anything else, and wanted it to be as authentic to my core self as possible. As a natural introvert, going along with a crowd and I would become little more than an observer. I needed to be *silent* (Maitland 2009: 25).

- **I would only sleep in the van, and not buy anything new that I didn't already own.**
  Even though I could draw on AHRC funding I didn't want to, and spent as little money as I could. I hired a van and that was my home for three days. I decided to only take with me things I already owned, not buying news things to make myself comfortable, which would be tempting, but inevitably new things go to waste.

- **I would only use my camera phone.**
  This was a difficult one. As a photographer I knew I would probably want to make a book from images of the journey. I also knew, though, that I didn't want it to be all about being behind a camera, and

I engage differently with my surroundings with a large camera in tow. On a more theological/philosophical level, limiting my camera use was a form of sacrifice: humbling myself. In the photographic world there's a masculine tradition of men undertaking long endurance hikes, large cameras in tow, akin to a hero's journey (Bate 2009: 98; Sontag 1977: 15, 90; Wells 2011: 110, 114; Taylor 1994: 61).

- **I would not plan anything or tell anyone I was arriving, and would leave everything to chance.**
  I am a big fan of serendipity. I didn't want anything to feel forced and wanted to remove any sense of expectation, my own and other people's. I was happy to see what naturally occurred, and be open to opportunity.

- **I would not broadly advertise that I was doing this, I would not tell social media I was doing this or post photographs during my journey.**
  I wanted to respect the solo element of this journey, and only tell a select few people who needed to know I'd be away, or who happened to have contacted me during that period. Updating my timeline as I went would break the silence, and arguably be too ego-orientated.

- **I would return to Bristol by 6 pm, Saturday 23 December, to look after my children.**
  It could be said that going on a pilgrimage is the most selfish of acts, as, unlike meditation, you are removing yourself physically as well as psychologically from those around you, albeit temporarily. I say this unapologetically, as I think it is everyone's right to have time alone, and fully respect that not everyone is in this privileged position. It had been in my diary that I was to be with the children that evening, so it was important to respect that.

As with all journeys there is so much that could be said. Here I have discussed key moments from Wemyss Bay under themes which I feel are pertinent to the topic here. I have subdivided the themes into chronologically ordered parts, so they make sense told in context. The themes are; being present, serendipity, vulnerability, and talking with strangers.

Being present

The van arrived in Bristol at 8 am, but before I could collect it, I first drove to Bath at 7.30 am to drive my parents back to Bristol for a hospital appointment. I drove them home to Bath afterwards, and then returned to Bristol to pick up the van. I mention this as I realised, while driving to Bath first thing in the morning, that I was not doing my family duty *before* my pilgrimage started; I realised that this *was* the beginning of my pilgrimage. One of my favourite quotes is 'The only Zen you can find on the tops of mountains is the Zen you bring up there' (Pirsig 1989: 248). I think of this quote often, and thought about it whilst embarking on this journey. It came into my mind again as I was driving to Bath to pick up my parents. They are my mountains.

I picked up the van, leaving Bristol at about 3 pm. Twenty minutes into my journey, I realised that I had forgotten my camera tripod. It was hidden under clutter in the boot of my car. Even though I had vowed to only use my camera-phone, I brought along my DSLR as a backup, and the tripod also had an attachment for my phone. So rather than being in the present and preoccupied with romantic thoughts about Rastafari philosophy and tracing the footsteps of Haile Selassie, I could only think about the missing tripod, and berating myself for being so stupid. After about an hour more of this distraction as I drove, I eventually decided to fix the situation, making a detour to the nearest city (Worcester) to buy a tripod. This added at least an hour and a half onto my journey, and I had already stopped at most service stations to write my journal along the way. By the time I bought the new tripod it was 6.08 pm, and in real terms

I was only about two hours outside of Bristol. As I was leaving Worcester my friend Rob rang me to check my progress. I complained to him I was going a lot slower than I had planned. He reminded me that the diversions, the pauses to write, and whatever other experiences I had were the point of the journey, and that any expectations of time were entirely self-imposed. It was a good reminder. After that I progressed as slowly as I wanted to.

Serendipity

I am interested in how, when you give certain ideas your attention, links can be made. Listening to BBC Radio 4 in the van, a programme comes on with comedian Susan Calman. She was trying her hand at baking, something she cannot do, being taught by a man called Selassie. She says he was on the show the *Great British Bake Off*, but as I have never watched it, I had no idea who he was. I later find out the spelling of his name is Selasi (surname Gbormittah), but it is pronounced Selassie.

A few hours later, at around 1 am, still driving, I turn to BBC Radio 1. Reggae comes out of the speakers. The presenter/DJ, (whose name I later discover as Toddla T), takes listeners to Jamaica as he speaks with veteran reggae singers, local Rastafari, and other Jamaican locals about certain sites that were important to the history of reggae music. They talk about trying to get certain buildings protected by UNESCO, visit an old record shop and talk about how they get visitors coming there specifically from all over the world in a form of pilgrimage to reggae. They mention roots reggae and the importance of Haile Selassie. I could not believe my ears. No one would believe this if it was scripted.

Vulnerability

Not long after passing Lockerbie I pull over on the side of the road to take a photograph of the landscape. I put the camera on the new tripod and try a few long exposures. I (uncharacteristically) decide to take a few self-portraits, but only want the back of me in the shot. I set the camera to timer to give myself chance to get in position, and still keep the setting to long exposure, so I have to stand in position for quite a long time. I hear a beep from a passing motorist, and it immediately dawns on me it must look like I am urinating on the roadside. I am mortified. My double-consciousness kicks in. I think the people out here probably do not see many people of African descent, let alone Rastafari, and when they do they think he urinates in full public view. Even though this is no doubt a comedic moment and I do chuckle to myself, I also feel a certain shame, and it stays on my mind. Such is the burden of the double-conscious mind. This episode confirmed to trust my instincts and not use my DSLR but only my camera phone as I previously vowed.

Talking with strangers

I arrive at Wemyss Bay, at the site where the castle once stood. I find the remaining castle flagpole and photograph it among the surrounding houses. It is now, by the looks of it, quite an upper-class housing area. I walk around and take some photos and, becoming slightly self-conscious, get back in the van. As I drive away, I suddenly become a bit braver and flag down a woman driving slowly towards me in a BMW. I ask her for some clarification of my surroundings. We speak through vehicle windows, and she voluntarily tells me about Haile Selassie's connection with the area, and of the tree he planted which is at the end of the road. I knew he planted a tree at the castle in 1936, but did not think I would find it, presuming it was impossible to try. I was now being shown it without asking. It is now just a stump, rather than a tree, but it is still there. The lady was very proud of it, as were other locals I spoke to. They all had fond things to say.

Figure 4.1 Front pages of *The Daily Record*, 1936 and 2017.

Being present

The reason I have come to Wemyss Bay is because it is the location of the first ever colour photograph in a newspaper in the world: *The Daily Record* in 1936 (Hutchinson 2003: 225). The photograph was a portrait of Haile Selassie, his daughter Princess Tsehai, and local lady Mrs. Olive Muir.³ As a Rastafari photographer this has special resonance for me, and was first pointed out to me by former researcher at Fairfield House, Kayley Porter. Whilst there I buy that day's edition of the paper. On the front page are Prince Harry and Meghan Markle, soon to be a Duchess: an African American. The front pages read (Figure 4.1)

'Another Colour Revolution: First News Photograph in Nature's Hue Ever Published.'
The Daily Record – Monday 22 June 1936.

'Harry and his Jewel in the Gown.'
The Daily Record – Friday 22 December 2017

The irony of the 1936 headline 'Another Colour Revolution', read in context of the 2017 headline story, was not lost on me.

Being present

I leave Wemyss Bay, visit Kelburn Castle, Malvern's Abbey Hotel, and Holy Trinity Church, with stop-offs to sleep along the way, and many more stories to tell. I reach Bristol at 6.15 pm on 23 December. I rang my daughters on the way and said I would be 15 minutes late; my daughters are teenagers, up in their bedrooms doing their own thing. I fall asleep on the sofa.

Conclusion

This chapter attempted to present an African-centric engagement with the British landscape through autoethnographic research, firstly through focus on a particular site and an exploration of what a body politic ownership of that space can look like, and then through a more personal embodied immersion into the landscape. My thoughts are still emerging on this topic; I think of both experiences as forms of pilgrimage. I think of accidental pilgrimage and how a state of true pilgrimage can begin even after you have physically arrived. How pilgrimage can be a form of archaeology, unearthing the significance of a place, and the importance of being fully present in that place, not just physically there.

This journey saw pilgrimage enacted as a form of creative expression, as much drawn and responding to a site of narrative, rather than primacy of the physical place itself. I look through my photographs from both journeys and begin to see each image as a pilgrimage in its own right. Photographer and writer Robert Adams (1996) tells us that good landscape photographs contain three elements: geography, landscape, and metaphor. I think he is absolutely right. I pledge to take full ownership of all three in life, and try to not let the weight of history get in the way.

Notes

1 Throughout this chapter I use the terms 'African descendent' and 'African heritage' as opposed to 'African Caribbean' or 'Black,' other than in direct quotations.
2 The Slave Trade Legacies is facilitated by project director Lisa Robinson for the Nottingham community organisation Bright Ideas.
3 In *The Daily Record* article, Olive Muir, from Glasgow, is mentioned as chaperone to Princess Tsehai, but she is an interesting person in her own right – an artist, author, and activist. After this day she painted a large portrait of Haile Selassie, which now hangs in the museum at Fairfield House.

References

Adams, R. (1996). *Beauty in photography: Essays in defence of traditional values*. London: Aperture.
Bachelard, G. (1994). *The poetics of space: The classic look at how we experience intimate places*. Boston: Beacon Press.
Bate, D. (2009). *Photography: The key concepts*. Oxford: Berg.
Behar, R. (1997). *The vulnerable observer: Anthropology that breaks your heart*. London: Beacon Press.
Bowers, K. (2016). *Imperial exile: Emperor Haile Selassie in Britain 1936–40*. Bath: Brown Dog Books.
Chang, H. (2008). *Autoethnography as method*. Walnut Creek, CA: Left Coast Press.
Coules, V. (2007). *The trade: Bristol and the transatlantic slave trade*. Edinburgh: Birlinn Publishing.
Du Bois, W.E.B. (1903 [2007]). The souls of black folk. In: B.T. Du Bois and W.E.B. Douglass, eds. *Three American classics*. Washington, DC and New York: Dover Publications, pp. 159–331.
Gilroy, P. (1993). *The black Atlantic: Modernity and the double consciousness*. London: Verso.
Hall, S. (1999). Un-settling 'the heritage', re-imagining the post-nation, Whose heritage? *Third Text*, 13 (49): 3–1.
hooks, b. (2009). *Belonging: A culture of place*. London: Routledge.
Hutchinson, I.G.C. (2003). Scottish newspapers and Scottish national identity in the nineteenth and twentieth centuries. In: H. Walravens and E. King, eds. *Newspapers in international librarianship: Papers presented by the Newspapers at IFLA general conferences*. Berlin: De Gruyter Publishers, pp. 215–228.
Jacobs-Huey, L. (2002). 'The natives are gazing and talking back': Reviewing the problematics of positionality, voice, and accountability among 'native' anthropologists. *American Anthropologist*, 104 (3): 791–804.
Maitland, S. (2009). *The book of silence: A journey in search of the pleasures and powers of silence*. London: Granta.
Perry, V. (2013). Slavery and the sublime: The Atlantic trade, landscape aesthetics and tourism. In: M. Dresser and A. Hann, eds. *Slavery British country house*. London: English Heritage, pp. 102–122.
Pirsig, R. (1989). *Zen and the art of motorcycle maintenance*. London: Vintage.
Sobers, S. (2017). Ethiopian stories in an English landscape. In: H. Roude-Cunliffe and A. Copeland, eds. *Participatory heritage*. London: Facet Publishing, pp. 143–152.
Sobers, S., and Mitchell, R. (2013). 'Re:Interpretation: The representation of perspectives on slave trade history using creative media. In: M. Dresser and A. Hann, eds. *Slavery British country house*. London: English Heritage, pp. 142–152.
Sontag, S. (1977). *On photography*. London: Penguin.
Tafari, D. (2016). *Revelation of an Emperor: Journey in exile*. Manchester: Devon Publishing.
Taylor, J. (1994). *A dream of England: Landscape, photography and the tourist's imagination*. Manchester: Manchester University Press.
Wells, L. (2011). *Land matter: Landscape photography, culture and identity*. London: I.B. Tauris Publishing.

5
MAPPING MEMORIES OF EXILE

*Sébastien Caquard, Emory Shaw, José Alavez,
and Stefanie Dimitrovas*

Introduction

Memories do not land particularly well on maps. Indeed, while memories are spatial, they are also fluid, evanescent, and ever-changing, and their geographies fluctuate along with the individual that bears them as well as with the context and form in which they are expressed such as oral life stories, memoires, or diaries. On top of being difficult to characterise qualitatively or quantitatively, they are challenging to circumscribe spatially. In other words, the elusive geographies of memories do not easily overlap with the rigid Euclidean structure of the conventional map. Thus, to map memories would inevitably require that memories be distorted in a way that fits a rigid cartographic structure, or to distort this structure in a way that would accommodate memories. In this chapter, we propose to explore these two modes of distortion by mapping the life stories of exiles.[1]

The transformation and visualisation of memories using conventional cartographic frameworks have been carried out in several ways. Geographers Kwan and Ding (2008) developed an approach using GIS tools to represent memories of daily life, collected through interviews, to study how the events of 9/11 affected the use of urban space by Muslim women in the United States. Historian Vincent Brown developed an interactive map to study the 1760–1761 Slave Revolt in Jamaica based on 'diaries, letters, military correspondence, and newspapers,' which enabled him 'to observe the tactical dynamics of slave insurrection and counter revolt' (Brown 2015: 136–137). According to historian Tim Cole and geographer Alberto Giordano, historical GIS can 'provide a crucial context for rereading and better contextualising one of the key sets of sources in Holocaust Studies: diaries, memoirs, and oral testimony' (2014: 151). Digital mapping technologies have also been used by communities in Spain to locate and represent mass graves and 'abandoned places of memory' related to the Spanish Civil War of 1936–1939 (Ferrándiz 2014). The lives of Spanish Republicans who left Spain after fighting Franco have also been mapped to study the types of emotions these exiles developed with places over time (Dominguès et al. 2017). Based on these examples, it is clear that conventional maps present a real potential to represent and study different aspects of memories, especially when they relate to exile.

Although, beyond its potential, the previously mentioned authors are all fully aware of the limits of this approach, as noted by the loss of the many collective and intimate experiences of exile and migration when plotted on conventional maps such as state-centric maps (Campos-Delgado 2018). Alternative forms of mapping have been developed to propose representations which more appropriately fit the geographies of these personal experiences and memories associated to exile: places of departure and destination, mobility and immobility, hope and despair, violence and relief. Artists and academics have worked with asylum seekers and refugees to facilitate the creation of their own maps of personal experiences of exile (Miller et al. 2011; Mekdjian et al. 2014), while historians have embraced inductive visualisation[2] as an alternative cartographic approach to mapping survivor accounts of the Holocaust based on the content and meaning that is unique to each testimony (Knowles et al. 2015). Meanwhile, cognitive mapping has been used as a way 'to challenge the invisibility of irregular migrants' stories' (Campos-Delgado 2018: 3). Conventional cartographic symbols have also been revisited by cartographers who were invited to transform memories of border crossings by Syrian migrants who had fled war into graphic symbols that captured the diversity of these human experiences (Kelly 2016). These alternative approaches to mapping illustrate the impulse and need for developing spatial representations that are better aligned with the experiences and memories of exiles, potentially at the expense of a precise alignment with absolute, Euclidean space.

In this chapter, we propose to further explore the potential that each of these two broadly identified approaches have for mapping the memories of exiles through a three-phase project. The first phase focused on methodological aspects of the transformation of memories – narrated as life stories – into conventional cartographic data structures and symbologies. In the second phase, these digital maps were put into conversation with the original authors (i.e. 'storytellers') of those memories: first as a means of validation, but also as a way of triggering further storytelling. The final phase consisted of a series of alternative mapping workshops in which exiles were invited to work in close collaboration with artists and cartographers to co-create their own life story map. We conclude by discussing the process and results of these three consecutive exercises and interpreting the observed experiences of storytellers and how they relate to broader, more collective spatial narratives about exile.

From memories to maps: charting the life stories of exiles

The *Mapping the Life Stories of Exiles* project began in 2013 with a collaboration between the Centre for Oral History and Digital Storytelling (COHDS) and the Geomedia Lab at Concordia University. In the context of an earlier project called the *Montreal Life Stories Project*, the COHDS collected over 500 video recordings of life story interviews from exiles living in Canada (High 2014). Ten interviews were selected to be mapped by the Geomedia Lab using conventional cartographic approaches: five from Montreal residents of Rwandan and Haitian origin respectively.

These interviews were not originally designed to be mapped in that explicit questions about places were not central nor were explicit spatial indices prioritised, but the ten selected stories were particularly rich spatially, with clear locations associated to specific events and detailed descriptions of geographical movement. To map these stories, we first developed a methodology that guided the transformation of narratives into geographical data. Two 'analysts' carried out a listening of each story to identify and characterise 'story units,' or what Gérard Genette (1972) called 'narrative segments,' interpreted here as sections of the interview that were spatiotemporally discrete. Each unit corresponded to an entry in a database with associated attributes such as a descriptive summary of the story unit's qualitative content, the geographic location to which it had been associated, the characters involved, a historical time period, as well as other binary indicators such as whether the unit involved violence, forced migration or was explicitly positive or negative.

Faced with how complex and ambiguous spatial and temporal references tend to be in narratives, it took nearly two years to develop a methodology that could consistently retrieve relevant information from more than a single story. Since each storyteller had different expressive styles, it was challenging to define the place and time of story units in a way that was consistent both within and between stories. For example, it was easy to associate a place and time to Alexandra Philoctète's traumatic memory of two prisoners dying in front of the library in her hometown of Jérémie. On the other hand, it was impossible to locate the segment in which Gisimba talked extensively about grandparents' personalities as well as their views on religion,

Figure 5.1 A screenshot of a map designed with Atlascine that combines the five Rwandan stories (each with its own colour) and differentiating places that were associated with violence (opaque) from places that were not (translucent) in the Lake Kivu region of East Africa.

kinship, and social class. Narrative segments of expression like this one were recorded in our database but not associated to any location and were therefore not mapped.

Geovisualising these data was done using Atlascine, an online mapping application developed by the Geomedia Lab at Concordia University in collaboration with the Geomatic and Cartographic Research Center at Carleton University using the open source software *Nunaliit* (see Caquard & Fiset 2014). This application allowed us to transform the story units into proportional symbols whose sizes reflected the overall importance of different places within a story and whose colours and opacity levels could represent any associated attributes. We produced at least four different maps per story, in addition to a series that combined stories to explore collective themes such as forced displacement and violence (Figure 5.1). Maps which aggregated memories from multiple stories enabled the identification of junctions between storytellers. For example, places known to be associated with violence by the respective diasporic communities emerged, such as Gikongoro in Rwanda and the Casernes Dessalines in Port-au-Prince, Haiti. There were also locations of intersecting positive memories, such as the Côte-des-Neiges neighbourhood in Montreal.

From maps to memories: conversing with exiles

In the second phase of this project, our goal was to solicit feedback from the storytellers themselves on the maps produced during the first phase. Not only was this a way to validate, question, and improve our interpretations and mappings of their stories, it was also an opportunity to use maps to facilitate more affective recollections of memories which may not have been captured by the discursively smoothed, narrated recollection of past events characteristic of spoken life stories (see Martouzet et al. 2010). In other words, this was an opportunity to move away from 'knowing from' storytellers to 'knowing with' them (Greenspan 2014).

This phase began by presenting the Atlascine maps to members of Maison d'Haïti and Page Rwanda, respectively: these were organisations which acted as key liaisons with the Rwandan and Haitian communities of Montreal. After incorporating feedback from these sessions, the individual storytellers were contacted for one-on-one meetings concerning the maps of their stories. Four of the ten storytellers agreed to meet with us: three of Rwandan origin (Gisimba, Emmanuel Habimana, and Emmanuelle Kayiganwa) and one from Haiti (Alexandra Philoctète).³ These meetings took place at Concordia University during the winter of 2016–2017 and lasted between one and two hours each.

Overall, feedback on the maps was positive. All storytellers easily recognised the mapped spatial trajectories as their own. Emmanuel Habimana and Alexandra Philoctète were pleased by the transparent, unapologetic precision of the maps, describing them as refreshing and unromanticised compared to other media. Alexandra Philoctète pointed out that her map accurately reflected the more detailed memories she had of life in New York and Quebec City, versus the much vaguer and distant ones she had of Jérémie, in Haiti. Gisimba recognised that, as a whole, his map represented his world quite well.

However, all the participants acknowledged difficulty in understanding the maps without guidance, especially regarding the categories that the cartographers had chosen to qualify story units. Although we chose categories that emerged heuristically from each story while loosely following the themes in the interviews such as family, education, work, and experiences of violence, Gisimba thought these were crude and oversimplified, while Emmanuel Habimana was surprised to not find certain elements he thought were significant to the collective memory of the Rwandan diaspora such as religion. Furthermore, our earlier meeting with members of Maison d'Haïti had also revealed a lack of satisfaction regarding the chosen categories: they commented that such a content analysis of memories, and their cartographic symbology, required direct collaboration with members of the diaspora. This greatly emphasised the need to create a dialogue with storytellers during the mapping of their life story.

What was most noticeable about these meetings was how the storytellers' interactions with these maps triggered further reflection and recollection. These usually began with clarifications about the spatiotemporal accuracy and precision of the events mapped, as well as the filling-in of new events that were not expressed in their original life story. For example, Alexandra Philoctète provided some clarifications regarding what she then saw as the correct sequence of events in her life by correcting the dates we had attributed to specific events and by providing more geographic precision to some of those that had been associated with course locations. This effort to correct and improve the map led to the divulgation of more memories from her life, further adding to the temporal scope that had theoretically already been covered by the map. For example, upon noticing that Vancouver was not featured anywhere, Alexandra elaborated on the dramatic story of her brother who had lived there (unmentioned in the original interview), and his recent passing. In this way, the maps served as catalysts for imaginative wanderings, associations, memories, and words (Ryden 1993). Alexandra had figuratively inscribed yet another narrative onto the map, a more contemporary iteration built off of her interaction with it. Another storyteller, Gisimba, was surprised to see that his story showed very little violence in Burundi. This inspired him to describe several shocking and violent incidents he had witnessed during his stay in Bujumbura whereas, in the interview, he had only talked about his experience as a student there in more general, neutral terms. In brief, the map, through its concretisation of spatiotemporal memories, provoked the reiteration of untold ones, confirming the idea that maps, like photos and videos, can serve to flush out and elucidate stories and memories (Martouzet et al. 2010; Palmer 2016).

These meetings also shed light on some aspects of the previous mapping phase. Emmanuelle Kayiganwa, who noted having difficulty reconstructing and communicating certain traumatic memories in consistent and precise terms, acknowledged that such abstracted cartographic symbols might appropriately communicate difficult and sensitive themes to audiences such as Rwandan youth or the general public, a comment which was also made during our meeting with members of Page Rwanda. While this heavily abstracting aspect of maps has been criticised for its power in facilitating dehumanised decision-making processes including militaristic planning and even genocides (Harley 1988), these remarks might suggest that such conventional mappings can also be used by individuals to communicate memories and experiences too difficult to fully express and convey through more realistic and evocative media. Yet Emmanuelle also shared a critique of this mapping method. When observing the trajectory of her journey from Kigali to Goma, she was reminded of how a friend of hers who had made the journey with her described her experience in a way that she did not remember at all, and thus Emmanuelle ended up questioning the relevance of spatial detail in her own story. For example, she could not even begin to remember the number of times she was forced to move during the crisis in Rwanda: 'I really try to remember, how many times we moved: I simply cannot. I cannot. How many times we moved in ten years, before my parents were deported to Nyamata, I have not a clue. I have not a clue' (Emmanuelle Kayiganwa, translated by authors). These remarks echo what Gisimba had also emphasised in his original life story recording about his own memories whereby he 'remembers people, not places.'

These observations point to the bias of spatial reification inherent in any attempt to map life stories with conventional cartographic methods. As described in this section, the difficulty in representing the less explicitly spatial moments in stories is also foreshadowed by the extraordinary variability in memory itself. Although this phase showed us how maps can reactivate memories, refine others, and trigger new ones (Martouzet et al. 2010), it also made it clear that the distance between cartographer and storyteller, which characterised the first phase's mappings, beckoned a resolution. Furthermore, these representations of memories remained constrained by the Euclidean structure of conventional maps which, while novel to some, might not be sufficient for expressing more personal and intimate relationships to places. This led to the third mapping phase focused on individualised mappings.

A participatory workshop on memory- and map-making

The third phase of the project was designed to enable storytellers to actively participate in the mapping process through the design of their own life story map. Although our original goal was to develop this phase in collaboration with the storytellers from the first and second phases, only one consented to joining us for this exercise. Thus, during the spring of 2017, we solicited artists interested in working with storytellers who identify as refugees or exiles to facilitate the development of their own life story map, as well as the said storytellers. Inspired by a series of mapping workshops organised in 2013 with asylum seekers in France (see Mekdjian et al. 2014), our intention was to facilitate a collaboration between artists, mapmakers, and storytellers to map life stories using the medium and method of their choice.

Participants ended up including four storytellers, three artists, and four mapmakers from the Geomedia Lab. Three meetings were organised throughout May and June of 2017 to provide a space for the participants to team up and develop their collaborative works. By the end of the first meeting, two of the four storytellers had voluntarily paired up with two mapmakers, while the two others bonded with two of the artists. These four duos then worked together during the two following meetings and throughout the summer and fall seasons to finalise their projects. Results demonstrate four very different ways of mapping the memories of exiles.[4]

Navigating memories with maps

Alexandra Philoctète was the only storyteller who participated in all three mapping phases. Having appreciated the Atlascine maps of her life story interview, especially for their spatial precision, she suggested to further develop them, thereby choosing online mapping as a medium for her project. She was therefore matched with Stefanie Dimitrovas, a mapmaker from the Geomedia Lab who had participated creating her web maps.

Alexandra is a writer whose work includes co-authoring the book *D'Haïti au Québec: quelques parcours de femmes*, text was therefore an important creative medium for her. Also, during her interview in phase two, she expressed an interest in incorporating music into her life story map. Thus, with the help of a comparative analysis of online story mapping applications (see Caquard & Dimitrovas 2017), ESRI Story Maps was selected due to its capacity to combine text, music, and images with an interactive map.

Throughout the process, text became the primary medium of the project; Alexandra ended up producing a book entitled *Rencontre avec l'autre: l'histoire de vie d'Alexandra Philoctète*, completed in February 2017, which chronicled events in her life from the time she left Haiti in 1956 to 2016. It also became clear that the multi-media web mapping application was no longer appropriate as the primary medium. As a result, the project evolved in two directions. The first consisted of the said book, with images and musical references in the form of footnotes and with all the places where Alexandra had either visited or lived in marked in colour. These demarcated place names linked to the second part: the ESRI Story Map which became secondary. Conceived as an abridged version of the book, each of the six chapters was represented by colour-coded points showing the geography of different eras in her life. These were associated with quotes from the text summarising the memories associated to each place (Figures 5.2a/b).

This collaboration revealed two key limitations with web-mapping life stories. For one, it became clear over time that Alexandra was motivated to reproduce her life story in a way that went beyond a spatial perspective, relegating the map to a supportive role. Second, while she expressed her enthusiasm for the project and its outcomes on several occasions, Alexandra's frustration with how the artefact created with ESRI Story Maps was only available online and could not be archived locally (except as a pdf, which implied losing the application's interactivity and most of its media) revealed how leaving a legacy to her friends and family was an important motivation for recording her memories.

Mapping to share memories

Another partnership took shape as a result of the workshop between C.K., a Montrealer of Rwandan origin who immigrated to Canada in 2008, and Emory Shaw, a student in the Geomedia Lab. C.K. came in with a very clear vision: he wanted his story digitised, made tangible and public, and he wanted to produce

Figure 5.2a Screenshot of the ESRI Story Map associated to her life story. The map can be viewed at arcg.is/0S1HP1.

Figure 5.2b Alexandra Philoctète (left) and Stefanie Dimitrovas (right) designing Alexandra's life story map.

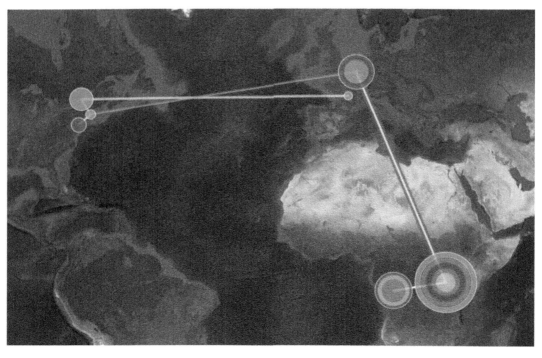

Figure 5.3a A screenshot of the final map at the global scale. A video recording of the map as it evolves over time with the recorded interview can be viewed online at vimeo.com/246040198.

Mapping memories of exile

Figure 5.3b C.K. (left) and Emory Shaw (right) designing C.K.'s story map.

a map similar to the ones designed during the first phase of the project using Atlascine. Though several other cartographic methods and tools were proposed to him, he had already taken a liking to Atlascine's minimalistic symbology and capacity for abstraction. Since he never recorded his life story, the first part of this collaboration consisted of this recording. Beyond his clear intention to share his story via maps, the process was also constructive for him personally: having mentioned that he wanted to write down his story at some point, this exercise started off the process.

Since C.K. told his story in order to create a map, it contained more spatial and temporal precision than the life story interviews from the Montreal Life Stories Project mapped previously. He navigated his memories in a careful, brief, and structured way, scaling from the joys of stable, daily life spanning months and years, to the very crucial minutes, hours, and days when in times of danger, deception, and dread. All the story units therefore comprised of locations which were almost universally at the city level, unless during key moments of movement when the story became most spatiotemporally precise. The mapping process also differed from the first phase since it was created collaboratively with the storyteller. After converting the content of two storytelling sessions into a story unit database, C.K. and Emory met again to review the database and attribute a generalised emotion to each story unit as well as colors to represent these qualities. In addition, while the result was a digital map that visually resembled the maps designed during the first phase (Figures 5.3a/b), the final product combined the map with the story's audio recordings to produce a video featuring the narrative accompanying the temporal unfolding of the map. In this collaboration, the conventional map was perceived as a powerful way of sharing and conveying individual memories that are part of a collective memoryscape.

Mapping as memory-catching

The third collaboration was carried out by the visual artist Lilia Bitar and Italian-born illustrator Nasim Abaeian. Nasim, whose father is Iranian and mother Italian, reflected on her identity as follows: 'I am not a refugee, but I have immigrated so much in my life that I don't really know where [home is] for me

Figure 5.4a Two screenshots of the 'Tree Map' viewable online at prezi.com/0pg3auer9jwc.

Mapping memories of exile

Figure 5.4b Lilia Bitar (left) and Nasim Abaeian (right) working on their tree map.

anymore.' Similarly, Lilia, who identifies as Russian-Syrian, has migrated multiple times throughout her life: from her childhood in Algeria, to other locales in the Middle East such as Abu Dhabi, to Montreal in 2001. The hybrid identities and displacements that characterised their lives, as well as their artistic sensibilities, gave them common ground for collaboration, enabling them to develop their map in metaphorical terms. As described by Lilia: 'The map is an emotional map. It's a map of memories. It took the shape of a living map, like a parabola or a dreamcatcher, so it works as a trigger. We [made] the map as our dreamcatcher or memory catcher, and we gave time [for] the memories to come.'

Nasim and Lilia conceived the map as a tree, with each branch representing one chapter and place of Nasim's life: Genoa, Tehran, Dubai, Savannah, and Montreal, each of which was illustrated by artefacts that were sentimental to Nasim. A key feature of this tree map (Figures 5.4a/b) was its easily editable and dynamic capacity: Lilia and Nasim produced a representation that provided space for, and could be adapted to, the changing nature of memories, serving as a creative and inspiring answer to a key challenge encountered in previous phases of the project. By approaching the mapping of memories through the metaphor of a living entity such as a tree, they addressed the issue of designing maps that can evolve with the memory of the storyteller. In doing so, Lilia and Nasim also illustrated one of the main challenges to such a process: the long-term commitment required to update and maintain a living memory map. Indeed, shortly after the end of the workshop, Nasim moved to Toronto, implying the eventual growth of yet another branch on her tree map.

Embodied memory mapping

Meghri Bakarian, a Canadian of Armenian-Syrian origin, came to the workshop with a clear idea of how she wanted to express her story world: 'I want to merge different materials ... I have photos that I would like to merge with dance. Also, I want to use my voice ... I'd like to use several languages.' During the first workshop, she discussed her project extensively with Khadija Baker, a multidisciplinary artist of Kurdish-Syrian origin. Her artistic background and personal experience of migration helped create a common language between storyteller and artist. Together, they conceived their map as a video representation of a performance melding collage and dance. Demonstrating how memory is inscribed on and within of the body of the dancer (West 2014), and the relations 'between embodied performance and the production of knowledge' (Taylor 2003: xviii–xix), Meghri hoped to depict the stories and places of her life through such performance. Khadija, who was receptive to this, saw dance as an effective way of expressing embodied memories, or what she called our 'second map': 'the map that we have within us, what we carry in terms of culture, memory, languages, and relations within communities that becomes more complicated to reflect on in our daily life.'

Meghri's dance was filmed and photographed by Khadija and served to illustrate important chapters of Meghri's journey from her hometown of Aleppo, her voyage to Armenia, her life in Lebanon, and finally her arrival in Montreal (Figures 5.5a/b). For Meghri, the process of choosing the photographs was meaningful but difficult because of what it triggered: 'It was like moving the memories all over again.' Yet she also emphasised the opportunity this exercise gave her to catch up with her past: 'Living here is too quick. You never have the chance to think or reflect on yourself. During this workshop, I had the chance to reflect on my journey that had been crazy over the last four years.' A short poem[5] lyrically framed Meghri's performance, which was recited three times: in Armenian, in Arabic, and finally in English.

> I'm going to say a poem that is four sentences long. The poem is about a bird who built her nest. Each time she builds her nest she builds it with joy and happiness. Each time she builds her nest she remembers the one she lost before. The whole story is about the poem. It's about moving and migration.
>
> *Meghri Bakarian,* Interview, 2017

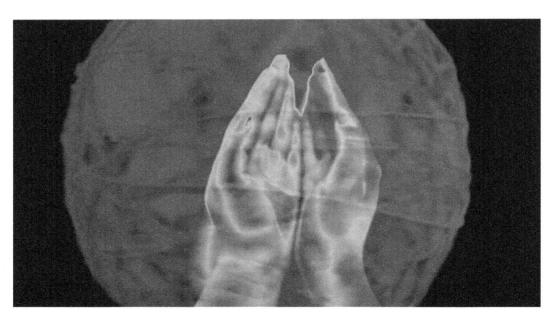

Figure 5.5a Three still frames of the Meghri Bakarian's life story made in collaboration with Khadija Baker. The video can be viewed at vimeo.com/244583982.

Figure 5.5 (Continued)

Figure 5.5b Khadija Baker (left) and Meghri Bakarian (right) developing their project.

Conclusion

Historians such as Vincent Brown (2015) and Anne Knowles (2015) have called for new ways of mapping and reading spaces of displacement and exile while emphasising the need for combining what Karen Bishop (2016: 3) calls 'formalist and non-formalist' mapping strategies. Both strategies were deployed in the project presented here, enabling us to reflect on each of their potentialities as well as their complementarity and relevant fields of application.

Process was an important element for both mapping strategies, yet these processes require fundamentally different approaches. Mapping memories of exile with conventional cartographic representations requires a rigorous and replicable methodology. Although such a methodology may not be required when the goal is to make memories more tangible, visible, and accessible to others, it becomes central to taking full advantage of cartography's power in aggregation and pattern identification: for example, that of combining multiple stories onto a single map to enable the identification of places of shared experience that are otherwise buried in individual memories. Beyond this interest in revealing geographic structures, a rigorous mapping of memories can be leveraged for place-related claims. A famous example is the 1970s Inuit Land Use and Occupancy Project (ILUOP) in Canada that relied on aggregating 'map biographies' (maps that locate and explicate Indigenous land use and occupancy based on memory) to define the total area used within living memory (see Tobias 2009) and which successfully enabled the Inuit community to claim ownership of its territory (i.e. Nunavut). Similarly, memories of exiles collected and mapped with such a methodology could be aggregated to provide tangible evidence of the existence of particular events in given places and of their spatial extent. These conventional maps of combined stories could then be leveraged by communities to argue for recognition of such events. In this way, conventional maps would be mobilised to do what they do best: emerge from the territory through the reification of past events in locales and precede it through their potential instrumentalisation towards place-related claims.

Yet, as argued by Karen Bishop, mapping memories of exile can be considered a personal act of leaving a visual trace of spatial experience as a way for the 'exile-turned-cartographer' to gain 'a certain control over the foreign environment she newly inhabits' (Bishop 2016: 9). This exercise is seen as a way of materialising the diversity of voices, identities, and experiences that are erased by the standardisation process characteristic of conventional maps. As illustrated in the third section of this chapter, these mappings reflect the range of intentions of each storyteller: from a very personal need to take the time to remember and reflect on events, places, and feelings related to exile, to a more collective statement to make these memories as public as possible. These processes as well as the cartographic outcomes in their multiple forms can allow and support these different intentions. The mapping therefore becomes an interface between the storyteller and their memories and between these memories and the map's audience.

Acknowledgments

We would like to thank all the individuals mentioned in this chapter who made this project possible either by sharing their life stories or by contributing to the mapping of these stories. Special thanks to Page Rwanda and Steven High for supporting the project and for actively contributing to its completion. This project was supported by the Fonds de Recherche du Québec Société et Culture (FRQSC) and by The Social Sciences and Humanities Research Council of Canada (SSHRC).

Notes

1. The term 'exile' is used to refer to individuals who have experienced forced migrations. As explained by Karen E. Bishop (2016: 2), based on Edward Said's work, 'exile includes, yes, those few who are still formally banished from their homes, but also the internal exile, the refuge, the asylum-seeker, the diasporic subject, etc.'
2. Inductive visualisation is an intuitive method which aims to collect and represent spatial and phenomenological data. It is a 'creative, experiential exploration of the structure, content, and meaning of the source material' (Knowles et al. 2015: 244).
3. The names of some participants have been changed to respect their request for anonymity.
4. These projects and their interactive media can be viewed online at geomedialab.org/artist_workshop.html.
5. *The Swallow* is a poem by Armenian writer Ghazaros Aghayan (1840–1911).

References

Bishop, K.E. (2016). The cartographical necessity of exile. *In*: K.E. Bishop, ed. *Cartographies of exile: A new spatial literacy*. London: Routledge, pp. 1–24.

Brown, V. (2015). Mapping a slave revolt: Visualizing spatial history through the archives of slavery. *Social Text*, 33 (4(125)): 134–141.

Campos-Delgado, A. (2018). Counter-mapping migration: Irregular migrants' stories through cognitive mapping. *Mobilities*, 1–17.

Caquard, S., and Dimitrovas, S. (2017). Story maps and co. The state of the art of online narrative cartography. *Mappemonde*, 121: 1–31.

Caquard, S., and Fiset, J.P. (2014). How can we map stories? A cybercartographic application for narrative cartography. *The Journal of Maps*, 10 (1): 18–25.

Cole, T., and Giordano, A. (2014). Bringing the ghetto to the Jew: Spatialities of ghettoization in Budapest. *In*: A.K. Knowles, T. Cole and A. Giordano, eds. *Geographies of the holocaust*. Bloomington: Indiana University Press, pp. 120–157.

Dominguès, C., Weber, S., Brando, C., Jolivet, L., and Van Damme, M.D. (2017). Analyse et cartographie des sentiments dans des récits de vie de migrants. *In: Spatial analysis and GEOmatics 2017*. Rouen, France: INSA de Rouen, https://hal.archives-ouvertes.fr/hal-01649150.

Ferrándiz F. (2014). Subterranean autopsies: Exhumations of mass graves in contemporary Spain. *In*: E. Schindel and P. Colombo, eds. *Space and the memories of violence*. Palgrave Macmillan Memory Studies. London: Palgrave Macmillan, pp. 61–73.

Genette, G. (1972). *Discours du récit. G. G. figures III*. (448 p.). Paris: Seuil.

Greenspan, H. (2014). Voices, places and spaces. *In*: S. High, E. Little and T.R. Duong, eds. *Remembering mass violence – oral history, new media and performance*. Toronto: University of Tortonto Press, pp. 35–48.

Harley, B. (1988). Silences and secrecy: The hidden agenda of cartography in Early Modern Europe. *Imago Mundi*, 40: 57–76.

High, S. (2014). *Oral history at the crossroads: Sharing life stories of survival and displacement*. Vancouver: UBC Press.

Kelly, M. (2016). Collectively mapping borders. *Cartographic Perspectives*, 84: 31–38.

Knowles, A.K., Westerveld, L., and Strom, L. (2015). Inductive visualization: A humanistic alternative to GIS. *GeoHumanities*, 1 (2): 233–265.

Kwan, M., and Ding, G. (2008). Geo-narrative: Extending geographic information systems for narrative analysis in qualitative and mixed-method research. *The Professional Geographer*, 60 (4): 443–465.

Martouzet, D., Bailleul, H., Feildel, B., and Gaignard, L. (2010). La carte: Fonctionnalité transitionnelle et dépassement du récit de vie. *Natures Sciences Sociétés*, 18 (1): 158–170.

Mekdjian, S., Amilhat-Szary, A-L., Moreau, M., Nasruddin, G., Deme, M., Houbey, L., and Guillemin, C. (2014). Figurer les entre-deux migratoires. Pratiques cartographiques expérimentales entre chercheurs, artistes et voyageurs. *Carnets de Géographes*, 7: 1–19.

Miller, E., Luchs, M., and Dyer Jalea, G. (2011). *Mapping memories: Participatory media, place-based stories and refugee youth*. Montreal: Centre for Oral History and Digital Storytelling.

Palmer, M.H. (2016). Kiowa storytelling around a map. In: C. Travis and A. Von Lünen, eds. *The digital arts and humanities*. Cham: Springer, pp. 63–73.

Ryden, K.C. (1993). *Mapping the invisible landscape: Folklore, writing, and the sense of place*. Iowa City: University of Iowa Press.

Taylor, D. (2003). *The archive and the repertoire: Performing cultural memory in the Americas*. Durham, NC: Duke University Press.

Tobias, T. (2009). *Living proof: The essential data-collection guide for indigenous use-and-occupancy map surveys*, www.ubcic.bc.ca/livingproof/index.html

West, T. (2014). Remembering displacement: Photography and the interactive spaces of memory. *Memory Studies*, 7 (2): 176–190.

PART II

Difficult memories

Sarah De Nardi and Eerika Koskinen-Koivisto

Introduction

Scarring places, people, and things at a variety of scales, conflict, colonialism, and trauma have variously led to physical and/or psychological damage, displacement and death (Kidron 2012); war and conflict have also more often than not engendered the destruction of memory as sites of historical significance are demolished or partly erased (Navaro-Yashin 2009; Logan & Reeves 2009; Macdonald 2009). Ethnic cleansing and wars on cultural property are often carried out hand in hand, '[d]emolition has often been deployed to break up concentrations of resistance among the populace' (Bevan 2016: 132). In such cases, history and the preservation of historical memory is deliberately targeted.

Contributors to this section attempt to make sense of the alternative narratives and themes emerging from veteran and witness storytelling, using what Alessandro Portelli sums up as the typical ingredients of much Second World War memorialisation: 'history, myth, ritual and symbol' (2003: 31). A key prism through which to conceptualise the emplaced memories of war, violence, deterritorialisation and conflict is the recognition of the more-than-representational politics of memory of these places, their non-cartesian topography and nature. Memories of territorial subjugation have longevity and the markers of war that remain on/in the landscape can be evident both materially and as ghosts of place in what Karen Till has called 'wounded cities' (2005). Cities, places, and people can be wounded by the ravages of conflict and social divisions, leading to fragmented memories that can be hard to reconcile.

The first chapter, by Sarah Gensburger, unpacks the legacies of the Austerlitz work camp in Nazi-occupied France through the optics of Halbwachsian memory and postmemory. The history and mythic qualities of this camp affects the positionality of the victims who were internees there. The site's 'liminal' nature in the wider context of the Holocaust has shaped the ways that internees have defined themselves and cultivate their memory and storytelling after the war, examined here through the lens of Maurice Halbwachs's sociology of memory (1992). Tracing the postmemory mobilisation work of an association seeking to preserve the materiality and memory of transit or work camps, Gensburger unpacks the ways in which this association's agenda spatialises the relationships between the Jewish 'collective milieu' (sensu Halbwachs) and the non-Jewish collective milieu.

Lilia Topouzova's haunting chapter on the Belene work camp on the Danube, Bulgaria, lays bare the secrets and myths developing around an infamous blackspot in post-Soviet memory, tied to countless traumas of political prisoners and persecution. In her text, Topouzova addresses the tension between the silence and absence of a post-communist state effort to address the memory of camp violence and the survivors' will to remember and bear witness. The two sets of memories exist side by side and resist negation. Place-making and memory-making blur into one process in this case study.

Melisa Salerno's chapter draws on ethnographic and historical-archival fieldwork in the Tierra Del Fuego, Argentina, to paint a vivid picture of a liminal place, a Salesian mission cemetery in Río Grande (Argentina), in which contested and liminal memories exist at the margins of competing belief systems: the indigenous Selk'nam or Ona people's cosmology, and the Catholic faith. Salerno explores the landscapes of death within the context of a religious mission which hosted the Selk'nam people in the late nineteenth and early twentieth centuries and puts this historical experience in dialogue with the Selk'nam tradition. Intriguingly, this case study also shows the ways in which an identity crisis may have taken place: as some individuals started to recognise themselves as Selk'nams, they claimed their ancestors had suffered genocide, while also negating their culture's extinction.

In the next chapter, Layla Renshaw also considers a tense deathscape of memory and interment. Drawing on decade-long research, she lays bare the complex and interwoven agencies and memory rehearsals at Spanish Civil War grave sites, urging us to think beyond the memorialisation and the tangible towards a reframing of memory and forgetting as emplaced, intimate, and often hidden practices. More specifically, Renshaw argues for the importance of situating the mass grave in its wider landscape, beyond the forensic excavation site edge, to incorporate other connected affects and memories that define more than the scale and complexity of a crime. Making this connection, she argues, allows us to understand how a mass grave is conceptualised within a network of sites with interconnected mnemonic and historical meanings.

The final chapter in this section takes us to London. In his revisiting of an urban bombsite in the city, Gabriel Moshenska opens up the entanglement of material destruction and social construction of a bombsite through the materiality of 'rubble'. He approaches bombsites as continuous sites of abjection and disorder that assert their haunting memories in the everyday lives of post-conflict environment. The core of the chapter is his reflection on a ruined building that is made even more visible and disturbing by the juxtaposition of surviving buildings nearby. Surveying the visible and invisible traces of the wartime past and their connections with living landscapes serves as a powerful ethnography of memory and a stratigraphy of the present-past.

Taken as a whole, the contributions in this section not only engage with the multiplicity of ways that people and place shape and diffract painful and controversial memory; they also open up new ways of conceptualising the traces that go beyond the discursive and the historical, beyond the tangible and the intangible, to create pervasive assemblages of affectual connections between how place and event are remembered when they carry hurtful legacies.

References

Bevan, R. (2016). *The destruction memory: Architecture at war.* Second Expanded Edition. London: Reaktion Books.
Halbwachs, M. (1992 [1925]). *Les cadres sociaux de la memoire.* Paris: Gallimard.
Kidron, C. (2012). Breaching the wall of traumatic silence: Holocaust survivor and descendant person – object relations and the material transmission of the genocidal past. *Journal of Material Culture*, 17 (1): 3–21.
Logan, W., and Reeves, K. (2009). Introduction: Remembering places of pain and shame. *In:* W. Logan and K. Reeves, eds. *Places of pain and shame: Dealing with 'difficult heritage.'* Abington: Routledge, pp. 1–14.
Macdonald, S. (2009). *Difficult heritage: Negotiating the Nazi past in Nuremberg and beyond.* London: Routledge.
Navaro-Yashin, Y. (2009). Affective spaces, melancholic objects: Ruination and the production of anthropological knowledge. *Journal of the Royal Anthropological Institute* (N.S.), 15: 1–18.
Portelli, A. (2003). The massacre at the Fosse Ardeatine: History, myth, ritual and symbol. *In:* K. Hodgkin and S. Radstone, eds. *Contested pasts: The politics of memory.* London: Berghahn Books, pp. 29–41.
Till, K. (2005). *The new Berlin.* Minneapolis: University of Minnesota Press.

6
MEMORY AND SPACE
(Re)reading Halbwachs

Sarah Gensburger

'Maurice Halbwachs founded the field of collective memory' (Schwartz & Schuman 2005: 183). The nineteenth-century French sociologist is indeed constantly described as a 'founding father' of the emerging field of memory studies (Olick 2009; Lustiger Thaler 2013). However, the reference to Halbwachs remains formal in nature.

What follows aims at introducing the French author's 'sociological theory of memory' (1994 [1925]: VIII). To do so, it will, firstly, stress the centrality of space in the process of remembrance, conceptualising memory dynamics as a process of localisation. Secondly, it will illustrate the empirical potentialities of Halbwachs's views on the relation between memory and space through the case study of the memory of the internment camps in Paris during the Holocaust. In doing so, this chapter may contribute to the current epistemological reflexions on the nature of memory. It advocates the importance not to mainly consider memory as 'a content' but as a relational process.

Memory as a localisation process

The work of Maurice Halbwachs is most often cut into two independent parts. The sociologist would have first studied urban spaces and city life, based mainly on statistics (Topalov 1999, 2006; Cléro 2008; Halbwachs 1960). At some point, he would have changed his interests, deciding to focus on the understanding of memory, relying then only on personal impressions and philosophical considerations (Namer 1983). This dichotomisation reflects a huge misunderstanding of the scope and aim of his study of memory.

In Halbwachs's mind, memory is not an issue of time but a matter of space and localisation (Jaisson 1999). As early as in the *Social Frameworks of Memory*, Halbwachs discusses 'the original society that every individual, in a way, forms with himself.' (1994 [1925]: 139). Memory does not have an origin, an initial moment. Almost 20 years later, in the *Legendary Topograph*, Halbwachs still writes: 'whatever epoch is examined, attention is not directed toward the first events, or perhaps the origin of these events, but rather toward the group of believers' (2008 [1941]: 234–235). In his view, the remembrance process is first and foremost an operation that localises. Memory is localised inter-subjectively, via mutual recognition between various individuals located in a structured space. The 'group' as such does not define collective memory; rather, it is the individual's position in a complex and structured social space as the evolution of the structure of this very space which does. The localisation of memory thus results just as much from a given individual's place in a structured ensemble of relations as from the modifications within this ensemble. By perceiving the space in which individuals are located as always relational,

Halbwachs is able to conceive of both the individual and the collective: in short, to conceptualise the social. Therefore, he forges jointly the concepts of 'social space,' a system structured by inter-individual relations, and 'collective memory.'

> That which we call our own feeling of oneness, wherein various states occasionally come together in unique cohesion, is essentially just our own constant consciousness of belonging to diverse milieus at all one time. But this consciousness exists only in the present: how would it come to persist for states stuck up in the past, where the pressures of social milieu are no longer exercised?
>
> *(1997 [1950 posthumous]: 89)*

From there it is possible to understand to what extent studying urban space and memory are interlinked in Halbwachs's analytical perspective. Studying memory was for him a way to deal with the central issue for the sociologists of his time: social morphology. In *La Morphologie sociale*, Halbwachs claims that,

> social morphology, like sociology, is above all a question of collective representations. We study these material forms so that we may discover the element of collective psychology beneath them. Society locates itself in the material world, and the group's thought emanates from these spatial conditions and proves regular and stable, just as individual thought needs to perceive its own body and space in order to preserve its equilibrium.
>
> *(1946 [1938]: 18)*[1]

Halbwachs's spatial theorisation of memory is useful for thinking about contemporary issues, such as the presence and transformation of the past in multiculturalist and global society. The anthropologist Roger Bastide, who put these kinds of cultural 'remodelings' at the crux of his work during the 1960s, was perhaps the first one to realise how Halbwachs's approach could be used for empirical work on memory (Lavabre 2004). Bastide studied the South American 'Black Americas,' born as a result of the slave trade (1996 [1967]).[2] For Bastide, Halbwachs provided a key conceptual framework (1970), for his ethnological study which dealt with the conservation, transformation, and 'gaps' in collective memory following the slave trade from Africa to Brazil. These memory reconfigurations are, for Bastide, fundamentally linked to transformations in the structured topography of the original groups.

> In short, collective memory can indeed be regarded as a group memory provided we add that it is a memory articulated among the members of the group.... It is the structure of the group rather than the group itself that provides the frameworks of collective memory; otherwise it would be impossible to understand why individual memory needs the support of the community as a whole. If we need someone else in order to remember, it is because our memories are articulated together with the memories of others in the well-ordered interplay of reciprocal images.
>
> *(2007 [1960], 342–343)*

This 'well-ordered interplay of reciprocal images' takes place between 'complementary actors' in a structured social space. Memory will be conserved all the better since the group's structure is preserved during the move to Brazil.

> It is this idea of structure, or this system of communication between individuals, which explains ... instances of selection and forgetting in collective memory, if group-blocks manage to explain the instances of conservation.
>
> *(Bastide 1970: 91–92)*

A Halbwachsian case study: the Parisian camps during the Holocaust

To illustrate forward the empirical potentialities of Halbwachs theorisation of memory and space, we will focus on a contemporary iconic case study related to the memory of a Holocaust related past: the existence of internment camps at the very centre of Paris between 1943 and 1944. While most of the former inmates survived the war and most of the buildings still exist, the memory of these places has long been forgotten.

The history of the Parisian camps (Gensburger 2015; Dreyfus & Gensburger 2011) is situated at the intersection of physical extermination and economic spoliation. At the beginning of 1942, the Nazi decision to implement the 'Final Solution to the Jewish Problem' led to the installation of a new type of plundering in Western Europe: the *Möbel Aktion* (Operation Furniture). The *Dienststelle Westen* (Western Service) was quickly created in Paris to house the furniture and objects from the apartments of Jews who had fled or had already been deported. The service had few military personnel. Thus, in July 1943, it hired a Jewish workforce from Drancy, the main *Sicherheitsdienst*'s (Security Service's) transit camp in France, located in a Paris suburb, where Jews were parked before the deportation to Auschwitz. The prisoners chosen for this work and the transfer to Paris belonged to one of three categories of the temporarily 'non-deportable' inmates: Jews who were 'spouses of Aryans', *Mischlinge* or 'half-Jews,' and Jewish wives of Jewish war prisoners. Occasionally, 'ordinary' Jews interned at Drancy succeeded in obtaining false documents which allowed them to pass as 'spouses of Aryans' or as 'half-Jews,' thereby temporarily staving off their deportation.

In July 1943, 180 internees left Drancy for the Lévitan furniture store in the 10th arrondissement. In November, the second satellite camp, called 'Austerlitz,' opened in the 13th arrondissement and it received a group of nearly 200 prisoners. In both locations, the internees were subjected to forced labour. All day long, they sorted, cleaned, and repaired objects and furniture pillaged from Jewish apartments. In March 1944, a third, smaller camp opened in an Aryanised private mansion at 2 rue de Bassano in the 16th arrondissement. There, a luxury fashion house catering to Nazi dignitaries was established. In all three locations, the prisoners participated in every step of Operation Furniture, which emptied 38,000 Parisian homes.

In principle, this work detail in the capital signified, above all, eluding the fatal deportation to the East. The prisoners, however, remained dependent upon the main camp. A hallmark of the Nazi administrative operation, arbitrariness was permanent. In several cases of attempted escape and insubordination, internees were taken back to the main camp to be deported. Nearly 800 people worked in the three camps between July 1943 and August 1944; 166 were deported to Auschwitz or Bergen-Belsen. The remaining 80% – or 600 internees – from the three camps stayed in Paris and survived the war.

Until recently, the very existence of these forced labour camps was almost ignored. A large majority of the potential witnesses remained mute, both privately within their families and publicly. For a long time, only fragments of the past lingered.

The Austerlitz camp was located in an industrial zone of *Magasins et Entrepôts Généraux de Paris* (Paris general stores and warehouses company). Among the buildings still standing in the early 1990s was a refrigerated warehouse that had been occupied by artists for 20 years. In the middle of the 1990s, an urban renewal plan threatened their presence. A rumour persisted that the building had been a camp during the war. Hoping to defend their building from destruction, several artists decided in 1997 to ascertain whether a 'Jewish camp' had indeed been on the premises.[3] They put out a call for 'witnesses' and published in the national press a notice for former internees. Some individuals responded to the notice and 'Amicale Austerlitz-Lévitan-Bassano,' a non-profit organisation gathering former inmates, family, and friends was created. Its mission was to 'retrace the history of Parisian internment, work and transit camps and to preserve their memory.'[4] Consequently, a group of individuals decided to recount, publicly, the history of these camps. Understanding how certain individuals remember, here publicly, illuminates how others long guarded and continue to guard their silence. In other words, it enables us to grasp how memory works.

'Each individual memory represents a point of view on the collective memory'

> If collective memory derives its force and duration from a group of individuals, these are after all individuals who remember *as* members of a group. The common memories in this mass are interdependent, and it is not always the same memories that will seem strongest to each group member. We suggest that each individual memory represents a point of view on the collective memory. This point of view changes depending on the place I occupy, and the place I occupy changes depending on the relations I pursue with other milieus. Thus, it is not surprising that not everyone makes the same use of a common tool. In trying to explain this diversity, however, we always return to a combination of influences that are, by nature, social.'
>
> *(Halbwachs 1997 [1950 posthumous]: 94–95)*

On 1 October 2001, at the peak of its mobilisation,[5] this 'Amicale Austerlitz-Lévitan-Bassano' counted 29 members, who were either former internees, the child or grandchild of an internee, or individuals interested in history. Yet, during the association's internal discussions, the relationships between the Jewish 'collective milieu,' according to Maurice Halbwachs's formula, and the non-Jewish collective milieu very clearly structured the space of memory's expression. Those who recounted their past and spoke publicly hold singular positions in the social space. The parallel between the social morphology of the organisation and the original one of the Parisian camps during the war is striking. While people 'fully' Jewish accounted for no more than 15% of the Parisian camps internees, and are over-represented among the people finally deported to Auschwitz with the highest death rates, they, nevertheless, constitute 65% of the association's members. Conversely, constituting 85% of the original population of the Parisian camps, internees having matrimonial or genealogical connections with non-Jews constitute only 35% of the organisation members. The observation of the organisation's general meetings confirmed the peripheral position of people situated at the margins of Jewish milieus. Yet the large majority within the Parisian camps barely spoke during the organisation's meetings. This was left to members situated clearly within a Jewish collective milieu.

> The dynamics of memory – even the most personal memories – can be explained by the changes taking place in our relations with various collective milieus: that is to say by the transformations taking place in each of these milieus and within these milieus as a whole.
>
> *(Halbwachs 1997 [1950, posthumous]: 95)*

Only a 'change' in the 'relationships with the diverse collective milieus' solicits the expression of memory and its transmission to others. The remarks of the girl of a first marriage of a former Parisian camps internee, interned as 'spouse of an Aryan,' revealed the difficulties of remembering with others while one is socially isolated, situated on 'an island.' Questioned about post-war family conversations, this woman explains:

> One couldn't say that we knew deported people or really old acquaintances. We were a little like castaways on an island in relation to what happened on the outside. I must say one thing. We were complete atheists. Neutral forever. And so, my sister's children married Catholics. So, only my sister and I remained Jews. We went home and then didn't deal with the rest. And by chance, not very long ago, I had a friend who was a friend with Denise Weill [the organisation's secretary-general]. She put me into connection.

In this case, the silence appears to be directly linked to the isolation, to the absence of relationships with 'deportees,' implicitly 'Jews,' or, in other terms, those of Halbwachs with 'collective milieus' in which the past may be likely to be meaningful and shared. Conversely, the concomitance between the 'connection' with others – the organisation's secretary-general a common contact – and the decision of remembering belong together. Memory, a social phenomenon, only forms through complementary links between members of a group, carriers of a common past.

> One only remembers on the condition of putting oneself in the perspective of one or several groups and resituating oneself in one or several collective intellectual currents.
> *(Halbwachs 1997 [1950, posthumous]: 65)*

The renewal of contacts, the connections allows one to again insert oneself in these 'collective intellectual currents.' This is even more improbable when the social position is situated at the group's margin or is really complex, the result of a number of distinct affiliations that the individual cannot proceed to a minimal identification. Another example reveals the relationship between localisation and remembrance, the modification of connections to collective milieus and the formation of memory. When a veteran of the Austerlitz camp, interned as a 'spouse of an Aryan' woman, recounts his internment, for the first time since the end of the war, he immediately and spontaneously evokes the recent modification of his relations to collective milieus. First of all, he shared with us the recent death of his wife and went on:

> She was not Jewish. So, she didn't want to hear any more about this story. She blamed me for having let myself, according to her, get arrested too easily. Besides, no one among my children was interested in this story. So, it is important to know that my wife, like me, were militant atheists.

This person's total rupture with the social group with which he shares a pertinent common past – the Jews' past before and during the Occupation – explains his long silence. Conversely, it was under the label of a recent change in his potential relationship to this milieu – the death of non-Jewish wife who did not want to speak of this past – that the man situates his testimony. If it enables him to testify, this transformation of relationships with different collective milieus remains, nonetheless, marginal. This man is not a member of the former inmates' organisation and does not want to become one. On the phone, before accepting to meet with us, he asked us, first of all, to assure him that the association had 'no denominational nature.' In order to accept witnessing of one of the places of the persecution of Jews, this man preferred a framework not linked to the Jewish identity, in this case the completion of the 'academic' research project I was working on.

The connections between collective milieus and memory explains that, in the case of the Parisian camps, the potential witnesses have rarely expressed themselves and that their memories did not consolidate as a unified memory. Their testimony falls into a social mechanism, a morphological one, Halbwachs would say. Several former internees, like their families, do not speak of this 'ever-present' past (Conan & Rousso 1994) because they do not find their place(s) in relation to it. The structuration of the social space and the position that the individual occupies within it explains the expression of the memory, both 'recognised and reconstructed.'

> In order for our own memory to be aided by others, listening to another's accounts is not sufficient.... It does not suffice to reconstruct, piece by piece, the image of a past event in order to obtain a memory. The reconstruction must take place on the basis of data or common notions present in our minds just as in the minds of others. These notions move continuously from our minds to others' and can only do so if both have taken part, and continue to take part in, the same society. Only this way may we understand how a memory can be simultaneously recognised and reconstructed.
> *(Halbwachs 1997 [1950 posthumous]: 63).*

Beyond forgetting and remembrance: memory as a structured relational process

So, the Parisian camps have for a long time formed the centre of what, based on his study of African religions in Brazil, Roger Bastide termed as a 'memory hole' (1970: 95): neither forgetting nor remembrance, but the memory of a void. While researching the history of these camps, it was consequently difficult for us to find and meet with former detainees. Their silence was linked to their sense of isolation, to the absence of any relations with 'deportees' who would, by implication, be 'Jews' and so stemming, in other words – the words of Maurice Halbwachs, to be precise – from their being cut off from the 'collective milieux' able to sustain this memory. 'A person remembers only by situating himself with the viewpoint of one or several groups and one or several currents of collective thought' (Halbwachs 1997 [1950 Posthumous]: 33).

The position occupied by most of these internees at the crossroads between two groups, 'Jews' and 'non-Jews,' has made it difficult to become a witness. Several former detainees, along with their families, have decided to say nothing about this past, not so much because it is 'ever-present,' traumatising them, because they themselves are unable to find their own place in relation to it.

> So that our memory aids that of others, it is not enough for some to offer their testimony: they must never cease to fit with their memories and there must be sufficient points contacts between one another so that the memory that they recall can be reconstructed upon a common foundation.
> *(Halbwachs 1997 [1950, posthumous]: 63).*

Moreover, because most of these detainees had ended up being sheltered from the danger of deportation, living in much better conditions than those in Drancy, let alone Bergen-Belsen or Auschwitz, there was a perception that many of the inmates of the Parisian camps had in this way been 'spared.' Following the Liberation, several detainees did in fact speak about their internment in some articles published in Parisian Resistance's newspapers in September 1944. This public airing of their experiences ceased with the return of the surviving deportees and the dawning awareness of the reality of the extermination camps. After April 1945, nothing more would be said about the satellite camps of Drancy, whether in the public arena or in more specialised publications. Filled with a vague feeling of guilt at having escaped deportation in exchange for forced labour in the context of the looting of Jewish property, in a marginal position with respect both to their identification with Judaism and to the group formed by the deportees, the majority of the former detainees of Lévitan, Austerlitz, and Bassano could find no suitable social framework within which to express their memories. No 'space for narration' (Pollak 1993: 201), no place to tell their story, was then available.

It was only after 1998, when the anti-Semitic looting during the war became a public issue and led to an official reparation process, that some of the former internees began to speak out. They had found a social place from which to tell their story. At a time when French state and secular social actors started gathering around the evocation of this looting operation, a space opened for them to speak out, this time as French citizens. In doing so, they did not tell their personal experience of suffering, anxiety, cold, and hope. They told what they saw from 1943 to 1944: how the looting operation worked and the amount of the plunder. In accordance to the social frameworks which had finally led them to tell their story, they became historical witnesses rather than autobiographical ones. They testified of what they saw rather than what they experienced. Here the localisation dynamics of memory directly frames the content of what is recalled from the past.

* * *

From the 1980s onward, an increasing number of researchers in the social sciences have become interested in the topic of 'memory.' Twenty years later, the interest has reached 'boom' status and has, as such, been critiqued variously (Confino 1997; Olick & Robbins 1998; Berliner 2005; Winter 2000). Memory studies as an integrated field should help to deal with one of the principal 'problems plaguing contemporary research on collective memory, namely, the tendency to reify it' (Olick 2007: 42; Klein 2000; Crane 1997). If Maurice Halbwachs' 'sociological theory of memory' is often referred to as inspirational, it turns out that this new research field largely oversees the potentialities of the French sociologist's conceptualisation of memory as a localisation process. Halbwachs's perspective is particularly important in a time where memory has become a field of intense dialogue, and sometimes conflict, between neurobiology and social sciences. The issues of network and localisation are indeed at the core of the social turn neurobiology has underwent in the recent years. Halbwachs's intuition opens a new direction to conceptualise memory as something other than content coming from the past, which is so often described as 'collective memory.' It calls for developing a spatialised, and then genuinely social, approach of memory dynamics.

Notes

1. Halbwachs evokes this point several times in *La Morphologie sociale*.
2. Some of this chapter was reproduced in an English translation in the *Collective memory reader* (2011).
3. There wasn't. The camp was elsewhere. Research of this location also inspired W.G. Sebald's eponymous work (2001), which situated the camp elsewhere.
4. Legal statutes, 14 January 1998, Paris Police Prefecture.
5. The historical sources include documents from public and private archives and 26 oral interviews conducted with former prisoners or their children. For the present sociological study of memory, a field work notebook mixing self-analysis with observations, as well as contemporary official documents, was also used.

References

Bastide, R. (1970). Mémoire collective et sociologie du bricolage. *L'Année sociologique*, 65–108.
Bastide, R. (1996 [1967]). *Les Amériques noires*. Paris: L'Harmattan.
Bastide, R. (2007 [1960]). *Les religions africaines au Brésil*. Paris: Presses Universitaires de France.
Berliner, D. (2005). The abuses of memory: Reflections on the memory boom in Anthropology. *Anthropological Quarterly*, 78 (1): 197–211.
Cléro, J.P. (2008). Halbwachs et l'espace fictionnel de la ville. In: M. Jaisson, ed. *Maurice Halbwachs: La Topographie Légendaire des Evangiles en Terre Sainte*. Paris: Presses Universitaires de France, pp. 43–72.
Conan, E., and Rousso, H. (1996 [1994]). *Vichy, un passé qui ne passe pas*. Paris: Gallimard.
Confino, A. (1997). Collective memory and cultural history: Problems of method. *American Historical Review*, 105: 1386–1403.
Crane, S. (1997). Writing the individual back into collective memory. *American Historical Review*, 102 (5): 1372–1385.
Dreyfus, J.M., and Gensburger, S. (2011). *Nazi labor camps in Paris*. New York: Berghahn Books.
Gensburger, S. (2015). *Witnessing the robbing of the Jews. A photographic album, Paris 1940–1944*. Bloomington: Indiana University Press.
Halbwachs, M. (1946 [1938]). *Morphologie sociale*. Paris: Armand Colin.
Halbwachs, M. (1960). *Population and society: Introduction to social morphology*. Glencoe, IL: Free Press.
Halbwachs, M. (1994 [1925]). *Les cadres sociaux de la mémoire*. Paris: Albin Michel.
Halbwachs, M. (1997 [1950 Posthumous]). *La mémoire collective*. Paris: Albin Michel.
Halbwachs, M. (2008 [1941]). *La topographie légendaire des évangiles en terre sainte. Etude de mémoire collective*. Paris: Presses Universitaires de France.
Jaisson, M. (1999). Temps et espace chez Maurice Halbwachs (1925–1945). *Revue d'histoire des sciences humaines*, 1: 163–178.
Klein, K. (2000). On the emergence of memory in historical discourse. *Representations*, 69: 127–150.
Lavabre, M.C. (2004). Roger Bastide, lecteur de Maurice Halbwachs. In: Y. Déloye and C. Haroche, eds. *Maurice Halbwachs: Espaces, mémoire et psychologie collective*. Paris: Éditions de la Sorbonne, pp. 161–171.
Lustiger Thaler, H. (2013). Memory redux. *Current Sociology*, 61 (5–6): 906–927.
Namer, G. (1983). *Batailles pour la mémoire. La commémoration en France de 1945 à nos jours*. Paris: Papyrus.
Olick, J.K. (2007). Collective memory and nonpublic opinion: A historical note on a methodological controversy about a political problem. *Symbolic Interaction*, 30 (1): 41–55.

Olick, J.K. (2009). Between chaos and diversity: Is social memory studies a field? *International Journal of Politics, Culture and Society*, 22: 249–252.

Olick, J.K., and Robbins, J. (1998). Social memory studies: From "Collective memory" to the Historical sociology of mnemonic practices. *Annual Review of Sociology*, 105–140.

Pollak, M. (1993). *L'expérience concentrationnaire: Essai sur le maintien de l'identité sociale*, Paris: Métailié.

Schwartz, B., and Schuman, H. (2005). Commemoration and belief: Abraham Lincoln in American memory, 1945–2001. *American Sociological Review*, 70 (2): 183–203.

Sebald, W.G. (2001). *Austerlitz*. London: Penguin.

Topalov, C. (1999). "Expériences sociologiques": Les faits et les preuves dans les thèses de Maurice Halbwachs (1909–1913). *Revue d'histoire des sciences humaines*, 1: 11–46.

Topalov, C. (2006). Maurice Halbwachs et les sociologues de Chicago. *Revue française de sociologie*, 47 (3): 561–590.

Winter, J. (2000). The generation of memory: Reflections on the memory boom in contemporary historical studies. *Bulletin of the German Historical Institute*, 27: 69–92.

7
REMEMBERING BELENE ISLAND
Commemorating a site of violence

Lilia Topouzova

On the Danube

In the late 1950s, the Bulgarian writer Georgi Markov, who was on his way to becoming one of the country's most acclaimed authors, travelled on a cruise boat along the banks of the Danube. At first, he revelled in the peaceful journey: 'The boat glided along with the current of the river, we sunbathed on the upper deck and cooled off under the showers, the radio played pleasant light music' (Markov 1984: 88). Markov was certainly not the first to evoke the serenity of a voyage across the European continent's second-largest river. Flowing through ten eastern and central European countries, the Danube stands as an important cultural and historic symbol of the region. Famously memorialised in the 1867 *Blue Danube* waltz by the Austrian composer Johann Strauss, the river has been, for centuries now, both a source of life for the communities through which it passes and a muse for artists who encounter it. In their creative homages, those who immerse themselves in the Danubian life world generally celebrate its splendour – but they also alert us to the river's turmoil. 'Troubled, wise was the Danube, mighty force,' wrote the Hungarian modernist poet Attila József in 1936 (Ersoy et al. 2010: 454). In 1986, the Italian scholar Claudio Magris described the river as 'a symbol for life and death and disappearance' (Magris quoted in Flanagan 2016). Markov's prose, historically situated halfway between József and Magris, conjures up similar imagery, at once blissful and ominous. His initial impressions of the tranquil trip darkened as the boat approached the largest island in Bulgarian waters, when someone uttered its name:

> On every journey, this ghostly voice speaks up . . . "Belene." All the Bulgarians on the boat turned and stared for a long time at the green island until it was left far behind. It was as if the music in the wireless had suddenly become the hooting of an owl, the whole innocent enchantment of the river had disappeared, and the sun had faded, darkened by frightening memories of so many tales and rumors about this island, Bulgaria's Calvary. There is hardly a Bulgarian living who has not known some inhabitant of this socialist showcase. . . . I remembered how, feet dangling over the edge of the boat, a youth with a guitar once sang a strange song:
>
> *Danube, white river, how quiet you flow*
> *Danube, black river, what anguish you know.*
>   *(Markov 1984: 88)*

Poignant but understated, Markov's words only hint at the muted terror experienced by Bulgarians on their encounter with the physical space of the island. In other passages of his posthumously published writing, the fear associated with Belene appears more stridently. In fact, it rears its head frequently in his descriptions of everyday life in Bulgaria in the 1950s. This is because Belene Island was the site of the largest and longest-running forced labour camp – prison complex of the communist era. To this day, it remains the most enduring symbol of the repressive legacy of the People's Republic of Bulgaria (Skochev 2017; Topouzova 2014; Luleva et al. 2012; Koleva 2010; Znepolski 2009). Yet what we now know about Belene, with the benefit of archival hindsight, back then existed only as a sinister rumour, never to be spoken aloud but certain to intimidate – much as it did the passengers on the boat. Prior to the collapse of the communist regime in 1989, Georgi Markov was one on the first to publicly name Belene as a site of violence.

A few years after his voyage on the Danube, the writer received the most prestigious Bulgarian literary award. Yet, despite rising to the top of the cultural ladder, Markov gradually became disillusioned with the regime. In 1969, he defected to the West. Markov's transformation from a socialist literary star to a prominent dissident cut it rather close: had he not escaped, he would likely have been sent to prison or a camp (Segel 2012). Instead, Markov's powerful reportages, 'political in tone and content,' regularly and clandestinely (in the Soviet satellites) aired on the BBC, on Radio Deutsche Welle, and Radio Free Europe (Kenarov 2014). His radio essays lent a very public dimension to the private fears of his compatriots and earned him a popular following. Eventually, however, his unrelenting criticism of communist Bulgaria's human rights abuses and corrupt officials cost him his life. In 1978, the Bulgarian secret police assassinated Markov as he was crossing Waterloo Bridge in London, England (Brunwasser 2008; Hristov 2006). While the Bulgarian state may have succeeded in cutting tragically short Markov's life, it did not manage to silence his voice. He left behind a rich body of writing: fiction, plays, literary criticism, and his crowning achievement, the *In Absentia Reports*, which many specialists consider one of 'the most important narrative nonfiction documents' of Bulgaria's communist period (Igov quoted in Kenarov 2014).

Nearly 30 years since the fall of the Berlin Wall and the disintegration of the Eastern European communist regimes, scholars across the disciplines continue interpreting the multifaceted legacy of the communist experience. Consensus on what exactly constituted twentieth-century European communism and how to remember it, however, is difficult to achieve. Simply put, this is because communism as a lived experience proved simultaneously repressive and emancipatory. Markov's insights seem especially pertinent, particularly for the Bulgarian case, where the debate on the communist past is deeply polarised and mired in the politics of a difficult transition to democracy. His writing, circumspect and self-reflexive, underlines the complexity of the communist world. In his attempt to render 'that life as it is,' Markov succeeds in recounting it 'with breathtaking honesty,' as the journalist Dimiter Kenarov put it (Kenarov 2014). So it seemed apt to begin the discussion on Belene with Markov's reminiscence of his journey on the Danube, which speaks of both the 'frightening' and 'the enchanting,' implicitly urging us to somehow make sense of both.

The Belene that emerges in the pages of Markov's dissident writing both overlaps and collides with present-day Belene. Belene Island constitutes one of Bulgaria's profoundly contested memoryscapes: infamous as the country's most notorious communist-era forced-labour camp and home to one of Europe's vitally important ecological regions, the island's past and present belie a harmonious historical narrative. Yet it is precisely Belene's paradoxical character that provides an opportunity to inquire into how lived experiences of violence become historical and public memoires. In what follows, I trace how such memories enter the commemorative record despite attempts to relegate them to oblivion.[1] Against the landscape of Belene Island, I map out the tension between the absence of a post-communist state effort to address the memory of camp violence and the survivors' will to remember and bear witness.

Belene: a historical overview

Located in northern Bulgaria, on the shore of the Danube, Belene is a quaint little town with a population of about 9,000 people. Old peasant houses and Soviet-style architecture absurdly blend with the serenity of the Danube and its archipelago of islands. Island Persin, or Belene Island, as it is more commonly called, is the largest of these islands, the fourth in size on the Danube, and the biggest in Bulgaria. Measuring 15 kilometres in length and six kilometres in width, it is considered one of the most ecologically important areas in the region, home to an idyllic nature reserve with more than 170 rare water bird species. Its unique environmental characteristics have been globally recognised, and the island constitutes a protected area, part of the European Ecological Network and a wetland of international importance. In addition to the nature park, the island currently houses a penitentiary, a small portion of a nonoperational nuclear power plant, some Roman ruins, and a medieval fortress. Ecologists and nature enthusiasts have expressed much admiration for the picturesque island, dubbing it the 'Bulgarian Pearl of the Danube' and urging visitors to discover its rich and captivating historical heritage. To this end, eco-trails, kayaking and cycling tours, and a special water route, 'In the Embrace of the Danube,' are offered to those interested in exploring the region. While these landmarks and facilities appeal to travellers' imaginations, references to Belene Island's more sinister historical past are commonly omitted and obscured.

That Belene Island became the site of the country's largest communist-era forced-labour camp and prison complex often remains unsaid. Starting in 1949, chosen especially for its remote location, Belene Island, alongside the smaller surrounding islands, was carved into five separate sites.[2] With a three-year interruption beginning in 1953 following early de-Stalinisation reforms, it actively operated from 1949 to 1959. In 1952–1953, sentenced prisoners were sent to the camp and when, in the fall of 1953, the camp closed down, only those sentenced prisoners stayed on the island. When the camp was reinstated at the end of 1956 in the aftermath of the Hungarian uprising, the prisoner population remained there. Starting in 1957, Belene Island thus housed both a prison and a camp (Figure 7.1). Authorities most actively used the camp until 1959, but it remained functional until 1987. The Belene prison continues to operate to this day; it is a maximum-security facility that houses some 500 inmates.

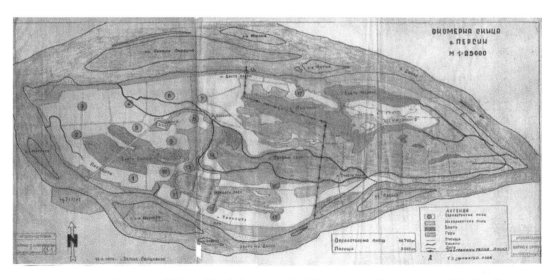

Figure 7.1 A secret police map of Belene Island, the site of the Belene camp and prison, from 1956. Site II, the location of the camp, is at the far left. The image is persevered in the archive of Bulgaria's former Ministry of the Interior. AMVR, f.23, op.1, a.u.149.

Both archival evidence and the oral-history testimonies of those who spent years of their lives in the Belene camp describe its regime as very severe (Topouzova 2014; Skochev 2017). Although the level of harshness tended to vary depending on the year of internment and on the camp commander, hard physical labour, impossible-to-fulfill work quotas, 14-hour workdays, disease, beatings, hunger, and other brutal punishments characterised life in Belene for its internees. Officially, the camp closed down in the summer of 1959, when inmates transferred to another site near the town of Lovech. The island archipelago system, however, remained operational, even though no camp internment took place there for a few years. Authorities transferred the dead bodies of inmates killed in the Lovech camp some 100 kilometres back to Belene. Sometimes transports arrived twice a day, once in the morning and once in the evening.[3] The corpses were either thrown into the Danube or buried in unmarked graves on one of the smaller islands.[4] The archipelago's remoteness and its natural frontier, the majestic river surrounding it, turned it into the ideal location both for the isolation of 'enemies' and for a silent graveyard of those who perished in the camps. In particular, the archipelago's smaller islands, which regularly flooded, became convenient, anonymous, and secret burial grounds. According to camp authorities, the arrival of transports from Lovech and the burials sometimes stirred a fuss among the village population, but Belene Island's remoteness nonetheless prevented 'the oncoming flood of relatives of the deceased, cries at the graves, and the possible theft of the corpses.'[5]

Barred from mourning or retrieving the bodies of their loved ones at the time of death, the families of victims attempted to reclaim their remains in the post-communist years. The search for unmarked mass graves in early 1990 proved unsuccessful, however, whether due to the 'unreliable evidence and slack investigation' conducted at the time, the uncertainty of eyewitness recollections, or the islands' geography (Znepolski 2009: 7). The Danube naturally exhumed and disposed of the bodies, ultimately fulfilling the camp system's goal of confining the killings to silence: 'Erasing the traces of the dead is also the erasure of death itself as a camp reality,' the cultural theorist Ivaylo Znepolski rightfully observed (Znepolski 2009: 7).

Belene Island was so well situated that despite official pronouncements of its closure in 1959, the camp never completely shut down and remained operational throughout most of the communist regime. As Bulgaria's head of state, Todor Zhivkov, remarked in 1959 at a Politburo session discussing the fate of the camp and the pressing need to close it, 'Belene should remain [open] for when times get complicated' (Znepolski 2009: 22). Times seem to have remained complicated until the end. In fact, as late as May 1989, the deputy minister of the interior insisted that camp internment in Belene continued as an option.[6] From 1964 until 1977, people were interned in Belene at a significantly reduced scale, though generally without a trial or a sentence (Skochev 2017). Between 1984–1986, during the communist government's final repressive operation on a mass scale – the forced-assimilation campaign against Bulgaria's Turkish and Muslim minority – the camp once again functioned at an increased pace. During the camp's most active phase from 1949 to 1959, an estimated 14,000 people passed through Belene without sentences; if we take into account internment from 1959 until 1989, there were around 15,000 inmates altogether (Skochev 2017). Another scholar estimates that perhaps as many as 30,000 people were sent to the camp and/or the prison between 1949–1986 (Znepolski 2009). Unfortunately, the available records from the purged secret police archive do not provide enough evidence to estimate with any reasonable certainty the number of those who died.

Making sense of Belene: a non-place or a *Lieu d'Oubli*?

The purging of the archives made difficult not only counting the victims but also writing the history of the Belene prison-camp complex. Shortly after the collapse of the Bulgarian communist regime, in late 1989, Belene made headlines as 'the Island of Death' in Bulgaria and abroad. Suddenly what had circulated as hushed but frightening rumours (what had been buried deep in archives) exploded onto the pages of newspapers and on TV screens. Notwithstanding all the media coverage and the sinister epithets the island received, research on the camp remained scant. Bulgarian pundits to this day characterise Belene as the embodiment of the Bulgarian Gulag, a symbol of the most nefarious years of the communist regime (Todorov 1992; Znepolski 2009; Methodiev 2009; Koleva 2010; Luleva 2012; Baichev 2014). Yet their scholarly investigations frequently ponder the dearth of research on Belene and lament the absence of any memory about its past. It was only in late 2017 that a 923-page monograph on the history of the Belene forced-labour camp appeared, penned by the independent researcher and civil society activist Borislav Skochev. His meticulous archival study of the establishment and operation of the Belene camp revealed the degree to which administrative repression had come to define the functioning of the Bulgarian communist regime. If the publication of the most recent studies has generated a much-needed historical narrative about the Belene prison-camp complex, the question of how we remember the lived experience of camp violence and, concomitantly, how we integrate its memory into our present also demands our attention.

Following Pierre Nora, it seems worthwhile to 'look beyond' Belene's "historical reality to discover the symbolic reality and recover the memory it sustained" (Nora 1996). Weaving together Belene's symbolic threads – as a *milieu de violence* (a physical site of violence) during communism, and a contested *lieu de mémoire* (a realm of memory) in the early 1990s, one that gradually transformed into a *lieu d'oubli* (a realm of forgetting) in the present day – reveals a memory that is 'absence in presence,' to use Paul Ricœur's observation (Ricoeur: 2004). In this trajectory of Belene as a memoryscape of atrocity and forgetting, grasping for the camp past in the present remains an impossibility because that past has been expunged. Znepolski in fact has argued that the erasure of the camp past has been so extreme that, borrowing a term from Marc Augé, he referred to all former Bulgarian communist forced-labour camp sites as a 'non-place' (*nemiasto*) (Znepolski 2009; Augé 1995). Augé's neologism, with which he describes places devoid of historical meaning, at first seems counterintuitive when used in relation to Belene Island, a place replete with historical memories. Yet I regard it as an apt metaphor (rather than an analytical category) through which we can begin thinking through Belene's silenced history of violence.

Belene's silence about the camp's past is also Bulgaria's silence. The hush is not confined to a particular place, but spread across the entire country, even if it manifests most evidently in this small town and the island across from it. In this sense, I see Belene as a microcosm of Bulgaria's difficult relationship with its communist past. It is ultimately a relationship fraught with contested memories and underlined by the absence of a state discourse on the violent past. The Bulgarian state's decision to maintain a prison on the island ominously echoes this silence. 'And when we die, which will be soon, who will remember what happened on that island in the 1950s, and will they know that people were sent there without a trial and sentence?' the camp survivor Krum Horozov poignantly asked in 2011 (Horozov, interview with the author, 2011). Nowhere are the silence and the absence of memory more evident than on the territory of the former camp. Yet it is also there, on the edge of the island, in the midst of the hollow but cumbersome silence, that the traces of the camp past are most visible.

Site II

The site of the former Belene camp, Site II, has now become integrated into the penal complex and is used as a prison farm. In 2006, the senior prison inspector in charge of it gave me a tour of the empty buildings, including the old house that served as the camp commandant's office and the watchtower. 'They say there used to be a concentration camp here,' the inspector told me; 'I began working here only recently and I don't know if it's true or not.'[7] He had worked at the site for ten years but was reluctant to answer questions about its past. 'I am a few years away from retirement and don't want any problems,' he explained. 'There are different stories about Belene. Some people exaggerate. Others tell the truth. It's a complicated thing.' The prison inspector remained opaque and decided to talk only about the site's present: 'Now, there are fifteen, twenty prisoners here, as well as goats, sheep, rabbits, a horse ... and the cross and the crescent.'[8] What of the last two? 'It seems that the cross and the crescent were put here in memory of the people who were in ... the camp?'[9]

The two modest wood-and-metal structures, no more than six feet tall, were a makeshift memorial erected by survivors' organisations in the 1990s and 2000s (Figure 7.2). Together with a small plaque hanging on the wall of one of the empty buildings, they were meant as temporary memorials until a permanent one was built. But the construction of an official monument never happened. In 2009, a concrete wall on top of a concrete pedestal was placed where the cross and crescent had been. The concrete slab was supposed to serve as the future memorial's foundation, and the names of former camp inmates were supposed to have been inscribed on the wall (Koleva 2010). To date, however, the concrete wall remains empty, resembling a segment from an unfinished construction site.

Other than that, the prison inspector's description of the site seemed fairly accurate. Several crumbling buildings, used for the internment of Bulgaria's Muslim minority in the mid-1980s, remained deserted, as did the watchtower erected in the 1950s, from when the camp had operated at full capacity (Figure 7.2). Although the buildings were in a state of neglect, they had not been fully abandoned: a young mare, some racing pigeons, a pet pig, and an encaged wild boar lived there. Such are the memory keepers of Bulgaria's longest-running site of violence. And such are the narratives produced by representatives of the government responsible for managing the site: vague, punctuated by doubt about the historical events that took place at the Belene camp.

Occasionally, people other than the 15 to 20 prisoners and the prison staff visit the former camp site. Since 1990, survivors have tried to gather there once a year. In the early post-communist years, the meetings were large, frequented by survivors, their families, and their friends, as well as journalists, filmmakers, priests, and politicians. Survivors delivered passionate speeches, priests read prayers and held mass, and sometimes imams joined them. With time, however, the numbers have begun to shrink noticeably: survivors have aged,

Figure 7.2 The camp watchtower and the memorial cross and crescent at Site II of the former Belene camp, Belene, 2006. Photo stills from the documentary film *The Mosquito Problem and Other Stories* (2007).

Source: Courtesy of Agitprop.

Figure 7.2 (Continued)

and many of them have died. Friends and families have moved on with their lives. Politicians no longer need the camp's past for election platforms. Journalists have more pressing stories to cover, and priests are busy holding liturgies elsewhere. In other words, many people formerly motivated by politics no longer have any investment in attending survivors' meetings. Yet for the few remaining camp and prison survivors, elderly men and women, the annual pilgrimage to the former Belene camp still proves extremely important. Visiting the island, however, is not easy: bureaucratic challenges abound, and accessing the former camp site proves rather difficult. Since the island still operates as a prison, a visit requires authorisation by the Bulgarian interior minister and approval by the prison governor. In 2006, for example, camp survivors did not obtain permission to hold their annual meeting there.

Offsite and the absent state

Instead of convening on Belene Island, they gathered in another part of the country, at the site of one of the first camps, the harsh coal mines of Bogdanovdol, just outside the city of Pernik. It was a casual meeting at an unofficial and unmarked site, on a grass field just off the highway. It was 'close enough to where the camp had been,' one of the survivors informed me.[10] The small group of senior citizens, around 30 of them, exchanged remarks about the weather and caught up with one another. There were no speeches or ceremonial procedures. Unless bystanders listened closely to their conversations, they would have found it difficult to guess that the group had congregated to commemorate their shared experience of political violence. One of them carried several self-published copies of his memoir; I was his sole customer. Many of them were upset that their gathering had been refused in Belene. Even though not all of them had been interned there, they all insisted, much like the other survivors I interviewed, that the Belene site had to be transformed into a memorial, one easily accessible to both survivors and future generations. 'It is a mockery that the state maintains a prison where once the camp had been,' contended a former political prisoner.[11]

Many in the survivor community argue that this state of affairs creates an intentional blurring of the lines by the Bulgarian government, one meant to whitewash the history of communist violence. Although it is difficult to ascertain whether intent or indifference have motivated the various post-communist Bulgarian governments to keep the Belene prison intact, the decision has certainly had a negative impact on the community of camp and prison survivors.

Not only are its commemoration efforts hampered by administrative challenges but, in Bulgarian public discourse, the site itself has also gradually been stripped of the meaning bestowed on it by survivors. History textbooks in Bulgarian secondary schools rarely devote more than a page to the history of the communist repressive apparatus, and the names and stories of camp and prison survivors are never incorporated (Topouzova 2014; Kelbecheva 2012). Unsurprisingly, in 2013, a widely circulated media headline disclosed that Bulgarian youth often confused the word 'Gulag' with 'Google.'[12] Mistaking the popular online search engine for a system of forced labour may well be a sign of the times, but it is nonetheless an unsettling one, especially for those whose lives were forever changed by the camps. If one of the main aims of communist camp internment, without a trial or sentence, was to render the interned non-existent to the outside world, then this goal has been fully realised at present. For the survivors' memories and their narratives have no bearing on the public or the social worlds. More important, they have also been largely excluded from educational and academic discourses. The absence of these life stories from the collective narrative about the socialist past constitutes yet another sign of what the Bulgarian scholars Alexander Kiossev and Daniela Koleva have identified, albeit in the form of a question, as 'the disintegration of basic social solidarities in Bulgarian society' (Kiossev & Koleva 2017: 413). In this multivalent process of disintegration, the former Belene camp has become a 'non-place.' Yet it remains a very real and significant one for the survivors.[13] How to reconcile this disjuncture between the traumatic but insular and increasingly fading memories of survivors and the swelling non-place status of the former Belene camp in Bulgarian society makes for a most pressing question. It is precisely what Krum Horozov meant when he warned: 'Our voices would not suffice; there need to be others' (Horozov, interview with the author 2006).

Certainly some others have taken up the challenge of commemorating the experience of camp violence and of establishing the historical narrative of the Bulgarian forced-labour camp system. The work of individual academics, civil society activists, archivists, NGOs, investigative journalists, filmmakers, and, in recent years, that of the Catholic priest Paolo Cortezi, based in the town of Belene, bear special mention.[14] The outcome of these individual and sometimes collaborative projects is important, but it remains fragmented. None of these efforts, and here I include my own work on the topic, has translated into a systemic and public engagement with the memory of violence. This is because the Bulgarian state continues to remain absent from the conversation. I do not wish to assert that state-sponsored commemorative initiatives would automatically bring forth genuine societal reflection or a meaningful engagement with the camp past. State-sponsored memorial events can be perfunctory in their execution, devoid of meaning, or even deployed as cynical instruments to gain political capital (Nora 1998; Minow 1998; Neumann and Thompson 2015). Yet the continued blanket absence of the state, as the Bulgarian case reveals, has devastating consequences for the lives of individuals who endured decades of political violence, as well as for their families. However limited, government responses – from the creation of memorials to public education programs – when done in collaboration with non-state-actors and institutions, can lead to a public dialogue in which even the existence of divergent perspectives does not negate the possibility of a shared and inclusive narrative on the recent past (Minow 1998; Barkan 2009). Such a process would allow for the survivors' narratives to enter the Bulgarian public space and for their memories to exit the marginal realm they currently occupy. Theirs will no longer be memories that resist representation (Ricoeur 2004). Yet at the moment, they remain just that: secluded memories relegated to the margins.

Remembering Belene Island

Individual remembrance

One attempt at commemoration, by the former political prisoner Krum Horozov, set itself apart by the variety of means he employed. In addition to writing a memoir and giving interviews, Horozov attempted to render the history of the camp and the experience of imprisonment through a number of drawings that depicted the camp in both abstract spatial terms and from the perspective of a person living on the island. These drawings included conventional maps of the various camp and prison sites on the island reconstructed from memory, plans of the different sites rendered in the spatially abstract style of architectural blueprints (Figure 7.3), more naturalistic colour drawings of the sites (Figure 7.4), and drawings in between the architectural and the naturalistic, such as the interior of one of the prisoners' barracks.

Because no photographs appear to exist of Belene Island as an active camp, Horozov's colourful naturalistic drawings may be the closest we can get to a visual representation of the camp. Yet the subtle visual distortions of perspective and scale attest to the gulf separating reconstructed representations and the historical experience of people in the actual space of the camps. Like all survivors' accounts, they commemorate both the experience of living inside the camps and the impossibility of rendering this experience with complete precision. They are expressions of what happened and what has been lost in commemoration.

We find another powerful symbol of the difficulties in expressing the lived experience of the camps in Horozov's self-published visual memoir. In a copy of his book that he gave me, he had pasted small pieces of paper on some of the pages (Figure 7.5). These additions to the book were photocopied printouts of English text, mostly translating the Bulgarian descriptions of the drawings. Although the English is awkward and full of grammatical errors, the translations powerfully symbolise Horozov's need to communicate. They also attest to the limits of language to express the violence of the camp. Much as language breaks down and reveals its weakness in the face of the trauma of the camps, so, too, Horozov's testimony breaks down as it

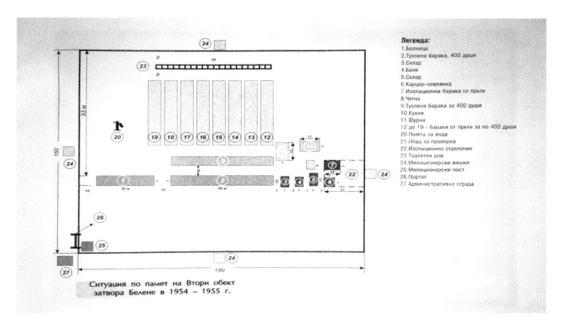

Figure 7.3 Site II of the Belene camp and prison (1954–1955), rendered from memory in 1977.
Source: Drawing courtesy of Krum Horozov. Ruse, 2003.

Figure 7.4 Site II of the Belene camp and prison (1952–1960; 3,500 to 5,000 people) rendered from memory in 1977.
Source: Drawing courtesy of Krum Horozov. Ruse, 2003.

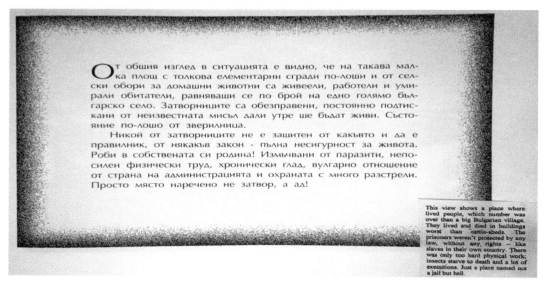

Figure 7.5 Description of the camp compound of Site II, the Belene camp and prison complex.
Source: Courtesy of Krum Horozov, Ruse, 2003.

tries to reach an audience beyond his mother tongue. Horozov spoke no English, so, he told me, a friend with only a basic grasp of the language did the translations for him.

Describing the first drawing in the book, a bird's-eye view of the camp compound, the translation reads, in full:

> This view shows a place where people lived people, which number was larger than a big Bulgarian village. They lived and died in buildings worst than cattle-sheds. The prisoners weren't protected

by law, without any rights – like slaves in their own country. There was only too hard physical work; insects starve to death and a lot of executions. Just a place named not a jail but hell.

Although the message is awkwardly conveyed, it speaks loudly. There is something deeply moving about the breakdown of language in the translation, as it powerfully expresses the strength of this survivor's desire to be heard and his will to communicate. With an almost poetic power, the troubled language expresses Horozov's determination to use whatever means are available to him to represent Belene. In this case, those means were scissors, paste, and a friend with a Bulgarian-English dictionary and a basic grasp of English. Horozov had already written about his experiences in Bulgarian, drawn images of the camps, attended conferences, spoken to journalists, and had conversations with as many Bulgarians as would listen. With his ad hoc English translations, we register his desire to go further and push into realms beyond his own ability to communicate.

Conclusion

In lieu of an adequate effort by the state to investigate and memorialise the history of the camps, we are left primarily with the isolated and sometimes idiosyncratic voices of survivors and select individuals to commemorate the lived-experience of camp violence. For despite all the headlines and judicial endeavors of the early post-communist period, none of the interest within Bulgaria and beyond translated into much in the way of systemic outcomes. The archives were purged, no guilty verdicts were achieved, a state memorial remains unbuilt, and current history textbooks barely mention the camp's existence. With varying degrees of significance, these multiple acts of erasure and denial continue to perpetuate the violence of the camp long after the removal of the barbed wire surrounding it. As a result, close to 30 years since news of the Bulgarian Gulag first struck the Bulgarian psyche, the memory of its troubling past has gradually receded into silence. Yet the camps' legacy continues to haunt those few remaining survivors unable to forget and unwilling to relegate their experiences to historical oblivion. With the passage of time, however, their voices are becoming less audible. Still, the written records, memoirs, testimonies, and Horozov's drawings have enabled the preservation of the memory of repression from Bulgaria's biggest and longest-running site of violence, Belene Island. By the time I completed the writing of this chapter, many of the survivors I interviewed, Horozov included, had passed away. Yet the pain of the camp's past does not cease to exist. Though repressed and unacknowledged, the survivors' pain has nonetheless found its way into the crevices of Bulgarian society. Its presence, traumatic and spectral, has interwoven itself with the rhythm of an everyday life that refuses to acknowledge it. But trauma stitched with silent threads always remains. With time, it begins to rear its head in that same everydayness, demanding of us to bear witness to it, to unravel the silence that envelops it.

Notes

1 This chapter draws on oral history interviews that I carried out between 2006 and 2016 with 40 survivors of political violence from Bulgaria's communist era, as well as archival research conducted in recently declassified files from Bulgaria's former Ministry of the Interior. My research on the Bulgarian gulag was generously supported by post-doctoral fellowships from Brown University's Pembroke Center for Teaching and Research on Women and from the Social Sciences and Humanities Research Council of Canada held at Concordia University's Centre for Oral History and Digital Storytelling. Unless otherwise indicated, all translations from Bulgarian are my own.
2 The shores of the village of Belene had also briefly been used for internment in the pre-communist years. Between 1942 and 1943, some soldiers and peasants were mobilised for national labour service (a form of compulsory labour) there.
3 Archive of the Ministry of the Interior, thereafter AMVR, f.23, op.1, a.u.109
4 AMVR, f.23, op.1, a.u. 109.
5 AMVR, f.23, op.1, a.u. 109.

6 See Grigor Shopov's remarks in AMVR, f.1, op.12, a.u.933, pp. 38–39.
7 Senior prison inspector of the Belene prison, interview with the author, Belene, August 2006.
8 Ibid.
9 Ibid.
10 Alexandur Nakov, interview with the author, Pernik, August 2006.
11 Todor Anastasov, interview with the author, Sofia, 11 March 2011.
12 *Sega*, 21 September 2013.
13 I am grateful to Steven High for making me rethink how survivors might relate to the concept of *non-place*.
14 In addition to the work of scholars and journalists discussed in this article, it is important to note the recent initiatives of the Sofia Platform NGO responsible for organising a summer school in Belene.

References

Augé, M. (1995). *Non-places: Introduction to an anthropology of supermodernity*, trans. John Howe. London: Verso.
Baichev, P. (2014). *Spomeni ot lagerite: Portreti na lageristi ot Belene 1948–1943*. Sofia: Institut za izsledvane na blizkoto minalo.
Barkan, E. (2009). Introduction: Historians and historical reconciliation. *The American Historical Review*, 114: 4.
Brunwasser, M. (2008). A book peels back some layers of a Cold War mystery. *The New York Times*, 10 September. Accessed 22 September 2018, www.nytimes.com/2008/09/11/world/europe/11sofia.html
Ersoy, A., Gorny, M., and Kechriotis, V. eds. (2010). *Modernism: The creation of nation states*. New York and Budapest: Central University Press.
Flanagan, R. Why Claudio Magris's Danube is a timely elegy for lost Europe. *The Guardian*, October 22 Accessed 17 January 2019, www.theguardian.com/books/2016/oct/22/why-claudio-magriss-danube-is-a-timely-elegy-for-lost-europe.
Hristov, H. (2006). *Ubiite "Skitnik" Bulgarskata i Britanskata Durzhavna Politika po Sluchaia Georgi Markov*. Sofia: Ciela.
Igov, S. (2009). *Istoria na Bulgarskata literatura*. Sofia: Ciela.
Kelbecheva, E. (2012). Istoriata kato pamet ili kato prikazka-lichen pogled vurhu komunizma v Bulgaria. In: M. Gruev and D. Mishkova, eds. *Bulgarskiat komunizum. Debati i interpretatzii*. Sofia: Tzentur za izsledovatelski isledvania, pp. 68–84.
Kenarov, D. (2014). A captivating mind. *The Nation*, 18 March. Accessed 20 November 2018, www.thenation.com/article/captivating-mind/.
Kiossev, A., and Koleva, D. eds. (2017). *Trudniat Razkaz: modeli na avtobiografichnoto razkazvane za sotzialisma mezhdu ustnoto i pismenoto*. Sofia: Institut za izsledvane na blizkoto minalo.
Koleva, D. ed. (2010). *Belene-miasto na pamet? Antropologichna anketa*. Sofia: Institut za izsledvane na blizkoto minalo.
Luleva, A., Troeva, E., and Petrov, P. eds. (2012). *Prinuditelniat trud v Bulgaria (1941–1962): Spomeni na svideteli*. Sofia: Akademichno izdatelstvo, "Profesor Marin Drinov."
Magris, C. (1986). *Danubio*. Milano: Garzanti.
Markov, G. (1984). *The truth that killed*. New York: Ticknor and Fields.
Methodiev, M. (2009). Bulgaria. In: L. Stan, ed. *Transitional justice in Eastern Europe and the former Soviet Union: Reckoning with the communist past*. New York: Routledge, pp. 152–173.
Minow, M. (1998). *Between vengeance and forgiveness: Facing history after genocide and mass violence*. Boston: Beacon Press.
Neumann, K. and Thompson, J. eds. (2015). *Historical justice and memory*. Madison: University of Wisconsin Press.
Nora, P. ed. (1996–1998). *Realms of memory: Rethinking the French past*, trans. A. Goldhammer. New York: Columbia University Press.
Ricoeur, P. (2004). *History, memory, forgetting*, trans. K. Blamey and D. Pellauer. Chicago: University of Chicago Press.
Segel, H. ed. (2012). *The walls behind the curtain: East European prison literature, 1945–1990*. Pittsburgh: University of Pittsburgh Press.
Skochev, B. (2017). *Kontzlagerut "Belene" 1949–1987. Ostrovut, koito ubi svobodniat chovek*. Sofia: Ciela.
Todorov, T. (1992). *Au nom du peuple: Témoignages sur les camps communistes*. Paris: Les éditions de l'Aube.
Topouzova, L. (2014). *Reclaiming memory: The history and legacy of concentration camps in Communist Bulgaria*. PhD Dissertation, University of Toronto.
Znepolski, I. (2009). *Bez Sleda. Lagerat Belene 1949–1959 i sled tova*. Sofia: Institut za izsledvane na blizkoto minalo.

8

THE LANDSCAPES OF DEATH AMONG THE SELK'NAMS

Place, mobility, memory, and forgetting

Melisa A. Salerno

Introduction

Traditionally, the Selk'nam or Ona people were nomads who lived by hunting and gathering on the Isla Grande de Tierra del Fuego (southernmost end of South America). For centuries, they had brief encounters with Westerners who visited the coasts of the island. However, cultural interactions became more intense in the late nineteenth and early twentieth centuries, when Argentina and Chile sought to consolidate their sovereignty, and entrepreneurs decided to exploit natural resources (Martinic 1973; Belza 1974, 1975; Borrero 2001). Within this context, some voices were raised expressing concern about the fate of the Selk'nams. The establishment of gold washings and sheep-farming 'estancias' unleashed numerous conflicts. Even though Nuestra Señora de La Candelaria, a Salesian mission in Río Grande (Argentina), sought to 'reduce' the Selk'nams with an aim to integrate them into a new social order, many indigenous people found death at the institution – mainly as a result of infectious diseases (Guichón et al. 2006, 2017; Casali 2011).

By the mid-twentieth century, the Selk'nam population had decreased significantly. From that moment on until recently, the media and the academic world spoke of 'extinction,' spreading news from time to time about the identity of the survivors. There always seemed to be a 'last ona' who had gone previously unnoticed (Méndez 2012; Peñaloza 2016). Some years ago, some began to recognise themselves as Selk'nams. They claimed their ancestors had been the victims of a 'genocide,' but they also stressed that the 'extinction' had not been real (Badenes 2016). Indigenous people had always been there, even though they had preferred to remain silent to avoid stigmatisation. In the 1990s, the Selk'nams began to organise themselves. Moreover, they demanded their constitutional rights to be respected, including the return of their ancestors' remains to Tierra del Fuego.

The subject of death has been always present in cultural studies surrounding the Selk'nams. The first systematic records were produced as a means to preserve information about practices which were thought to be doomed to oblivion. In the late nineteenth and early twentieth centuries, researchers and other Westerners made written and photographic surveys, and occasionally drew together large collections of objects to shed light on the material world of the Selk'nams (Gallardo 1910; Beauvoir 1915; Gusinde 1982). Human remains were not beyond their reach, and some skeletons were taken to museums to study human diversity (under a racial paradigm) (Gusinde 1951). As time went by, different projects attempted to study the impact that the modern world had on the Selk'nams. However, few researchers sought to discuss how Selk'nam people had traditionally understood death, and how these understandings could have changed over time.

My own research on the subject began when I started working with a group of bioarchaeologists (led by Riardo Guichón) who were interested in studying health in the context of La Candelaria. Between 1897 and 1947, more than 340 people were buried at the mission cemetery. Almost two thirds of the burials corresponded to indigenous people, while the rest included members of the religious community and other settlers (Salerno et al. 2016; Salerno & Guichón 2017). Fieldwork focused on the cemetery. Counting on the support of local groups, 33 bodies were recovered and taken to research facilities in the Province of Buenos Aires for further study. Bioarchaeological analyses led researchers to discuss the impact of tuberculosis on the indigenous population of the mission (Guichón et al. 2006, 2017; García Laborde et al. 2010).

Some time later the project decided to move beyond its original purpose by studying the disposition of the bodies. Considering my experience in historical archaeology, I developed a new line of enquiry focused on mortuary practices. Even though my research took into account multiple social actors, my attention was placed on the Selk'nams. At the beginning, I did not consider the possibility of studying what I later understood as the 'landscapes of death' among these people. My intention was to discuss how the mortuary practices involving the Selk'nam deceased and their survivors (within the framework of undeniable relationships with other social actors) could have helped (re)produce certain aspects of the cemetery space (Salerno et al. 2016; Salerno & Guichón 2017).

However, during my research it became clear that in order to understand what had happened at the cemetery, it was necessary to go beyond its limits. At first, I decided to broaden the spatial scale of my work, considering practices that – while connected to the burials – indigenous people could have carried out in other spaces of the mission or outside the institution. Later I thought it convenient to broaden the timescale of my research in an attempt to provide some sort of historical sense to the aforementioned practices. Therefore, I started reading ethnohistorical records trying to understand the relationship between the deceased and their survivors in Selk'nam tradition (Salerno & Rigone 2017). The preparation of the remains recovered at the cemetery for their return to Tierra del Fuego and eventual restitution to the Selk'nam Community (in the case of the individuals genetically identified as Native Americans) opened up new questions regarding their final disposition (Guichón et al. 2015).

Only then did I become aware of the relevance that the spatial dimension of certain practices could have had in the cultural understanding of death among the Selk'nams. Moreover, I had a chance to perceive that the spatial dimension of these practices could have undergone changes over time. Following this perception, one of the goals of this chapter is to explore the features defining the landscapes of death among the Selk'nams. In particular, I am interested in discussing the flows and tensions between places and mobilities, and memory and forgetting, all while considering the practices that (re)produce relations of proximity/distance between the dead and their survivors.

The chapter does not follow a chronological sequence conforming to Selk'nam history (even though it involves a multi-temporal analysis and the re-ordering of historical processes in the Final Words). On the contrary, it follows the changes in the landscapes of my research. Therefore, after presenting some concepts that I find to be relevant, I will consider the landscapes of death within the context of 1) La Candelaria mission, 2) the Selk'nam tradition, and 3) the current discussions surrounding the restitution of Selk'nam remains. This sequence allows me to discuss the places and movements of my impressions and interpretations, and the specific way each context challenged me and made me think about the others. Considering this, I also attempt to critically reflect on my own work and assumptions.

Some concepts

The notion of landscape has been understood in multiple ways. Taking up elements of different proposals, here I define the landscape as the way in which people understand and engage with space – not as an abstract container, but as an array of material elements including topographical features, buildings, objects, etc. (Thomas 1996; Ingold 2000; Bender 2001). Considering that our being-in-the-world is corporeal, an important part of our engagement with the landscape is intertwined with bodily practices and experiences (Tilley 1994). There is not a single way of dealing with the landscape, but multiple sociocultural understandings. Furthermore, landscapes do not represent static realities, but ever-changing processes responding to historical contingencies (Bender 1993, 1998).

In this chapter I focus on the relationship between landscape and death, considering that death may become a focus of extreme concern under certain historical circumstances and that the decision of studying the connection between both of these terms can provide insight into some aspects of the spatial dimension of death that would otherwise go unnoticed. The idea of the landscapes of death is not new. While some authors have associated them with theaters of conflict where death has affected large segments of a population (as in the case of wars, epidemics, etc.), some others have focused on specific places whose materiality has been devoted to the disposition and memory of the deceased (necropolis, memorials, etc.) (Dov Kulka 2013; Hannum & Rhodes 2018).

My research took up some elements of previous definitions, as it stemmed from the analysis of a historical context (a religious mission) where the death of the Selk'nams not only became massive, but also materialised in a cemetery. Notwithstanding this, here I understand the landscapes of death in a broader sense, as the ways in which people grasp the surrounding space within the framework of death-related practices and experiences. Death as a cultural process can take various forms. It may begin in the moments before clinical death, and it may go far beyond that (Martínez 2013). In this chapter I focus on the spatial dimension of the practices and experiences connected with the disposition of the bodies and mourning, stressing the relationship between the deceased and the survivors.

At the level of landscape, the disposition of the bodies and mourning might involve places and mobilities. These are concepts in tension and dialogue. On the one hand, places are the result of focusing on certain spaces within more extensive horizons (Bender 2001). On the other, mobility requires the deployment of horizons as well as broader spatial scales (Ingold 1997, 2000). Thanks to movement it is possible to leave behind a certain place and engage with another, while places can be thought of as stopping points within the framework of movement. Death-related dynamics of place and mobilities can be varied. For instance, the disposition of the bodies can be carried out at the same place of death, or imply the transport of the bodies to new locations which can be revisited or not by the survivors.

Death-related dynamics of place and mobility are associated with the cultural forms that mobility (in a general sense) takes up in a given group. La Candelaria mission intended to show the Selk'nams the virtues of sedentism. However, missionary documents recorded abundant entries and exits, which were either resisted or tolerated by the religious community (Marschoff & Salerno 2016; Salerno & Marschoff 2017). As mentioned before, the Selk'nams were traditionally nomadic. At present, people who recognise themselves as Selk'nams relate to the landscape in a variety of ways – as it is with any other group of contemporary society showing diversity in people's life contexts and occupations.

The material forms that the disposition of the bodies and mourning can take maintain close ties with the practical senses associated with the memory and forgetting of the deceased, (re)producing relations of proximity or distance between the dead and their survivors (Salerno & Rigone 2017). Just as place and mobility, and memory and forgetting, are terms in tension and dialogue (Buchli & Lucas 2001). To remember is to focus on the past in the present. However, it is only possible to remember certain things to the extent that some others are forgotten (Ricoeur 2004). Memory and forgetting are not inherently positive or negative, but their value needs to be evaluated in context (Weinrich 2004). Survivors can resort to the potential of the material world to recall the dead or erase their traces to cope with mourning. However, some groups may try to erase the material traces of other people's dead to delete them from history (Salerno & Zarankin 2015).

Even though I have raised these issues at the beginning of the chapter, they were only strengthened as I developed my research. My reflection on historical and archaeological evidence allowed me to defy certain assumptions that I had accepted before the beginning of the work or during certain moments of the investigation (for instance, an emphasis on place at the expense of mobility, or vice versa). In the following sections, I present the course of my research. I believe that, only after traversing certain places and movements, I could not only broaden the landscapes of my thoughts, but also grasp some elements of the changing landscapes of death among the Selk'nams.

The religious mission of Río Grande

As mentioned earlier, my research initially focused on La Candelaria, discussing how the mortuary practices that involved the Selk'nam deceased and their survivors could have (re)produced certain aspects of the cemetery space (Salerno et al. 2016; Salerno & Guichón 2017). Thus, and without fully intending to, I ended up circumscribing mortuary practices to a place understood in a particular way; in other words, as a self-contained space, associated with routine activities, and capable of (re)producing a sense of identity and memory. The association between death and this idea of place was probably the result of different assumptions (some of which had been naturalised during my professional training):

1 the cemetery was the limited space that the project had chosen to carry out fieldwork. When I made the decision to study mortuary practices, I took it for granted that it would be capable of providing me with enough evidence. In contexts we define as 'colonial enclaves' (such as religious missions), archaeologists frequently resort to site perspectives, fixed to certain places that are deemed to be the result of a sedentary experience, at the cost of mobility dynamics (Marschoff & Salerno 2016). Therefore, though not always deliberately, we narrow or restrict the understanding of landscape to the (re)production of certain senses of place as those referred to previously;
2 the mission cemetery was designated as a national historical monument in 1999 (Decreto 64/99). In Argentina, a national historical monument describes 'an immovable [object or property] of material existence, either erected or built, where events of historical, institutional or ethical-spiritual character took place, whose transcendental consequences are valuable for the cultural identity of the Nation' (Comisión Nacional de Museos y de Monumentos y Lugares Históricos 1991, Disposición 5/91). As can be seen, heritage discourses refer to monuments as immovable realities: as rooted or fixed spaces. At the same time, they point to the historical relevance of these places, as they are thought to maintain close ties with identity and memory; and
3 the cemetery was limited by its own materiality, as it was surrounded by a perimeter wall. The religious community had intended to demarcate the community of dead Christians who had been part of the history of the mission. For the Congregation, the cemetery combined senses of identity and memory (Salerno & Guichón 2017). But up to this point, I have referred to the perspective of archaeologists, heritage discourses, and the missionaries. What can I say about the Selk'nams? Even though they had been buried at the cemetery, did they also understand it as a place, with all the senses of identity and memory that the other references assigned and still assign to the graveyard?

When I first visited the cemetery, I felt surprised. Most of the space within the perimeter walls had no visible signs which could be attributed to burials. Furthermore, the few structures and gravestones still standing were associated with members of the religious community and other settlers who had died between the 1920s and the 1950s, rather than with indigenous people (Salerno & Guichón 2017). More than bound to memory, the cemetery appeared to be a slightly forgotten place. The general condition of the graveyard accounted for these circumstances. There were fallen sections of the perimeter wall and no indication of a well-maintained path leading to the cemetery. This did not seem to have a relationship with the declaration of the cemetery as a national historical monument. Such a declaration had not been enough to protect the graveyard nor to make some people associate it with identity and memory.

I started wondering which practices could have been the product or producers of the cemetery over time (Salerno & Guichón 2016). I looked for answers in the practices of the religious community and the other settlers, but I especially focused on the Selk'nams. Missionary documents did not explicitly refer to the existence of significant grave markings until the 1920s. Before that date (when most deaths corresponded with indigenous people), only wooden crosses had been apparently placed over the head of some burials (Beauvoir 1915). By the 1920s, the cemetery practically lacked grave markings, making it difficult to recognise who had been buried where (Gusinde 1920a). Following some references, the cemetery had been unsystematically excavated by Westerners looking for the remains of indigenous people (Gusinde 1920b). Some of these remains could have been taken/sold to museums (Gusinde 1951).

These circumstances seemed to change between the 1920s and the 1940s when the indigenous population of the mission decreased, and the number of settlers buried at the cemetery became larger. At that moment, missionary documents claimed that some structures and gravestones had been placed to mark new burials. However, it is possible that the structures associated with indigenous people could have been different to those associated with settlers; the first ones being made of wood, and the latter of metal, concrete, and marble. As time went by, wooden structures could have been affected by weather conditions, but also by an extreme remodelling of the cemetery that took place between the 1970s and the 1980s. This remodelling only left standing those structures made of metal, concrete, and marble.

More clues were eventually found in missionary documents. For the members of the religious community, the material presence of the cemetery was necessary to maintain the ties between the deceased and their survivors. When a death occurred, the priests insisted on asking Selk'nam people to accompany the body to the cemetery. The same thing was done every 2 November, a day in memory of the dead. Even though the deceased had to be remembered, the Salesians emphasised that mourning should not last too long, as indigenous people had to trust in the existence of a better life in the hereafter. At the mission, the observance of some Christian principles could have been respected (or at least, that was what some priests wanted to stress).

However, in the stories surrounding the first deaths of Selk'nam people in the institution, there are mentions of practices associated with the disposition of the dead or mourning that are obviously different. There is a description of an indigenous man trying to run away with the body of his son (Beauvoir 1915). On several occasions, the death of one or more persons was followed by the decision of some indigenous people to leave the mission for mourning (after trying to burn down the house and the artefacts of the deceased). Some references indicate that, at the mission, some indigenous people painted their bodies, cut their hair, and performed lamentations and self-flagellation during mourning (Fernández 2014). Finally, some documents suggest that, as time went by, some Selk'nams started associating La Candelaria with death and decided to avoid it (Gusinde 1982).

Although I had initially focused on the cemetery as a place having unequivocal senses, the evidence made me pose new questions. Could the cemetery effectively operate as a place of identity and memory for all the Selk'nams? Could some people not identify themselves with the cemetery, or understand it as a place to be avoided or forgotten? Could some indigenous practices be related to the (re)production of some sort of material distance between the dead and their survivors? The landscapes of death eventually took me out of the cemetery and to consider Selk'nam tradition.

The Selk'nam tradition

Following ethnohistorical sources, when a Selk'nam felt about to die, she or he decided to rest in her or his tent (Gusinde 1982). In the moments before death, the members of the group surrounded the dying person, limiting their mutual interaction to a 'hidden or veiled attention' (Salerno & Rigone 2017). The imminence of death produced a general state of excitement. The dead body was prepared some hours later. The body was not painted or dressed in a particular way. The guanaco skin that the person had worn in life was spread out on the floor, and the body was placed on the skin together with some twigs to facilitate transportation. Finally, the body was wrapped in the skin, and bound with leather straps (Gusinde 1982).

The burial was not carried out where death had occurred. A group usually made up of men took the body to a location which they deemed appropriate for the inhumation. The Selk'nams resorted to individual burials but not to cemeteries. Those who acted as 'undertakers' were obliged to remain silent about the location of the grave. They had to erase any signs of it, and even their own footprints so no one could follow their path (Gusinde 1982).

While the 'undertakers' carried out their duty, those who remained at the camp devoted themselves to burning all things that once had been used by the deceased. Survivors wanted to remove everything that could trigger the memory of the dead (Gusinde 1982). Finally, the group abandoned the place where death had occurred and resumed the cycle of mobility. The men who had acted as 'undertakers' did their best to keep the group at a proper distance from the burial. Some time later, some of them secretly returned to the grave to check if it remained hidden (Gusinde 1982). The Selk'nams were horrified by human remains, and when they found them they felt the urge to rebury them and clean themselves immediately (Gallardo 1910).

The Selk'nams understood mourning as an extended and particularly anguishing moment. The practices of mourning demanded the survivors to attend to their bodies in particular ways (Salerno & Rigone 2017). Firstly, they painted themselves with certain colours and designs. Secondly, they cut their hair, creating a tonsure. Thirdly, the relatives performed lamentations at different times of day. Fourthly, during these lamentations, some people scratched or cut their own bodies with rocks or shells, occasionally drawing motifs on their skin with blood (Gusinde 1982).

The present

The archaeological work at the mission sought to study the sociocultural context of that institution and some aspects of the Selk'nam tradition. But it also required maintaining a relationship with the Selk'nam people who still live in Tierra del Fuego. As mentioned earlier, some time ago a number of people began to openly recognise themselves as Selk'nams (Méndez 2012). Following the words of a Community leader,

> [b]eing Ona is our original, and also chosen, identity. Too often people are discriminated against. Who has the right to measure a person's blood, a person's degree of purity.... To say that the 'last pure Ona' died can only lead to confusion and do harm to the Community.
>
> *(Maldonado in Iparraguirre 2009: 161)*

The research project counted on the support of the Rafaela Ishton Community, some members of which had their ancestors buried at the mission. This Community was granted juridical status and obtained a provincial law granting them communal lands in the rural area of Tolhuin, Department of Río Grande. The Community also requested that the human remains of Selk'nam people, which were in the hands of the Museum of Natural Sciences of La Plata (Buenos Aires Province), were restituted in compliance with Ley Nacional 25,517 (National Law 25,517). Even though negotiations took some time, the remains eventually returned to Tierra del Fuego after more than 100 years.

The archaeological work at the mission cemetery, as well as the moving of the bodies to Buenos Aires for further analysis, distanced the remains from Río Grande. At present, the research project is organising the return of the bodies to the heritage authorities of Tierra del Fuego. Following Ley Nacional 25,517 (2001), 'The mortal remains of aboriginal people . . . that are part of museums and/or public or private collections, should be made available to the indigenous people and/or communities of origin that request them.' With an aim to fulfill legal and ethical responsibilities, after or together with the return of the remains to the provincial authorities, the bodies identified as Native Americans will be restituted to the indigenous Community.

Considering my interest in the landscapes of death, I started wondering about the final destination that contemporary Selk'nams will choose for the bodies. Will they resort to the old mortuary traditions, burying the bodies at secret locations in the lands that the Community was granted (as did other indigenous groups in South America)? Will they prefer to leave the bodies at the mission cemetery, where they had rested for decades? Will they resort to another option, not necessarily based on ancestral tradition or missionary practices? Even though Selk'nam people have not yet reached a decision, some details concerning the restitution of the remains kept by the Museum of La Plata could provide relevant information.

In 2014, the members of the Rafaela Ishton Community and the research project participated in a series of informal meetings to reflect on the possible destination of the bodies that had been exhumed at the mission (Guichón et al. 2015). Considering previous experiences involving other indigenous communities, the meeting considered the possibility of creating a temporary reservoir for human remains. The restitution of the bodies once kept by the Museum of La Plata was delayed until 2016 as the Community could not reach a decision whether to take them to Ushuaia, the city of Río Grande, or Tolhuin. Finally, the Community brought them to the communal lands, with a view of building a mausoleum. A Selk'nam leader pointed out: 'In the same place that these Onas rest, we will place some other brothers that we will bring from different parts of the country, and the world, so they could finally rest in their land' (Maldonado in Télam, 21 April 2016).

The landscapes of death that presently involve the bodies of ancient Selk'nams frequently imply exhumations, the transport of remains out of Tierra del Fuego, research facilities, the return and restitution of the bodies, community decisions regarding their final destination, etc. Statements surrounding the restitution of the bodies previously kept by the Museum of La Plata indicated that they would be preserved in a mausoleum or a space for memory. This would be different from the traditional form of disposing of the dead among the Selk'nams, as ancient practices defied the senses of identity and memory that are now attached to collective mausoleums. However, unlike the mission cemetery, a mausoleum could effectively mobilise both of these notions. The decision to create a space for memory could be connected with a world tendency that has allowed minorities subjected to massacres, genocides, and discrimination to recover visibility and their own voices after years of marginalisation (Salerno & Zarankin 2015).

Final words

In Selk'nam tradition, the landscape of death was bound to a general form of mobility. Since death found a group in different points, the disposal of the dead did not occur in a single location. The moving of the body created a distinction between the places and journeys 'of life,' known to all the members of the group, and the places and journeys 'of death,' only known to the 'undertakers.' Both realities were eventually combined, as the 'undertakers' defined certain aspects of the cycle of mobility in order to avoid the spaces of death. Secrecy surrounding the burials prevented these places from being associated with senses of identity or memory. The decision to abandon the camp, the practice of burning the things the deceased used, etc., strengthened the material distance from death. This was considered relevant to break the constant recall that prevented people from closing mourning. In the long term, the places of burial became lost in the landscape, and the memory of single actors fused into the horizon of the ancestors.

The mission attempted to create another landscape of death, connected with the principles of Christian and modern thought. Under the project of the Salesian Congregation, the cemetery was transformed into a place capable of gathering the community of Christians. At the same time, it was associated with notions of identity and memory, stressing the possibility of maintaining a material relationship with the dead. However, while most of the Selk'nams tolerated burying the deceased at the cemetery, many of them could have maintained traditional mourning practices. This made them abandon temporarily or permanently the mission, having an impact on the decline of the institution (without denying the effect of infectious diseases).

Today the situation is different. Unlike the mission context, where the destination of the bodies seemed to be controlled by the members of the religious community, the people who recognise themselves as Selk'nams discuss what they will do with the remains of their ancestors. As empowered actors, the opportunities that open up before them are numerous; and they could resort to certain elements of Selk'nam tradition, recent history, and/or the community's present understanding of death (considering that they are part of an inter-cultural society).

Regarding my own understanding on the landscapes of death, the research confronted me with a changing array of places and mobilities. Starting with La Candelaria, I initially emphasised a notion of place, conceived of as a fixed category and associated with unequivocal senses of identity and memory. Following some practices made me go beyond the institution and reach a series of places and mobilities that I had not considered earlier. Approaching Selk'nam tradition was relevant to understanding what had happened at the mission. Thus, I had the chance to explore a different context, where mobilities gained strength, and the places connected with death accounted for different senses, attempting to reinforce – at least to some extent – distance and forgetting.

Reflecting on the present was part of the research project agenda and allowed me to close the temporal exercise. The new landscape of death involved different places and mobilities. But the possible final destination of the remains made me think of the relevance that certain places, associated with senses of identity and memory, can have for contemporary groups. All of this, despite of the past and the interests that we, as researchers, can have in overcoming what we sometimes understand as fixed categories (Caftanzoglou 2001).

Acknowledgments

I would like to thank the Multidisciplinary Institute of History and Human Sciences, of the National Council for Scientific and Technical Research (IMHICIHU-CONICET, Argentina), Ricardo Guichón (director of the research project at La Candelaria), Romina Rigone, the other members of the team, and the Selk'nam Community Rafaela Ishton. The ideas presented here are my sole responsibility.

References

Badenes, D. (2016). Crónica de una restitución. La devolución de restos como parte del reconocimiento del genocidio Selk´man. *Aletheia*, 6 (12). Accessed 12 December 2017, Retrieved from www.aletheia.fahce.unlp.edu.ar/numeros/numero-12/dossier/cronica-de-una-restitucion.-la-devolucion-de-restos-como-parte-del-reconocimiento-del-genocidio-selkman

Beauvoir, J. (1915). *Los Shelknam*. Buenos Aires: Librería del Colegio Pío IX.

Belza, J. (1974). *En la isla del fuego. 1. Encuentros*. Buenos Aires: Instituto de Investigaciones Históricas Tierra del Fuego.

Belza, J. (1975). *En la isla del fuego. 2. Colonización*. Buenos Aires: Instituto de Investigaciones Históricas Tierra del Fuego.

Bender, B. (1993). Introduction: Landscape: Meaning and action. In: B. Bender, ed. *Landscape: Politics and perspectives*. Oxford: Berg, pp. 1–17.

Bender, B. (1998). *Stonehenge: Making space*. Oxford: Berg.

Bender, B. (2001). Introduction. In: B. Bender and M. Winer, eds. *Contested landscapes: Movement, exile and place*. Oxford and New York: Berg, pp. 1–18.

Borrero, L. (2001). *Los Selk'nam*. Buenos Aires: Galerna.

Buchli, V., and Lucas, G. (2001). Between remembering and forgetting. In: V. Buchli and G. Lucas, eds. *Archaeologies of the contemporary past*. London: Routledge, pp. 79–83.

Caftanzoglour, R. (2001). The shadow of the sacred rock: Contrasting discourses of place under the Acropolis. In: B. Bender and M. Winer, eds. *Contested landscapes: Movement, exile and place*. Oxford and New York: Berg, pp. 21–35.

Casali, R. (2011). *Contacto interétnico en el norte de Tierra del Fuego: La misión salesiana La Candelaria y la salud de la población selk'nam*. PhD dissertation, Mar del Plata, Universidad Nacional de Mar del Plata.

Comisión Nacional de Museos y de Monumentos y Lugares Históricos. (1991). *Disposición, 5/91*. Accessed 12 February 2017, https://static.cpau.org/mp/publicaciones/patrimonio/anexo/files/assets/basic-html/page96.html

Decreto 64/99. (1999). *Monumento y lugares históricos. Declárense a edificios y sitios históricos de la Provincia de Tierra del Fuego, Antártida e Islas del Atlántico Sur*. Accessed 12 February 2017, http://servicios.infoleg.gob.ar/infolegInternet/anexos/55000-59999/55843/norma.htm

Dov Kulka, O. (2013). *Landscapes of the metropolis of death: Reflections on memory and imagination*. Cambridge, MA: Harvard University Press.

Fernández, A. (2014). *Con Letra de Mujer: La Crónica de las Hijas de María Auxiliadora en la Misión Nuestra Señora de la Candelaria (Tierra del Fuego – Argentina)*. Buenos Aires: EDBA.

Gallardo, C. (1910). *Los Onas*. Buenos Aires: Cabaut y Cía.

García Laborde, P., Suby, J., Guichón, R., and Casali, R. (2010). El antiguo cementerio de la misión de Río Grande, Tierra del Fuego. Primeros resultados sobre patologías nutricionales-metabólicas e infecciosas. *Revista Argentina de Antropología Biológica*, 12 (1): 57–69.

Guichón, R., Casali, R., García Laborde, P., Salerno, M., and Guichón, R. (2017). Double coloniality in Tierra del Fuego, Argentina: A bioarchaeological and historiographical approach to Selk'nam demographics and health (La Candelaria mission, late 19th and early 20th centuries). *In*: M. Murphy and H. Klauss, eds. *Colonized bodies, world transformed. Toward a global bioarchaeology of contact and colonialism*. Gainesville: University Press of Florida, pp. 197–225.

Guichón, R., García Laborde, P., Motti, J., Martucci, M., Casali, R., Huilinao, F., Maldonado, M., Salamanca, M., Bilte, B., Guevara, A., Pantoja, C., Suarez, M., Salerno, M., Valenzuela, L., D'Angelo del Campo, M., and Palácio, P. (2015). Experiencias de trabajo conjunto entre investigadores y pueblos originarios. El caso de Patagonia Austral. *Revista Argentina de Antropología Biológica*, 17 (2). Accessed 5 October 2017, www.scielo.org.ar/pdf/raab/v17n2/v17n2a05.pdf

Guichón, R., Suby, J., Casali, R., and Fugassa, M. (2006). Health at the time of Native-European contact in Southern Patagonia. First steps, results and prospects. *Memoria Instituto Oswaldo Cruz*, 101 (2): 97–105.

Gusinde, M. (1920a). Expedición a la Tierra del Fuego. *Publicaciones del Museo de Etnología y Antropología de Chile*, 9–44.

Gusinde, M. (1920b). 2do Viaje a la Tierra del Fuego. *Publicaciones del Museo de Etnología y Antropología de Chile*: 133–164.

Gusinde, M. (1951). *Fueguinos*. Sevilla: Escuela de Estudios Hispano-Americanos.

Gusinde, M. (1982 [1939]). *Los Indios de Tierra del Fuego*, Part I, Vol. 1 and 2. Buenos Aires: Centro Argentino de Etnología Americana.

Hannum, K., and Rhodes, M. (2018). Public art as public pedagogy: Memorial landscapes of the Cambodian genocide. *Journal of Cultural Geography*, 13 (7): 675–682. Accessed 29 April 2018, https://doi.org/10.1080/08873631.2018.1430935

Ingold, T. (1997). The picture is not the terrain: Maps, paintings, and the dwelt-in world. *Archaeological Dialogues*, 1: 29–31.

Ingold, T. (2000). *The perception of the environment: Essays on livelihood, dwelling and skill*. London: Routledge.

Iparraguirre, S. (2009). *Tierra del Fuego. Una biografía del fin del mundo*. Buenos Aires: Del Nuevo Extremo.

Ley Nacional 25,517. (2001). *Comunidades Indígenas*. Accessed 16 October 2017, http://servicios.infoleg.gob.ar/infolegInternet/anexos/70000-74999/70944/norma.htm

Marschoff, M., and Salerno, M. (2016). ¿Sedentarios vs. Nómades? Repensando la movilidad en el marco de proyectos reduccionales (Esteco, s. XVIII; Tierra del Fuego, fines s. XIX – principios s. XX). *In*: V. Aldazábal, L. Amor, M. Díaz, R. Flammini, N. Franco and B. Matossian, eds. *Territorios, memorias e identidades. Actas de las IV jornadas multidisciplinarias*. Buenos Aires: Instituto Multidisciplinario de Historia y Ciencias Humanas, CONICET, pp. 231–241.

Martínez, B. (2013). La muerte como proceso: Una perspectiva antropológica. *Ciência & Saúde Coletiva*, 18 (9): 2681–2689.

Martinic, M. (1973). Panorama de la colonización en Tierra del Fuego entre 1881–1900. *Anales del Instituto de la Patagonia*, 4: 5–69.

Méndez, P. (2012). La extinción de los selknam (onas) de la Isla de Tierra del Fuego. Ciencia, discurso y orden social. *Gazeta de Antropología*, 26 (2). Accessed 12 December 2017, http://hdl.handle.net/10481/22063

Peñaloza, F. (2016). Transpacific discourses of primitivism and extinction on 'Fueguians' and 'Tasmanians' in the nineteenth and twentieth centuries. *In*: E. Dur and P. Schorch, eds. *Transpacific Americas: Encounters and engagements between the Americas and the South Pacific*. New York and London: Routledge, pp. 89–109.

Ricoeur, P. (2004). *Memory, history, forgetting*. London: University of Chicago Press.

Salerno, M., García Laborde, P., Guichón, R., Hereñú, D., and Segura, M. (2016). Prácticas mortuorias, dinámicas de poder e identidad en el cementerio de la misión salesiana Nuestra Señora de la Candelaria (Río Grande, Tierra del Fuego). *In*: V. Aldazábal, L. Amor, M. Díaz, R. Flammini, N. Franco and B. Matossian, eds. *Territorios, memorias e identidades*. Buenos Aires: Instituto Multidisciplinario de Historia y Ciencias Humanas, CONICET, pp. 305–318.

Salerno, M., and Guichón, R. (2017). Sobre la memoria y el olvido: Los difuntos selk'nam y el cementerio de la misión salesiana Nuestra Señora de La Candelaria (Río Grande, Tierra del Fuego). *Magallania*, 45 (2): 135–149.

Salerno, M., and Marschoff, M. (2017). En el camino: Misioneros, recorridos y paisajes (Esteco, s. XVI-XVII; Tierra del Fuego, s. XIX-XX). *Revista Memória em Rede*, 10 (17): 116–129. Pelotas, Brasil.

Salerno, M., and Rigone, R. (2017). Atendiendo a la muerte: Experiencias intersubjetivas en contextos tradicionales selk'nam y la misión salesiana Nuestra Señora de La Candelaria (Tierra del Fuego, fines del siglo XIX-principios del siglo XX). *In:* J. Pellini, A. Zarankin and M. Salerno, eds. *Sentidos indisciplinados. Arqueología, sensorialidad y narrativas alternativas.* Madrid: JAS, pp. 45–76.

Salerno, M., and Zarankin, A. (2015). Discussing the spaces of memory in Buenos Aires: Official narratives and the challenges of government management. *In:* A. González-Ruibal and G. Moshenska, eds. *Ethics and the archaeology of violence.* New York: Springer, pp. 89–112.

Télam – Agencia Nacional de Noticias. (2016). *Los restos de cuatro onas restituidos ya descansan en su tierra ancestral*, 21 April. Accessed 12 December 2017, www.telam.com.ar/notas/201604/144442-pueblos-originarios-restitucion-restos-selknam-ona-tierra-del-fuego.php

Thomas, J. (1996). *Time, culture and identity: An interpretative archaeology.* London: Routledge.

Tilley, C. (1994). *A phenomenology of landscape: Places, paths and monuments.* Oxford: Berg.

Weinrich, H. (2004). *Lethe: The art and critique of forgetting.* New York: Cornell University Press.

9
FORENSIC ARCHAEOLOGY AND THE PRODUCTION OF MEMORIAL SITES

Situating the mass grave in a wider memory landscape

Layla Renshaw

Introduction

The period since the Second World War has seen a sustained increase in the use of archaeological and forensic techniques to investigate the material traces of both historic and recent episodes of mass violence (Rosenblatt 2015; Moon 2014; Ferllini 2003). The exhumation of mass graves can enable the identification and eventual return of the dead to their families or communities for reburial and commemoration. Furthermore, the detailed reconstruction of the events surrounding death and burial can support the prosecution of perpetrators, or other forms of transitional justice, such as truth and reconciliation hearings. The physical evidence obtained from mass burials can also inform future historical accounts and counteract revisionism or denial (Saunders 2002). Focusing on the mass grave as a spatial feature, and the wider notion of the forensic landscape, this chapter will highlight some of the ambivalent properties of mass grave sites as places of horror, abjection, pollution, absence, and haunting, but also, conversely, as potential indices for ancestral bonds, collective sacrifice, and shared suffering, and a connection to the land. The highly charged psycho-geography of missing and concealed bodies in the landscape, which remain out of place and unshriven by normative burial rites, exert a powerful hold on the individual and collective imagination. This chapter will also look at the affordances of the forensic process itself, how a forensic investigation radically reframes a site of mass violence, changing the narrative that surrounds it, and the reconstituting of the grave as a site of scientific enquiry (Renshaw 2017a).

This chapter underscores the importance of situating the mass grave in its landscape, both forensically to understand the scale and complexity of a crime, and culturally to understand how a mass grave is conceptualised within a network of sites with interdependent mnemonic and historical meanings. Aspects of the forensic process, reconstructing the particularities of each individual death, and uniquely identifying each body, whilst vital to the investigative goals, can risk de-historicising a grave or untethering it from its wider cultural and symbolic context. The evolution of forensic archaeology as a discipline, stressing objectivity, impartiality, and an ultra-empirical paradigm have arguably compounded this untethering of the mass grave from its multi-layered cultural and historical associations, engendering a 'radical distance,' as Domanska (2005) describes it (see Crossland 2009, 2013 for a deeper discussion of the development of this evidentiary paradigm and its implications). In defining itself as a new field with the authority to speak in a juridical setting, and produce evidence that passes stringent legal standards, forensic archaeology has arguably divorced itself from more socially informed, theoretically engaged, and reflective areas of archaeological work (Steele 2008). The reinsertion of the mass grave into its wider cultural landscape is effectively an argument for a mutual engagement between forensic archaeology and the approaches of funerary archaeology, conflict archaeology, and landscape archaeology in their broadest manifestations (Renshaw 2013).

This contribution will draw on a number of geographical and historical contexts. Illustrative examples are taken from my own participation in exhumations in a number of settings, particularly my participant-observation conducted in the excavation of Republican civilian mass graves from the Spanish Civil War, as well as extensive ethnographic interviews conducted with both the investigative teams and the relatives of the dead in rural communities in Spain (Renshaw 2011). It will also draw on ethnographic work conducted in Australia and Northern France with relatives of Anzac soldiers recovered from a Second World War mass grave on the Western Front (Renshaw 2017b). In addition to this, some of the considerable body of literature concerning the wars in former Yugoslavia (Wagner 2008) will be drawn upon as a key example of the enduring power and significance of mass graves in the landscape.

Exhumation produces new representational spaces and new memorial sites

The dead in mass graves resulting from conflict or violence can be characterised as bodies out of place. The fact that they lie in a mass grave means, by definition, that these are individuals who have not gone through the normal rites of passage that follow death. They are often not fully accounted for by state bureaucracies, and the deaths have not been subjected to medical, scientific, or legal scrutiny. They have not been the subject of normative burial rituals, nor received the common kinds of collective commemoration by their family and community. Although many will be missed and mourned by loved ones for years, without a body or burial site as a focal point, this may be a very private and atomised form of grief. This means the biographies of the dead, and particularly the circumstances of their deaths, are often shrouded in uncertainty. Accounts of these deaths may be repressed in both public and private if they are politically dangerous or psychologically painful. This may result in silence surrounding the dead in mass graves, or, conversely, the absence of certainty is filled by multiple competing and unstable narratives.

Opening up a mass grave for exhumation creates a new space in which representations of the past are created and contested. The open grave as a representational space functions on multiple levels. On a fundamental level, opening the grave demarcates a new physical space, a new geographical location which is a focal point and destination for those seeking both the remains of the dead and evidence of the past. This seems like a simplistic observation, but the creation of a destination, and, above all, a public space, can be central to the impact of an exhumation in those societies where the past has been strongly repressed. A strong illustration of this is the exhumation of Republican civilian victims of the Spanish Civil War, which has grown into a popular social movement since the inception of these investigations in 2000. The repressive dictatorship that followed this conflict achieved such a profound domination of public space and over the collective commemorations of the war that it was politically and socially unthinkable to publicly mourn the Republican dead (Bevernage & Colaert 2014). In the post-war period, many communities had a surveillance culture so strong that it even inhibited the memorialisation of the dead within the home and family. In small communities that experienced traumatic levels of violence during the war, the practices of self-censorship were deeply entrenched and conditioned in representations of the past in all public spaces, such as the street, village square, town hall, church, school, or bar. The creation of a brand new public space cuts through these entrenched prohibitions. This is a manifestation of Weizman's usage of the term 'forensis' as a form of 'public truth' (2014), tracing the etymology of the word 'forensic,' to its roots in the forum of the ancient world. For those who have experienced the kind of grassroots, community-led exhumations that occur in Spain, the image of a classical forum is very resonant. Relatives of the dead and community members of all ages visit the grave to share testimony and anecdotes, to observe the progress of the archaeological work, to offer commentary, manual labour, equipment, and food. These are dynamic social spaces where the mood and focus can shift rapidly, from scientific analysis, to mourning, to humour. As a participant-observer, it was clear that the unstructured space was extremely helpful in breaking down prohibitions, with some survivors or descendants visiting multiple times before gaining the confidence to give testimony or engage with the investigation (Renshaw 2011).

The more conventional image of the forensic mass grave exhumation, with highly controlled access to the site, and expert practitioners in protective clothing, using specialist equipment, can also create new representational spaces. The paraphernalia of the forensic process dress the site and create a new category of space, namely, the crime scene. This produces an accompanying category shift in the events that took place there. They are no longer simply events of the traumatic past, but deaths and burials that have finally arrived at the point of recognition that warrant expert investigation, collective attention, and resourcing. Paradoxically, even as the exhumation process exposes human remains and buried objects in all their horror or pathos, it also reframes the sites by situating it in the redemptive narratives of scientific investigation and historic justice (Renshaw 2017a). The exhumation process has many benign associations, as a rational, sense-making activity, with the potential to bring order to the chaos of violent death and a jumble of bodies in a grave. If the grave and the dead inside it have been the subject of denial or forgetting, this is dramatically reversed by the intense scrutiny inherent in the forensic process, and changes the status of the dead profoundly, even before they have been excavated and analysed. The evidence gathered from exhumed bodies and objects engenders new representations about the past in terms of scientific reports, or legal hearings, which in time may inform academic or historiographical understandings of these events. But running in tandem to these kinds of empirical findings is the production and circulation of a wealth of stories and images in news coverage and online sources. As a rich source of striking visual images, human stories, and metaphors about the past, mass grave exhumations often elicit a huge creative response by artists, writers, filmmakers, and photographers (Ferrándiz 2006). Where access is allowed, the graves themselves can become a magnet for these kinds of representational activities. These images and stories, spread via the media and the arts, can have a much more profound impact on the collective or popular understanding of the traumatic past than a scientific or legal report.

The forensic landscape

The search for mass graves, crime scenes, and missing persons all engender a particular way of understanding and evaluating the landscape. This means seeking out the potential for risk, wrongdoing, and concealment in an environment, and anticipating how both victim and perpetrator might behave within it. Forms of spatial analysis have long been used in police investigations (Rossmo 2000; Canter 2003), and the accrued knowledge of trends and patterns in human spatial behaviour regularly inform the search for both missing persons and clandestine burials (Killam 2004). Congram et al. (2017) assert the necessity of more sophisticated spatial thinking, more refined and systematic spatial recording, and the use of GIS in post-war investigations. They cite an example of how seemingly benign places in the landscape can inform the search for mass graves, referring to a map presented in evidence against the Bosnian Serb former General Ratko Mladic. It was demonstrated that during the Yugoslav wars, school buildings were repurposed as detention centres and furthermore, that the location of these schools had a predictable proximity to the mass execution sites where detainees were killed. Congram et al. advocate the use of software to generate layers of data such as soil type, gradient, and proximity to roads to map the probability of a potential grave location. Temporally specific data relating to the unfolding conflict can also be inserted into the map to visualize the situation on the ground, including the last reported sightings of victims, military control of territory, and the destruction of access routes such as roads and bridges. The ability to combine data sources, and assign a weighting of significance to the different factors that determine the location of a mass grave, makes this a powerful tool in locating burials and execution sites.

The necessity of this kind of spatial data becomes apparent when considering how mass killings and burials can be crimes of such scale and complexity that they spread to occupy great swathes of the landscape, and ultimately come to define that landscape. As Cyr phrases it, 'crimes of scale have a way of invoking landscapes' (2014: 90). The forensic landscape resulting from the Srebrenica massacre exemplifies this phenomenon, comprising of a web of interconnected crime scenes (Jugo & Wagner 2017). These range in nature from holding centres, to execution sites, to the escape routes and hiding places of the survivors, which were also places of great hardship and trauma. Aside from the horrifying scale and brutality, a distinctive feature of the Srebrenica massacre was the post-mortem movement of bodies from primary, secondary, and even tertiary burials as the perpetrators attempted to disperse the evidence of mass killings and confound future investigations. The detailed analysis of environmental evidence relating to the Srebrenica massacre conducted by Brown (2006) gives some insight into the macabre churn of these graves.

The evidence from pollen, soil, and sediment was used to map a series of links between grave sites and reconstruct the post-mortem movement of human remains through the landscape. Brown describes his analysis of seven primary mass graves and a further 19 secondary sites. The difference in underlying geology, and variable plant cover such as meadow, arable, woodland, and orchard, allowed a distinctive signature for each. The dead bodies transferred trace evidence from the execution sites into the primary graves, and then further traces were transferred between primary and secondary graves. The movement of decomposing bodies, often using heavy mechanised diggers, caused dismemberment and the disassociation of body parts, clothing, and possessions. This meant that family groups and even the parts of a single individual could be scattered through the landscape, perpetrating a further outrage on the dead and further obstructing their recovery and identification. The effort expended in the dispersal of the dead speaks of the planning, resourcing, and knowing culpability of the perpetrators. In her ethnographic work conducted in Bosnia-Herzegovina and Serbia, Petrović-Šteger (2009) captures the emotional impact of the shattered and scattered bodies of the Yugoslav wars experienced across all communities, exploring the dilemmas faced by those who receive partial human remains but know that the rest of their loved one is still out in the landscape.

The agony of these processes, as disparate sources of evidence combine into an assemblage that proves beyond doubt that a relative is dead, is also described powerfully by Wagner (2008) and Stover and Peress (1998). Cyr writes movingly of the affective resonance of this landscape, knowing that so many thousands could disappear into it. In a haunting phrase, she describes these missing men as 'becoming landscape' in that they were swallowed up by the earth and concealed within it. In the absence of human remains, the landscape becomes an index for the missing. Cyr describes how these places have been altered forever by 'the precarious but nevertheless enduring experience of looking to landscape for missing men,' or as one of the thousands of bereaved women of Srebrenica expressed it, 'we turn to our empty forests' (2014: 89). The landscape exerts a malign power because it will not relinquish these bodies to their families, but is simultaneously a site of memory and commemoration, sanctified by the presence of so many dead.

Beyond the graves: warscapes and the wider memorial landscape

The preceding section emphasises the mass grave as a focal point and as a bounded site in the landscape that can become emblematic of a complex sequence of events. The highly elaborated nature of the archaeological and scientific work carried out at the grave reinforces this focus, as does the extreme emotional and visual power of exposed bodies and objects. However, the risk of this bounded view is that the grave site becomes untethered from its wider spatial context, and stripped out from the temporal sequence that led to its formation. An exclusive focus on the mass grave can preclude an analysis of the way the grave is really understood and experienced by witnesses, survivors, descendants, and surrounding communities.

The predominance of the grave, as emblematic of the traumatic past, can also create an implicit hierarchy of loss or suffering which overshadows a whole suite of experiences during war, including displacement, destitution, imprisonment, forced labour, battle trauma, and gendered or sexual violence, all of which may be associated with other locales in the wider warscape (González-Ruibal 2016). An investigative paradigm that focuses on the mass grave, to the exclusion of other potential sites of memory, risks missing a close attention to vernacular memorial practices and the affective or symbolic significance of other sites, which may pre- or post-date the exhumation process. The meaning of a mass grave can change when it is brought into tension with these other locales. Vernacular or emic memorial practices can reveal the importance of these associations and highlight the wider context of the grave.

It is readily apparent that the Western Front represents a warscape, both in the large-scale monument building and concerted memorialisation of the dead, and more subtly in the way settlements and natural landscapes were transformed by processes of destruction and renewal in the war and its aftermath, and the physical traces of these forces can be discerned everywhere. The interconnected nature of the sites is reinforced by the form of modern pilgrimage that has emerged, in stages, over the last century (Beaumont 2015). The visitor reinforces the relationships between these sites by moving bodily between them. Initiatives to mark the centenary of the First World War include the regeneration of sites and their repackaging as the 'Australian Remembrance Trail' (Sumartojo 2014). Various stretches of the trail can be followed depending on the time, resources, and interests of the visitor. Lesser-known sites have explicitly had their status raised by forging a connection with the most popular and heavily visited locales. Visitors can move through the landscape to experience an unfolding narrative, mirroring the historical sequence of events. This is an intensive and immersive period of engagement with the past, and with the dead. Each site visited informs the experience and understanding of the next.

In this kind of transnational commemoration typified by the Western Front, there is clearly an impetus to experience multiple sites and maximise one's affective and sensory engagement with the landscape. For many Australian, Canadian, or even British and German visitors, the trip may be characterised as a once-in-a-lifetime experience. The warscape is consumed to a saturation point, the work expended in moving through the landscape reaffirms familial or national bonds with the dead, memories and mementoes are taken home to share with others who cannot make the trip, and once home, the impressions gathered in these sites of memory furnish further personal acts of memorialisation of the dead. The commemorative activities on the Western Front, at Gallipoli, or on the Normandy beaches represents a very conscious and concerted engagement with a warscape, sometimes making a pilgrimage of many thousands of miles to be there. On the other end of the spectrum are the clandestine graves of Spain's Civil War. These mass graves are spread throughout small communities in Spain, where many witnesses, survivors, and descendants live in intimate familiarity with the landscape in which past horrors have occurred. They navigate their warscape on a daily basis and negotiate the memories and emotions elicited by these places in order to be reconciled to the continuation of normal life.

It is worth reflecting in detail about the other sites that constitute the landscape of a Civil War and the sites that have a locally meaningful connection to the mass graves of the war dead (González-Ruibal 2016), as these may in fact be more significant as sites of memory than a grave site, which may not always be apparent on first sight. These places may be modest or mundane sites, because of the nature of the conflict (with victim and perpetrator often known to each other), with highly localised forms of violence running alongside national events, which are understood and remembered through the lens of tensions and grievances particular to each community. Taking the example of my fieldwork in two small villages in Burgos Province, Castile and Leon, the sites that came up in ethnographic interviews about the war were highly varied. Sites of political and class conflict that predated the war came up, as did local and regional sites of political authority and repression, such as the local police cells, or the central prison in Burgos where prisoners were concentrated prior to being handed back to militias and death squads for extrajudicial executions. In both villages, the church at the heart of the village and a nearby monastic complex were also cited as centres of authority, surveillance, and where decisions were made to blacklist Republican civilians. After the mass killings that occurred in these communities, both the empty homes of the dead, some of which fell into ruins, and the homes of the perpetrators, suspected of carrying out the murders, became psychically charged places for the relatives of the dead. Private property such as businesses, farmland, gardens, and homes changed hands during the war, frequently appropriated by the Francoist authorities, or more directly taken by the killers, and therefore function as enduring reminders of injustice.

Many shared public spaces in these small villages have vastly different meanings and associations, depending on one's experience of the Civil War. For the defeated, the public monuments to Franco or other Nationalist military leaders, the post-war decision to change key street names to honour Franco's victory, and the monument to those villagers who died fighting for Franco, 'fallen for God and for Spain,' reinforced the experience of defeat on a daily basis. Other, seemingly neutral, spaces could be associated with highly traumatic events. For example, in both these Burgos villages, survivors and descendants of Republican families gave accounts of extreme gendered and sexual violence that occurred in the main village square and throughout the streets, with its public and ritualised nature a key strategy in the humiliation of Republican women. The village cemetery was also a very charged space, especially for the spouses and mothers of the dead, as it came to signify the absence of the loved ones' bodies and the impossibility of burying them and mourning them in the socially prescribed place (Renshaw 2011).

The example of the mass grave at Pheasant Wood, Fromelles, is very useful to consider, particularly as it reveals how this context changes through time, and different locales in the warscape are foregrounded in different periods. In the case of Fromelles, there are a number of significant locations in close proximity that constitute a dense memorial landscape. As the site of a brief but horrific First World War battle, with an Australian casualty rate unsurpassed in any other conflict, visitors and relatives of the dead had long engaged with this cluster of sites (Lindsay 2008). With the realisation that not all the dead from this battle had been recovered and reburied after the war, and the subsequent discovery of mass graves nearby (Loe 2010), the character of Fromelles as a memorial site has changed dramatically, but many of the key sites remain the same. Relatives of the dead and visitors to Fromelles highlight the importance of the village itself, which has diligently honoured the memory of the battle, and the sacrifice of the Anzac forces there. Some visitors highlight the original cemetery constructed for the Anzac dead from this battle, known as V.C. Corner and the 'Cobber's' statue nearby which has become an icon of a particularly Australian martial heritage. Other visitors emphasise the collection of materiel and personal possessions collected over decades from the battle sites and the environs of the grave which give a very tangible and intimate insight into the dead soldiers. For some, the most significant part of visiting Fromelles is the phenomenological experience of the battle terrain itself, retracing the lines of trenches, and understanding the topography of the opposing positions.

These features are all in immediate proximity to the both the Commonwealth War Grave Commission (CWGC) Cemetery and the mass grave site that resulted from the battle (Summers 2010). Visitors' affective responses and understanding of the grave are strongly informed by their experiences of these other locales, and the immersive layering of different indices for the battle, and for the dead. However, a wider memorial landscape can operate on multiple scales, and, in the case of the Fromelles, the sites of memory that interconnect around the dead are truly transnational. A considerable proportion of the vernacular memorial practices and invented traditions that have grown up around the Fromelles dead are explicitly focused on connecting the graves back to Australia and reasserting the bonds between the dead soldier's origin and his final resting place (Scates 2016). In the ethnographic interviews I conducted, a broad range of objects and images were used to achieve a form of recursive binding between Fromelles and home in Australia, an assertion of affective and symbolic connections that transcended the obvious temporal and spatial distances (Renshaw 2017a; Scates 2016; and see Pinney 1997, 2005, for an exploration of temporality and the recursive). These included bringing offerings or deposits to the graves such as rocks, soil, and leaves taken from the informant's own home; from favourite places; from local war memorials; from the graves of other relatives; and from the house in which the dead soldier had been born and raised, even if that house was now empty or in ruins. Other practices clearly intended to bridge these distances included bringing photographs to Fromelles – of the family home and photographs of the graves of now long deceased relatives, particularly the parents or children of the dead soldier – bringing together these important locales in family history in a virtual form.

Bodies, blood, ancestors, and ownership

Mass graves, like any type of human burial, are intimately connected with ancestors, and the physical emplacement of those ancestors is one of the ways contemporary societies trace their historical connection to a landscape. These ancestral associations may, in fact, be intensified in the case of mass graves. If the graves result from violent deaths, they become indices of sacrifice and blood that has been shed in a particular location, strengthening the psychic and moral investment in a particular locale. If the graves relate to a conflict over territory, attempted ethnic cleansing, genocide, or forced displacement, they serve as shorthand for this attack against the connection between people and place. Following displacement from ancestral land, the buried bodies of the dead may often be the most enduring tangible link between a cultural group and their erstwhile home, even after other forms of cultural heritage have been erased and living populations have been forced to leave. In this way, the dead can act as a placeholder, staking a claim for the historical presence of a particular group and can also exacerbate the pain of exile, as living communities are no longer in proximity to their dead and not able to recover or care for the remains. Sant Cassia's (2007) ethnography of the divided communities of Cyprus powerfully illustrates the obstacles to mourning engendered by forced displacement. Even though they are in close geographical proximity, many of the war dead of the 1974 conflict are missing on the other side of the political divide, and, for decades, the families of the deceased could not undertake the search and recovery of these remains, except by clandestine means.

There are many examples of the connection between mass graves, missing bodies, and contested historical claim to territory, with the conflicts in former Yugoslavia being a paradigmatic case (Denich 1994). Verdery's analysis of the complexity of dead body politics in the region notes that, even prior to political instability, there was an intense funerary culture and 'burial regime' which made the dead a likely source of symbolic capital for Nationalist movements. 'People hold strong ideas about proper burial and about continuing relations with dead kin; frequent visits to tombs are common; and violence against enemy graves has a history at least as old as World War II' (1999: 97). Bax also analysed the role of historic graves in resurgent Balkan nationalism and competing territorial claims: 'in the Bosnian countryside the deceased continue to be part of a kinship group. Via them, their progeny can lay claim to the use of land and water and to the produce of fruit and olive trees' (1997: 17). Bax reports hearing the oft-voiced suspicion in divided communities that bodies were being secretly removed from cemeteries, a type of ethnic cleansing of the dead which they referred to as killing the dead again.

The historic mass graves that become pivotal to the breakdown of Yugoslavia in the 1990s contained the dead resulting from the extreme violence of the Second World War in which competing ideological and ethnic groups committed countless atrocities. 'Mass slaughter occurred on all sides, the victims being thrown into caves or buried in shallow mass graves or simply left to rot' (Verdery 1999: 99). A contemporary commentator noted how the great swathes of unstable and heaving graves made it appear that 'the very earth seemed to breathe' (ibid., 99). The Yugoslav communist regime managed to obscure the details of much of these killings within a narrative of national struggle, but from the 1980s onward moves were made on a community level to locate and recover these bodies dating from the Second World War massacres. In 1991, a series of mass exhumations and reburials of Serbian victims of Croatian Ustaše militias were the focus of intensive media coverage and public participation. 'Retrieving and reburying these nameless bones marked the territory claimed for Greater Serbia.... We might say that these corpses assisted in reconfiguring space by etching new international borders into it with their newly dug graves' (Verdery 1999: 102). The movement of dead bodies prefigured the movement of troops, leading some commentators to describe them as a 'vanguard of bones.' Skinner et al. (2002) remark on how this preoccupation with ancestry, kinship, and human remains persisted even during conflict and societal breakdown, with a large volume of 'body trading' reportedly occurring between opposing forces, and even the planned exhumation and relocation of ancestral graves, as families and communities permanently relocated due to violence and the redrawing of national borders. This demonstrates how even whilst being sites of abjection, mass graves can be inextricably bound up with narratives of group identity, collective sacrifice, territorial possession, and inheritance, with bones functioning both as physical place markers in the ground, and as indices of the blood that has been shed into, and for, the homeland.

Conclusion

Forensic exhumation, like all archaeological excavation, destroys a feature at the same time it brings it into being as a site of enquiry and locus of evidence. Throughout this chapter, there are repeated examples of the overwhelming human impetus to find the dead, gather them in, analyse, commemorate, and grieve for them. However, the psychic power of mass grave sites is strong. The associations between the grave site, the memory of conflict, and the memory of the dead are multi-layered and can exert a strong pull even after the grave has been emptied. In my own ethnographic work concerning graves in rural Spain and Fromelles in Northern France, before the exhumations occurred, the mass graves were predominantly characterised as sites of abjection, signifiers of insult and injury to the dead. But after exhumation, with the bodies formally reburied, the grave site changed. It retained much of its psychic hold, but in a more benign form. Informants in rural Spain talked about the solidarity of the dead lying together, that they had 'sanctified' the soil with their bones (Renshaw 2011), and that the empty grave held more memories than the village cemetery, as a place to go and think about the dead, (Renshaw 2010). Although not expressed in these same explicit terms, the relatives of the Fromelles dead repeatedly expressed a desire for strands of continuity between the old mass grave and the newly built CWGC cemetery. The majority commented positively on the close proximity, even inter-visibility, of the old and new graves, with the same community, ambience, and familiar sounds surrounding the dead. All approved of the decision to order the graves in the new cemetery in the same sequence as the bodies had lain in the mass grave, and also strongly approved the decision to preserve the empty grave site in perpetuity, and not to farm or build on it. Several commented that the now empty mass grave was 'a kind of monument.' This illustrates that mass graves are inherently ambiguous and unstable sites, and that forensic intervention radically alters not only the physical site itself but the narrative surrounding it, both in its affect and meaning. To more fully understand the significance of a grave site, and therefore the impact of exhuming it, the grave must be brought into dialogue with a network of significant sites that make up the wider warscape or memorial landscape.

References

Bax, M. (1997). Mass graves, stagnating identification, and violence: A case study in the local sources of "The War" in Bosnia-Herzegovina. *Anthropological Quarterly*, 70 (1): 11–19.

Beaumont, J. (2015). Australia's global memory footprint: Memorial building on the Western Front, 1916–2015. *Australian Historical Studies*, 46 (1): 45–63.

Bevernage, B., and Colaert, L. (2014). History from the grave? Politics of time in Spanish mass graves exhumations. *Memory Studies*, 7 (4): 440–456.

Brown, A.G. (2006). The use of forensic botany and geology in war crimes investigations in NE Bosnia. *Forensic Science International*, 163: 204–210.

Canter, D. (2003). *Mapping murder: The secrets of geographic profiling*. London: Virgin Publishing.

Congram, D., Kenyhercz, M., and Green, A.G. (2017). Grave mapping in support of the search for missing persons in conflict contexts. *Forensic Science International*, 278: 260–268.

Crossland, Z. (2009). Acts of estrangement. The post-mortem making of self and other. *Archaeological Dialogues*, 16 (1): 102–125.

Crossland, Z. (2013). Evidential regimes of forensic archaeology. *Annual Review of Anthropology*, 42: 121–137.

Cyr, R.E. (2014). The "Forensic landscape" of Srebrenica. *Kultura*, 4 (5): 81–91.

Denich, B. (1994). Dismembering Yugoslavia: Nationalist ideologies and the symbolic revival of genocide. *American Ethnologist*, 21: 367–390.

Domanska, E. (2005). Toward the archaeontology of the dead body. *Rethinking History*, 9 (4): 389–413.

Ferllini, R. (2003). The development of human rights investigations since 1945. *Science and Justice*, 43 (4): 219–224.

Ferrándiz, F. (2006). The return of civil war ghosts: The ethnography of exhumations in contemporary Spain. *Anthropology Today*, 22 (3): 7–12.

González-Ruibal, A. (2016). Beyond the mass grave: Producing and remembering landscapes of violence in Francoist Spain. *In*: O. Feran and L. Hilbink, eds. *Legacies of violence in contemporary Spain: Exhuming the past, understanding the present*. London: Routledge, pp. 93–118.

Jugo, A., and Wagner, S. (2017). Memory politics and forensic practices: Exhuming Bosnia and Herzegovina's missing persons. *In*: Z. Dziuban, ed. *Mapping the 'Forensic Turn': Engagements with materialities of mass death in holocaust studies and beyond*. Vienna: New Academic Press, pp. 195–214.

Killam, E.W. (2004). *The detection of human remains*. Second Edition. Springfield, IL: Charles C. Thomas.

Lindsay, P. (2008). *Fromelles: Australia's darkest day and the dramatic discovery of our fallen World War One diggers*. Prahran: Hardie Grant Books.

Loe, L. (2010). Remembering Fromelles. *British Archaeology*, 111 (March–April): 36–41.

Moon, C. (2014). Human rights, human remains: Forensic humanitarianism and the human rights of the dead international. *Social Science Journal*, 65 (215–216): 49–63.

Petrović-Šteger, M. (2009). Anatomizing conflict – accommodating human remains. *In*: H. Lambert and M. McDonald, eds. *Social bodies*. New York: Berghahn Books, pp. 47–76.

Pinney, C. (1997). *Camera Indica: The social life of Indian photographs*. London: Reaktion Books.

Pinney, C. (2005). Things happen: Or from which moment does that object come? *In*: D. Miller, ed. *Materiality*. Durham, NC: Duke University Press, pp. 257–271.

Renshaw, L. (2010). Missing bodies near-at-hand: The dissonant memory and dormant graves of the Spanish civil war. *In*: M. Bille, F. Hastrup and T. Flohr Sorenson, eds. *An anthropology of absence: Materializations of transcendence and loss*. New York: Springer, pp. 45–61.

Renshaw, L. (2011). *Exhuming loss: Memory, materiality and mass graves of the Spanish civil war*. Walnut Creek, CA: Left Coast Press.

Renshaw, L. (2013). The exhumation of civilian victims of conflict and human rights abuses: Political, ethical and theoretical considerations. *In*: S. Tarlow and L. Nilsson Stutz, eds. *The Oxford handbook of the archaeology of death and burial*. Oxford: Oxford University Press, pp. 781–799.

Renshaw, L. (2017a). The forensic gaze: Reconstituting bodies and objects as evidence. *In*: Z. Dziuban, ed. *Mapping the 'Forensic Turn': engagements with materialities of mass death in holocaust studies and beyond*. Vienna: New Academic Press, pp. 215–236.

Renshaw, L. (2017b). Anzac anxieties: Overcoming geographical and temporal distance in the recovery and reburial of Australian war dead at Fromelles. *Journal of War and Culture Studies*, 10 (4): 324–339.

Rosenblatt, A. (2015). *Digging for the disappeared: Forensic science after atrocity*. Stanford, CA: Stanford University Press.

Rossmo, D.K. (2000). *Geographic profiling*. Boca Raton, FL: CRC Press.

Sant Cassia, P. (2007). *Bodies of evidence: Burial, memory, and the recovery of missing persons in Cyprus*. New York: Berghahn Books.

Saunders, R. (2002). Tell the truth: The archaeology of human rights abuses in Guatemala and the former Yugoslavia. *In*: J. Schofield, W.G. Johnson and C.M. Beck, eds. *Matériel culture: The archaeology of twentieth century conflict*. London: Routledge, pp. 103–114.

Scates, B. (2016). The unquiet grave: Exhuming and reburying the dead of Fromelles. *In*: K. Reeves, G. Bird, L. James, B. Stichelbaut and J. Bourgeois, eds. *Battlefield events: Landscape, commemoration and heritage*. London: Routledge, pp. 13–27.

Skinner, M., York, H., and Connor, M. (2002). Postburial disturbance of graves in Bosnia-Herzegovina. *In*: W. Haglund and M. Sorg, eds. *Advances in forensic taphonomy: Method, theory, and archaeological perspectives*. Boca Raton: CRC Press, pp. 293–308.

Steele, C. (2008). Archaeology and the forensic investigation of recent mass graves: Ethical issues for a new practice of archaeology. *Archaeologies*, 4 (3): 414–428.

Stover, E., and Peress, G. (1998). *The graves: Forensic efforts at Srebrenica and Vukovar*. New York: Scalo.

Sumartojo, S. (2014). Anzac kinship and national identity on the Australian remembrance trail. *In:* S. Sumartojo and B. Wellings, eds. *Nation, memory and great war commemoration: Mobilizing the past in Europe, Australia and New Zealand*. Oxford: Peter Lang, pp. 291–306.

Summers, J. (2010). *Remembering Fromelles: A new cemetery for a new century*. Maidenhead: CWGC Publications.

Verdery, K. (1999). *The political lives of dead bodies: Reburial and postsocialist change*. New York: Columbia University Press.

Wagner, S. (2008). *To know where he lies: DNA technology and the search for Srebrenica's missing*. Berkeley, CA: University of California Press.

Weizman, E. ed. (2014). Introduction: Forensis in forensic architecture. *In: Forensis: The architecture of public truth*. Berlin: Sternberg Press, pp. 9–34.

10
URBAN BOMBSITES

Gabriel Moshenska

Introduction

The Neues Museum on Berlin's Museum Island, first opened in 1855, was gutted by bombs in the Second World War. In 2009 it reopened after a lengthy restoration, displaying its extraordinary prehistoric and Ancient Egyptian collections, including the famous bust of Nefertiti (Barndt 2011). The architect David Chipperfield, who oversaw the lengthy and troubled rebuilding of the museum from what he described as a 'Piranesian pile' into an award-winning recreation, retained large areas of bare brick, burned stone columns, fragmented frescoes, bullet holes, and the scars of shell and bomb fragments (Kimmelman 2009). In this unflinching exhibit of its architectural wounds, together with displays of artefacts burned in the bombing, the contemporary Neues Museum is as much a museum of the urban bombsite as a museum of archaeology.

The life history of the Neues Museum from its first (of several) bombings in 1943 through to the present illustrates many of the themes surrounding urban bombsites. In its partial destruction, it was one amongst millions of buildings across the world left scarred by the Second World War, and its ruins remained a blackened reminder of the horrors of that conflict for another half century in the heart of divided Berlin. Chipperfield's reference to classical ruins marks another theme in the lives of urban bombsites: attempts to find comfort, meaning, and a sense of healing in the evocation of more romantic and politically sterile monuments of the reassuringly distant past. Unusually, the Neues Museum manages to reconcile two generally distinct pathways in the later life histories of bombsites: the first being rebuilding and the comfortable amnesia of material erasure; the second being the preservation and presentation of architectural ruins as sites of commemoration.

In this chapter I want to present some of the principal themes, trends, practices, adhesions, representations, and confusions surrounding urban bombsites. To weave together these disparate threads, I will consider the bombsite as a place that is created, inhabited, used, transformed, annihilated, and represented across a range of trajectories. One of my over-arching aims is to dismiss any conception of the life history of bombsites or any ruins of violence as a straightforwardly linear progression from the violent wound of destruction through a gradual process of 'healing' through, variously, rebuilding, clearing and redeveloping, or the softening processes of time and nature (Moshenska 2015). This too simplistic and teleological approach to bombsites as 'healable' is often implicit in the notion of architectural ruins as 'wounds.'

I am an archaeologist, anthropologist, and historian of bombsites writing these words in an office built upon a Second World War bombsite, from where I can watch *YouTube* videos of Predator drone missile strikes reducing buildings to rubble on the other side of the world. In my journey to work today I have travelled past the sites of bombings in the 1880s, 1940s, and 1970s in a city – London – whose bombed ruins have witnessed social cleansing, modernist dreaming, cinematic absurdity, children's play, ecological diversity, archaeological discovery, death, haunting, and wave upon wave of rebuilding and re-rebuilding. The life histories of bombsites between place, memory, and haunting absences is a part of everyday life for many people across the world in societies ripped apart by, or healing from the violent conflicts of the twentieth and twenty-first centuries.

Coming into being

In the modern city built of brick, stone, steel, and glass, the bomb presents an existential threat in the way that fire did (and still does) for cities of wood and thatch. My focus in this chapter is the individual or bounded bombsite: not the remains of a city flattened by an atomic bomb, artillery barrage, or a firestorm of incendiary bombs, but the ruined building made notable and disturbing by the juxtaposition of surviving buildings nearby. The opposite case can also be seen: the survival of St Paul's Cathedral during the London Blitz was made particularly striking by the near-total destruction of every building surrounding it (Allbeson 2015).

The urban bombsite is the remains of a building marked out for destruction, either by guided missile or smart bomb in some more recent conflicts, or by the more inhumanly random finger of fate directing a bomb, rocket, or missile from the sky. Or it could be seen as the absence of that building in the space where it had previously stood.

The creation of a bombsite might appear instantaneous, but it is a combination of several distinct events taking place within moments of detonation (Cullis 2001). The bomb, typically a high explosive charge inside a metal casing, might detonate on impact, close to the surface, or after burying itself some distance into the ground. The initial blast of hot, high-pressure gas will destroy structures close by and cause diminishing damage further away. The fragments of bomb casing will cut into walls and act as projectiles over longer distances, while the secondary fragments – pieces of nearby structures and other objects carried by the blast – will also be propelled at high velocity across the blast area and beyond. Shockwaves after the initial blast will cause further damage, and seismic shockwaves will damage or even demolish buildings further from the detonation point, literally shaking them to pieces. The heat of the detonation can char or ignite flammable materials such as timbers and soft furnishings, and fires from broken gas pipes can also spread through the wreckage. Many of the stranger effects of a bomb are caused by the blast wind, in which the low pressure caused by the initial blast wave creates a powerful suction, drawing wreckage and fragments back towards the point of detonation. All of this occurs within seconds, leaving a building transformed into a bombsite.

The social construction of a bombsite continues after this point. In many conflict areas such as London during the Blitz or contemporary Syria, dedicated teams of Civil Defence workers will arrive at the site. Fire crews will work to extinguish fires and turn off gas and water mains, while rescue crews will make the wreckage stable and begin to search for survivors and casualties. Police, military, or paramilitary troops might be deployed to guard the site against looters or to create a protective cordon, and efforts will be made to clear the wreckage and debris from the street back into the footprint of the destroyed or damaged structure (O'Brien 1955). Unstable structural elements such as standing walls might be torn down to prevent their subsequent uncontrolled collapse and possible harm to those nearby. In some cases these rescue efforts might themselves be targeted for attack, either by a continued blind or random bombing of the area, or by deliberate targeting: some contemporary drone operators have been accused of targeting rescuers who rush to the scene with follow-up missile strikes. When the fires have been extinguished, the wounded rescued, the dead removed, and the site made safe and bounded, the next stages in the life of a bombsite can begin.

Rubble

The defining artefact of an urban bombsite is rubble. Rubble has a curious life of its own, and one worth following briefly. Some of the rubble of bombed British cities in the Second World War was carried to America as ballast by the empty convoy ships that had brought war materials, food, and other resources to Britain. Shipped from Bristol and dumped in New York, this rubble formed the foundations of the waterfront in parts of Queens, and is marked today with a memorial stone (Lstiburek 2014). The rubble from other cities was shipped to the east of England and formed the foundations for the miles of runways and taxiways for the new airfields from which the bombers of the British and American air forces reduced German and continental cities to rubble. The work *What Dust Will Rise?* by artist Michael Rakowitz included a piece of rubble from a bombed British building that had itself been dropped by bombers, labelled 'Stone fragment dropped by the Royal Air Force over Essen, Germany, 1942. British bombers released the rubble of English buildings destroyed by the Luftwaffe during the Battle of Britain on German cities as a prelude to actual bombs' (Rakowitz 2012).

The vast heaps of rubble in bombed cities in post–Second World War Germany presented logistical problems. Winston Churchill described Europe in this period as a 'rubble heap,' and Bertold Brecht called Berlin 'the heap of rubble near Potsdam,' but gargantuan efforts were made by teams of workers, mostly women, to sort and clear the rubble. Much of it was dumped out at sea, while more was formed into mounds, artificial embankments or hills, and covered with earth (Sebald 2003). Nevertheless, an object-agency perspective on warfare might conclude that it is a means by which rubble reproduces itself.

What kind of place?

Rubble, together with the weeds that grow quickly on bombsites, contribute to a sense of bombsites as 'fuzzy' spaces, in contrast with the hard-edged spaces of the urban built environment. Bombsites can be considered 'fuzzy places' in a number of different ways, and it is worth considering some of these briefly, again drawing primarily on accounts of the bombsites of Second World War Britain.

The rich and complex ecology of bombsites fascinated horticulturalists and others in wartime Britain. As Richard Mabey noted, the ruins of London including those around St Paul's became a rich carpet of colour with the rosebay willowherb and buddleia that grew enthusiastically on burned and ruined sites (1996: 236). The director of Kew Gardens gave a lecture on the ecology of London's bombsites in 1945, noting the prevalence of windborne seeds as well as others liberated from the long-buried urban earth, amongst the more than 150 species of plant found growing on bombsites. Alongside the flora came the fauna: birds nested amid the rubble, stray cats found quiet spots to birth and house their kittens, and stray dogs foraged for food. Where standing water could be found – including the huge emergency water tanks erected on some bombsites during the Blitz – a wealth of insect life could be found on and below the surface (Moshenska 2014).

The chaos and mess of bombsites attracts more than wild animals. In conflicts across the world and into the present, children and young people have found the disorderly spaces of bombed buildings to be irresistible play spaces, with near endless potential for imaginative play, transformation and modification, escape from adult supervision, risk-taking, privacy, fighting, hiding, and illicit sexual activity (Moshenska 2014) (Figure 10.1). Geographers Cloke and Jones (2005) have traced both the longstanding attraction of disordered spaces to children, but also the cultural history of children in ruins as an expression of their wildness and closeness to nature. The 'adventure playground' movement from the 1940s onwards drew heavily on the wartime experiences of educational reformers such as Marie Paneth who had observed children's play on bombsites and recognised the potential it held for free expression, creative play, and working through traumatic experiences (Kozlovsky 2007). The typical adventure playground of this period with open spaces and construction materials including scrap timber, bricks, and scrap metal bears a clear resemblance to the

Figure 10.1 Boys digging an allotment on a bombsite in London, 1942.
Source: Ministry of Information/Wikimedia Commons.

bombsites in Denmark and Britain that inspired educationalists. Meanwhile, first person accounts of wartime childhoods reveal the social complexity and cultural richness of children's encounters with bombsites, including the physical dangers, joys, complexities, competitiveness, possessiveness, and imagination that characterise children's play in free environments (Cranwell 2003).

The marginality of bombsites extends beyond children and stray cats. In many conflicts such as wartime Europe bombsites, alongside air raid shelters, became hiding places for deserters, refugees, criminals and others on the margins of society. Many people whose homes had been destroyed built modest shelters in the wreckage of their homes, as social services struggled to find replacement housing in the face of heavy demand, or totally collapsed as the war took its toll. In many cases people were reluctant to leave the wreckage of their homes due to psychological trauma, or due to the not wholly irrational fear that their possessions would be looted or the land taken away from them. The fuzziness of the bombsite blurred boundaries of possession as well as concepts of indoors and outdoors, the home and the street. In the reverse of looting, bombsites – and particularly isolated ones – often become sites for the illicit or official dumping of rubbish.

Transformations

How are bombsites transformed by human intervention? What 'next chapters' are available in their life histories? The primary theme here is development, of bombsites put to work as productive spaces within society. Here, what they become – a public park, a new block of flats – is almost as important as what they are not – an eyesore, a patch of waste ground, an urban wilderness. This sense of absencing, of bombsites as absences that must themselves be made absent, highlights an important aspect of their being: that they are not wanted, not valued in their own right, not regarded as an end-point or a legitimate form of urban space by anybody with any power to decide. Wild children and stray cats do not constitute stakeholders. What else must be made absent? The redeveloped bombsite is an erasure of the past in numerous ways. In many places the destruction of war, like destruction wrought by natural disaster, creates opportunities for social engineering: slums have been swept away, and now slum-dwellers, vulnerable and transient communities can be exiled or further marginalised through the redevelopment of their former homes.

Archaeologists have long treated redevelopments as keyhole views into deep urban pasts, and the development of bombsites is no exception. In post–Second World War London archaeologists such as W.F. Grimes, Ivor Noël Hume, and their teams worked in advance of the construction crews to retrieve and record the traces of prehistoric, Roman and medieval London before they once again disappeared from view beneath office blocks and homes: Rose Macauley referred to these diggers as 'a civilised intelligence . . . at work among the ruins' (quoted in Mellor 2004: 86). These buried pasts mean little to developers and planners with ambition and a vision of the future, for whom the opportunities embodied in a bombsite are endless. In the ruins of the Second World War, Britain planners such as Patrick Abercrombie saw the foundations of a modernist future urbanism, a harmonious and humane creation, rational and totalising, built on the ruins of the messy, palimpsest city of the past (Tiratsoo 2000). Abercrombie's plans for London, Greater London, Plymouth, and other cities are astonishingly detailed, grand, beautiful, and cold. Only the poverty of post-war Britain kept them from reality, and the absence of these grand modernist reconstructions becomes just another absence haunting the bombsites until their eventual redevelopment.

Memorials

A small subset of bombsites are destined not for redevelopment but for preservation in situ: the Neues Museum mentioned earlier is a rare example of compromise between these two forces. The use of bomb ruins as war memorials was discussed in Britain as early as 1943, when detailed plans were put in place to preserve the ruins of Christ Church Greyfriars as a memorial to the victims of the Blitz (Casson 1944). The Kaiser Wilhelm Memorial Church in Berlin is one such structure, preserved in a state of ruin as a reminder of the horrors of war, and accompanied by a memorial hall with interpretation materials, and a new church building next to it constructed in the 1960s. Bombed churches and cathedrals left in ruins as war memorials can be found in Liverpool, Hanover, Bristol, Cologne, Coventry, and other cities.

One of the notable things about bombsites used as war memorials is that their preservation puts them outside the timeframes of entropy and decay that other bombsites must obey. These memorial ruins are cleared of rubble and weeds and fitted with flower gardens or sombre plain stone interiors. Their crumbling walls are propped with beams and capped with cement to prevent collapse, while moss, ivy, and other signs of romantic, abandoned, gothic ruin are periodically purged (Moshenska 2015). They are not allowed to fade or mellow or slowly collapse over the centuries, like a ruined monastery, until only a few lumps and bumps of stone can be seen amid the grass and weeds and bushes. The bombsite preserved as a war memorial is a sterile space, its fuzzy edges made hard again by cement and a heritage management plan. But despite their sterility these managed ruins still maintain an atmosphere of abjection and disorder, particularly as modern buildings of glass and metal rise around them, and the sense of being frozen and unchanging in time cuts both ways.

Representations

Bombsites are culturally generative spaces. Scholars and writers including W.G. Sebald, Leo Mellor, Rose Macauley, and others have queried this richness and examined the legacies of film, literature, and art that thrived in the ruins (Dillon 2014; Mellor 2011). In post-1945 Germany, 'Rubble Films' portrayed life in the broken cities with an unflinching realism, and are now recognised as a distinct genre of the post-war years. Mellor has argued that literary modernism in Britain, including the works of T.S. Eliot, Elizabeth Bowen, and Louis MacNeice, were strongly influenced by the idea and the reality of bombsites, and that similar influences can be found in the works of British surrealist artists (Mellor 2011). In different media, bombsites serve different purposes: they set the scene of stories set in wartime, they signify violence in their burned, jagged imagery, they provide a glimpse or possibility of social chaos amidst physical urban order, or they offer a stage for a story to play out on the margins of society. As Mellor notes, 'Reading – and writing – the ruins of war requires the material spaces cut violently into the city fabric to be acknowledged and understood' (2011: 203).

Conclusion

Urban bombsites are conceptually and culturally rich places that are good to think with. The tearing open of the domestic sphere and exposing it to the outside world is *unheimlich* in the full Freudian sense: uncanny, unhomely, unsettling, obscene, fascinating. As material traces of extreme violence, bombsites sit uncomfortably in civilised society, and where possible they are razed, turfed over, or developed into oblivion. Those that remain are domesticated and sacralised into clean and tidy memorial spaces. But the sense of uncanniness lingers. Bombsites are haunted by absences: of the building that is gone, the dead, the absent living, and of buildings unbuilt in its footprint.

Perhaps the most famous bombsite in Europe was the ruined Frauenkirche in Dresden, shattered by fire in the infamous bombing of the city in February 1945. For decades the blackened ruins of the church, surrounded by rubble, were the subject of political tensions between the East German government and the population of Dresden. Variously inscribed as a memorial of the Western Allies' aggression against Germany or as a focus for anti-regime peace protests, from the 1980s the ruins of the Frauenkirche were also the focus of a growing reconstruction campaign (James 2006; Moshenska 2015). Following German reunification this campaign grew rapidly, and an international fundraising campaign focused in the UK and the United States raised the necessary amount. In 2005 the reconstructed church re-opened: like the Neues Museum in Berlin a few years later, it integrated burned and broken structural elements into its reconstruction, so as to preserve in its walls the form, fabric, and intangible heritage of the iconic bombsite it had been.

References

Allbeson, T. (2015). Visualizing wartime destruction and postwar reconstruction: Herbert Mason's photograph of St. Paul's reevaluated. *The Journal of Modern History*, 87 (3): 532–578.
Barndt, K. (2011). Working through ruins: Berlin's Neues Museum. *The Germanic Review: Literature, Culture, Theory*, 86 (4): 294–307.

Casson, H. (1944). Ruins for remembrance. *In: Bombed churches as war memorials*. Cheam: Architectural Press, pp. 5–22.
Cloke, P., and Jones, O. (2005). Unclaimed territory: Childhood and disordered space(s). *Social and Cultural Geography*, 6 (3): 311–333.
Cranwell, K. (2003). Towards a history of adventure playground 1931–2000. *In:* N. Norman, ed. *An architecture of play: A survey of London's adventure playgrounds*. London: Four Corners Books, pp. 17–25.
Cullis, I.G. (2001). Blast waves and how they interact with structures. *Journal of the Royal Army Medical Corps*, 147 (1): 16–26.
Dillon, B. (2014). *Ruin lust: Artists' fascination with ruins, from Turner to the present day*. London: Tate.
James, J. (2006). Undoing trauma: Reconstructing the Church of our lady in Dresden. *Ethos*, 34 (2): 244–272.
Kimmelman, M. (2009). For Berlin Museum, a modern makeover that doesn't deny the wounds of war. *New York Times*, 12 March. Accessed 18 December 2017, www.nytimes.com/2009/03/12/arts/design/12abroad.html
Kozlovsky, R. (2007). Adventure playgrounds and postwar reconstruction. *In:* M. Gutman and N. de Coninck-Smith, eds. *Designing modern childhoods: History, space, and the material culture of children: An international reader*. New Brunswick: Rutgers University Press, pp. 171–190.
Lstiburek, J.W. (2014). Luftwaffe, ballast and shipping containers. *ASHRAE Journal*, 56 (7): 46–50.
Mabey, R. (1996). *Flora Britannica*. London: Sinclair-Stevenson.
Mellor, L. (2004). Words from the bombsites: Debris, modernism and literary salvage. *Critical Quarterly*, 46 (4): 77–90.
Mellor, L. (2011). *Reading the ruins: Modernism, bombsites and British culture*. Cambridge: Cambridge University Press.
Moshenska, G. (2014). Children in ruins: Bombsites as playgrounds in second world war Britain. *In:* B. Olsen and Þ. Pétursdóttir, eds. *Ruin memories: Materiality, aesthetics and the archaeology of the recent past*. Abingdon: Routledge, pp. 230–248.
Moshenska, G. (2015). Curated ruins and the endurance of conflict heritage. *Conservation and Management of Archaeological Sites*, 17 (1): 77–90.
O'Brien, T.H. (1955). *Civil defence*. London: Longmans/HMSO.
Rakowitz, M. (2012). *What dust will rise? Bamiyan travertine, glass, vitrines, bullets, shrapnel, meteorites, Libyan desert glass, trinitite, fragments of the destroyed Buddhas of Bamiyan, books burned during the Second World War*. Accessed 18 December 2017, www.contemporaryartdaily.com/2012/06/documenta-13-michael-rakowitz/.
Sebald, W.G. (2003). *On the natural history of destruction*. London: Penguin.
Tiratsoo, N. (2000). The reconstruction of blitzed British cities, 1945–55: Myths and Reality. *Contemporary British History*, 14 (1): 27–44.

PART III

Memoryscapes

Sarah De Nardi and Steven High

Introduction

The conceptual portmanteau memoryscape is more than just a melding of the ideas of memory and landscape; it effectively renders the ideas of place and remembrance as interdependent, entangling their manifestations and significations in one epistemological whole. In thinking of this concept, of this mutual assemblage of meanings, authors in this section have used 'memoryscape' more or less explicitly in unravelling and peeling back the stories, mythic narratives, and materialities of space. In doing so, these authors, and others in this volume, push past the criticism of Mark Riley and David Harvey (2007: 1–4), who wrote that place has been 'largely treated by oral historians [and other qualitative researchers outside of geography] in a superficial, Euclidian, manner – a frame for research rather than an active part.'

By contrast, humanities and social science researchers have long engaged with the world-making role of memory as a building block of social identities and cultural enactments. Memory may be a social binder, asserting and keeping groups together. Then again memory may split open a group's fractured identity through mnemonic dissonance and dissent. At any rate, for Misztal, 'the reconstruction of the past always depends on present-day identities and contexts' (2003: 14). The memory of a past and the perception of a present are interdependent, caught up in the same lived assemblage, enacted in the present in big and small ways. Moreover, '[t]he only past we can know is the one we shape by the questions we ask, and these questions are moulded by the social context we come from' (Kyvik 2004: 87). The context in which and through which we make, share, or abandon memories is the framework with which we interpret events from the past, our own experiences, and the memories of others around us. These entities form networks of past and present affects that aggregate in space to form, as it were, memoryscapes.

The premise of this section is that emplacement and negotiation of memories depend on perceptual and culturally specific modes of being in the world. It may also be that, in Paul Thompson's words, 'The memory process depends on that of perception. In order to learn something, we have first to comprehend it' (1978 [2000]: 129). In other words, we cannot remember something we have not assimilated, metabolised and made our own.

The five chapters in this section speak to complex patterns in the historical fabric as well as the centrality of political contestation and structures of power in determining what is visible to others. Jiwan Yoon and Derek H. Alderman show the symbolic importance of statues of Korean 'Comfort Women' in North American cityscapes, eloquently bearing witness to the abuse and state-sanctioned rape perpetrated against women and girls by the Imperial Japanese Army during the Second World War. From the vantage point of North America today, these statues also prompt passersby to consider the wider issue of violence against women and girls in global conflicts. Making visible what is otherwise rendered invisible is also the subject of the chapter on revolution and reconciliation in Hungary by Anett Árvay and Kenneth Foote. The mnemonic nature of the Hungarian past reveals hidden 'moods' in the 'quilting' of national memory. Iconic sites like Parcel 301 in the Rákoskeresztúr Cemetery (Budapest), where the then Prime Minister and others were secretly buried after their executions in the aftermath of the 1956 uprising, encapsulate the counter-memory of anti-communist upheaval. These fraught memories exist at the edge of vibrant everyday cityscapes, not always visible to the naked eyes as they are incorporated and sometimes silenced by official memory politics in the urban fabric.

The next chapter, by Keith Thor Carlson and Naxaxalhts'i (Albert 'Sonny' McHalsie), on the memoryscapes of the Stó:lō peoples of Canada, invites us to decolonise and re-Indigenise our understanding of history and geography. There is a fundamental division between settler and Indigenous relationships to the land and its history. Different worldviews produce different traces and craft different ways of remembering place, diffracting the mnemonic landscape in multiple directions. So, for example, fixed stones that are associated with certain 'transformer stories' transcend the human and the non-human, and these are specifically meaningful to certain places and to specific Indigenous ways of knowing. For the authors, the specific memory practices of this Indigenous group overturn the logics of colonialism, and their cosmologies of time-space can yet open hitherto unthought ontological possibilities for the wider study of memory and remembrance.

Moving forward in the section, Ashton Sinamai explores the powerful sense of place of Great Zimbabwe in a compelling ethnography focusing on the local, a perspective which, he posits, is often subjugated to the primacy of macro-narratives in constructing national identities. His chapter weaves together materiality, identity, ancestral memory and contemporary politics to demonstrate the fundamental role that the local, in its myriad affects, plays in how the Zimbabwean Shona society perform, rather than discursively define, their relationship with this sacred site that to them means much more than an iconic UNESCO asset-heritage site.

Finally, the section concludes with Toby Butler's pedagogical work on memoryscapes in East London. An early proponent of the memoryscape idea in the context of his own memory-infused audio walks, Butler maps out the learning potential, as well as some of the challenges, involved in training and mentoring students in the construction of memoryscapes. In doing so, he draws on different projects to foreground and open up the politics of working-class memories and representation through multi-media engagements. Taken together, these chapters raise productive questions about the contemporary politics of memoryscapes. They show the potential of collaboration and co-production to better understand and frame the pervasiveness of the past in the present.

Taken as a whole, these contributions demonstrate how memory's critical engagement with contemporary politics of place and the anti-memory of conflict, uprising, and exploitation of sacred grounds and sacred sites re-writes a mnemonic geography across the globe in more inclusive, fairer paradigms of vigilance and care. These projects may be channelled and enabled by spatially coalescing memories that speak to a past, present, and future sense of place; and by emplaced, dynamic, and inclusive social and cultural programmes of remembrance and social action.

References

Kyvik, G. (2004). Prehistoric material culture, presenting, commemorating, politicising. *In:* F. Fahlander and T. Oestigaard, eds. *Material culture and other things: Post – disciplinary Studies in the 21st Century*. Gothenburg: University of Gothenburg, pp. 93–108.

Misztal, B. (2003). *Theories of social remembering*. London: Routledge.

Riley, M., and Harvey, D. (2007). Talking geography: On oral history and the practice of geography. *Social and Cultural Geography*, 8 (3): 1–4.

Thompson, P. (1978 [2000]). *The voice of the past*. Oxford: Oxford University Press.

11
WHEN MEMORYSCAPES MOVE
'Comfort Women' memorials as transnational

Jihwan Yoon and Derek H. Alderman

Introduction

The study of memory and place has tended to emphasise the weighty materiality and fixed nature of memorials and monuments, often focusing on single memorial sites grounded within a specific social construction and contestation of place. Such an approach remains valid, but recent scholarship also focuses on the mobile and spatially interconnected nature of commemoration, and in particular, this work points to a 'transnational turn' in memory studies (Bond et al. 2017). This transnational turn seeks to understand how people move memory making beyond national borders and reconfigure the spatial extent and influence of commemorative politics to create 'new forms of belonging, solidarity and cultural identification in a world characterised by streams of migration and the lingering impact of traumatic and entangled pasts' (Assmann 2014: 546). Other scholars (e.g., Levy & Sznaider 2002: 92) write about the transition from nationally bound collective memories to globalised, cosmopolitan memory formation, stressing that the 'meanings [of these commemorations] evolve from the encounter of global interpretations and local sensibilities.'

Importantly, a transnational perspective prompts scholars to examine instances of globalising commemorative politics, and to understand the social actors and processes driving the expansion of memory and attendant memoryscapes, and how these commemorative circulations become embedded within, shaped by, and adapted to local needs, identities, and affects. The purpose of our chapter is to explore these complex dynamics by carrying out a brief case study of the 'Comfort Women' commemorative and human rights campaign, which addresses the abusive sexual treatment of Korean women under Japanese imperialistic rule. The campaign is increasingly moving across international borders and, at the same time, moving people emotionally and politically to identify with and support the memory of victimised 'Comfort Women,' but it has received limited attention within memory studies generally and, specifically, the geographical study of memorialisation. Our chapter seeks to offer insight into how and why the practice of building memorials to these traumatic histories migrates transnationally, focusing on the way memory becomes 'reterritorialised' within the socio-spatial relations of receiving communities and 'quilted' into their discourses or ways of thinking and talking about the politics of the past.

The 'Comfort Women' cause

'Comfort Women' is a euphemistic term for the women and girls forced into sexual slavery by the Imperial Japanese Army before and during the Second World War. The term also mistakenly suggests that these female victims participated in voluntary prostitution despite incontrovertible evidence otherwise (Choi 1997; Chung 1997; Hicks 1997; Yang 1997). At least 200,000 women endured daily, state-sanctioned rape, with many of them coming from occupied Korea. Many of these women spent the rest of their lives suffering from serious levels of mental and physical illness. Until the 1990s, a conservative Korean society institutionally suppressed this history of trauma and violence, forcing 'Comfort Women' to hide their personal feelings and memories for fear of stigmatisation and exclusion within their own country. According to critics, the 'Comfort Women,' with less than 50 still alive, remain unsatisfactorily addressed by a Japanese government more concerned with moving beyond the issue than issuing strong official apologies, full reparations, and revising their own history books.

Over the past few decades, surviving 'Comfort Women', their allies, and other civic activists have used testimony, art and performance, public protests and rallies, and monument and memorial building to raise public attention to these historical and contemporary injustices. In 1992, they began holding rallies called 'Wednesday Demonstrations' in front of the Japanese Embassy located in Seoul. By 2011, the 'Comfort Women' campaign had buttressed these weekly protests by constructing the 'Statue of Peace' memorial, which shows a barefoot Korean teenage girl sitting in a chair with her eyes squarely fixed upon the embassy in calm defiance, in effect keeping a vigil even when protestors are not present. The crowd-attracting statue (Figure 11.1) and the Wednesday Demonstrations, which continue to be held today, can be understood as

Figure 11.1 Statue of Peace in Seoul, Korea is an important focal point for public memorial services in support of 'Comfort Women' and Wednesday Demonstrations against the Japanese government.

Source: Photograph by Jihwan Yoon.

'memory work' (Till 2012), the activist and artistic practices of spatially materialising memories of trauma and discrimination, enacting a place-based ethics of care for victimised groups, and demanding that a state comes to terms with its perpetuation of violence.

While the 'Statue of Peace' has drawn considerable opposition from the Japanese government, it has gathered considerable support and inspired other cities inside and outside of the country to engage in similar monument building. By 2018, one could find replicas of the famous Seoul statue and other 'Comfort Women' memorials in over 100 South Korean administrative districts. In addition, at least ten US cities have memorials honoring the victims of sexual violence at the hands of the Japanese. Korean-American immigrant communities are primarily (but not entirely) responsible for this memory movement in the United States, working without visible connections to civic organisations or activist political bodies in South Korea. 'Comfort Women' memorials are also in Sydney (Australia), Toronto (Canada), Okinawa (Japan), and the Chinese cities of Nanjing and Shanghai. Many of these geographically dispersed memorials face resistance from pro-Japanese interests, including the Japanese government itself, which have challenged the establishment of these memorials or demanded their removal. In San Francisco, for example, the construction of a 'Comfort Women' statue overlooking a small downtown park prompted the mayor of Osaka, Japan, to end its 60-year-old 'sister city' relationship with the City by the Bay (Fortin 2017). Atlanta's Center for Civil and Human Rights decided to cancel plans to install a 'Comfort Women' memorial, citing logistical problems but also influenced by pressure from Japanese political and commercial interests in the city. Brookhaven, Georgia, a smaller city in the northeast suburbs of Atlanta, later welcomed and served as the site of the one-ton bronze copy of the memorial in South Korea (Croxton 2017).

Reterritorialising and quilting transnational memory

We focus on the 'Comfort Women' commemorative campaign and the appearance of its memorials outside of Korea, particularly in the United States, as an important theoretical and empirical moment to reflect on the transnational migration of memoryscapes and their accompanying identity politics. The mobility of 'Comfort Women' monuments works to re-scale the geography of remembering war crimes and female subjugation, transforming it from a national issue or a strictly binary geopolitical struggle (Korea versus Japan) to a multinational campaign. Several scholars note that human rights movements manipulate and modify geographic scale as a strategy for redefining the spatial and social extent to which their cause resonates (Alderman 2003; Kelly 1997; Post 2011). Scaling up the resonance of and public investment in identity politics requires more than just widely disseminating news of social struggles and injustices but also building, affectively, a solidarity between these historically marginalised groups and other social and spatial communities.

Our analysis of 'Comfort Women' public commemoration demonstrates that transnational memorial politics and monument making is not a monolithic process of simply relocating or merely transplanting human rights and commemorative concerns. Rather, transnational memory is actively made and our chapter illustrates how the mobile memoryscape undergoes a reterritorialisation and discursive quilting specific to the social groups and places receiving and adopting the commemorative cause. Quilting and reterritorialisation are the conversion processes by which 'Comfort Women' memory politics move – physically, politically, and emotionally – outside the region of origin and come to be incorporated within a different culture and landscape and thus matter and make sense to a receiving community. A transnational perspective of 'Comfort Women' memorials recognises that these global movements of memory open up new lines of activism, identity, and debate within spaces beyond Korea even as they give public recognition to atrocities that are key to nation building and memory work within South Korea and geopolitics between Korea and Japan.

Drawing from Gilles Deleuze and Felix Guattari (see Holland 2002), we use 'reterritorialise' to capture the manner in which the dispersed and migrating memory of 'Comfort Women,' upon being deterritorialised from its immediate geographic context in Korea, becomes reproduced in spatial form and adapted to the United States by Korean-Americans. With reterritorialisation, memoryscapes are claimed geographically and thus come to project and serve the place-specific political experiences, practices, and needs of the adopting community. Quilting, drawn from Jacques Lacan's (see Lee 1991) idea of *point de capiton* (quilting point), is helpful for understanding how memories and memorials develop a cross-cultural meaning and resonance as they cross national boundaries (Alderman et al. 2013). Quilting recognises the way in which cultural meanings – in this case, memories – become anchored to and converge with other images, causes, and popular discourses that enhance the acceptance and efficacy of the mobile memory at its destination. Building monuments and memorials to Korean 'Comfort Women' within the United States, while appearing to be simply a matter of diffusion, is a complex process that requires understanding how the commemoration of sexual slavery in Asia becomes emplaced within Korean immigrant community identity politics in the United States and connected to wider American discourses about human rights and social justice.

The remainder of our chapter delves into the processes of reterritorialisation and quilting that characterise the transnational mobility and efficacy of 'Comfort Women' memory work within the United States, focusing particular attention on monument building in Glendale, California (in 2013), and Bergen County, New Jersey (in 2013). Our discussion draws from semi-structured interviews conducted by one of us (Jihwan) from 2015 to 2017 with civic activists working with and for 'Comfort Women' commemorative campaigns, artists who created comfort women memorials, and politicians who supported commemorative projects in the United States.

Mobile memory and immigrant communities

To understand how and why the building of 'Comfort Women' memorials has been able to move into the United States (and move certain US groups to support the cause), it is necessary to reflect on the role these memorials play within the social lives and integration experiences of Korean-Americans, and more generally, the relationship between memory mobility and immigrant communities. The transnational migration and adoption of 'Comfort Women' memory within the United States relies upon, more broadly, a theoretical understanding of the mobility of memory and its significant contributions not only to human rights movements at the initial place of origin but also to the empowerment of an ethnic immigrant groups who re-territorialise or claim the memory for their own objectives. Indeed, our research suggests that Korean-American civic activists have viewed and treated 'Comfort Women' memorials and commemorative advocacy in terms of addressing their own social marginalisation, developing a belongingness and encouraging political participation within their own ethnic communities.

Importantly, the 'Comfort Women' commemorative movement, as it circulates into other regions and countries, develops and takes on meanings and political value beyond its immediate location- and time-specific context within Korea. This is due to the more universally important role that memory politics plays in the struggles of marginalised social groups to address and advocate for human rights (Alderman 2003; Bosco 2004; Levy & Sznaider 2010). Marginalised subjects are reliant on alternative ways, such as materialising corporeal senses and subjective emotions, to raise public attention of unjust social conditions they have suffered from and sought to overcome (Foucault 1980; Lefebvre 1991). As a critical foundation for symbolically materialized practice, landscape provides a visible, permanent arena for the historical and ongoing struggles of marginalised subjects to be remembered through symbolic images and meanings produced in space and thus inspire people to have a more socially responsible sense of history (Till 2008, 2012).

Given the current prevalence of diaspora around the world, ethnic groups tend to bring and transplant memories of their relatives and families into their new local destination communities (Jacobson 2002; Novick 2000; Shirinian 1998). A number of studies have discussed the essential role of multiculturalism in constructing and strengthening the cultures of ethnic groups within immigrant societies (Glazer 1998;

Howe 1999; Nash et al. 2000). According to Brubaker (2001), forms of multiculturalism – which include diverse forms of memorialisation – contribute to social collaboration and the assimilation of ethnic immigrant communities into a host society while also helping excavate and develop a common immigrant identity rooted in their homeland. This hits upon the duality in immigration, that immigrant populations can and often do take root in another country without losing their distinct sociocultural identity. Indeed, as Fortier (2000) argues, migrant groups do not abandon a collective identity and memory of homeland but rather utilise them for preserving their social status in the process of integrating themselves or carving out a place within the prevailing political and economic system.

In addition, we should consider the importance of 'moral integration' in understanding immigrants moving their home country's memory into the host society (Honneth 2004: 354). By bringing memories, especially emotionally and politically charged ones, into current living spaces, immigrant community members actively reshape their group identity, appealing to a public affect and sense of justice from other local people and political groups. However, because the transnational expansion of 'Comfort Women' memory inevitably connects with issues of social justice and sexual violence, it has an even broader range of implications for understanding the socially contested nature of space. When these traumatic commemorative narratives are materialised and visualised in various forms of public recognition, different social classes and ethnic groups come into conflict and debate over hegemony and ideology (Bosco 2004; Robertson & Hall 2007). Memoryscapes evolve into 'symbolic capital' that can be utilised for heterogeneous sociopolitical objectives (Rose-Redwood 2008). More specifically, 'places of memory . . . may be symbolic spaces where officials and other social groups express their contemporary political agendas to a larger "public"' (Forest et al. 2004: 358).

Though much academic work on memory politics concentrates on immigrants versus nativists (Anderson 2006; Roudometof 2002; Weiner 2015), an immigrant community often needs to work with other local groups and participate in the political activities of the host society to boost public recognition and draw collective support for their own immigrant memory politics. As indicated in several studies of Jewish populations, the transnational movement of the Holocaust memory – the classic example of a globalised commemorative cause – has progressed via mass media, tourism, shared symbolic forms, and the simultaneous diaspora of Jewish migrants. These migrants have worked within their own communities but also externally within a wider society to construct public understanding of their victimisation (Assmann 2010). As descendants of those immigrants have learned and internalised the history of victimhood, the traumatic memory experienced by their ancestors can be inherited and take root within immigrant and non-immigrant communities from generation to generation through 'sites of memory' such as museums and monuments, within the everyday social and spatial remains of past history (Blickstein 2009; Fogel 2000). These movements aroused by traumatic memory resonate with and affect a larger public, as well as the immigrant group, when they are quilted into broader discussions of social justice and morality and adapted to and emplaced within local social environments and territories.

Making a place for 'Comfort Women' memorials within the United States

As explained earlier, identity formation among immigrant populations inevitably results in appropriating, deploying, and re-territorialising collective memory of the homeland to strengthen a common sense of belonging (Fortier 2000). This evocation of belonging with shared memory is especially needed among marginalised social groups as they react to and challenge their exclusion from the mainstream host society. In exploring the movement of 'Comfort Women' memory into the United States, we found evidence of Korean-American activists claiming and actively interpreting the history of Japanese-imposed sexual slavery to assert a common sense of ethnicity and legitimacy in the face of marginalising social and spatial conditions in the United States. For example, one of the major Korean-American civic activists who initiated the nation's first 'Comfort Women' memorial described how the trauma of the 1992 Los Angeles Riots, specifically the violence against Korean-Americans and the looting and destroying of their businesses, greatly informed the decision to build the monument almost two decades later.

For this activist, named D.S., the emplacement of 'Comfort Women' memory politics within the United States was not just about honoring victims in Asia but also the need for Korean immigrants to build strong connections with US society and strengthen their political influence, especially relative to other historically marginalised groups such as African Americans and Hispanics (Lien 2010: 183).

> For the last 30 years, we have constantly made efforts to strengthen political power of Korean-Americans. But as Koreans got obsessed with internal issues of the Korean community ignoring participation in US politics, we decided to bring the contested issue, the history of Comfort Women, to raise public attentions to a common sense of ethnicity among Koreans and US citizens. We expected that if this is successful, then Koreans will be politically involved in civil society of the US.
>
> *Interview with D.S., transcribed and translated by Jihwan Yoon*

Several studies of political activities among ethnic groups in the United States show that Korean-Americans have until recently been less active in voter registration than other ethnic communities (Lien et al. 2001; Uhlaner et al. 1989; Wong et al. 2005).

Yet, the memory of 'Comfort Women' was much more than a convenient and evocative signifier; it was an especially affective and highly charged one for Korean-Americans. People who have undergone their 'own marginalizing experiences' can feel an emotional shift toward the identity of 'others' who have gone through similar types of disenfranchisement and discrimination within the same imagined social community (Villenas 1996). In this respect, the transnational mobility of 'Comfort Women' memory politics was not merely the process of transplanting a global concern for acknowledging victimisation within Asia. Rather, it was also a reterritorialising and reinterpretation of that trauma in terms of the social and spatial inequalities and political experiences of Korean-Americans. Korean immigrants such as D.S. sought a potent symbolic resource for building solidarity, advocating for greater participation in American politics, and publicly recognising the vulnerabilities and injustices that have long characterised the Korean experience, not just in Asia but also in the United States.

While Korean-American activists hoped that a 'Comfort Women' memorial would affect their ethnic immigrant community, they also realised that a wider transnational acknowledgment of this trauma required 'collaboration, integration, and coalitions' among marginal subjects and other populations (Bradshaw 2000: 134; Hopkins 2006). Social groups can align themselves with other racial, ethnic, and religious communities if there are 'collective attitudes to the extent to which various subjects become aware of the commonality of their social status' (Honneth 1996: 165). Morality can be the most common and acceptable matter around which social groups can build up a collective sense of social justice and work towards a common goal of human rights (ibid). Yet, for the 'Comfort Women' memory politics to benefit from this wider social collaboration and to have resonance beyond the Korean-American community, the trauma of sexual slavery had to be quilted or joined with larger American understandings and discourses about injustice and victimisation.

An elected official in Bergen County, New Jersey (named J), defended his support for a 'Comfort Women' memorial, even in the face of Japanese resistance, by quilting or knitting together the Korean issue with expansive discourses about human rights, including memories of and empathy for the victims of the Holocaust and African enslavement.

> People said to me, 'J. [the interviewee's pseudonym], you know . . . you're a politician. Why would you want to alienate? You know, because we do have a pretty good population here of Japanese people there in Bergen County as well.' And I said, 'That doesn't matter. You know,

this is not about blanketing, [Interruption] this is about doing right thing.' It's human rights, and particularly, the issue of sexual slavery. Uh . . . you know, people do that what the Jewish Holocaust, people do that well . . . slavery wasn't that bad. Those were bad things. Those were very bad things.

Interview with J, transcribed by Jihwan Yoon

This interview and another one conducted with a second Bergen County elected official who support the 'Comfort Women' monument suggest that political leaders made sense of the appropriateness of the Korean memorial by employing a notion of morality and justice based on achieving 'reciprocal recognition' within a multicultural America (Honneth & Farrell 1997: 17). In particular, these officials frequently heard stories about human rights abuses from the various immigrant groups who settled in Bergen County. As both interviewees mention, the experience of 'Comfort Women' is not the sole concern among them or other US citizens. Rather, the power and mobility of this transnational memory comes from being quilted or discursively connected with a wider American and Bergen County understandings and experiences of human rights, including the struggles of other racial and ethnic groups. Indeed, this quilting is present within the spatial narrative of the New Jersey memorial; the 'Comfort Women' memorial sits alongside those for recognising African American slavery, the Irish Famine, the Holocaust, and the Armenian Genocide (Figure 11.2).

Figure 11.2 Comfort Women memorial in Bergen County, New Jersey quilted, discursively and spatially, with and alongside memorials to other human rights struggles.

Source: Photograph by Jihwan Yoon.

Glendale, California, was another city in which 'Comfort Women' memory politics became reterritorialised or adapted to the US multicultural social environment. According to one of the Korean-American activists who led the construction of the 'Comfort Women' statue in Glendale, the strong emotions held about injustices endured by Korean women could easily penetrate other ethnic communities who had also experienced human rights abuses. In particular, the activist (named P.) describes how the shame felt by Korean sex slaves and the Japanese government's denial of responsibility resonated with and was quilted onto the genocidal histories of Armenian-American supporters.

> One thing notable in Glendale is that there are many Armenians living in the town. Armenians had experienced the ethnic cleansing . . . but the Turkish government still denies the crime [like the Japanese government]. So, Armenians understand the sadness of Comfort Women better than anyone. [Interruption] Armenians had gone through the time of silence for decades too [like Comfort Women] because the survivors could not explain the genocide to their children as it was too horrible to tell. But their descendants started to report the genocide to remember the history. They said, 'If we had let the world realise the inhumanity of ethnic cleansing in a louder and clearer voice, the Holocaust would have never occurred.' [Interruption] So, when we brought this issue [of Comfort Women] to city council members and Armenian-Americans, they understood very well. [Interruption] When they [freeholders] received many letters of complaint from Japanese politicians and citizens, they told them, 'Enough of your nonsense.'
>
> *Interview with P., transcribed and translated by Jihwan Yoon*

This reminds us that public memory can function as 'performative acts' to refresh the collectivity of an ethnic community through 'a repetition of [symbolic] images and words' (Levy & Sznaider 2002: 92, 2011; Fortier 2000; Levy & Sznaider 2002). This in turn encourages us to consider how the transnational movement of memory politics is not a direct copying of politics from Korea but a complex translation of that memory within the receiving community. The memorial creation process has the power to draw emotional and political supports from those who share the same kind of traumatic memory. This, in turn, can lead to 'scaled performances of memory' that further expands the resonance and meaning of 'Comfort Women' beyond just the ethnic boundaries of Korea and Korean-Americans to include solidarity among greater range of global and transnational collaborators (Bosco 2004; Kelly 1997; Post 2011).

Concluding remarks

In advancing the general study of memory and place, we have sought to introduce a transnational dynamism and mobility within the study of memoryscapes that has been previously underdeveloped within the geographic literature. In doing so, we shed light on the inter-scalar connections and movements – spatial and politico-emotive – between commemoration in different places and 'collaborative works' among different social groups. Transnational memorialisation cannot be reduced to a simple model of time-space diffusion or wholesale transplantation. The concept of reterritorialisation allows us to think about the social actors and processes that accompany and shape the meaning of commemoration as it moves across space and becomes adapted to different sociocultural environments of receiving communities. While Korean-American activists have adopted and transformed 'Comfort Women' memorials as part of the politics of ethnic identity formation and political empowerment in the United States, transnational commemoration is not the exclusive property of a certain social group but requires a memory politics that appeals to the moral sense of other cultural and ethnic communities in US society. The concept of quilting prompts us to pay attention to the collaborations and conflicts that characterise the conversion of 'Comfort Women' into a transnational memory, and how moving memories become knitted with various discourses, identities, and ideas of morality as people interpret, internalise, and redefine the scaled meanings and politics of 'Comfort Women' memory and make it their own.

References

Alderman, D.H. (2003). Street names and the scaling of memory: The politics of commemorating Martin Luther King, Jr within the African American community. *Area*, 35 (2): 163–173.

Alderman, D.H., Kingsbury, P., and Dwyer, O.J. (2013). Reexamining the Montgomery bus boycott: Toward an empathetic pedagogy of the civil rights movement. *The Professional Geographer*, 65 (1): 171–186.

Anderson, B. (2006). *Imagined communities: Reflections on the origin and spread of nationalism*. London: Verso Books.

Assmann, A. (2010). The holocaust – a global memory? Extensions and limits of a new memory community. In: A. Assmann and S. Conrad, eds. *Memory in a Global Age*. Basingstoke: Palgrave Macmillan, pp. 97–117.

Assmann, A. (2014). Transnational memories. *European Review*, 22 (4): 546–556.

Blickstein, T. (2009). Forgetful "Sites of memory": Immigration museums and the uses of public memory. *The New School Psychology Bulletin*, 6 (2): 15–31.

Bond, L., Craps, S., and Vermeulen, P. (2017). *Memory unbound: Tracing the dynamics of memory studies*. Oxford: Berghahn Books.

Bosco, F.J. (2004). Human rights politics and scaled performances of memory: Conflicts among the Madres de Plaza de Mayo in Argentina. *Social & Cultural Geography*, 5 (3): 381–402.

Bradshaw, T.K. (2000). Complex community development projects: Collaboration, comprehensive programs, and community coalitions in complex society. *Community Development Journal*, 35 (2): 133–145.

Brubaker, R. (2001). The return of assimilation? Changing perspectives on immigration and its sequels in France, Germany, and the United States. *Ethnic and Racial Studies*, 24 (4): 531–548.

Butler, J. (2003). Performative acts and gender constitution. In: P. Auslander, ed. *Performance*, Vol. IV. New York: Routledge, pp. 97–110.

Butler, J. (2011). *Bodies that matter: On the discursive limits of sex*. London: Taylor & Francis.

Choi, C. ed. (1997). *The comfort women: Colonialism, war, and sex*. Durham, NC: Duke University Press.

Chung, C.S. (1997). The origin and development of the military sexual slavery problem in Imperial Japan. *Positions*, 5 (1): 219–255.

Croxton, K. (2017). Brookhaven reveals controversial comfort women statue. *Marietta Daily Journal* (Atlanta, Georgia USA) 30 June. Accessed 12 October 2017, www.mdjonline.com/neighbor_newspapers/dekalb/brookhaven-reveals-controversial-comfort-women-statue/

Fogel, J.A. ed. (2000). *The Nanjing Massacre in history and historiography*, Vol. 2. Berkeley, CA: University of California Press.

Forest, B., Johnson, J., and Till, K. (2004). Post-totalitarian national identity: Public memory in Germany and Russia. *Social & Cultural Geography*, 5 (3): 357–380.

Fortier, A.M.F. (2000). *Migrant belongings: Memory, space, identity*. Oxford: Berg.

Fortin, J. (2017). Comfort women statue in San Francisco leads a Japanese city to cut ties. *New York Times*, 25 November. Accessed 1 December 2017, www.nytimes.com/2017/11/25/world/asia/comfort-women-statue.html

Foucault, M. (1980). *Language, counter-memory, practice: Selected essays and interviews*. Ithaca, NY: Cornell University Press.

Glazer, N. (1998). *We are all multiculturalists now*. Cambridge, MA: Harvard University Press.

Hicks, G. (1997). *The comfort women: Japan's brutal regime of enforced prostitution in the second world war*. New York: WW Norton & Company.

Holland, E.W. (2002). *Deleuze and Guattari's Anti-Oedipus: Introduction to schizoanalysis*. London: Routledge.

Honneth, A. (1996). *The struggle for recognition: The moral grammar of social conflicts*. Cambridge, MA: MIT Press.

Honneth, A. (2004). Recognition and justice: Outline of a plural theory of justice. *Acta Sociologica*, 47 (4): 351–364.

Honneth, A., and Farrell, J. (1997). Recognition and moral obligation. *Social Research*, 64 (1): 16–35.

Hopkins, G. (2006). Somali community organizations in London and Toronto: Collaboration and effectiveness. *Journal of Refugee Studies*, 19 (3): 361–380.

Howe, S. (1999). *Afrocentrism: Mythical pasts and imagined homes*. New York: Verso.

Jacobson, M.F. (2002). *Special sorrows: The diasporic imagination of Irish, Polish, and Jewish immigrants in the United States*. Berkeley, CA: University of California Press.

Jelin, E. (1994). The politics of memory: The human rights movement and the construction of democracy in Argentina. *Latin American Perspectives*, 21 (2): 38–58.

Kelly, P.F. (1997). Globalization, power and the politics of scale in the Philippines. *Geoforum*, 28 (2): 151–171.

Lee, J.S. (1991). *Jacques Lacan*. Amherst, MA: University of Massachusetts Press.

Lefebvre, H. (1991). *The production of space*, Vol. 142. Oxford: Blackwell.
Levy, D., and Sznaider, N. (2002). Memory unbound: The Holocaust and the formation of cosmopolitan memory. *European Journal of Social Theory*, 5 (1): 87–106.
Levy, D., and Sznaider, N. (2010). *Human rights and memory*, Vol. 5. University Park, PA: Penn State University Press.
Lien, P.T. (2010). *Making of Asian America: Through political participation*. Philadelphia: Temple University Press.
Lien, P.T., Collet, C., Wong, J., and Ramakrishnan, S.K. (2001). Asian Pacific-American public opinion and political participation. *PS: Political Science and Politics*, 34 (3): 625–630.
Nash, G.B., Crabtree, C.A., and Dunn, R.E. (2000). *History on trial: Culture wars and the teaching of the past*. New York: Vintage.
Novick, P. (2000). *The holocaust in American life*. Boston: Houghton Mifflin Harcourt.
Post, C.W. (2011). Art, scale, and the memory of tragedy: A consideration of public art in Pleasant Hill, Missouri. *Material Culture*, 43 (2): 43–58.
Robertson, I., and Hall, T. (2007). Memory, identity and the memorialization of conflict in the Scottish Highlands. In: N. Moore and Y. Whelan, eds. *Heritage, memory and the politics of identity: New Perspectives on the cultural landscape*. Aldershot: Ashgate, pp. 19–36.
Rose-Redwood, R.S. (2008). From number to name: Symbolic capital, places of memory and the politics of street renaming in New York City. *Social & Cultural Geography*, 9 (4), 431–452.
Roudometof, V. (2002). *Collective memory, national identity, and ethnic conflict: Greece, Bulgaria, and the Macedonian question*. Westport, CT: Greenwood Publishing Group.
Shirinian, L. (1998). Survivor memoirs of the Armenian Genocide as cultural history. In: R.G. Hovannisian, ed. *Remembrance and Denial: The case of the Armenian genocide*. Detroit: Wayne State University Press, pp. 405–423.
Till, K.E. (2008). Artistic and activist memory-work: Approaching place-based practice. *Memory Studies*, 1 (1), 99–113.
Till, K.E. (2012). Wounded cities: Memory-work and a place-based ethics of care. *Political Geography*, 31 (1), 3–14.
Uhlaner, C.J., Cain, B.E., and Kiewiet, D.R. (1989). Political participation of ethnic minorities in the 1980s. *Political Behavior*, 11 (3), 195–231.
Villenas, S. (1996). The colonizer/colonized Chicana ethnographer: Identity, marginalization, and co-optation in the field. *Harvard Educational Review*, 66 (4), 711–732.
Weiner, M. (2015). *Sons of the soil: Migration and ethnic conflict in India*. Princeton, NJ: Princeton University Press.
Wong, J.S., Lien, P.T., and Conway, M.M. (2005). Group-based resources and political participation among Asian Americans. *American Politics Research*, 33 (4), 545–576.
Yang, H. (1997). Revisiting the issue of Korean "military comfort women": The question of truth and positionality. *Positions*, 5 (1), 51–72.

12
THE SPATIALITY OF MEMORYSCAPES, PUBLIC MEMORY, AND COMMEMORATION

Anett Árvay and Kenneth Foote

Introduction

Memoryscapes and traumascapes have emerged in recent years as important issues of debate and research. A key focus of attention has been the embodied experiences of victims, their families, and members of traumatised communities. The political and social dynamics of memoryscapes have also been the subject of considerable research. Indeed the 'politics of place' and 'contested sites of meaning' are dominant themes in research on memoryscapes, public memory, and commemoration. In this chapter we concentrate on one aspect of these sites of memory and meaning – their *spatial* dimension (Foote & Azaryahu 2007; Foote et al. 2000). Our concern is how the location and positioning of monuments and memorials in public and private spaces affects their meaning and interpretation at the international, national, regional, and local scales.

In certain cases, for example, it is important when a memorial is placed on the exact site of a historical event while in other cases it is positioned off site. Memorials are sometimes located in close proximity one to another to draw symbolic parallels between the events and people being honoured. In other settings distance is maintained between memorials to emphasise their differences, perhaps to mark a political or historical disjuncture. Our point is that, just as the positions and movements of chess pieces on the game board are keys to strategy, the positions and movements of memorials are one aspect of understanding their meanings across a range of scales.

This idea of spatialising public memory is not new. Halbwachs, one of the earliest and most influential writers on collective memory, considered localisation to be an important dynamic for sustaining these cultural practices (Halbwachs 1992). This close mnemonic relationship between space and memory is an essential element of his *La Topographie légendaire des Evangiles en Terre sainte* (Halbwachs 1941) in which he details how the passion of Christ was spatialised by Christians within the Old City of Jerusalem. This same sensitivity to space, place, and location is apparent in a number of subsequent works on public memory. Lowenthal's note that 'features recalled with pride are apt to be safeguarded against erosion and vandalism; those that reflect shame may be ignored or expunged from the landscape' (1975: 31) addresses one key relationship between memory and place discussed in more detail in some of his other writings (Lowenthal 1985). Many of the essays edited by Nora for *Les lieux des mémoire* (1984) argue for a close connection between space, place, and region in defining a wide range of French historical traditions. Bodnar (1992: 13) is more explicit in this connection, arguing that 'The shaping of a past worthy of public commemoration in the present is contested and involves a struggle for supremacy between advocates of various political ideas and sentiments' and that these contests often play out in the renaming of a city street, the dedication of a new memorial in local park, or the creation of a museum focusing on a event or era.

Although space, place, and location are not always foregrounded in the rapidly growing literature on public memory (Olick 2007), a number of researchers are sensitive to this relationship. Linenthal (1995), Hartman (1994), Young (1993), and Marcuse (2001) among others have focused on the spatiality of Holocaust memory. Farmer (1999), Tumarkin (2005), Doss (2012), and Kelman (2013) are further examples of the writers who have made additional contributions to this literature. In this short chapter, we provide a synopsis of some of ways space and place are interwoven with memoryscapes and plays roles in contemporary memorial practice.

Some examples from Hungary

In this chapter we use examples drawn primarily from our current research in Hungary. Our intent is not to downplay the growing body of work on public memory worldwide, but only to offer examples that illustrate the major points of our argument. In Hungary, as in other countries, questions of *where* are of great importance in understanding the creation of monuments and memorials as well as contestations over their symbolism and meaning. Our analysis is part of a growing literature on public memory and commemoration in Central and Eastern Europe (Andersen & Törnquist-Plewa 2016; Bernhard & Kubik 2014; Dobre & Ghita 2017; Krasnodębski et al. 2012; Lebow et al. 2006; Luthar 2012; Todorova et al. 2014; Törnquist-Plewa 2016) with some of these researchers focusing considerable attention on issues of space and place (Mark 2010; Mink & Neumayer 2013; Rampley 2012). Excellent studies on Hungary include, among many others, Apor (2014), Boros (1997), James (2005), Pótó (1989), Rév (2005), Wehner (1986), Seleny (2014) and Jakab (2012).

By way of a background, the Hungarian state traces its origins to the conquest of Europe's Carpathian Basin by Magyar tribes arriving from Central Asia at the end of the ninth century. Ever since, Hungary has played an important role in Central European history sometimes as a buffer between the East and West. Yet Hungary's experience of many pivotal events of the nineteenth and twentieth centuries has been far different from those of even its closest neighbours in Central Europe. The revolution and the War of Independence of 1848–1849 against the Habsburg monarchy, though unsuccessful, are still commemorated in Hungary. Commemoration of the losses of the First World War, so important in many nations, was used in Hungary to rally opposition to its territorial losses under the Trianon Treaty of 1920. Hungary was the only nation, apart from Russia, to have a communist government after the First World War. Suppressed in less than five months, this Council Republic (1919) was replaced by a conservative regime (1920–1944) that allied itself with Germany during the Second World War. This interwar and wartime government was, in turn, replaced by a still-Hungarian puppet government (1944–1945) that oversaw the deportation and killing of most of Hungary's Jews during the spring and summer of 1944. After the war, Hungary's devastating military losses were not acknowledged until after the fall of the communist government in 1989.

These numerous turnovers of government – among regimes of highly divergent ideologies – are not unique to Hungary. But such rapid changes of political culture do affect commemorative traditions and national narratives (Roudometof 2002). Each regime had its own vision of the significant moments in the Hungarian past and each regime attempted to inscribe its vision of the national past on the landscape, with social and religious divides also playing a role in debates. The result is that Hungary provides good examples of the roles of space and place in the development of commemorative traditions.

'Here is the place': locating memory on site

In many cases space is important in an absolute sense: the placement of a memorial at the actual geographical location of an event, the orientation of monuments one to another, as well as the distances and directions between can all impact their meaning. Memorials that have been erected on the sites of a particular battle,

The spatiality of memoryscapes

the outbreaks of revolution, or the death sites of national heroes. These places are often sacralised and serve as pilgrimage sites, either spontaneously or through grassroot efforts initiated by veterans, supporters, family, and local communities. Some of these may eventually be incorporated into a national canon of memory sites (Foote 2003: 265–292). These places are also important as loci of national and local holidays – memorial services and annual commemorations with speeches delivered by politicians, survivors, and other community members. These types of 'non-representation' embodiments of memory in ritual and ceremony are often anchored in 'representational' forms, such as memorial tablets and physical monuments. These places may also serve as sites of protest and resistance, when commemoration of an event or individual is banned. This was the case in Hungary with a major forced labour camp operated between 1950–1953 at a stone quarry near the village of Recsk in northern Hungary. Former prisoners organised to reclaim and mark this site after the fall of the communist government in 1989 (Figure 12.1).

Another example is Parcel 301 in the Rákoskeresztúr Cemetery on the outskirts of Budapest, where Prime Minister Imre Nagy and other martyrs were buried in secret after their executions in the aftermath

Figure 12.1 Examples of site-specific memorials where the location of memorials is tied to particular places. These include (clockwise from top) a reconstructed barracks at the Recsk prison camp honouring victims of political repression during the communist period; grave poles in Budapest's Parcel 301 cemetery honouring victims of the 1956 Hungarian Uprising; and a memorial at Pákozd honouring one of the greatest victories of the Hungarian revolutionary army during the 1848–49 War of Independence.

Source: Photos by authors.

of the 1956 uprising. Demands to honour the heroes of 1956 were critical in contributing to the downfall of the communist regime in 1989. The bodies were exhumed and reburied in the cemetery, with Parcel 301 becoming a national memorial site (György 2000, Rainer 2001). On the day of the reburial, a ceremony with six empty coffins was held in Budapest's Heroes' Square, a major spatial spectacle in which hundreds of thousands of people paid tribute to those killed for resisting communist rule.

Relative location and symbolism

The location and placement of memorials can also be important in a relative sense. The placement of memorials with respect to one another and in relation to significant civic buildings or public spaces can also be important: are they sited in a central square, adjacent to a town or county hall, or in a distant, hard-to-reach place? In these cases, the relative location of a memorial can take on allusive, figurative, and connotative meanings. For example, in many Hungarian cities and towns, memorials to the 1956 uprising have been placed close to those commemorating the 1848–1849 War of Independence. This positioning draws a symbolic parallel between these two events as representing Hungarian resistance to outside domination. Sometimes memorials for 1956 and 1848–1849 are also placed close to those for the losses of the First and Second World Wars. These are often gathered together in 'remembrance' or 'martyrs' squares emphasising the community's losses in the service of the nation.

Relative location can even be important in terms of spaces that are public or private. In the aftermath of the Holocaust, surviving members of Hungary's Jewish communities sometimes created memorials in private, protected places – inside of synagogues, on the grounds of synagogues, or in cemeteries. It was not until 1986 that a public memorial was raised to victims of the Holocaust. At the same time, this memorial to the 'Hungarian Martyrs' was sited at the northern edge of what had been the International Jewish Ghetto during the Second World War. This was a place of significant meaning, but is also quite distant from the government quarter of central Budapest and other major public spaces near Parliament and Castle Hill.

In recent years this memorialisation of the Holocaust – as an example – has gradually moved inward toward central Budapest, and from private to public spaces. Now an important Holocaust memorial, 'Shoes on the Danube Bank' is located along the riverfront side of the Parliament building. This is a memorial that marks the place where Jews were told to kick off their shoes before being shot, then pushed into the river in the winter of 1944–1945. In this central area of Budapest, roads along the banks of the Danube have also be renamed to honour 'rescuers,' people who helped save Jews from death during the Holocaust. Apart from these memorials, several public museums focusing on Holocaust remembrance have since been opened in Budapest and other cities.

Symbolic assemblages and the accretion of meaning across scales

Another spatial process is the accretion of multiple memorials in one place to draw symbolic parallels between events. These assemblages may be located on the site of a particular historical event or at a different, relative location such as those discussed in the previous section. This pattern can be seen in a number of Hungarian town centres and even more frequently in a wide range of small towns and villages. There are cases where the extent of these accretion is so large that entire memorial parks may develop through time.

Examples are the memorials in Pákozd, a village southwest of Budapest and the site of an important victory for the Hungarian army during the unsuccessful 1848–1849 War of Independence. The first memorial to the battle was erected in 1889 from public donation, a martial memorial honouring the troops that fought in the battle. In 1951, the communist government sought to celebrate the importance of its own Peoples' Army and decided to create a major new memorial on the edge of Pákozd on a hill overlooking the battle site. By choosing this location, the government seemed to be trying to establish a symbolic connection between the struggles of 1848–1849 and the creation of the communist Peoples' Republic a hundred years later, in 1948–1949. Since then, the area around the latter memorial has developed into a

The spatiality of memoryscapes

memorial honouring Hungary's military losses in the First and the Second World War, the 1956 revolution, and during recent Hungarian peacekeeping missions with the United Nations. This park was one of the first places a memorial was created to honour the loss of the Second Hungarian Army during the Second World War in the retreat from Stalingrad.

When observing assemblages of memorials in close proximity, the Liberty Square (Szabadság tér) in Budapest offers another good example (Figure 12.2). The square implied a strong, homogenous

Figure 12.2 A view of central Budapest from Parliament to Freedom Square [*Szabadságtér*] indicating a few of the many important sites and memorials in this area. The spatial positioning and alignment of these memorials illustrates the process of symbolic accretion in which the positioning of one memorial leads to the siting of others to create assemblages of public art and architecture in public spaces.

Source: Google Earth and photos by authors.

irredentist historical narrative between 1921 and 1945, with the huge flowerbed depicting the map of prewar Hungary, with the territories it lost after the First World War represented metonymically in the form of human figures, and flanked by a flag-standard for flying the national banner. All these were removed during the communist period. Today the square presents multiple narratives of Hungarian collective memory embodied by the Soviet Liberation memorial (1945), the German occupational memorial (2014), a bust of Regent Miklós Horthy the interwar leader (2013), a statue of Ronald Reagan (2011) facing the Soviet memorial, and Imre Nagy's memorial (1998–2018) placed at the edge of the square facing the Hungarian Parliament building. The first three express unresolved, competing memories and have generated heated public arguments. The most striking example of the public disagreement was the spontaneous protest against the state-sponsored German occupational memorial during its construction and the creation of a permanent counter-memorial which consists of documents and commentaries on the Hungarian Holocaust. The official memorial was completed in 2014, nevertheless it was never dedicated. Now it faces the private memorial right across, thus implying a counter-narrative in Hungarian collective memory (Erőss 2016).

Locality and scale

Location is also important in terms of what cities and communities are studied. Perhaps too often studies of public memory have focused on capital cities or other major cities and have tended to valorise the development of national memoryscapes rather than local variations in commemorative traditions. We do not deny the major cities provide excellent case studies, but we argue that it is just as important to look beyond the capital to consider how memoryscapes are created in smaller cities, towns, villages, and rural areas throughout particular countries and regions.

Such considerations are important for three reasons. Firstly, political, social, cultural, and economic values usually vary considerably across most states. It is risky to assume that political debates about memoryscapes in the capital or other major city are nowadays representative of trends at regional or local levels. Secondly, local communities have their own heroes, causes, and issues that may be more important than those debated at the national level. Finally, looking broadly at a commemoration across an entire nation and region can reveal the spatial patterning in how memoryscapes develop and change through time. Did the first memorials of a particular type emerge in rural or urban areas? Were they created first in a particular region of a country, or everywhere at once? Our point is that locality matters to meaning. Memoryscapes may vary across local, regional, national, and international settings.

Sometimes these differences appear around sensitive issues. For example, the devastating loss of the Hungarian Second Army (discussed in the next section) during the Nazi retreat from Stalingrad is still not widely commemorated. There was a large upwelling of commemoration of Hungary's wartime losses after the fall of the communist government in 1989, since these had been banned from the late 1940s onward. But the scope of the losses in Russia have only gradually been incorporated in the memoryscapes of Hungary in cities such as Kaposvár (1993), Nagykanizsa (1993), Siófok (2003), Szeged (2009), and Balatonfenyves (2010), but not yet in Budapest. The legacy of political oppression during the Cold War period is now more widely acknowledged, particularly after the opening of the House of Terror in Budapest in 2002. This building in central Budapest served as the headquarters of the Nazi Gestapo during the Second World War then, in the post-war period, of Hungarian secret police. Yet apart from this major memorial, sites of political oppression between 1949–1989 are sparsely marked, with only a few other markers in somewhat out-of-the-way places in Budapest and elsewhere in Hungary, for example at the Recsk labor camp itself in northern Hungary and a memorial in Mórahalom (2003), or a relocation memorial in Budafok (2011).

Meanings across boundaries and barriers

From a spatial point of view, many issues of memory stretch across borders and barriers, resulting in complex negotiations over the creation and symbolism of memoryscapes. Paramount among these are episodes of violence from the First and the Second World Wars including not only events of warfare, but atrocities, expulsions, genocides, political purges, sieges, and other events that caused the death and suffering of millions of people. These have been commemorated in and outside of their home countries, however, often not without conflicts.

Countries that used to be enemies during the war have to cope with the memory of aggression and killings that is very hard to overcome even after decades. During the Second World War period, for example, Hungarians and Serbians committed brutal massacres against each other. The so-called Újvidék [Novi Sad] 'raid' in southern Bačka of January 1942, 3,000–4,000 civilians were killed as part of the military operation by the Hungarian Army in a number of locations over a period of weeks with the goal of stopping partisan attacks on Hungarian forces, but focusing especially on Jews. In 1944–1945, as Serbian forces pushed Axis forces northward, the partisans massacred ethnic Hungarians, Germans, and Croatians who had not supported them, killing as many as 60,000 civilians (A. Sajti 2004). Josip Tito's partisans deported and exterminated almost the entire Hungarian population of this region.

This period from the Nazi/Hungarian invasion of Serbia in 1941 to the end of the Second World War long remained an issue of tension between Serbia and Hungary. For example, an on-site memorial to the killing of ethnic Hungarians was created in 2006, but has been vandalised many times since. Although such local conflicts over memory may continue for some time, some conflicts can be resolved at the diplomatic level. In this case, the Serbian and the Hungarian prime ministers dedicated a museum and a memorial together in 2013 to serve justice for both countries, thus expressing the common wish for reconciliation over the events of the Second World War.

Another important issue that involves every country that participated in the World Wars is the commemoration of their war dead. For Hungarians, the largest military cemeteries are in Russia, since Hungary supported the Nazi invasion of the USSR and suffered as the invasion was crushed. In 1943, the Hungarian Second Army 'disappeared' in the retreat from Stalingrad along what is sometimes called the 'Don-Bend.' This was the front stretching northwest from Stalingrad to Voronezh along which the Don River makes a sharp turn toward the Black Sea. The early efforts of commemoration started only in 1989, and finally the Rudkino cemetery was created for the 30,000 Hungarian victims. The project was sponsored by the Hungarian State in 2001.

Conclusion

We have used examples from Hungary in this chapter, but our argument about the spatiality of memoryscapes is applicable in many other contexts. Our point is that the spaces and places where commemoration takes place are closely tied to the meaning and symbolism of memoryscapes. It is important, however, to qualify our argument in at least four ways. Firstly, memoryscapes are not stable forms that come to be marked permanently on landscapes and cityscapes. They change and evolve through time. Memoryscapes may be anchored in assemblages of stone, bronze, and concrete, but even these durable forms change through time as do their meanings and symbolic connotations. Decades or generations may pass for these changes to occur, but they do happen. Some of these changes are quite dramatic in the aftermath of wars or revolutions, or in the wake of major political upheavals, such as the fall of Hungary's communist government in 1989 and of the collapse of the Soviet Union two years later, but, at other times, changes can extend over much longer periods.

Secondly, memoryscapes are shaped by social, cultural, and national traditions that vary from place to place. We have used examples from Hungary where issues of public art, memoryscapes, and politics are subjects of much public discussion and where many traditions of public commemoration are deeply rooted in European precedents. Memoryscapes that develop in other settings are likely to be shaped by very different values and processes. These differences may involve the spatial dynamics of memoryscapes – where they are created – as well as the symbolism embodied in their forms.

Thirdly, as a corollary to the previous point, our analysis has not addressed a range of ways in which memoryscapes are gendered, racialised, and shaped in ways that privilege one perspective over another. The cases we have used in this chapter highlight the gendered representations of the past that are current in Hungary. Apart from a few queens, saints, and brave women heroes, the contributions of women to Hungarian history is underrepresented in the landscape. Though Hungary was once the part of Dual Monarchy, one of the largest multi-ethnic states to exist in Europe up to the First World War, few elements of contemporary memoryscapes credit the diverse peoples and populations that were long part of Hungarian life. This tendency slowly changes, a few memorials have appeared in recent years to honour ethnic groups (for example Armenians, Swabians) that suffered, were displaced, or expelled from Hungary after the Second World War. The losses of the Roma population of Hungary in the Holocaust is only faintly represented in the national commemorative landscape.

Finally, focusing the material and spatial is neither to dismiss the temporal aspects of memoryscapes nor to see them as the sole, or even primary means of conserving memory. Other chapters in this *Handbook* approach public memory, commemoration, and memoryscapes from very different perspectives including their expression in myth and ritual and their embodiment in day-to-day experience. Our point is that space, place, and location are often interwoven with these other ways of expressing and representing the past in contemporary life.

References

Andersen, T.S., and Törnquist-Plewa, B. (2016). *Disputed memory: Emotions and memory politics in Central, Eastern and South-Eastern Europe*. Berlin: De Gruyter.
Apor, P. (2014). *Fabricating authenticity in Soviet Hungary: The afterlife of the first Hungarian Soviet Republic in the age of state socialism*. London: Anthem Press.
Bernhard, M.H., and Kubik, J. eds. (2014). *Twenty years after communism: The politics of memory and commemoration*. Oxford: Oxford University Press.
Bodnar, J. (1992). *Remaking America: Public memory, commemoration and patriotism in the twentieth century*. Princeton, NJ: Princeton University Press.
Boros, G. (1997). *Emlékművek '56-nak* [Monuments to 1956]. Budapest: 1956-os Intézet.
Dobre, C., and Ghita, C. eds. (2017). *Quest for a suitable past: Myths and memory in Central and Eastern Europe*. Budapest: Central European University Press.
Doss, E. (2012). *Memorial mania: Public feeling in America*. Chicago: University of Chicago Press.
Erőss, Á. (2016). "In memory of victims": Monument and counter-monument in Liberty Square, Budapest. *Hungarian Geographical Bulletin*, 65 (3): 237–254.
Farmer, S. (1999). *Martyred village: Commemorating the 1944 massacre at Oradour-sur-Glane*. Berkeley, CA: University of California Press.
Foote, K.E. (2003). *Shadowed ground: America's landscapes of violence and tragedy*. Revised Edition. Austin: University of Texas Press.
Foote, K.E., and Azaryahu, M. (2007). Toward a geography of memory: Geographical dimensions of public memory and commemoration. *Journal of Political and Military Sociology*, 35 (1): 125–144.
Foote, K.E., Tóth, A., and Árvay, A. (2000). Hungary after 1989: Inscribing a new past on place. *Geographical Review*, 90 (3): 301–334.
György, P. (2000). *Néma Hagyomány* [Mute tradition]. Budapest: Magvető Kiadó.
Halbwachs, M. (1941). *La topographie légendaire des Evangiles en Terre Sainte: Etude de mémoire collective* [Legendary topography of the Gospels in the Holy Land: A study in collective memory]. Paris: Presses Universitaires de France.
Halbwachs, M. (1992). *On collective memory*, trans. Lewis Coser. Chicago: University of Chicago Press.
Hartman, G.H. (1994). *Holocaust remembrance: The shapes of memory*. Cambridge, MA: Blackwell.

Jakab, A.Z. (2012). *Emlékállítás és emlékezési gyakorlat. A kulturális emlékezet reprezentációi Kolozsváron* [Memory construction and practice. representations of cultural memory in Cluj-Napoca]. Cluj-Napoca, Romania: Kriza János Néprajzi Társaság-Nemzeti Kisebbségkutató Intézet.

James, B. (2005). *Imagining postcommunism: Visual narratives of Hungary's 1956 revolution*. College Station: Texas A&M University Press.

Kelman, A. (2013). *A misplaced massacre: Struggling over the memory of Sand Creek*. Cambridge, MA: Harvard University Press.

Krasnodębski, Z, Garsztecki, S., and Ritter, R. eds. (2012). *Politics, history and collective memory in East Central Europe*. Hamburg: Krämer.

Lebow, R.N., Kansteiner, W., and Fogu, C. eds. (2006). *The politics of memory in postwar Europe*. Durham, NC: Duke University Press.

Linenthal, E.T. (1995). *Preserving memory: The struggle to create America's Holocaust Museum*. New York: Viking.

Lowenthal, D. (1975). Past time, present place: Landscape and memory. *Geographical Review*, 65 (1): 1–36.

Lowenthal, D. (1985). *The past is a foreign country*. New York: Cambridge University Press.

Luthar, O. ed. (2012). *The Great War and memory in Central and South-Eastern Europe*. Leiden, The Netherlands: Brill Academic Publishing.

Marcuse, H. (2001). *Legacies of Dachau: The uses and abuses of a concentration camp, 1933–2001*. Cambridge: Cambridge University Press.

Mark, J. (2010). *The unfinished revolution: Making sense of the past in Central-Eastern Europe*. New Haven, CT: Yale University Press.

Mink, G., and Neumayer, L. eds. (2013). *History, memory and politics in Central and Eastern Europe: Memory Games*. Basingstoke: Palgrave Macmillan.

Nora, P. (1984). *Les lieux de memoire* [Realms of memory], 2 vols. Paris: Gallimard.

Olick, J.K. (2007). *The politics of regret: On collective memory and historical responsibility*. New York: Routledge.

Pótó, J. (1989). *Emlékművek, politika, közgondolkodás: Budapest köztéri emlékművei, 1945–1949 így épült a Sztálin-szobor, 1949–1953* [Monuments, politics, public opinion: Public art memorials of Budapest, 1945–1949 and the Stalin statue, 1949–1953]. Budapest: Magyar Tudományos Akadémia (MTA), Történettudományi Intézet.

Rainer, M.J. (2001). Nagy Imre újratemetése – a magyar demokratikus átalakulás szimbolikus aktusa [The reburial of Imre Nagy – the symbolic act of the Hungarian democratic transformation]. *In*: B. Király and L.W. Congdon, eds. *A magyar forradalom eszméi. Eltiprásuk és győzelmük (1956–1999)* [The ideas of the Hungarian revolution. Their treading out and victories (1956–1999)]. Budapest: Atlanti Kutató és Kiadó Társulat-Alapítvány, pp. 240–258.

Rampley, M. ed. (2012). *Heritage, ideology, and identity in Central and Eastern Europe: Contested pasts, contested presents*. Woodbridge: Boydell Press.

Rév, I. (2005 [2001]). *Retroactive justice: Prehistory of post-communism*. Stanford, CA: Stanford University Press.

Roudometof, V. (2002). *Collective memory, national identity, and ethnic conflict*. Westport: Praeger.

Sajti, E.A. (2004). *Impériumváltások, revízió és kisebbség* [Replacement of empires, revision and minority]. Budapest: Napvilág Kiadó.

Seleny, A. (2014). Revolutionary road: 1956 and the fracturing of Hungarian historical memory. *In*: M.H. Bernhard and J. Kubik, eds. *Twenty years after communism: The politics of memory and commemoration*. Oxford: Oxford University Press, pp. 37–59.

Todorova, M., Dimou, A., and Troebst, S. eds. (2014). *Remembering communism: Private and public recollections of lived experiences in Southeast Europe*. Budapest: Central European University Press.

Törnquist-Plewa, B. (2016). *Whose memory? Which future? Remembering ethnic cleansing and lost cultural diversity in Eastern, Central and Southeastern Europe*. New York: Berghahn Books.

Tumarkin, M. (2005). *Traumascapes: The power and fate of places transformed by tragedy*. Melbourne: Melbourne University Press.

Wehner, T. (1986). *Köztéri szobraink* [Our public statues]. Budapest: Gondolat Kiadó.

Young, J. (1993). *The texture of memory: Holocaust memorials and meaning*. New Haven, CT: Yale University Press.

13
STÓ:LŌ MEMORYSCAPES AS INDIGENOUS WAYS OF KNOWING

Stó:lō history from stone and fire

Keith Thor Carlson with Naxaxalhts'i (Albert 'Sonny' McHalsie)

Despite the complex and deeply personal ways that individuals interact with their environs to create memoried places, in North America all memoryscapes can be divided into two categories – Indigenous and settler. Settler memoryscapes by definition are part of the ongoing settler colonial process of displacing Indigenous people from lands and resources. Indeed, it is largely through the process of creating memories on and about geographies from which Indigenous people have been displaced that settlers come to regard themselves as belonging on a particular place and to a particular polity (i.e. Canada or the United States, but likewise in settler colonial spaces in New Zealand, Australia, northern Norway, Finland, and Sweden, and Argentina, among others). Beyond this, settler and Indigenous memoryscapes additionally differ in that while both settlers and Indigenous people attach and invest memories onto landscapes (so as to build and sustain associations with the land), Indigenous people also regard landscapes as sentient and as having memories of their own independent of living humans. In this chapter we focus on one particular Indigenous community, the Stó:lō Coast Salish of the lower Fraser River watershed in western Canada. Importunately Stó:lō knowledge-keepers reject the suggestion that they are anthropomorphising a landscape. Rather, seen through a Coast Salish cosmology and ways of knowing, the landscape animates humans, for it is in the stones, plants, and animals that memories and knowledge nest. And in the right circumstances, the landscape shares these memories with people and in so doing invests new historical understandings into human society.

Remembering and forgetting Indigenous and settler landscapes

A growing body of international interdisciplinary scholarship is seeking increasingly sophisticated ways to pull back the veil of settler colonialism to better understand, interpret, and articulate historic and current Indigenous relationships to the land (Harrison 2005; Brunn & Springer 2015). To varying degrees this historiography is informed by and reflects Indigenous ways of knowing, and even those that are less informed are aspiring to engage with Indigenous ways of knowing to better understand the consequence of the creation of colonial memoryscapes (Ingold 1993; Kelly 2015).

While Indigenous people needed no assistance in coming to recognise that pioneer settlement in North America was actually a 're-settlement' facilitated by an earlier process of Indigenous depopulation, non-Indigenous academics interested in Indigenous memoryscapes owe a debt of intellectual gratitude to scholars like Keith Basso, Cole Harris, Julie Cruikshank, and Lynn Kelly, for bringing this insight into the academic mainstream (Basso 1996; Harris 1997, 2003; Emberley 2007; Kelly 2015). The mythologies informing the celebratory narratives of American Manifest Destiny and Canadian Dominion which depict re-settlement as 'natural' and 'inevitable,' likewise normalise the historic displacement of Indigenous people from their land and resources. The process underlying the ways in which settlers came to regard these spaces as their own thus reflects a convenient amnesia that memory studies has the potential to disrupt (Carlson 2011).

While academic and political attention has generally focused on rural and 'wilderness' spaces, historians such as Coll Thrush have highlighted the ways in which the creation of North American urban spaces likewise continue to be predicated upon the erasure and eclipsing of earlier Indigenous spaces. The settler politics of forgetting through the selective creation of settler memoryscapes, as Thrush has shown, was often facilitated through the creation of an urban settler aesthetic that appropriated Indigenous symbols and imagery to commemorate an Indigenous past through art that implied contemporary Indigenous consent (Thrush 2009).

Indigenous people together with allied scholars have in recent decades produced remarkable collaborations aimed at alerting settler society to the significance of Indigenous peoples' historical presence and ongoing special relationships with the lands and waters of their ancestors. Hugh Brody's insightful and accessible *Maps and Dreams* (1992) revealed how Indigenous memories of past activities on the landscape could serve as a counterbalance against the exploitative capitalist resource extraction that Canadian provincial legislation had facilitated over a territory that settlers widely regarded a *Terra Nullius* prior to their arrival. In the seminal *Wisdom Sits in Places* (1996), Keith Basso worked with Apache knowledge-keepers to interpret a landscape that was an archive of memory and a generator of meaning. Apache memoryscapes, he argued, are created through a process in which 'people are forever presenting each other with culturally mediated images of where and how they dwell. In large ways and small, they are forever performing acts that reproduce and express their own sense of place – and also, inextricably, their own understandings of who and what they are' (Basso 1996: 57).

In a more northern landscape, Julie Cruikshank's collaborative scholarship with Tlingit and Athabascan people (*Do Glaciers Listen?* 2005) contrasted science's materialist view of glaciers with those of local Indigenous people who see them as sentient beings. Northern Indigenous voices revealed for Cruickshank the ways in which landscapes become markers of human history, as well as of 'memory, stability, and change in human affairs' (Cruikshank 2005: 11).

The resistance to Cruikshank's analysis by empirical scientists provides a context for understanding the contestations that continue to capture global media headlines over clashes between Indigenous people and setter societies on the question of how to understand and manage human 'uses' of mountains across cultural divides (Kraft 2010; MacKinnon 2017). These discussions reveal that progress in building understanding and respect between Indigenous people and those who venture into their traditional lands is progressing at a rate that Indigenous people often find frustratingly slow.

When the Stó:lō leadership authorised a team that included the authors to produce *A Stó:lō Coast Salish Historical Atlas* in 2001 (Carlson et.al. 2001), we strove to not only create a reference tool that could help local Stó:lō community members and resource managers in the process of reclaiming land by restoring memories, but a work of artistic historical geography that could unsettle settlers by alerting the more than two million people living in Canada's Vancouver Lower Mainland region that the landscape that appeared familiar, that they thought they knew, that they claimed to remember, could in fact be re-situated so as to be unfamiliar, unknown, and not remembered. We sought, in other words, to disrupt a colonial memoryscape by addressing what Lorenzo Veracini later identified as settler colonialism's efforts to sustain structures of oppression designed to be invisible so as to provided settlers with a degree of comfort – and by extension privilege (Veracini 2014). We regarded this as a contribution to decolonising and re-Indigenising Canadian space and history.

Transformer sites

Indigenous people have profoundly local, deeply historical ways of remembering, interpreting, and understanding the creation of the places they call home. One of the Stó:lō Atlas plates (McHalsie et al. 2001) within the *Stó:lō Atlas* in particular was inspired by the *sxwoxwiyá:m* (ancient narratives of creation and transformation) that Stó:lō Chief George Chehalis (Sts'ailes First Nation) shared with anthropologist Franz Boas in 1884. In his conversations with Boas, Chehalis' described how the world that people recognise today was in large part the product of the work of Xe:Xá:ls, three brothers and a sister, who in ancient times traveled throughout the Coast Salish world permanently transforming people, animals, plants, stones, and water into their currently recognisable form.

The division that Westerners who arrived in the Stó:lō world drew between humans and everything else was not a part of Stó:lō epistemology. Prior to Xe:Xá:ls' arrival nothing had an entirely predictable physical appearances. People sometimes transformed themselves, and sometimes others, into animals or objects, and vice versa. Sometimes these transformations were consensual, sometimes not; sometimes they were morally guided by a desire to reward or punish, sometimes not. In making the contemporary physical expression of the formerly chaotic ancient world predictable, Xe:Xá:ls fixed spirits into a variety of physical forms. The result are ongoing kin ties between people and the environment. The heroic founders of Stó:lō tribal communities were transformed into particular animals and plants. At Leq'á:mel, for example, Xe:Xá:ls transformed a man into sturgeon, and as a result, all members of the Leq'á:mel First Nations regard sturgeons as their relatives (their ancestor's spirit is retained in all sturgeon). The ancestor of the Matsqui tribe was transformed into a beaver, and so on along the lower Fraser River.

Not everyone Xe:Xá:ls transformed founded a tribal community, nor were all transformations into animals or plants. A great elk hunter was turned to stone, located in the Fraser River near Yale, as was a great seal hunter on a lower stretch of the Fraser. A dangerous witch with a toothed vagina was likewise transformed into stone located along the edge of the Harrison River. And one of the tallest mountains in the region was formerly a kind and protective Stó:lō woman who had been married to Mt. Baker. When she returned to the Fraser River from her husband's home in the south, Xe:Xá:ls transformed her and her daughters into stone.

Among other things, Xe:Xá:ls created a distinctly local Indigenous memoryscape where plants, animals, and stones serve as mnemonic devices to remind Stó:lō people of both the ancient histories of transformation and the more recent histories (*sqwelqwel*) that have occurred on the landscape subsequent to the work of the transformers. Thus the landscape is populated with storied creatures and storied plants and animals who serve to remind people of their ancestral connection to the land.

These stories cut across one another and informed one another in complicated ways. The tribal origin stories that tie one group of Stó:lō people to a particular sub-region of the broader Stó:lō landscape serve to reinforce the autonomy of tribal collectives just as the fact that the animals and plants which carry the spirits of particular tribal founders grow and live throughout the broader region serves to remind Stó:lō people from all tribes of their interconnections to one another. Moreover, geographically fixed stones associated with certain transformer stories are particular to certain places. Unlike a beaver that regardless of its location throughout the broader Stó:lō territory can serve as a mnemonic of the origin story of the Matsqui tribe, the Elk Hunter and the Seal Hunter who were turned to stone need to be visited in situ for their anchored story to be shared in a fulsome way. Historically these stones' stories served to encourage and facilitate visits between in-laws and friends. The story of the kind protective mother mountain and her daughters, by way of contrast, is universally regarded as a shared super-tribal Stó:lō story not specific to any particular tribe, for on clear days they can be seen from anywhere in Stó:lō territory. The memory associated with that mountain and its peaks transcend particular tribes, for she is regarded as having been placed in that prominent position by Xe:Xá:ls so that she could see and look over *all* the Stó:lō people as well as the annually returning salmon of the Fraser river. Her presence and visibility are constant sources of comfort.

Knowing one's history

Knowing from where you come, and from whom you come, is a defining way that Indigenous people build identity in themselves and acknowledge it in others. Within the Stó:lō community, Elders describe people from high-status families by the Hal'qeméylem term *smelalh* – a word that translates as 'to be worthy/worthy people.' In 1995 Elder Rosaleen George, when asked to elaborate on the meaning, answered, 'to know your history is to be smelalh – that's "worthy". If you don't know your history (if you've lost or forgotten it), well then you are stexem – and that's "worthless."'[1]

Before colonial settlement in Stó:lō territory and the establishment of laws and practices aimed at alienating Indigenous land and resources, 'knowing one's history' was largely tied to knowing where one had a hereditary right to fish, hunt, and gather. At potlatch naming ceremonies, families transferred hereditary names that were linked to such sites across generations. High-status visitors validated the host family's rights to such valuable resources sites by acknowledging the correctness of the genealogical stories shared at the potlatch that tied people to places on the landscape. Likewise, in the past as today, worthy families share with their guests and relatives the stories of transformation that make them who they are as individuals, as members of tribal collectives, and as members of the broader super-tribal Stó:lō community.

Where memories rest

Indigenous historical narratives may function in ways differing from those circulating in Western society (Deloria 1994; Fixico 2003; Chamberlin 2003). Among the Stó:lō, for example, stories about transformation are not merely mnemonic narratives that people use to attach memories to the land. In 1995 Elder Bertha Peters shared with us a legendary transformer story that highlighted this fact, while simultaneously challenging Western chronologies associated with literacy's connection to colonialism and Western scientific ways of understanding non-human life:

> The Great Spirit travelled the land, sort of like Jesus, and he taught these three *siyá:m*, these three chiefs, how to write their language. And they were supposed to teach everyone how to write their language, but they didn't. So they were heaped into a pile and turned to stone. Because they were supposed to teach the language to everyone and because they didn't, people from all different lands will come and take all the knowledge from the people – because they wouldn't learn to write they lost that knowledge.
>
> When the first white people came, a white man raped this Indian woman. And she got syphilis. Then, when her husband *went* with her, he caught syphilis too. But they didn't know about these sicknesses, and so the man went up the mountain to die. He was laying there naked and a snake came up to him and ate all the sickness off his penis, then wiggled away. Then it ate three types of plants and got well. So the man went and ate the three plants and got well. So they knew a cure for this sickness, but they couldn't write it down, so they lost it.[2]

Beyond its explicit lessons about the need for Indigenous people to find ways of preserving traditional knowledge in the face of colonial incursions, and what it reveals about Mrs. Peter's historical consciousness pertaining to an earlier time when an Indigenous literacy preceded European contact (Carlson 2011), her story principally highlights memories of violent trauma and cultural loss in the history of Stó:lō people's relations with settlers. In terms of its potential to reveal insights into Stó:lō memoryscapes, what strikes us is that in telling the narrative Mrs. Peters did not share information that would allow us to locate the sites on the landscape where rapes had occurred, where diseases like syphilis had been caught, or the location of medicinal plant gardens such as that where the woman's husband had been cured following lessons learned from the snake. Rather, she seems to be suggesting specific ways of understanding the significance and implications of colonial erasure and its role in collective forgetting by highlighting a particular Stó:lō way of remembering.

What was implicit in Mrs. Peter's telling of the narrative is explicit in Elder Aggie Victor's explanation of the rock's significance. When visiting the transformer stone on the occasion of an announcement that it was being protected from the threat of development destruction, Mrs. Victor explained to the gathered crowd that 'I want you young people not to forget that the spirits of those three chiefs is still in that rock.'[3] For Mrs. Victor, people who took the time to learn how to properly listen would still be able to hear from within the stone the voices of the three *siyá:m*.

The idea that memories and history can exist autonomously of human agents was additionally the focus of a conversation between Elder Peter Pierre (Katzie First Nation) and anthropologist Diamond Jenness in 1936. According to Pierre, seven spirit entities reside within each Stó:lō person. One of these is *smestíyexw* ('vitality' or 'thought'), which is responsible for a person's conscious thought (Pierre 1955). Pierre explained to Jenness that the sun provided *smestíyexw*, and unlike the spirits associated with people's soul or shadow, whose departure brings instant death, *smestíyexw*/vitality could leave the body for short periods.

Importantly for the purpose of this chapter, Old Pierre elucidated that *smestíyexw* is responsible for not merely vitality and thought, but also memory. An individual's *smestíyexw* literally has the ability to travel through the spirit world to locations that are considered *xá:xa* (sacred or taboo) and there acquire knowledge, or reacquire lost knowledge (forgotten memories).[4] A person's *smestíyexw*/vitality could also acquire power or talent (such as knowledge and lost memories) during its travels.

Elder Jimmie Charlie (Sts'ailes First Nation) explained in 1996 that knowledge and memories sometimes rested within stones and plants and other objects, and sometimes memories could additionally float about in areas frequented by a person who had carried them during his or her life.[5] People with the right gifts from the Creator and appropriate training from knowledge-keepers within their families acquired the ability to listen in special ways that would allow them to hear what others had forgotten. Jimmie's grandson Kelsey Charlie is one such person, widely renowned for his ability to find and bring back songs that have been lost or forgotten for decades and even generations.[6] He also has the gift of being able to remember songs after they come to him, and a reputation for generously sharing those songs back with the relatives and descendants of the people who originally sang them in earlier times.

In conversations Kelsey explains that he does not compose these songs, nor does he consider there to be a process of 'inspiration' at work in what he does in the way Western singers and songwriters often describe their own process of first articulating a song. Rather, when in the proper place on the landscape, and when listening the proper way, the songs come to him. He remembers them even though he has never heard them before, because the memories are nested in the landscape where he has 'picked them up.'[7]

Bending time and space

In Indigenous societies, time sometimes bends spaces in ways that settlers struggle to perceive, let alone appreciate. Most Stó:lō spirit energies vary in strength depending upon how far away they are acquired. Spirit memories are often acquired by Stó:lō people from locations that are near at hand for the simple reason that these are the places their ancestors knew and frequented; the memories and knowledge there are familiar and knowable. But as Peter Pierre explained to Jenness in 1936, and as remains a common understanding within the Stó:lō community today, memory and knowledge that are acquired from great distances hold the potential to be especially powerful, relevant, and meaningful, if for no other reason than that they are initially less predictable and thus potentially more dangerous.

Indeed, this is one of the main reasons that northern 'coastal raiders' used to be such a concern for Stó:lō people. These hostile strangers entered the Stó:lō riverine world without kinship ties, without local spirit guides, and with warrior spirit power that was unfamiliar and strange. If successful they stole away young girls as slaves and took them to distant places to be exploited without the benefit of family and ancestor spirits to comfort and guide them. Today, contemporary Elders and knowledge-keepers raise a concern over the Canadian government's child foster and adoption systems wherein Stó:lō children deemed by social workers to be 'at risk' are apprehended and placed in the care of non-Indigenous families. Among the objections Stó:lō people raise is the concern that children relocated outside of the Stó:lō homeland are by definition being denied the ability to interact with the locally grounded ancestral spirits that populate the Stó:lō memoryscape.

Sonny McHalsie, one of the authors, explains that 'When I leave the territory I don't feel my spirits are with me.'[8] When required to journey beyond the borders of the geographic range of his ancestors, McHalsie is careful to inform the spirits of his intentions through prayers which explain why he is going, where he is going, and when he intends to return. While away, the disconnection causes him to feel vulnerable. The return home is always a cause for rejoicing and relaxing. Then McHalsie engages in further conversations with his ancestral spirits to remind them of where he has been, why he was absent, and the purpose and value of his journey.

The further away one has to travel to obtain *swia'm* spirit power (the type Peter Pierre described as 'talent' and explained was often indistinguishable from the memory spirit of *smestíyexw*) the stronger that spirit power's manifestation. While all animals, plants, and stones have *smestíyexw* and *swia'm* that they might decide to share with humans, certain spirit entities exist only in the mystic *xá:xa* (sacred or taboo) realms, where they reside at varying distances from humans. A person (or at least a person's *smestíyexw* spirit) has to make the perilous journey into the *xá:xa* dimension in order to encounter such a spirit.

However, a person who has undergone intense and prolonged purification and received proper hereditary training need not necessarily travel bodily across a physical geography in order to find and acquire memories and knowledge. Dotting the Stó:lō landscape are invisible mystic portals that render spaces immediately adjacent to one another that a settler Canadian might consider distant and apart. Although they cannot be found on government-produced maps, for those who know the history, have the training, and know where to look, the tunnels are real. Successful travel through these dangerous tunnels is almost instantaneous. As such, the tunnels bend time and space to make locations, spirits, and memories that might otherwise appear far apart actually close together.

In the nineteenth and twentieth centuries, technologies such as steam locomotives, automobiles, airplanes, telegraphs, and ultimately satellites and the internet transformed the way humans traveled and how they acquired knowledge from distant places. People no longer needed to travel great distances to communicate and share knowledge face to face or even to wait weeks or months while written text were transported from one locale to another. Instead, knowledge increasingly enabled someone to acquire something from strangers from distant locations in ever shorter periods of time.

Such changes were interpreted by nineteenth-century Stó:lō people through the cosmology discussed earlier. Consider, for example, the oral histories still circulating in the Stó:lō community that describe one of the two colonial-era governors (either James Douglas or Frederick Seymour) having made a promise in his capacity as representative of the Crown that the Stó:lō would be compensated for the alienation of lands outside of their reserves. Further, consider that a delegation of Coast Salish Chiefs led by Joe Capilano of Squamish to London in 1906 to visit King Edward VII reported that they had secured the King's confirmation of the Governor's earlier promise to address the issue of Aboriginal title. As we have argued elsewhere, the fact that no archival documents have been found confirming the promise need not imply that the promise was never made (Carlson 2005). The fact that members of the 1906 delegation were trained in communicating with ancestral spirits may well mean that, to an extent that is not appreciated by Westerners, the Coast Salish received confirmation of the Crown's promise not from Edward VII but from Edward the Confessor (the first Edward) while visiting his tomb in Westminster Abbey – contemporary newspaper accounts describe how affected the delegates were upon discovering Edward's tomb within the Abbey (Carlson 2005).

Seen through this lens, the Salish Chiefs sojourn to London by steam engine and steamboat constituted a great journey across a vast landscape to a strange and foreign place where *smestíyexw* could be acquired from appropriate ancestral spirits. This, we think, provides a context for Chief Capilano's exasperation when speaking with a journalist from the Vancouver *Province* newspaper four years later when he observed that just because Westerners observe Indigenous people does not mean that they understand them: 'They tell you things they have heard, but they do not understand them. If they have seen them they do not understand them, for white men go about with a veil over their eyes and do not think as we think' (Morton 1970: 36).

Memories through fire

Indigenous memoryscapes can be both dangerous and comforting. In the early 1990s, Elders and education workers in the Sumas First Nation were seeking to better understand recent youth suicides. They were aware of oral traditions that identified a large cedar tree in their community as the 'hanging tree.' Contemporaneous archival research that the authors of this article were then conducting had helped bring back into people's consciousness the story of a 14-year-old Stó:lō boy named Louie Sam who in 1884 had been abducted from provincial police custody by a mob of Americans and lynched on Canadian soil (Carlson 1996). Archival evidence generated by Canadian undercover detectives suggested the boy had been framed for the murder of an American shopkeeper and that the lynching was orchestrated by the real murderer to prevent Louie Sam from testifying through a translator in the Canadian courts. People in the Sumas community wondered if the 'hanging tree' was the site of the lynching, and if so was Louie Sam's spirit potentially lingering in the vicinity? Could his spirit's ongoing sorrow and suffering as the victim of settler colonial violence be causing inadvertent harm to contemporary youth and perhaps contributing to suicides?

What also emerged from the archival records, however, were detailed coroner reports describing the precise location where Louie Sam had been lynched – at a site 152 metres north of the Canada/United States border, and a full eight kilometres away from the 'hanging tree' at Sumas. The apparent contradiction between the oral history and the archival records was a concern. For answers the Stó:lō turned to their ancestors. A ritual burning ceremony was organised with two purposes: firstly, to contact the spirits of Louie Sam and his mother to let them know that the contemporary community had not forgotten them; that the living cared for them, and would continue to help ensure that they rested in peace. The first goal, thus, was providing knowledge to the spirits. The second objective was to request knowledge and memories from the spirits. Was there a connection between Louie Sam and the 'hanging tree'? The Stó:lō ritualist started a sacred fire just before dawn. Cedar was used to kindle the fire, the crackling designed to drive away spirits so as to create an open spiritual space. Coniferous wood (that does not crackle) was then placed on the fire to create a portal through which the specific spirits of Louie Sam and his mother could be contacted. Eventually food and clothing were placed on the fire to feed and comfort the spirits. The ritualist then silently communicated with the spirits.

Such ceremonies take several hours, after which a meal is shared where the ritualist conveys the information learned from the spirits with the community. In this case several messages were received. Among them was confirmation that Louie Sam and his mother felt comforted, and they expressed that no one needed to worry that they might be causing harm to contemporary youth. Another was that the hanging tree was indeed associated with Louie Sam. But not as the site where the boy's body had been hanged, but rather where Louie Sam had been *hung*. The spirits confirmed that Louie had indeed been lynched on a tree just north of the American border line (as the archival records described) but that immediately afterwards his body had been cut down and returned to his family. It was late February, the ground was frozen, and so the family was not able to immediately perform the Christian-style internment ceremony that Stó:lō people had recently adopted. So, following ancient tradition, Louie's body was wrapped in blankets and hung in the branches of the old cedar tree in the Sumas community where it would be safe from animals until the ground had thawed sufficiently to allow for a burial.

Conclusion

Despite the alienation of lands and the imposition of government regulatory practices and policies, Indigenous people like the Stó:lō continue to 'make memories' on the landscape of their ancestors in ways that are intimate and profoundly meaningful. As with all people everywhere, they come to know places in complex ways. They go to places, do things, and then associate memories with those places. Such memories play an important role in grounding them and connecting them to place. Indeed, Indigenous memoryscapes are not radically different from the ones that numerous settlers Canadians and Americans have also been making over the same North American physical geography for the past two centuries.

But Indigenous memoryscapes differ in important ways from all the memoryscapes that settlers have draped over the land. The differences are both epistemological and ontological, and while they assume intimately local expressions, the conflicts that emerge have global commonalities. In Ireland, people with deep ancestral memories continue to clash with developers and archaeologists over the issue of how best to interpret and protect 'fairy forts' – rock formations which locals believe were created by spirit entities, archaeologists assert were built by ancient humans, and developers see as impediments to highway and urban development (Cheallaigh 2012). In Canada, similarly, the highly publicised conflict between the Ktunaxa First Nation and developers over whether or not a mountain was just a geological rock formation with suitable slopes for the development of a 6,000 unit ski resort or a sacred site that would be rendered spiritually vacuous by economic development reached the Supreme Court (MacKinnon 2017). And recently in Malaysia, Canadian tourists intentionally violated Indigenous customs and beliefs when they stripped naked and urinated on the peak of a sacred mountain. In this instance, occurring as it did in a country where Indigenous people were not a marginalised minority but the governing majority, the settler tourists found themselves jailed for 'desecrating a holy site, insulting a culture and – last but not least – causing a deadly earthquake' (Miller 2015).

Within the Stó:lō world, people recognise that they are related to the landscape by virtue of their ancient common origins with the region's stones, plants, and animals. In the Stó:lō cosmology, therefore, these 'things' are not merely places that can trigger memories, but rather places where memories and knowledge nest independent of human agency. Places and things, in the Stó:lō cosmology, are sentient; but they are not sentient in a way in which settler Canadians can ever fully share. Settlers may come to love the geography and call it home, but they will never have the ancestral connections to the land that Stó:lō enjoy for the simple fact that they arrived too late. Xe:Xá:ls had already completed his work long before the fur traders, gold miners, farmers, longshoremen, stockbrokers, and vegan gluten-free bakers arrived.

Settler colonialism has the power to eclipse Indigenous memoryscapes by challenging and contesting Stó:lō ways of knowing as well as by alienating lands from Stó:lō people through the seemingly never-ending expansion of simple title holdings and government regulation. What a settler society might regard as a process of building places for creating memories often poses challenges to Stó:lō people because of their tendency to damper Indigenous memory transfers. Stó:lō Grand Chief Steven Point, for example, explains that ancestral spirits sometimes have difficulty communicating with the living because of the interference caused by asphalt and concrete. It is for that reason (among others) that big cities like Vancouver can be dangerous places for Stó:lō people. Prolonged visits to such environments, Point explains, can result in Stó:lō people becoming disorientated, confused, and susceptible to unhealthy temptations and malevolent forces. 'In Vancouver you don't have connection with the earth; with the soil,' Point explains, and so 'the spirits have a hard time talking to you. As Native people we need that connection.'[9]

Settler colonialism creates settler memoryscapes in numerous ways, some of which involve industrial workspaces associated with such activities as blasting rocks, consuming mineral resources, harvesting trees, and paving streets. Such actions inevitably make it harder for Indigenous people like the Stó:lō to hear their ancestors' voices and to find and retrieve ancestral memories. This highlights for many settler Canadians what are perhaps largely invisible implications of the ongoing alienation of Indigenous people from their land and resources. As settlers and as scholars, our aspirations are, therefore, perhaps best directed at working with Indigenous people to try and better understand the complexity of the Indigenous relationships with the land so that we can position ourselves to more properly respect them. That is one way in which the study of memoryscapes can contribute directly to settler colonial decolonisation.

Notes

1. Rosaleen George in conversation with Keith Carlson, May 1995.
2. Bertha Peters in conversation with the authors, 20 September 1995.
3. Aggie Victor in conversation with Sonny McHalsie, July 1992.
4. Peter Pierre explained that if not returned within a reasonable period, a person without *smestíyexw* goes crazy and dies.
5. Jimmie Charlie in conversation with Keith Carlson, November 1996.
6. During the years when the potlatch and tamanawas dance were banned, songs that had been prominent with families sometimes ceased to be sung until they slipped from human memory.
7. Kelsey Charlie in conversation with Keith Carlson, May 2011.
8. Sonny McHalsie in conversation with Keith Carlson, 31 October 2003.
9. Steven Point in conversation with Keith Carlson, July 1995.

References

Basso, K. (1996). *Wisdom sits in places: Landscape and language among the western Apache*. New Mexico: University of New Mexico Press.
Brody, H. (1992). *Maps and dreams*. Vancouver: Douglas and McIntyre.
Brunn, S., and Springer, L. (2015). *The changing world religion map: Sacred places, identities, practices and politics*. New York: Springer.
Carlson, K.T. (1996). The lynching of Louis Sam. *British Columbia Studies*, 109: 63–79.
Carlson, K.T. (2005). Rethinking dialogue and history: The King's promise and the 1906 Aboriginal delegation to London. *Native Studies Review*, 16 (2): 1–38.
Carlson, K.T. (2011). Orality about literacy: The 'black and white' of Salish History. In: K.T. Carlson, K. Fagan and N. Khanenko-Friesen, eds. *Orality and literacy: Reflections across disciplines*. Toronto: University of Toronto Press, 43–70.
Carlson, K.T., Schaepe, D., and McHalsie, A. (2001). *A Stó:lō coast Salish historical atlas*. Vancouver: Douglas and McIntyre.
Chamberlin, J.E. (2003). *If this is your land, where are your stories? Finding common ground*. Toronto: Alfred A. Knopf.
Charlie, J. In conversation with Keith Carlson, November 1996.
Charlie, K. In conversation with Keith Carlson, May 2011.
Cheallaigh, M.N. (2012). Ringforts or fairy homes: Oral understandings and the practice of archaeology in nineteenth- and early twentieth-century Ireland. *International Journal of Historical Archaeology*, 16 (2): 367–384.
Cruikshank, J. (2005). *"Do glaciers listen?" Local knowledge, colonial encounters and social imagination*. Vancouver: University of British Columbia Press.
Deloria, V. (1994). *God is red: A native view of religion*. Golden, CO: Fulcrum Publishing.
Emberley, J.V. (2007). *Defamiliarizing the Aboriginal: Cultural practices and decolonization in Canada*. Toronto: University of Toronto Press.
Fixico, D. (2003). *The American Indian mind in a linear world*. New York: Routledge.
George, R. In conversation with Keith Carlson, May 1995.
Harris, R.C. (1997). *The resettlement of British Columbia: Essays on colonialism and geographical change*. Vancouver: UBC Press.
Harris, R.C. (2003). *Making native space: Colonialism, resistance, and reserves in British Columbia*. Vancouver: UBC Press.
Harrison, R. (2005). Dreamtime, old time, this time: Archaeology, memory and the present- past in a northern Australian Aboriginal community. In: J. Lydon and T. Ireland, eds. *Object lessons: Archaeology and heritage in Australia*. Melbourne: Australian Scholarly Publishing, 243–264.
Ingold, T. (1993). The temporality of the landscape. *World Archaeology*, 25 (2): 152–174.
Kelly, L. (2015). *Knowledge and power in prehistoric societies: Orality, memory and the transmission of culture*. New York: Cambridge University Press.
Kraft, S.E. (2010). The making of a sacred mountain: Meanings of nature and sacredness in Sápmi and northern Norway. *Religion*, 40 (1): 55–61.

MacKinnon, L. (2017). SCC Picks B.C. Ski Hill Project over Indigenous spirituality. *IPolitics*, November. Accessed 31 May 2018, https://ipolitics.ca/2017/11/02/scc-picks-b-c-ski-hill-project-over-indigenous-spirituality/

McHalsie, A. In conversation with Keith Calrson, 31 October 2003.

McHalsie, A., Carlson, K.T., Schaepe, D., and McHalsie, A. (2001). Making the world right through transformation. *In:* K.T Carlson, D. Schaepe and A. McHalsie, eds. *A Stó:lō coast Salish historical atlas*. Vancouver: Douglas and McIntyre, pp. 6–7.

Miller, M.E. (2015). Tourists strip, a sacred mountain shakes and Malaysia gets very angry. *Washington Post*, 9 June. Accessed 31 July 2018, www.washingtonpost.com/news/morning-mix/wp/2015/06/09/tourists-strip-a-sacred-mountain-shakes-and-malaysia-gets-very-angry/?utm_term=.07eb2389f22f

Morton, J. (1970). *Capilano: Story of a river*. Vancouver: McClelland & Stewart.

Peters, B. In conversation with the authors, 20 September 1995.

Pierre, P. (1955). Katzie book of genesis. *In:* D. Jenness, ed. *Faith of a coast Salish Indian*. Victoria: British Columbia Provincial Museum, pp. 10–34.

Point, S. In conversation with Keith Carlson, July 1995.

Thrush, C. (2009). *Native Seattle: Histories from the crossing over place*. Seattle: University of Washington Press.

Veracini, L. (2014). *The settler colonial present*. London: Palgrave Macmillan.

Victor, A. In conversation with Sonny McHalsie, July 1992.

14
POTS, TUNNELS, AND MOUNTAINS
Myth, memory, and landscape at Great Zimbabwe, Zimbabwe

Ashton Sinamai

Introduction

Archaeology interprets the past through materials, but people use all their senses to understand the pasts and the environments. Narratives are one way they try to understand the layers of their landscapes. They are memory tools that communities use to understand their environment and shape their present and futures. Recovering these memories in narratives and utilising them to enhance archaeological interpretations of place has, however, been difficult in archaeology and its related disciplines. However, listening to community narratives and analysing them through narrative inquiry has many benefits for the discipline. Narrative inquiry creates collaboration between the researcher and the people that he or she is researching. It thus makes people, and not artefact, central to archaeological inquiry (Clandinin & Rosiek 2007). Memory, restructured as myth, contains insights into how people think about their environments. This information can be used to define research agendas beyond the limits of archaeological boundaries and also force archaeologists to think of every place as a part of a wider landscape and not just a 'site.'

Narratives are also used to question the dominant ideologies which seem to focus on materiality in telling national stories at the expense of local narratives. This chapter examines the cultural landscape around Great Zimbabwe using community narrative to understand perceptions of heritage and landscape among the communities that created it. I argue that myths are metaphors of a cultural thought system that translates the intangible to the cultural and environmental present. Using Great Zimbabwe, I ask pertinent questions about how the focus on monumentality abbreviated traditional cultural landscapes and how these can be reconstituted through understanding landscape narratives in the form of legends.

In Africa, allowing people to tell their stories is also allowing multiple interpretations and recognising the various social layers that a place has. Memory has always been a way of representing the past as evidenced by *griots* and family keepers of history found in African communities. When used in heritage management, these narratives correct the epistemological injustices of archaeology, which was developed outside Africa, practiced for much of the colonial period by non-Africans, and is still practised with theories created outside Africa. The purpose of narrative inquiry is not to prove the efficacy of the stories, but to find their performative contexts and also read differently so that they feed into archaeological inquiry as well as heritage management. It is these contexts that can provide information on people's struggles to understand and appreciate the environment in which they find themselves in. Stories are built around and people and places and can be used to map cultural landscapes that have gone through political raptures.

There is a collective social process in telling the story and as a collective process the story captures intimate connections between the community and their landscape.

Cultural landscapes are not only a result of ingraining of culture on a landscape but also mark the development of an environmental literacy in a new environment (Basso 1996). The layers of environmental information collected, and the struggles to establish a society on that landscape, creates the invisible sacred bond between land and people. The stories of attaining environmental literacy and of the struggle to understand the landscape sometimes come down as myths and legends. By their nature, sacred landscapes cannot have boundaries. Narratives are therefore part of the memory of that struggle to understand a new environment and can give us an understanding of the old landscapes before a rapture caused by the colonial experience, traumatic events or natural disasters. A cultural landscape is therefore encoded with events, personalities, and institutions and passed on through myths, performance, and other forms of narratives.

A cultural landscape provides a 'geopsyche' in which people feel secure both physically and emotionally. The level of absorption, understanding, and memorialisation of the cultural landscape determines where 'home' and 'house' are located; home being a place with deeper personal and emotional connections and house being an emotionless space you occupy away from 'home.' Every person outside his or her familiar landscape does not feel secure enough and is also unable to read the cultural information within that landscape. The landscape is not simply a repository of human achievements as suggested by other definitions, but is also an actor in itself, influencing both the living and the dematerialised forces (Kohn 2013). It is thus not a passive commodity that can be conquered, used, and sold, but is an actor who can 'avenge' with brutal force a man's transgressions (Basso 1996). Western philosophies find difficulties in dealing with these concepts because of the inability to understand cultural landscapes beyond the 'use-value framework' (McFarlane 2014) and to understand that value is not always measurable.

People understand the landscape first before embedding it into collective memory, through using it initially as a source of food and security, and later, in enhancing their identity. That landscape is named with what is experienced in each particular feature. The process of narrating and naming the landscape is the initial stage of understanding that landscape (Spirn 1998). Placemaking is not only cultural; it is also an environmental activity in which the elements of the surroundings are recorded as both resources and mapping tools. A combination of the inserted culture and environmental literacy in a place creates cultural and social identities. Social and cultural identities linked to landscapes are only expressed once environmental literacy has developed. This connection to the land, once established, creates an ontological security for people living in it. The rupture of this security can be devastating for societies, but it also creates a nostalgic need which can create new narratives on top of the older narratives. The colonial experience in Africa is an example of such ruptures.

Among the Shona of Zimbabwe, a place acquires a human personality and behaves according to how it is treated by people. An individual learns of his or her landscape from childhood starting with where your umbilical cord is buried. The soil that it is buried in is imprinted into your being and will control and draw you back even when you move away from this environment. When an individual is asked where they come from, the answer is usually where their umbilical cord was buried and not where they have a house. This burial is symbolic of the intermingling of the human with the soil from which he is literally 'born' (Chitakure 2016). In this way, the landscape has a permanence that people and animals do not have and, in a way, it owns all these and controls their fate and not the other way round. Indeed, isotope studies have shown that the environment that nurtures us has a permanent print of in our bones. This should make us understand that focusing on human exceptionalism within cultural landscapes will not make us understand the connection that people have on land and landscapes. (Kohn 2013: 89). It is the mechanistic rationalism that makes it difficult to understand the intangible within the tangible without having to separate the two. In reality, the intangible is the reason we understand the tangible. In heritage studies, however, the intangible element is a surrogate of monumentality. Understanding the 'intangible' as metaphors of a cultural landscape can equip archaeologists and heritage managers with landscape literacy to read landscapes in the same way as those that societies that have experiences in them perceive them.

The Great Zimbabwe landscape

Currently being managed by the National Museums and Monuments of Zimbabwe (NNMZ), Great Zimbabwe (Figure 14.1) is listed as a 'World Heritage Site' not a cultural landscape. This categorisation of Great Zimbabwe as a 'site' has been problematic as narratives about the place show an intimate relationship between the wider landscape and the people who live in it. The word 'site' is used by archaeologists to show the presence of cultural heritage in a place. A cultural landscape, on the other hand, has no boundaries and considers both culture and nature as components of place. In a description given to UNESCO, Great Zimbabwe 'existed between 1000 and 1450 AD' (UNESCO, 2018). In other words, Great Zimbabwe has scientifically established boundaries both physically and chronologically. The major areas of interest include the Hill Complex, where stone walls easily blend into the boulders which are used as part of the architecture. The Hill Complex receives at least 70% of all visitors to Great Zimbabwe.

Umberto Eco (1998) argues that new worlds are not only discovered; they are read through 'background books' which determine how you see and interpret that new world. Archaeology, as a part of the colonial experience in Africa, is a part of those 'background books' that determine the perception of the new

Figure 14.1 The Great Enclosure at Great Zimbabwe.
Source: Photograph by A. Sinamai.

Pots, tunnels, and mountains

world. The colonial experience in Zimbabwe as a form of 'discovery' reflects these 'background books' in the way that Great Zimbabwe was researched, used, named, interpreted, and managed (Figure 14.2). The superimposition of the European experience at Great Zimbabwe and the surrounding landscape was not only a deconstruction of the native cultural landscape but a production of new stories used in claiming the new space understood through cultures of the 'old world.' The processes that the Great Zimbabwe has gone through, from a sacred landscape to a global heritage 'site,' reflect the use of these 'background books.' It was appropriated, mapped, renamed, and its uses changed. What the communities called *Dzimbabwe*, the hilltop royal residence, became the 'Acropolis,' the ancient citadel of Athens.

The new names become woven into new stories of colonial adventure, gallantry, and discovery and erased the previous narratives about the landscape. This, however, is the nature of landscapes – they are etched into the mind and are transferable to new territories. As 'existential spaces,' they 'are difficult to experience if you are not a part of that culture' (Dodge 2007: 8) and hence one has to find new ways to understand new landscapes. The colonial state inscribed new names to places, not only for appropriation purposes, but to also insert new meanings on these landscapes and create a new sense of belonging. Indeed, a part of Nyuni Mountain, north east of Great Zimbabwe, was renamed Glen Livet because it was a 'duplicate of the River Spey, caressing Glen Livet Mountain in my own county of Banffshire . . . Scotland' (Sayce 1978: 112).

Figure 14.2 The Allan Wilson crypt (1902) is still in its original place just outside the Site Museum at Great Zimbabwe.
Source: National Archives of Zimbabwe.

The landscape was developed for the enjoyment of the European settler population. The first hotel was constructed in 1898 followed by a golf course, which cleared much of the vegetation and created the open areas from the hotel to the Great Enclosure. Vegetation was also thinned within the stone-built area and, in some cases, it was replaced with exotic trees (jacaranda, eucalyptus) imported from elsewhere. A eucalyptus plantation was also planted in an area southwest of the Great Enclosure just above a sacred spring. It is this eucalyptus plantation that affected the water table, resulting in the drying up of the sacred *Chisikana* spring mentioned later. A camp for prisoners was also constructed south of the Great Enclosure to help with conservation work in the times when the Historical Monuments Commission did not have enough manpower to run to the site (Ndoro 2011: 61). Accompanying all these changes were the new narratives about the landscape. New stories of Great Zimbabwe being a city built by King Solomon or Queen of Sheba emerged in contrast to the local narratives told by communities. The original 'myths,' however, have been resilient and have continued to be narrated by elders within the Mugabe and Nemamwa communities living near Great Zimbabwe.

In current politics, Great Zimbabwe is the primordial source of modern Zimbabwe and part of the narrative of nation. It directly connects the modern state to the ancient polity and nullifies the existence of colonial Rhodesia (Fisher 2010). As a World Heritage Site, the site is also a preferred tourist destination for many foreign visitors to Zimbabwe. Indeed, before the current political and economic problems, Great Zimbabwe attracted a modest 150,000 visitors annually. Its potential to attract foreign and domestic visitors makes it a valuable site for the government, which collects revenue from hotels and the National Museums and Monuments of Zimbabwe (NMMZ), which collects entrance fees. It is, of course, still a valued archaeological resource as many researchers have used it to launch their careers. As an archaeological site, Great Zimbabwe is regarded as an archaeological island surrounded by rural areas where archaeology is absent. Archaeological maps end at the boundary of the estate as if archaeology is determined by the estate boundary.

Community memory and the landscape at Great Zimbabwe

For local communities, however, 'Dzimbabwe' is not a bounded piece of land but a centre of a cosmological network that creates an ontologically secure environment for people, nature and the non-human elements. Though it has monumental stone walls, it is not only the monumental stone walls that are valued. It is the totality of the landscape including mountains, springs, streams, and rivers not within the demarcated estate. To locals 'Dzimbabwe' is only a centre of that wider cultural landscape and is not more special than the other sacred parts of this landscape. These sacred places are said to be connected with Great Zimbabwe through various ways. Each of them thus suffers when one of them is negatively impacted by, for instance, development. One community story narrates of two pots that 'walked' from Great Zimbabwe to mountains and springs that surround Great Zimbabwe (Interview: Participant 3 and 5: 2016). The pots (a female and a male) visited Chepfuko spring, Bingura Hill, the Mutirikwi River before proceeding to prominent mountains, like Boroma, Nyuni, Beza, Ruvhure, Nyanda, Chamazango, and Mupfurawasha, which surround the Great Zimbabwe. One of the stories about the vessels (*Pfuko ya Kuvanji*) was recorded quite early (1871) by a German explorer, Karl Mauch. When he climbed up the mountain, his porters refused to accompany him (Burke 1969: 139). Great Zimbabwe is also said to be connected to the nearby hills of Mupfurawasha, Ruvhure, Nyuni, and Beza through tunnels from the Hill Complex.

According to the narratives, Great Zimbabwe also has 'entrances' (interviewees called them 'doors') that can be opened and locked through performances of certain rituals. One of these 'doors' is to southeast of the Hill Complex (*Mujejeje*). *Mujejeje* is 'a linear intrusive vein of quartz exposed on solid granite making a narrow bar ridge' (Summers 1965), situated 400m in the eastern limits of the Valley Ruins (East Ruin). For those travelling into or through Great Zimbabwe, this threshold is not to be crossed without carrying out a ritual or the traveller would suffer some misfortune. Individuals reaching this point in the course of

Pots, tunnels, and mountains

Figure 14.3 *Pfuko yaKuvanji* one of the two vessels that are said to have 'walked' in the Great Zimbabwe landscape.
Source: Photograph by A Sinamai.

their travel to Great Zimbabwe have to pick up a pebble and tap it along the line, murmuring a prayer to the ancestral spirits of the area and God and ask to enter into sacred space, which still has an ancestral presence. Bent also records this narrative during his excavations in the early 1890s after he observed his porters performing this ritual when they entered Great Zimbabwe from the east (Bent 1893: 75).

There is also a scared spring, *Chisikana*, that developed where a young girl who was kidnapped by a mermaid emerged from it after a long absence. This story has two versions: interviewees from the Nemamwa clan say that the little child was sent to fetch water from the sacred spring where she was kidnapped by mermaids. She disappeared for a very long time, after which she emerged from the spring as a very powerful spirit medium and healer. Participants from the Mugabe clan also claim her as a member of their clan. Both clans, however, agree that the spring became perennial and a stream developed. Along this stream were sacred pools in which fish were caught at certain times of the year. It is also from this spring that water for all religious ceremonies was collected.

The difference between the stories that emerge about Great Zimbabwe from these two clans is not about who has the truth as Fontein (2015) seems to suggest, but it is about what is the most sacred feature at Great Zimbabwe. How the two groups connect to the sacred sites is not important. The Chisikana Spring master-story has its key elements: a proficient natural spring, a girl, a kidnapping, and the empowerment of the spring and the river and the ceremonies that were carried out there. What is important is not whether the girl came back or belonged to the Mugabe or Nemamwa clan, but the centrality of water resources for people who have lived in this environment. The stories are thus accumulated wisdom collected through centuries of interaction with the environment around Great Zimbabwe. Their knowledge of water sources and the climate that creates this environment, for example, is apparent in these stories. The stories identify the perennial springs and rivers within the basin that supplied water to Great Zimbabwe.

Great Zimbabwe lies in the middle veld altitude 1,150 to 1,250 metres above sea level. The geology around the Great Zimbabwe basin is dominated by granites to the south with the north dominated by metasediments and felsic metavolcanics with greenstone belts. These northern hills also contain a few gold-bearing veins, which were probably exploited during the occupation of Great Zimbabwe. The ancient city itself is on the southern edge dominated by granite which is useful in building the stone walls. The granites of the south were a valuable source of the building materials needed in the construction of the city. The northern ranges are much steeper and higher than the granites, rising above 1,544 metres in the case of Nyuni and, together with the granite hills to the south, they create a sheltered basin with its own micro-environment. Local people recognise that the area enclosed by these mountains has a micro-environment which is only maintained by respecting the presence of sacred places like Great Zimbabwe. Their interpretation of the basin is thus not based on the rivers as suggested by Pikirayi et al. (2016) but on the sheltering mountains that create this micro-environment. Their recent hydrological description of the Great Zimbabwe area gives a northwest-south direction of the basin and focus on the rivers, streams, and springs (Pikirayi et al. 2016). The local communities, however, have a different perception of the landscape focusing on an east-west basin sheltered by these mountains, recognising the role of the mountains in creating this unique climatic environment.

In the *Chisikana* legend, the young girl becomes a relative of a recent, known ancestor. The storylines are however hardly changed and are actually strictly maintained, and in some cases, the same words are used to tell the stories. One reference to a voice calling someone to bring a milking can (*hwedza*) was quoted verbatim and this may be emphasising the centrality of cattle in the Great Zimbabwe culture. Great Zimbabwe being the centre of the cultural landscape is linked to all narratives told about sacred places around it. This description of Great Zimbabwe is not packaged for tourists and many walk past or through the very sacred areas without any knowledge of their existence. These narratives are not material evidence and so they remain unrecognised as they are difficult to quantify. They are also not monumental and are therefore not of interest to the researchers and tourists who emphasise the visual experience over the abstract connections of heritage and the mind.

Re-mapping the cultural landscape

Narratives and material culture are both equally tools of memory. Where material culture depends on sight and touch, immaterial heritage depends on experiences of the mind. Sight does not record an experience; it is the mind that keeps records of what is seen. Space cannot be a cultural landscape unless there is an effort to record it and etch it into the mind of individuals and communities and claim it. The narratives are a way of recording and remembering, and naming is the claiming of space and can be used to identify the most important components of an existing or past landscape. The current Great Zimbabwe 'site' is a place conceived through western conservation philosophies of monumentality. Supported by the national legislation governing heritage, these concepts have restricted what can be used to define and interpret the cultural landscape. The narratives told by communities about Great Zimbabwe show that it is not a 'site' but a part of a much wider cultural landscape. The narratives emphasise 'connection' through tunnels, the vessels that 'walked' to certain places as well as the fires that burnt the mountains towards the rain season. This connection highlights what is important around Great Zimbabwe and provides valuable information about the environment which archaeologists should have used to understand Great Zimbabwe.

The narratives collected highlight the centrality of water in the location of Great Zimbabwe. In these myths, water is represented by the mysterious mist that often covers the hills, the drizzling rain (*guti*) that falls often around Great Zimbabwe, the sacred springs, streams and rivers. When one looks closely at the mountains that are mentioned in the narratives, it is easier to see that Great Zimbabwe is located in a hydrographical basin. The sacred springs and streams and rivers within this area could withstand the serious droughts that the area often experiences and hence became important aspects of the landscape and hence their appearance in the legends.

The sacredness of Great Zimbabwe is also brought out through stories about sightings of lions within Great Zimbabwe. Lions, however, have been eliminated from this environment for over a hundred years. All participants, however, mentioned a cave on the northeastern slopes of the Hill Complex called the Cave of Lions where lions can be seen at a certain time of the day. Two elderly participants claim to have seen these lions when they were younger (Interviews Participant 2 and 4, 24 March 2016). Recent sightings have been reported by workers of the National Museums and Monuments of Zimbabwe and at least one tourist (Memo NMMZ, Interview Participant 12). For the local communities, the appearance of these lions is usually a warning of an impending problem that the community could face. Last year's sighting by a local tour group was therefore taken as a warning for the drought and the heatwaves that the area faced over a period of seven months. Lions are associated with royalty and represent the royal spirits which are said to still live at Great Zimbabwe. As important intercessors to God for issues that affect all the communities living near Great Zimbabwe, these spirits are important for ensuring rain not only around this area but for the rest of Zimbabwe as well. It is not surprising that the sightings of lions have been linked to the drought that area faced in 2015/2016 season.

Conclusions: landscape mapping, narratives, and the archaeological sub-disciplines

As Schmidt (1983: 75) posits, it is only 'when cultures in Africa participate in the interpretation of their own past, we can begin to build a self-enriching tradition of archaeology free from the domination of Western paradigms.' Using narratives inquiry in trying to understand the meaning of folklore, myths, and legend about a place is one way which the African voice can be present in archaeological enquiry. At Great Zimbabwe these voices identify the most important components of the landscape, many of which had been ignored by heritage managers before. Studying these narratives bring traditional knowledge systems to the foreground – which could be a way to identify the intangible outside of Western philosophies where they are taken as 'values.' They can help those of us trained in western philosophies to read landscape through the lenses of people that created them. The voices, the animal sounds that feature in community narrative are a memorialisation of the soundscape. Not only do the voices represent the wisdom from ancestors, they also record the animals that are within the cultural landscape.

The narratives are preserved for a purpose: they are occasionally used to subvert mainstream narratives that are sponsored by the political power of the state. State power threatens not only to change culture through single narratives but also through the denial of rights to performances of religion. In many postcolonies, archaeological research has dogmatically privileged state's needs against those who demand cultural rights. The discipline of archaeology thus militates against local sensibilities and aspirations in that it denies the presence of their religion and perpetuates the colonial presence in heritage management. Archaeologists have shaped cultural landscapes with what they know best: cultural/archaeological remains (Spirn 1998: 23). When archaeologists carry out archaeological surveys, the aim is to identify those elements of the landscape that are archaeological and they hardly invest time in narrative inquiry to acquire an African reading of heritage places they study (Schmidt 2014: 173).

Elsewhere archaeologists have argued that archaeological collaborations with the public have failed (Lane & Mapunda 2004) and this has been attributed to archaeologists failing to communicate the benefits of archaeology (Pikirayi 2011). The failure, however, lies not in failing to communicate but in failing to listen. Archaeologists and heritage managers have a top-down approach which does not recognise other forms of knowledge; they consider themselves to hold superior knowledge. Cultural landscapes should be read through artefacts, monuments, and sites, but also through an understanding of the minds of people who created it, lived in it, and have told stories about it. Telling a story about a place is a social process and understanding that social process will enhance landscape literacy among archaeologists.

Understanding the myths as metaphors of the landscape creates opportunities for meaningful collaboration between the researcher and the people that he or she is researching on. This makes people, and not artefacts, central to archaeological inquiry. The narratives also contain very crucial environmental information which can be used to define research agendas beyond the limits of archaeological boundaries. It forces archaeologists to think of every place as a part of a wider landscape and not just a 'site.' Narratives can also be used to question the authorised discourses which often focus on national agendas at the expense of local concerns. By allowing people to tell their stories one is also allowing multiple interpretations and recognising the various social layers that a place can have. Stories have places and people in them and can be used to map cultural landscapes.

Heritage cannot only deal with the 'premier' monumental achievements but must be reflective of all that societies deem important in a place including how it is memorialised (Breen Rhodes 2010: 34). Understanding this will give us tools to seek an understanding of other knowledges beyond archaeology and create an understanding of other people and how they relate to land and the past on that land. Very important sacred places at and around Great Zimbabwe were desecrated (and sometimes destroyed) through a landscape illiteracy promoted by Western philosophies of conservation. Some like the *Chisikana* spring were desecrated by practices which would be soundly condemned as environmentally unsustainable today. To dry out the marshes created by the spring and stream, eucalyptus trees were planted at the headwaters of the spring. In another sacred area with a spring, a public toilet has been constructed.

Recently the communities around Great Zimbabwe complained at the vandalism at *Mujejeje* (the 'threshold'). The stones that had piled at each end of the linear extrusion of quartz were removed by unknown individuals. In fact, workers at Great Zimbabwe had seen people in this part of the monument, but bothered to check what they were doing because they thought that part of the monument was not important. Indeed, it has not been important enough to monitor it in the same way that walls at the site are. The importance of these parts of Great Zimbabwe is clearly shown in the myths of place, but, through landscape illiteracy, these places have been physically damaged and desecrated. The inherited colonial legislation that archaeologists use to manage the landscape today defines what is to be preserved and celebrated, but it is apparent from problems experienced that there is a need to listen to community stories in managing and interpreting this cultural landscape. The stories about voices and animal sounds being heard from the Hill Complex point to the loss of a soundscape, but this aspect of heritage at Great Zimbabwe has been consistently eroded through various activities that communities have complained about in this landscape (Sinamai 2017). The narratives therefore assist heritage managers in identifying the important elements of a cultural landscape without separating them from the material culture. These narratives of place show that the landscape is not memorised by seeing it. It is an emotive experience that requires the use of all senses in creating intimacy of land. Current landscape studies seem to focus on current relationship between the land and the people, but, in actual fact, it is created by layers of ancestral experiences passed from one generation to the next through narratives. Without understanding community narratives as memory tools, we will continue to misread cultural landscapes.

Acknowledgements

Many thanks to communities living around Great Zimbabwe who allowed me to collect their stories and use them for this chapter. The research was funded by the EU's Horizon 2020 programme (Project: METAPHOR No. 661210).

References

Basso, K. (1996). *Wisdom sits in places*. Albuquerque: New Mexico University Press.
Bent, J.T. (1893). *The ruined cities of Mashonaland*. Books of Rhodesia (Reprint) Bulawayo, (Rhodesia).
Breen, C., and Rhodes, D. (2010). *Archaeology and international development in Africa*. London: Gerald Duckworth and Co. Ltd (Duckworth Debates in Archaeology).

Burke, E.E. (1969). *The journals of Carl Mauch, his travels in the Transvaal and Rhodesia, 1869–1872*. Salisbury (Rhodesia): National Archives of Rhodesia.
Chitakure, J. (2016). *The pursuit of the sacred: An introduction to religious studies*. Eugene: Wipf and Stock Publishers.
Clandinin, D.J., and Rosiek, J. (2007). Mapping landscape of narrative inquiry: Borderland spaces and tension. *In:* D.J. Clandinin, ed. *Handbook of narrative inquiry: Mapping a methodology*. Thousand Oaks: Sage Publication, pp. 35–76.
Dodge, W.A. (2007). *Black rock: A Zuni cultural landscape and the meaning of place*. Jackson, MS: University Press of Mississippi.
Eco, U. (1998). *Serendipities: Language and lunacy*. New York: Columbia University Press.
Fisher, J.L. (2010). *Pioneer, settlers, aliens and exiles: The decolonisation of white identity in Zimbabwe*. Canberra: Australia National University Press.
Fontein, J. (2015). *Remaking Mutirikwi: Landscape, water and belonging in Southern Zimbabwe*. London: James Curry.
Kohn, E. (2013). *How forests think: Towards an anthropology beyond the human*. Los Angeles: University of California Press.
Lane, P.J., and Mapunda, B. (2004). Archaeology for whose interests – archaeologists or the locals? *In:* N Merriman, ed. *Public archaeology*. London: Routledge, pp. 211–223.
McFarlane, R. (2014). *Landmarks*. London: Penguin Books.
Ndoro, W. (2011). Managing and conserving archaeological heritage in Sub-Saharan Africa. *In:* S. Sullivan and R Mackay, eds. *Archaeological sites: Conservation and management*. Los Angeles: Getty Publications, pp. 561–571.
Pikirayi, I. (2011). *Tradition, archaeological heritage protection and communities in the Limpopo Province of South Africa*. Addis Ababa: OSSREA.
Pikirayi, I., Sulas, F., Musindo, T.T., Chimwanda, A., Chikumbirike, J., Mtetwa, E., Nxumalo, B., and Sagiya, M.E. (2016). Great Zimbabwe's water. *Wires Water*, 3 (2). doi:https://doi.org/10.1002/wat2.1133
Sayce, K. (1978). *A town called Victoria*. Salisbury: Books of Rhodesia.
Schmidt, P.R. (2014). Hardcore ethnography: Interrogating the intersection of disease, human Rights and heritage. *Heritage and Society*, 7 (2): 170–188.
Schmidt, P.R. (1983). An alternative to a strictly materialist perspective: A review of historical archaeology, ethnoarchaeology, and symbolic approaches in African archaeology. *American Antiquity (Review of Old World Archaeology)*, 48 (1): 62–81.
Sinamai, A. (2017). Melodies of God: The significance of the soundscape in conserving the Great Zimbabwe Landscape. *Journal of Community Archaeology and Heritage*, 5 (1): 17–29.
Summers, R. (1965). *Zimbabwe: A Rhodesian mystery*. Cape Town: Thomas Nelson and Sons.
Spirn, A.W. (1998). *The language of landscape*. New Haven, CT: Yale University Press.
UNESCO. (2018). *Great Zimbabwe national monument*. Accessed 24 March 2018, https://whc.unesco.org/en/list/364.

15
LEARNING BY DOING
Memoryscape as an educational tool

Toby Butler

I have long been excited by the potential of memoryscapes, a term I originally applied to multi-media, oral history–based experiences designed to be undertaken in specific places. Creating a typical memoryscape requires a working knowledge of a range of practical skills including applied historical research, interviewing, digital recording, sound editing, creating websites and interactive maps, and developing an understanding and critical appreciation of user-friendly tour and trail design.

This chapter will explore the learning potential and some of the practical and pedagogical issues I have encountered in training and mentoring students in the construction of memoryscapes. I have been running courses on this theme for ten years in East London with a range of participants, from teenage youth groups to postgraduate students at Master's level, including some over retirement age.

The courses were all taught from the field of history and heritage studies in the United Kingdom, in a period of profound change in higher education. In the last decade there has been increasing pressure for degree programmes to deliver vocational skills and improve student employability.[1] Combined with the digital publishing and information revolution, many history programmes now provide applied digital skills training alongside the traditional research, critical thinking, and essay and report writing. The public-facing nature of web-based work is also ideal for showcasing student's achievements and the best projects were promoted on university and research centre websites.

Such a course could just have easily been part of a geography or creative media programme or used in an adult education or community development context. With this in mind, I hope this chapter might be useful for trainers, teachers, and lecturers interested in using memoryscapes to help students explore issues around place, public history, oral history, and local history; and along the way develop and apply skills in creating accessible and engaging digital multi-media.

My main aim is to share a range of experimental, exciting, and hopefully inspiring student project work, which, in the field of history at least, so rarely sees the light of day in academic publishing. Alongside this, I also give some context to the trails and give a sense of the nature and development of the courses that brought them about. This chapter is best read with a web browser to hand, so you can follow the links in the footnotes to the various websites and explore the work discussed.

My first attempt at a memoryscape building course was a free public workshop programme entitled *Making Community History Trails* focused on the communities around the Royal Docks in East London in 2007. The aim was to give local people the opportunity to develop multi-media and research skills and be involved in the creation of several artist-led trails around the docks (Figure 15.1). The aim of the project

Figure 15.1 The Ports of Call public workshop programme.

was to map and interpret something of the community history of North Woolwich and Silvertown, two communities that had been hit hard by the closure of the docks in the 1980s (Ports of Call 2008).

The programme attracted several older members of the community who were invaluable in helping us understand the historical geography of a deindustrialised area that had undergone an extraordinary amount of change. A community mapping day, in which photographs and memories were attached to a large map alongside talks on local history, planning, and mapping was particularly successful, as was a workshop on tour guiding (with increasing tourist interest in the locality following the success of the London 2012 Olympic Games bid). But it proved difficult to attract many attendees to skills workshops in digital recording or Google mapping, and in the end workshop participants featured in the trails as interviewees rather than participated as interviewers. On reflection a public workshop programme was perhaps too loose an organisational structure for sustained engagement; later projects had much more success with overtly recruiting volunteers with a clear written understanding of how much time would be expected to attend a training programme and work on project tasks. However, the open activities attracted a number of key members of the local community who were invaluable in helping the project staff to better understand the local history and politics of the dock communities.

The issue of authorship and control is intrinsic to any community project involving professional workers (composers, artists, digital producers, academics).[2] There are two quite distinct forces, or tensions, that run through projects that have public-facing media as the outcome. The first is the over-arching desire to produce something that is the best possible quality – it might not be broadcast standard, but it needs to be understandable, legible, and usable by someone unfamiliar with the project. In publicly funded projects like *Ports of Call* there is necessarily pressure on project managers and professionals involved to make the outcomes as accessible as possible, and those with extensive media expertise are often highly adept at producing content that works in terms of a comfortable and engaging listening experience. But there is also the knowledge expertise, energy, desire, and enthusiasm of volunteers to draw upon, which may not be aligned to the intended project objectives, but may well help the memoryscape become more grounded and meaningful.

The project manager must be mindful that usually the power relationship is heavily stacked in the artists or producers' favour – they have the technical expertise and often a very clear idea of the achievable form of the final product, particularly if the project is short term with a detailed budget, dreamed up months or even years before in a funding application. In initial meetings with community workers and potential volunteers, the intended scope of the product (the memoryscape) is naturally explained, perhaps with examples from other projects. The problem is how to build in enough genuine flexibility and scope for the local experts – usually volunteers and unpaid interviewees – to shape and participate in the project in ways that meet their own desires and aspirations. Without this the memoryscape might be technically impressive but have no meaningful heart or soul in terms of the *community* it is situated in (and should ideally serve in some way).

In these terms the *Ports of Call* workshop programme was not terribly successful – it engaged with some community members and established some relationships, but had a relatively shallow impact in terms of volunteer engagement with the trail production process, and two of the trails were ultimately heavily artist-led in terms of choice of content.

In this regard we had much more success with the *Asta Trail*, delivering a workshop programme for an existing youth group of 11 to 17 year olds. They met at the Asta Centre, a small youth centre after school in Silvertown in the shadow of London City Airport that had both a small recording studio and Jason Forde, a talented and enthusiastic youth worker with DJ mixing skills who used the studio for music recording and editing workshops. This meant we could run a more focused workshop programme from the Asta Centre in the heart of Silvertown rather than at the University of East London, and the three-hour sessions ran almost every evening over a two-week period in the summer holidays.

It also soon became clear that finding out about local history was going to be a tough to sell to a diverse group of young people; but there seemed to be a very strong desire to learn any skill that would be required to make rap music. Jo Thomas, a composer and university lecturer brilliantly built up the workshop programme around this interest and worked on lyric writing and multi-track recording, composition, and editing over two weeks in the summer holidays. A walk around the area with the young people revealed an array of issues that concerned them, ranging from fear of gangs in neighbouring communities to anger that a graffiti mural they had painted with a community artist was going to be removed by the council following a complaint by a local resident. Their concerns inspired their rap lyrics and composition that later featured very prominently in the trail.

After some training in recording and interviewing, the young people also recorded interviews with local experts including a property developer, museum curator, an airport representative, and the owner of the local grocery store. This led to some really interesting opportunities for the young people to question people who were prominent in the community (and in turn the interviewees had the opportunity to hear some of the current concerns from a younger generation).

The *Ports of Call* project was my first adventure in community-based project work. The project highlighted a number of engagement issues. The most successful work came from engagement with existing groups, brokered with experienced community workers who already had the trust of the young people concerned. This was not without difficulties; the summer period meant that some of the group left for holidays, or drifted away to do other things, leaving a core of a few young people to help Jo in the latter production stages. Some had moved to London from other countries recently, and we quickly realised that some who seemed quiet or reticent were actually coping with English as a second language. At times the issue of power and control over content had to be sensitively but overtly negotiated; some violent language and lyrics around gun crime made a great track for teenagers in a youth centre, but eventually were dropped by discussing their relevance to the locality and thinking more deeply about the potential audience – which included parents and carers.

However, the material was not all 'safe'; for example, critical content over the issue of the mural erasure was retained. The result was a rap song and interview material that entwined the present and future in a memoryscape around Silvertown entitled the *Asta Trail: Planes, Trains and Graffiti Walls* and some of the interview material was used in a another trail Jo constructed in West Silvertown (Asta Centre & Thomas 2008; Thomas 2008). The trails were put on to 1,000 CDs (Figure 15.2) and given away to community groups, local residents, and newcomers to the area. Brenda Ridge, one of the participants who had put days of effort into singing and song writing, featured on the CD cover, and the group all received MP3 Players as a reward for their work and were collectively acknowledged as co-authors (Asta Centre & Thomas 2008).[3]

Two years after the *Ports of Call* project I had an exciting opportunity to develop a course from these experimental workshops in partnership with Birkbeck College, a co-partner in the Raphael Samuel History Centre.[4] Birkbeck is well known for providing flexible and part-time courses for adult learners and there was a good fit with the ethos of the University of East London. The Centre was keen to jointly develop innovative history courses and events that might attract students to both institutions. At Birkbeck Mike Berlin taught London history and used both historical maps and walking as teaching methods; he was an ideal match to complement my experience in teaching oral history and digital media. The course we devised was entitled *Exploring London's Past: Archives, Architecture and Oral History*, part of Birkbeck's Certificate in History programme, a pre-undergraduate programme which accepted mostly mature applicants, some with few or no qualifications. Mike gave talks on the politics of heritage, specifically of twentieth-century London, and used maps and archives to research London history. I spoke on the theory and practice of walking trails, cultural geographies, and mapping experiences. In the final week students presented their proposals and their work in progress. I also gave practical workshops in oral history recording, sound editing, and Google map making.

The coursework assignments were to provide a short, 500-word trail plan, to include potential archives, locations, and interviewees; followed by a digital project – a *Google* map, website, or documentary equivalent. We asked that this should contain:

- written material on at least three locations, including references to sources;
- at least one oral history excerpt, either from an archive or recorded by the student;
- an image for each location; and
- a reflective account of around 800 words on the theoretical, artistic, and practical justification for the trail design (locations, content, what the trail aims to achieve).

For some students used to conventional, essay-based assessments, this was a forbidding assignment. It was important to be up front about the demands of the assessment in the course description and we found that students needed regular reassurance and where necessary technical support. We also made it possible to submit an almost entirely paper-based assignment for those who struggled with the web-based media. More user-friendly iterations of digital map and web platforms such as Google Maps, ZeeMaps, WordPress, and Wix.com have made matters much easier. It also helped to show and experience examples of memoryscape-based work; this was easier in later years as student projects could be used for inspiration and as a reassurance that a similar project could be created in the time allowed.

Many students focused on the local history of London's suburbs, far away from the more usual guided walks and tours around central London. Linda Davies took her home community, South Acton, and teased out hidden histories from her own memories and knowledge of how the locality had changed. She mapped lost laundries (of which there were once over 200), converted pubs, and demolished streets. In her reflective

Figure 15.2 A poster advertising the trails featuring Brenda Ridge outside the Asta Centre, Silvertown in East London.

account she explained that she found two approaches particularly useful in focusing her work; the first was to 'walk around the map mentally to see which locations suggested absences or memories'; the second was to use a more playful, surrealist approach that we had explored in the course, based loosely on Walter Benjamin's Convolutes method of using an A to Z list of keywords to provoke thought about specific aspects of place (Davies 2011a). Her experimental map, which is colour coded by historic period, included interview recordings with local people including her mother who had lived through the Blitz. This has now had over 2,200 views, impressive for a student history project (Davies 2011b).

Perhaps the most impressive of these early experiments in terms of scale was *47 Bus: a linear history project* by Jonathan Bigwood. Jonathan had worked professionally, producing public transport information, and he was struck by the way that a London bus route was largely unchanged for almost a century. He explored the idea that the bus route linked people and places historically, as well as geographically:

> Route 47's links include main line stations with City offices, and suburban homes with stations and shops. Today's cleaners and security officers, yesterday's 'char-ladies' and 'night-watchmen,' travel from the suburbs to The City.
>
> *(Bigwood 2011)*

The project was astonishingly ambitious; the original bus route ran from Farnborough near Bromley on the edge of suburbia in South London and took around 90 minutes to go through numerous communities until it reached Shoreditch in central London. Jonathan incorporated hundreds of points of interest, many with links to relevant websites, recordings, and newsreel clips, and eventually decided to create three maps; one is focused on transport history with historical routes mapped alongside 145 transport-related points of interest, another devoted to relevant voices, sound recordings, and film (from online archives), and a final 'other history' map with a further 93 points of interest, mostly focused on noteworthy buildings, markets, parks, and housing estates along the route (Bigwood 2011). This work demonstrated the potential of Google Maps for people to share histories, not just of their local communities but of routeways and networks that shape our understanding of places too. At the time History Pin was in its infancy, but it has since developed into a well-established international platform for hosting place-based collections and tours. Students have also experimented with app-based trail delivery platforms such as Woices (which recently shut down).

47 Bus also began to explore the public transport journey as a potential platform to curate memoryscape experiences via the mobile smartphone. The length of the journey, the relatively low speed of the London bus, and the elevated viewing position had advantages that are well known by bus tour operators in capital cities, but the smartphone was beginning to open up the public transport network as a very cheap platform for anyone to curate tour-like or gazetteer-based experiences. A much more oral history–based bus experience was developed later by a postgraduate student, Richard Turner, who interviewed 85-year-old author and journalist Bill Mitchell (a well-known chronicler of life in the Yorkshire Dales). The recordings were made into *A Daleman's Journey*, a memoryscape designed for a bus route from Skipton to Settle through the Yorkshire Dales, the bus route Bill used to take every day to get to work (Turner 2013).[5]

Postgraduate trails: place, oral history, and digital heritage

The following year (2012) the Raphael Samuel History Centre launched a new MA entitled *Heritage Studies: place, memory and history*.[6] The programme was designed to take a broad and inclusive approach to heritage, embracing the street and the internet as much as the museum and gallery, and give students the opportunity to develop digital skills and practice-led research projects. I led the programme and developed a core module that would explore the relationship between heritage, memory and place (particularly in relation to the urban and post-industrial landscape of East London). The module considered how places are conserved and 'made' in terms of popular memory, heritage, and artistic and historic interpretation. A more ambitious memoryscape assignment was designed to give students hands-on experience of researching and producing historical material for a public audience. Students came from a variety of disciplines, so the course also had to ensure that students developed appropriate skills in historical research, oral history recording, sound editing, web page design and considerations of audiences and accessibility.

The module was taught over 12 three-hour sessions followed by a three-week production period for students to work on their final projects. The assignment ran along similar lines to the History Certificate course detailed earlier with a web-based memoryscape project, but this time it had to be accompanied by a much more demanding 2,000-word reflective essay, with instructions that it should be a theoretical, artistic, and practical justification for the trail design and methodology and refer to examples and relevant literature to situate the work. The course covered theoretical and artistic ideas and case studies relating to the cultural geographies of place alongside skills workshops and field trips to experience and consider specific memoryscapes as a group. The challenge of providing the range of skills and expertise necessary for multi-media practice was helped enormously by involving UEL colleagues who could support the course with practice-led experience.[7]

Over the years, students experimented with app-based platforms for their project work. Students have also experimented with app-based trail delivery platforms such as *Woices* (there are dozens of others), although the latter recently shut down. Using any free platform has an element of risk in this regard, so we taught well-supported website services that stood most chance of survival (e.g. WordPress; Google Maps), and as a result most student websites have survived and are still operational at the time of writing (early 2018).

The involvement of the Rix Centre was invaluable in terms of exploring some of the major accessibility issues inherent in both walking trails and online and mobile platforms. The Rix Centre was established in 2001 as a charity based at UEL to develop new technologies to transform the lives of people with learning disabilities. They had already developed the *RIX Wiki* platform, an online website authoring tool specifically designed for use by and for people with learning difficulties and used by service providers, schools, local authorities, carers, and families. The ease of use made it a useful tool for our students to quickly create and gain confidence authoring online multi-media at the beginning of the module, and we hoped the accessibility issues explored in the process would be applied in the final projects to make their projects as accessible as possible. The platform itself was developed as a result of the programme, so Google Maps could be incorporated, and some students used it to deliver their entire project.

The first memoryscape devised specifically for people with learning difficulties was created by Sarah Mees for a seaside trail at Dover in Kent. Her trail *Channel Heritage Way* (2016) (Figure 15.3) was developed with a considerable amount of user testing on location in Dover with people with learning difficulties from the Tower Project in London. The trail incorporates a range of memorials, sculptures and statues along the seafront. The web platform is highly visual with minimal text and image led. The experience is extremely multi-sensory and the listener is encouraged to touch, feel, listen, smell, spot, and find things. The physical and sensual experience is augmented with music and video, and all elements are accompanied with audio narration.

The most sophisticated project in terms of production was the *Brixton Munch* (2013a) by Laura Mitchison. The trail is based in a food market in Brixton, South London, at the centre of the Caribbean community, amongst others, that settled here. Laura was influenced by Doreen Massey's idea that places are not fixed and bound but constellations of relationships and networks, and what better place than a market, where almost everything on sale has been brought from another place, to explore how the idea might play out in terms of local heritage.

In this trail Laura moves away from using conventional oral history recordings to more crafted stories and memories co-scripted with stall holders and locals. She explains:

Learning by doing

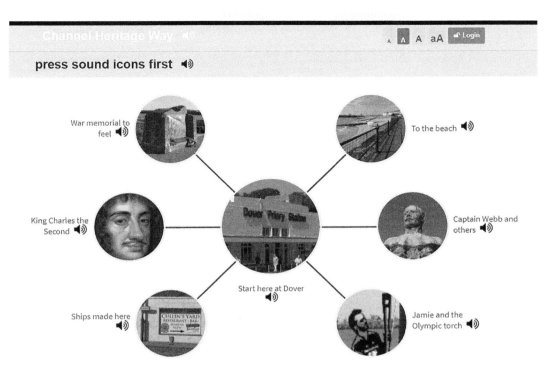

Figure 15.3 The *Channel Heritage Way RIX Wiki*.
Source: Mees (2016).

> In an early incarnation I tried 'straight' documentation – vox pops in situ – but they sounded flat and linear at home. They did not do justice to the emergent magic of the place as I know it; an emergence that, by definition, is difficult to pinpoint.
>
> *(Mitchison 2013b)*

The process by which she does this first involves creative preparatory work to stimulate the historical imagination alongside a process of co-producing the memories she collects along with the interviewees:

> Creating digital stories required intense preparation. I was intrigued by Heike Roms rhetorical question: can we stage conversations in different ways to generate different types of historical insight? First, I prepared foods and drinks relating to the story to create a rich mise-en-scène for the historical imagination. Cooking is akin to 'counsel woven into the fabric of real life', which Walter Benjamin's finds in the storyteller's art. Second, I took inspiration from other sources (archive cuttings, photos, vox pops, literature). The third phase was visualization exercises and re-enacting the story. For example, Shirley played her Tanty (aunt) getting hysterical while I played the youthful Shirley. . . . Finally, loosely scripted co-authored audio was recorded. The tellers became performers playing themselves – a strangely looped vantage point which enabled them to access descriptive details that would get lost in a casual retelling. I also encouraged them to channel other important people by implanting dialogue into the story. In the process, personal and collective significance was traded back and forth, written and spoken language saturated one another.
>
> *(Mitchison 2013b)*

The resulting stories are impressionistic rather than representational and are complemented with a superb local Rastafarian narrator, Shango. These voices are entwined with the smell, taste, music and sound relating to the foods and locations on the trail and I think very successfully capture what Laura terms 'the magic of the place.'

Some students experimented with fictionalised elements or characterisations. Halima Kanom's *Wander East Through East* (2013) of the first Chinatown in London, situated in Limehouse near the West India Docks from 1890 to the 1940s. Halima struggled to find accounts of the community from a Chinese migrant perspective, so she created an imaginary character, Chung Li, a recently arrived sailor in 1919, to imaginatively guide the listener around the old streets of Chinatown, drawing on research and archive images and oral history recordings (Kanom 2013). The most dramatic trail was created by an ex-television documentary maker, Bruce Eadie. He was an experienced interviewer but wanted a new challenge of writing what he termed 'aural drama.' Bruce's approach was inspired by *And While London Burns* (Platform 2007), a fictionalised audio trail concerned with the financial district of London's role in climate change by art activists Platform. He decided not use interview recordings, but create his own fictional characters to explore imagined memories of the inmates of Bethlam Royal Hospital for the mentally ill (or Bedlam as it was known, said to be the world's oldest mental hospital). This is not strictly a memoryscape in terms of using oral history, but I include it to illustrate how dramatically the concept was pushed. He uses the device of a conversation between two characters to explore ideas of madness around Liverpool Street Station near Spitalfields, London – the original site of the hospital. His two characters are personifications of *Raving and Melancholy Madness*, two statues displayed at the entrance of the hospital from 1676 to 1815 personified as brothers by Alexander Pope in a vitriolic poem *The Dunciad* in 1743. Bruce imagines the brothers as ghostly inmates of the hospital, discussing, arguing, and raving about the history of the hospital and the perception and treatment of madness:

> I was able to develop a tone of voice in which to deliver the audio drama: Raving Madness (or 'Ravi') restless, manic, disturbed, chaotic and with pressured speech; Melancholy Madness (or 'Mel'), pedantic, thoughtful, depressed, unable to let go of the past, trapped in a process of perpetual mourning, unable to move on. The two narrators personified the two forms of madness widely recognised in their day (the 17th and 18th centuries) and, to some degree, embodied the attributes ascribed to 'oral' and 'literate' man: Ravi inaccurate, fluid, subjective, rhythmic, participatory, adrift in time; Mel quantifying, fixed, objective, ordered, detached, obsessed with history and dates.
>
> (Eadie 2012)

The resulting trail *A Bowl around Bedlam with the Brainless, Brazen Brothers* is imaginative and engaging, and successfully conveys a large amount of historical information alongside other symbolic and psychological dimensions to the experience (Eadie 2012).

Students were also forced to reconcile an array of ethical, participatory, and representational issues that are bound up with locational public facing work. Rosa Vilbr's *The Way Home: a walk around Hackney's housing history* (2013) drew on a collection of oral history interviews collected by a campaign group, Hackney Housing for an oral history project in 2011. The aim of the project was to 'investigate and share personal and collective attempts to make housing better in the past. By doing this we hope to start conversations about what can be done to solve housing problems experienced today.' Rosa's trail includes a visit to the Hackney Service Centre, where Hackney residents came to get help from the Council with housing issues. The listener takes a seat in the waiting area and joins people queuing to see officials. The narrator explains the main principles of housing policy the officials are applying to decide who can be helped with their housing needs. This is followed by memories of the stressful interview process from those applying for council housing. It is an unsettling experience and some might feel uncomfortable in terms of the arguably voyeuristic nature

Learning by doing

Figure 15.4 The *Brixton Munch* food-related memoryscape.
Source: Mitchison (2013a).

of the experience. But this locational immediacy gives the memories a deeply impactful new life as you are witness the enactment of the housing politics of the present being played out before your eyes.

Another more politically orientated project by Sam Smith and Cindy Boga collected 17 interviews from the Occupy camp that had recently been established near the London Stock Exchange outside St Paul's Cathedral. The students had to tackle complex issues of representation, both in terms of reflecting the spectrum of opinion and perspectives they discovered in the camp and more widely in terms of the how the camp would be remembered. In her reflective essay, Cindy explained:

> As a heritage student I was concerned about the way such an event, possibly "the first major public response to thirty years of class war" (Chomsky 2012: 9), would be remembered. Mainstream media appeared either dismissive or antagonistic . . . oral history interviews seemed to offer the potential to ensure that 'the countless small actions of unknown people' (Howard Zinn in Chomsky 2012: 105) would enter the historical record – not to replace, but as a supplement to other, perhaps more dominant voices.
>
> *(Boga 2013)*

The students set up an Oral History Working Group in the camp and established a permanent presence in the camp itself. They had to navigate challenges from other researchers over their research strategy and consider questions of access and ownership alongside the fast-moving realities of recording in a protest camp (the interviews recorded for the project were archived and made publicly accessible at the Bishopsgate Library in London). The resulting trail *Occupy LSX* (Figure 15.5) (Boga & Smith 2013) is a good example of how memoryscape can capture and interpret temporary and transient histories, even in the shadow of huge historical edifices like St Paul's Cathedral that command the attention of the casual visitor.

Sam reflected:

> Whilst trying the trail, the open space where these tents once were feels desolate as you hear the connected voices from another time with a different use and meaning for the space. The occupiers remain almost as ghosts, of a time maybe already largely forgotten by the public and general media (but which may come back to haunt them). The close physical and emotional proximity of the audio stories to the locations and the longer than usual length of time required to wait in each area, adds a lingering moment of contemplation. The user can reflect on a space no longer occupied physically in the same way, but still available in virtual form.
>
> *(Smith 2013)*

There has not been the space to include many other impressive student projects, but I hope the range of trails discussed will inspire other educators to experiment with the medium. Creating memoryscapes is a uniquely challenging and exciting means of engaging students with place-based issues, using memories and oral histories in the public sphere and creating engaging and stimulating experiences. Over the years feedback suggests that students particularly enjoyed the variety of the sessions, the creative practice and the opportunity to develop practical multi-media skills and create something public facing. From the lecturer's perspective it has been immensely rewarding and at times exhilarating to teach, both in sharing the creative adventure with students and in giving me the opportunity to work with colleagues from different disciplines to create a supportive and stimulating workshop environment.

Figure 15.5 Occupy LSX Audio Trail web page.
Source: Boga & Smith (2013).

Notes

1 In this time, undergraduate course fee levels have risen from £3,000 a year to £9,250, some of the highest in the world, which means that UK students will graduate with an average debt of more than £50,000. As a result, the concept of higher education as worthy pursuit in and of itself has been overshadowed by enormous pressure to make undergraduate degree courses more vocational and utilitarian.
2 By professional, I mean someone who is usually paid for specific skills to successfully deliver a project outcome. I do not wish to imply that project volunteers are not professional or necessarily lack technical skills.
3 In retrospect it is a shame that the young people who participated weren't listed individually in the trail notes.
4 A research centre, originally called the East London History Centre, first established at the University of East London by Samuel. The centre was renamed in his honour after he died in 1996. By 2009 the centre had become a partnership between UEL, Birkbeck College, and the Bishopsgate Institute.
5 Buses have been utilised in exciting memoryscape participatory group-based work mapping and presenting refugee memories in Montreal (Miller 2011: 51–88).
6 Unusually this was devised as a joint programme between the University of East London and Birkbeck, and students from both institutions attended programme modules. UEL offered courses in heritage, memory studies, migration history, visual culture and heritage, and digital heritage; Birkbeck had a recently established MA Museum Cultures course taught from their Department of History of Art which offered courses in museum studies. The programme would not have been possible great deal of generosity and goodwill from colleagues at UEL, Birkbeck, Bishopsgate Library, and the London Metropolitan Archives.
7 Artist and theatre practitioner Luis Sotelo Castro gave a workshop on walking trails and collective memory; Andy Minnion and Ajay Choksi from the Rix Centre gave sessions on inclusive heritage and multi-media practice, and tour guide course director Caroline Dunmore gave a workshop on creating place-based narratives.

References

Asta Centre and Thomas, J. (2008). *The Astra Trail, Silverton*. Accessed 21 December 2018, www.portsofcall.org.uk/astatrail.html

Bigwood, J. (2011a). *47 bus*. Accessed 21 December 2018, https://47bus.wordpress.com/

Boga, C. (2013). *A reflective essay on Occupy LSX, an audio tour of the Occupy Camp at St Paul's Cathedral* (Unpublished essay).

Boga, C., and Smith, S. (2013). *Occupy LSX*. Accessed 21 December 2018, http://cindyboga.wixsite.com/occupy-london-lsx

Chomsky, N. (2012). *Occupy*. London: Penguin.

Davies, L. (2011a). *Finsbury work and play: An audio-walk around Finsbury and Shoreditch* (Unpublished essay).

Davies, L. (2011b). *South Acton hidden histories*. Accessed 21 December 2018, http://goo.gl/maps/B9ab

Eadie, B. (2012). *A bowl around Bedlam with the brainless, brazen brothers*. Accessed 21 December 2018, www.flanr.net/essay/

Kanom, H. (2013). *Wander east through east*. Accessed 21 December 2018, http://halimakhanom41.wixsite.com/wandereast

Mees, S. (2016). *Channel heritage way*. Accessed 21 December 2018, https://wiki.rixwiki.org/Default/home/channel-heritage-way-clone/

Miller, L. (2011). Going places: Creating a memoryscape of Montreal. In: L. Miller, M. Luchs and G. Dyer Jalea, eds. *Mapping memories: Participatory media, place-based stories and refugee youth*. Quebec: Marquis.

Mitchison, L. (2013a). *Brixton munch*. Accessed 21 December 2018, http://brixtonmunch.wixsite.com/brixtonmunch

Mitchison, L. (2013b). *An audio tour of food stories in Brixton market: Reflective essay*. Accessed 21 December 2018, http://brixtonmunch.wixsite.com/brixtonmunch/essay

Platform. (2007). *And while London burns*. Accessed 21 December 2018, www.andwhilelondonburns.com/

Smith, S. (2013). *A reflective essay on the creation and testing of the Occupy LSX audio trail: Virtually re-occupying London Stock exchange* (Unpublished essay).

Thomas, J. (2008). *The West Silvertown Trail*. Accessed 21 December 2018, www.portsofcall.org.uk/westsilvertown.html

Turner, R. (2013). *A Daleman's journey*. Accessed 21 December 2018, www.rixwiki.org [login with username: Bill-mitchell and password: Bill].

University of East London. (2008). *Ports of call*. Accessed 21 December 2018, www.portsofcall.org.uk

Vilbr, R. (2013). *The way home: A walk around Hackney's housing history*. Accessed 21 December 2018, https://hackneyhousingtour.wordpress.com/about/

PART IV

Industry

Steven High and Hilary Orange

Introduction

It is no coincidence that four of the six contributors to this section of the *Handbook* focus on abandoned deindustrialised landscapes in what was formerly known as the 'industrialised world.' That nobody was drawn to still-active industrial sites speaks volumes about how 'industry' is viewed today in much of Europe, North America, and Australia. For many people, industrialism belongs to the past, not the present – even though the world as a whole has not deindustrialised. Everything around us is made somewhere – just not where it used to be. Jackie Clarke has written about the new forms of 'working-class invisibility' in a country like France, reminding us that it is not a matter of outright disappearance, as there are still hundreds of thousands of industrial workers in that country, but rather one of occlusion and cultural displacement (2011: 446). Or, to borrow a phrase from anthropologist Kathryn Marie Dudley, those who 'once stood at the center of things now seem out of place' (1994: 161).

To lament these changes is to risk being charged with the crime of 'smokestack nostalgia' (Cowie & Heathcott 2003). But nostalgia need not be uncritical, it can challenge notions of progress and serve as a 'defense mechanism' against 'historical upheavals' (Boym 2001: xiv). As a historic emotion, nostalgia is a 'longing for a lost time and a lost home,' but there is no going back (Boym: xiii). This is the source of the heartache, and the political possibility. Several of the contributors suggest that the class politics of place represents 'radical spatial remembering' (Pleasant and Strangleman, Chapter 17) and 'memory as resistance' (MacKinnon, Chapter 16). Memoryscapes are political because they 'incite dissent,' insist Pleasant and Strangleman. Collective efforts to remember can counter cultural erasure, or even forced forgetting.

The contributors respond to post-colonial scholar Ann Laura Stoler's appeal to shift our gaze from highly aestheticised ruins of industry. To ruin, Stoler adds, is a 'violent verb.' It is 'an act perpetrated, a condition to which one is subject and a cause of loss' (2013: 9) In so doing, the contributors consider industry through various environmental, political, and cultural lenses. Here, the lived experience of work and encounters with places at work reveals much of their personal significance, their change over time including loss. However, this is not to deny that aesthetics can play a critical role in studies of industrial 'relations.' One of the ways in which people related to industrial sites within their neighbourhoods was through their particular aesthetic, and, even when factories represented 'forbidden cities,' their image still loomed large in place identity and collective memory. Indeed, certain views of sites are often noted within tales of homecomings and childhood remembrances, as well as operating as important city landmarks (Conlon, Chapter 18).

In the first chapter, Lachlan MacKinnon offers us a transatlantic perspective on deindustralisation's 'half-life' (Linkon 2018) grounding his analysis in distinct urban and rural post-industrial realities. Places are hollowed out as industry leaves, but what happens next is important. In some cities, abandoned spaces are quickly recontextualised as something new – filling the economic and cultural vacuum. Industrial buildings are torn down, the land then decontaminated and redeveloped. Or the factories are gutted and redeveloped as a condominium or arts space. To be sure, gentrification and culture-led 'revitalisation' often bring with them hidden hardships for working-class people who face residential displacement years after experiencing economic displacement. It usually plays out differently in rural areas and small towns where nothing immediately fills the economic or cultural vacuum. On this point, MacKinnon speaks to how the language of 'remediation' and 'reclamation' serve to green older brownfield narratives of industrial work, obscuring working-class histories in the process.

At the outset of the next chapter, sociologists Emma Pleasant and Tim Strangleman pose a key question: 'what is the relationship between memory and the world of work?' They note that the affective bonds built up at work over time go largely unnoticed until people suddenly find themselves forcibly disembedded, at which point the sense of loss becomes intensely painful. To understand the ways that loss can be very much present, Pleasant and Strangleman bring us along with them into the interview conversations with English shipyard and brewery workers. We hear from people who grapple with its significance. Asked if they returned to their old worksites to see them demolished, former workers did not respond uniformly: some were drawn to the sight of ruination, returning often, while others were repelled by the very idea. But we begin to see it as they do, thanks to these interviews.

The monumentality of lost industry is then explored in Martin Conlon's chapter on the destruction of the shipyard cranes in and around Glasgow, Scotland. For decades, these giant cranes were landmarks in the historic skyline of the area, not unlike other tall industrial structures such as blast furnaces, water towers, mine-frames, and smokestacks in other towns and cities. Over time, these structures become highly emotive places, and their loss is keenly felt. Conlon is particularly interested in the materiality of the shipyard cranes as well as their symbolic importance to area residence. He explores their significance in poetry and song as well as in the workplace recollections of those who once worked high above the city. The counterpoint between these different 'vehicles of memory' are not as great as one might think. Deindustrialisation is as much a cultural and political process as it is an economic process.

Next, we turn away from abandoned or demolished industrial structures, and consider the value placed on industrial heritage. Earlier Pleasant and Strangleman noted how official industrial heritage sites sanitise these physical spaces, clearing them of their ghosts and cobwebs. Lucy Taksa has studied one such place for much of her career. For a century, the Eveleigh Railway Workshops in Sydney, Australia, was one of that country's main sites of industry. Its closure in the 1980s led to its preservation. But what is being preserved precisely? In her previous work, Taksa has raised tough questions about the limitations of the preservation impulse. Buildings and machinery may be preserved, and their functioning interpreted, but not the shop-floor realities or the history of working-class struggle: public and labour history are 'like the two faces of Janus – the Roman deity of gates and doorways, of beginnings and endings – whereby one looks forward and the other looks backwards at the same time' (Taksa 2009, 89). In her contribution to this volume, Taksa examines the ways that masculinity and femininity are inscribed in the site interpretation plans for the site. In doing so, she finds that women's history of work is made visible in the Second World War Munitions Annex. However, the history of masculinity remained 'hidden in plain sight' as heritage interpretation opted for gender neutrality otherwise. It is thus a 'paradoxical space.'

The final two chapters in this section come at these issues from different vantage points. Each author poses an intriguing question, then spends the rest of the chapter answering it. If the earlier chapters explore the meaning of lost industry, Jeff Benjamin asks why the Delaware River never industrialised at all? Absence, not loss, is thus the focus of his attention. This contemplative chapter thus serves as a counterpoint to Theodore Steinberg's influential book, *Nature Incorporated* (1994), which examined the industrialisation of a river in New England. An industrial archaeologist and artist, Benjamin is interested in the materiality of the landscape and his own relationship to this place is central to his analysis. In many ways, the chapter is an extended meditation, written as if the author was writing these words from river's edge – which indeed he might have been.

The final chapter in this section, written by Samuel Merrill, brings us to the industry of transportation. Merrill's work centres on the spatial, political, and technological qualities of underground space and in this chapter he addresses transport systems in London and Berlin. Encouraging us to think vertically and volumetrically about memory in relation to space, he asks, what can we learn from these cities' sub-surface? In one example, we learn about the activist memory work around a far-right killing in a U-Bahn station in Berlin. In the other, we hear about the heritagisation of an abandoned Underground station in London. The chapter encourages us to look in another direction for new insights. It is a three-dimensional chapter that, in some ways, complements the flow evoked in Benjamin's contribution. While Benjamin sees the absence of industrialisation as a virtue, Merrill points out that population growth will stimulate 'the construction of an ever-growing number of new spaces beneath the city's streets,' suggesting that subterranean memory work will play a larger role in disciplinary futures.

Together, the chapters in this section demonstrate that there are manifold ways to approach industrial memoryscapes. Oral history provides several of the contributors with ground-floor access to hidden or buried histories. As one former dockyard worker told Emma Pleasant after he saw his old workplace stripped and people outside gawking: 'they're not seeing what I saw.' They don't see the ghosts he did, how could they? Elsewhere, Conlon reminds us of the value of visual and sonic mnemonics, while Merrill suggests we look beneath our feet. 'Memory itself is a *riparian* activity,' Benjamin writes, and he finds his answers in the land and the water as well as in books and his own thoughts. Place is approached as an archive, poked and prodded for what it can tell us about the history of this place. At one point, almost as an afterthought, Benjamin observes that absence was a virtue in this case as industrialism has been a catastrophe for humanity. We doubt that those interviewed by the other contributors would have agreed, even though industrial work may have left them injured or sick, and ultimately out of a job. Accordingly, there is a productive tension between these chapters.

References

Boym, S. (2001). *The future of nostalgia*. New York: Basic Books.
Clarke, J. (2011). Closing Moulinex: Thoughts on the visibility and invisibility of industrial labour in contemporary France. *Modern & Contemporary France*, 19 (4): 443–458.
Cowie, J., and Heathcott, J. eds. (2003). *Beyond the ruins: The meanings of deindustrialization*. Ithaca, NY: ILR Press.
Dudley, K.M. (1994). *The end of the line: Lost jobs, new lives in postindustrial America*. Chicago: University of Chicago Press.
Linkon, S. (2018). *The half-life of deindustrialization: Working-class writing about economic restructuring*. Ann Arbor, MI: University of Michigan Press.
Steinberg, T. (1994). *Nature incorporated: Industrialization and the waters of New England*. Amherst: University of Massachusetts Press.
Stoler, A.L. ed. (2013). *Imperial debris: On ruins and ruination*. Durham, NC: Duke University Press.
Taksa, L. (2009). Labor history and public history in Australia: Allies or uneasy bedfellows? *International Labor and Working-Class History*, 76: 82–104.

16
POST-INDUSTRIAL MEMORYSCAPES

Combatting working-class erasure in North America and Europe

Lachlan MacKinnon

Introduction

The intersectional relationships between landscape, place identity, historical memory, and individual histories have informed much scholarship in recent decades. These concepts are layered; spaces are composed of more than physical landscapes or boundaries that can be visualised or represented on a map. They are, in essence, 'relationally constructed' (Massey 2005: 9). This is as true of small, personalised spaces as it is of streetscapes, neighbourhoods, or large cities – and *memory* is foundational in maintaining these conceptual constructions. Our place-memories correspond with sets of relationships, each of which give meaning to the landscapes with which we are most familiar.

Martha Norkunas provides an excellent anecdote to explain this concept in her ruminations on working-class place identity in Lowell, Massachusetts. She describes her return to the home of her deceased mother only to have experienced a profoundly alien sensation. Her connection to that place, mediated as it was through her memory of her mother and their relationship, was disrupted by her mother's palpable absence (2002: 11). This sentiment, one of disruption and of being out-of-place, is one that speaks to many of us who have had the emotional experience of re-visiting an important landscape from our personal past. While this example is illustrative, if we re-direct our attention towards the places that we experience and construct collectively – neighbourhoods or workplaces – the socio-political aspects of memory and place identity are made clear.

In this chapter, the class dimensions of place identity are examined in the context of deindustrialised and regenerated work sites and communities in Canada, the United States, and the United Kingdom. Drawing upon several case studies, the transition of these spaces from places of work to working-class *memoryscapes* is compelling. Former industrial workers and displaced working-class community members perceive changes in the physical landscape as a corollary to a broader sense of powerlessness that emerges alongside deindustrialisation. Environmentalists or urban developers, often directly implicated in the remediation process, hold divergent and often conflicting perspectives. This focus corresponds with recent scholarship on urban change and gentrification, working-class erasure and displacement, and memory-as-resistance. It reveals the overarching power dynamics that inform class-based relationships through place, space, and environment and reflects the processes by which 'some things are more 'absorbed' . . . or incorporated into the place than others' (Massey 1995: 186).

Ruination accompanies the displacement of working-class men and women that has occurred in many former single-industry towns. In such landscapes the remnants of industry might exist, *ruined*, as visible markers of past relationships that have been torn away by the vicissitudes of global capitalism. Images of these places are familiar; abandoned cathedrals of industry, rusted machinery, and the empty shop floor are all hallmarks of the middle-class *ruin porn* aesthetic. These places, aestheticised or commodified as they may be, are produced by the rank materialism of capital mobility and the never-ending corporate search for higher rates of profit, lower rates of unionisation, and greater efficiencies (Mah 2012: 10). The politics surrounding these sites have been debated extensively (High 2007, 2013; Edensor 2005; DeSilvey & Edensor 2013), but it remains important to recognise them as temporal representations that retain great meaning among those who have been displaced.

The strength of these connections is made clear in Steven High's work on Sturgeon Falls, Ontario. High traces mill workers' experiences of the decline and closure of the town's major employer – a paper mill (High 2010, 2017, 2018). Ruination in this rural resource frontier became starkly visible on 2 December 2002, when Weyerhauser closed the mill after its two-month notice to employees ended. For its displaced workers, this announcement represented an immediate threat to their sense of place – one that did not go unchallenged. As High describes:

> Most of those interviewed, however, wanted nothing more than to return to the closed mill. Eventually, these hopes faded as union efforts to reopen the mill petered out, as did the daily protests at the front gate. Weyerhauser's decision to quickly demolish the mill confirmed the new reality. Until then, Randy Restoule [a former worker] still held out hope of its reopening.
>
> *(2017: 274)*

The demolition of on-site structures is one of the final and most insurmountable hurdles faced by displaced workers in retaining place identity. For workers in basic manufacturing, among many of whom constant fears of downsising and closure have taken hold since the 1970s, the physical remains of the mill are imbued with some sense of hope for the return of industrial employment. In Sturgeon Falls, employees created a 'mill history binder' – a portable record of the history of their workplace – that could continue to be used as a memory aid even after the destruction of the mill's physical structures.

Cultural practices like the mill history binder reveal how displaced peoples attempt to connect themselves, through memory, to the landscapes of the past. Popular interest in how such processes unfold has only increased in the years and decades after industrial crisis in the 1970s and 1980s. Indeed, as High et al. discuss, many of us who work in this field have either experienced industrial collapse first-hand or came of age in the shadow of defunct industry (2017).

Meanwhile, changes in post-industrial landscapes have provoked questions not only about ruination, but also about multiple forms of erasure (Clarke 2011). Changes in the present-day landscape are often accompanied by a set of explanations that rely upon popular – if circumscribed – ideas about the market, business cycle, environment, and economic or social development. Too often, the popularisation of these narratives serves to write-off working-class concerns about the end result of deindustrialisation as misguided or backward looking. How can positive memories of the industrial past co-exist, for instance, with development language that re-frames deindustrialised landscapes in terms of 'remediation' or 'reclamation?' And what about the men and women who retain connections to the un-reconstructed memory-landscapes of their own class experiences? In deindustrialised places, the memories of the displaced raise uncomfortable questions about the inevitability of industrial closure and the changing class boundaries of the present day. Working-class memories and bodies are not as neatly 'remediated' as their neighbourhoods or workplaces, after all.

Geography matters. The experiences of displaced workers in Pittsburgh, Montreal, or London differ from those in single-industry, rural resource frontiers – though they respond to the same general stimuli.

In North America, post-industrial urban transformation has been explored as 'an act perpetrated' (Stoler 2008) – in the cases of Hamilton, Ontario, and Pittsburgh, Pennsylvania, as the direct result of policies pursued by middle-class growth coalitions (Neumann 2016). Perhaps nowhere are the tensions between memory and the post-industrial city more visible than in Detroit, Michigan. Detroit has been simultaneously framed as the emblematic example of industrial collapse and – increasingly – as a space for middle-class regeneration or re-development. Aesthetic representations of *ruined* Detroit have certainly been, in the words of Dora Apel, positively post-apocalyptic, but the city has also been positioned as a supposed success story of neoliberal governance and middle-class re-development (2015: 132).

Challenging the development ethos: urban memoryscapes in the United States and Canada

In 2015, the Detroit Historical Museum began transcribing and conducting interviews as part of the Detroit 1967 Oral and Written History Project. The intention of this project was to collect the memories of Metro Detroiters, unpack changing perceptions over the past half-century, and reflect upon how these memories relate to the infamous 1967 Detroit riot. The riot erupted in late July, following several arrests made by police during a raid of an unlicensed drinking establishment. Racial enmity within the city erupted as African American residents protested over five days, which included a series of confrontations with the Detroit Police Department, the National Guard, and army paratroopers. There were 44 recorded deaths, millions of dollars in property damage, and hundreds of injuries as the result, with the majority of these affecting black residents of the city.

While many of the project's interviews include recollections of 1967, several also reflect explicitly upon the changing borders of class, racial, and ethnic geographies in Detroit alongside deindustrialisation and economic collapse. Lucille Schaffer, a daughter of German immigrants who was interviewed as part of this project, describes growing up in low-income, working-class neighbourhoods of Detroit. She remembers a sense of shared class identity, but also a city that was deeply divided along racial lines. As she recalls:

> I saw tensions all through my life, because I lived in Detroit.... It was like Miller and Mt. Elliot, by the Hamtramck Assembly, in that area. There, too, it was mostly segregated where, I would say the blacks were not treated as, well, you know. When you're low income you're not treated well anyway, so it was kind of like we were being treated the same, but the blacks still weren't treated fairly, I didn't think. The police would be a little more rougher, they'd be rough with them, kind of thing.
>
> *(Schaffer 2016)*

Her memories of the industrial city are conflicted. While basic manufacturing and the blue-collar middle class that it sustained helped to produce a sense of community, these circumstances were also rooted in deeply entrenched racial boundaries that were constantly policed not only by the state, but also by white, working-class 'neighbourhood organisations.'

Despite the divisions, Lucille's recollections imbue the city's past with an air of nostalgia not afforded to the deindustrialised present. She reflects:

> I love this place. If anyone talks about it, right away I am standing up for it. How it's changed: because of our economy and all the things that went on in the city, everything is vacant. It's a shame. It's dirty, broken, that type of thing, but I really feel like it's starting to come back.
>
> *(Schaffer 2016)*

Joyce Ross expressed similar concerns in 2017. For Joyce, the city has changed significantly since the 1960s and 1970s – but emerging changes may signal something positive on the horizon. She states, 'I think we're on the right track. I just hope that with the [Trump] administration in Washington, I hope it doesn't set us back.... If I were younger, I would move down here.' (Ross 2017).

The complex themes that inform these and other memories of the twentieth-century city – race and class chief among them – are central to dozens of interviews from the Detroit 1967 project. White and black residents of the city, spanning from working-class autoworkers and labour organisers to police officers and white-collar workers, recall an urban locale predicated upon the economic benefits of industrial production. While these memories are oriented towards a particular historical moment – the 1967 rebellion – they reveal a city of contradictions. Racial animus exacerbated by the great wave of black migration from the American South existed within an historical moment where the blue-collar middle class remained economically ascendant. While Detroit had lost thousands of manufacturing jobs prior to 1967, the years afterwards witnessed a surge in deindustrialisation, white flight to the suburbs, and periodic foreclosure crises. Each of these proved especially devastating to the city's black residents (Sugrue 2005: 5). As a result of these multiple and protracted forms of disinvestment, the city's population has declined from approximately two million in 1950 to just below 700,000 residents in 2017 (Herstad 2017: 85).

Urban developers have been quick to reorient working-class landscapes towards entirely new configurations of capital and social relationships as the result of the disintegration of the city's industrial base. Billionaires like Quickens Loan founder Dan Gilbert have been snapping up real estate in a small, waterfront-adjacent part of the downtown (Peck & Whiteside 2016: 236). While such acquisitions have opened spaces for small-business incubation and rentier capital, they have also deepened forms of geographical and class-based exclusion. Moreover, they have hardened the dividing lines between those within the class and racial boundaries of such spaces and those who remain on the outside. This occurs not only in Detroit, but also in deindustrialised urban centres throughout the United States (Slater 2009; Zukin et al. 2009: 47; Walley 2013).

The gentrification of downtown Detroit and the concurrent emptying out of working-class 'blighted' neighbourhoods provokes a sort of cognitive dissonance between the post-industrial landscape and the relational memories of current and former residents. While gentrification and urban regeneration is predicated upon imagining such spaces as blank slates, displaced workers, racialised peoples, and other marginalised groups have historically acted in resistance to this tendency. As Stefan Berger and Christian Wicke argue in a recent special edition of *The Public Historian*, heritage and memory activism have frequently been marshalled to assert the continued relevancy of these forms of identity in the face of peripheralisation (Berger & Wicke 2017: 10).

These processes, too, are often deeply contested. Companies have attempted to 'reclaim' Detroit's former working-class identity through various schemes – including one that re-uses material from demolished structures to produce $200 sunglasses for sale to the city's nascent urban elite. These approaches are fundamentally flawed in that they do not engage with either the trauma of deindustrialisation or the enduring forms of racial exclusion that continue to shape Detroit in its modern renaissance. On the other hand, as Kaeleigh Herstad notes, projects like Detroit '67 help to commemorate and foster discussion about Detroit's painful decline and history of racial exclusion – connecting past to present through oral history and ethnography. Such projects allow for residents of deindustrialised urban spaces to work through their contested pasts without appealing to the language of regeneration, which too-often obscures lived experiences of urban change and displacement (2017: 113). They provide a radical memoryscape of the city using the voices of working-class residents and challenge the pervasive idea that any positive recollection of the working-class city must, by definition, be nostalgic.

In Canada, where we have sometimes been content to see ourselves as an international resource hinterland in our own right, we have resisted the impulse to explore urban transformation in our largest cities as a function of deindustrialisation. This, as has been noted elsewhere, is perhaps because financialisation and the transition to service and knowledge industries obscures the disparity that erupts in the aftermath of workplace closures (High et al. 2017). In other words, post-industrialism within urban centres also relies upon the physical displacement of working-class men and women – not only from their workplaces, but also from their communities.

As we see in Griffintown, the Irish-Canadian working-class neighbourhood that long existed adjacent to Montreal's downtown core, there are similar incongruities. For urban developers, Griffintown was a poverty-ridden slum that was best 'regenerated' into a bedroom community for the downtown core. For its working-class residents, it was home. As condo-isation and urban growth encroached upon the area, residents launched a concerted effort to underpin their sense of place through oral history storytelling and various other *lieux de mémoires* (Barlow 2017: 16; Nora 1989). Similar forms of contestation continue to unfold in other threatened working-class areas of Montreal. These efforts rely upon the vibrant, living memory of residents and demand a careful re-calculation of the impact of gentrification as a function of deindustrialisation (Vickers 2013; Chatterjee & High 2017).

The tensions that arise from large-scale real estate investments or other efforts to reimagine working-class spaces as '*postindustrial*' occupy a significant place within the literature on deindustrialised urban landscapes, but these are not the only vectors of displacement. Commercial and retail spaces can also become sites of class struggle over the right of residents to stay put in the face of smallholder capital. In their study of Mount Dennis in Toronto, a former manufacturing hub and working-class neighbourhood, Katharine Rankin and Heather McLean examine how attempts to court incoming affluent residents may be more palatable to the liberal middle class than high-rise luxury apartments, but are no less damaging in the long term for the working-class residents that they come to displace. They write that such efforts constitute 'a new urbanist resistance favouring "green" and "creative" economic development' (2015: 217). These processes bring cultural pressures to bear on existing residents even before economic pressures begin to mount. As Starbucks and chain fitness studios begin to replace diners and local taverns, blue-collar residents feel their sense of place start to slip even prior to the first inklings of a rent hike or increased police presence (Lehrer & Wieditz 2009: 140; Harvey 2012).

Local memory in these areas consistently reveals one significant reality: deindustrialisation and class displacement are not natural processes. Nor are they neutral. They unfold with purpose, corresponding to both market forces and state policy. That workplace and neighbourhood landscapes are vessels for occupational and collective identity is clear, and the creation of memoryscapes is but one of the ways that these have been rigorously defended from within. As historians, our role in the production of such sites is political; this is inescapable. In foregrounding the memory of life before disruption, we necessarily question and challenge the structures that have emerged in the aftermath. This is especially important in urban areas, where skyscrapers and condos fit – too frequently – within uncritical, exclusionary assessments of the changing landscape.

Resistance in the aftermath of closure: rural deindustrialisation in Scotland and Canada

Unlike in Montreal or Detroit, where displacement is obscured by the citywide turn to other economic sectors, rural areas faced with deindustrialisation find themselves continuously defined in terms of the defunct industries that came before. This creates challenges for working-class public history initiatives or memoryscapes. While gentrification in urban locales pushes working-class residents to other neighbourhoods or suburbs, the absence of new orientations of capital in rural areas mean that residents are only infrequently directly displaced. Rather, the shape and sense of community changes around its residents. Some men and women choose to leave for economic opportunities elsewhere, which often requires moving away from an area entirely instead of simply relocating nearby. This results in a significant number of remaining residents who continue to recall the benefits of living in a community based around industrial production, the jobs it provided, and the relationships that it sustained (MacKinnon 2016; Donatelli et al. 2017). Such memories are re-produced in museums, theatre performances, documentary films, and monuments.

There are three general tendencies that must be navigated in the creation of working-class memoryscapes in such places. The first is the tendency towards *nostophobia*; or, representations of the industrial past that are uncritical or wholly celebratory, that distance the past from the present, or that ignore conflicting memories (Smith & Campbell 2011: 88). The second is the tendency to adopt development language that highlights the 'successes' of post-industrial initiatives, however constrained or ineffectual, in contrast to the dangerous, dirty work of the past. Finally, the recognition that industrial production has resulted in a host of negative bodily and environmental aftermaths is incredibly important and remains necessary in both scholarship and public history. But this also risks producing heritage efforts that focus wholly on such outcomes and read them backwards into history, colouring past experiences and imbuing narratives with a sense of impending, inevitable doom.

In the United Kingdom, Scotland has been the staging ground for a number of studies on rural deindustrialisation (Perchard 2012; Phillips 2012). Ewan Gibbs and Andy Clark devastatingly critique what they have termed the 'New Scotland thesis': that the decline of 'dinosaur' heavy industries has prompted the rise of a 'brain-intensive economy' based upon service and knowledge that is now poised to launch rural Scotland to hitherto unknown heights of prosperity (2017: 6). Gibbs and Clark are quick to note that such proposed benefits are not to be accrued by deindustrialised locales themselves, but by nearby urban centres to which rural residents will apparently be content to commute for work. At their most insidious, such narratives hold up those who remain outside of the new, 'brain-intensive' economy as being worthy of ridicule for their out-of-place working-class habits and cultures (Jones 2011).

The same holds true in the Canadian province of Nova Scotia, where workers on the rural island of Cape Breton produced both coal and steel before these industries were shuttered in 2000. The poverty and depopulation that have since plagued Cape Breton are not mentioned in a 2014 report to the provincial government, which instead praises the growth of the province's urban centre, Halifax. 'Economies grow around cities,' the report assures readers:

> Today Halifax [Nova Scotia] is emerging as the largest and most influential urban centre in the Atlantic region with new investment flowing into dynamic sectors such as aerospace and defense, ICT and financial services. If wisely planned and managed, this trend will generate positive linkages and spread effects for other areas of the province.
>
> *(Now or Never 2014: 27)*

The message is for residents of deindustrialised Cape Breton Island to tighten their belts in the face of austerity in the hope that some of the newly generated wealth of distant cities will make its way back in one form or another.

Memoryscapes help to challenge these exclusionary narratives by drawing upon and presenting the experiences of those who have been directly affected. The Scottish Oral History Centre and the Paisley People's Archive, for example, collaborated in the development of a Paisley Thread Mills Memoryscape. The hour-long documentary-style video, available online and through each partner organisation, draws upon oral history testimony of former employees, family members, and other residents to illustrate the deep sense of community that emerged from work at the mill. This is not a nostalgic representation; indeed, the difficulties of working-class life, strict regiments of gendered behaviour and sexuality, and the necessity of labour organising are all foregrounded. Nor does it conform to the New Scotland thesis; rather, the Paisley Memoryscape belies the idea that the movement away from blue-collar work can be framed entirely in progressive terms. As one female respondent recalls:

> I'm glad that you're, um, havin' this archives because . . . maybe through the years to come people will go back and think what it was like. And it was a good life if you could go and work . . . go home at days end with the money that you needed. That's – that's what my job was for . . . and

I've got good memories, because I always had money. And it was nae a lot of money, it was just ... but I've, I've got good memories.

(Paisley Peoples' Archive 2018)

Memoryscapes are a form of resistance. They complicate narratives that assign very particular, uncomplicated roles to working-class men and women who have been displaced or marginalised by deindustrialisation. They are not simply a response to a shift in economy, but to a threatened disconnect between place and identity. As Pierre Nora describes, 'Memory is a perpetually actual phenomenon, a bond tying us to the eternal present' (1989: 8). Such bonds have taken many forms, from local collections of documents and records to public history commemorations, audio walks, and digital projects (Dabakis 1999: 35–36; Heron 2000; Bradley 2016). Whether on the site of former workplaces or in working-class neighbourhoods under threat of gentrification or depopulation, these efforts frequently harken back to the sensory landscapes of the industrial past through the spoken word, ambient noise, or background textures (Butler 2007).

For labour historians, the reliance upon public history and memoryscapes as a form of resistance is perhaps jarring. Indeed, the very argument presumes that existing methods of combatting deindustrialisation have fallen short. They are frequently, after all, ways of maintaining a connection between place and identity in an environment where industrial production has already ceased. Organised labour has responded in a variety of ways, from the Crafted with Pride campaign of the 1980s to the UNITE effort against sweatshop labour in the 1990s, but none have achieved the type of sustained, systemic success that would be required to truly challenge deindustrialisation (Minchin 2012: 2; Johns & Veral 2000: 1193). While debates continue, we should not downplay the role of heritage and memory in maintaining a sense of power for affected communities.

Memory and environment in post-industrial places

Deindustrialisation has created contested spaces throughout urban and rural areas around the world. Whether at former workplaces or in working-class neighbourhoods facing erasure, residents of deindustrialised spaces find themselves at the forefront of a struggle for power over the shape of their future. Memory becomes a mechanism for resistance in urban and rural areas alike and helps to challenge essentialist notions about the positive nature of industrial decline. Such notions emerge – as we have seen – from within prevailing neoliberal conceptions of economic development, but so, too, can they arise out of an increasing awareness of ecology and the environment. In both urban and rural deindustrialised areas, eco-friendly repurposing of industrial spaces have been discursively positioned in opposition to the concerns of existing working-class communities. On such occasions, it is worthwhile to consider rhetoric about greening or cleaning up such spaces as a corollary to the gentrification process (Dooling 2009).

Rural environmental remediation efforts have frequently been necessary. In Anaconda, Montana, for instance, the operations of the Washoe Smelter produced pollutants that affected the health of workers and residents and damaged the local environment. (Bryson & Wyckoff 2010: 66) Love Canal, New York, and the Sydney Tar Ponds in Nova Scotia are two other examples; in each, working-class communities came together, in the face of deindustrialisation, to challenge both their economic dispossession and fight for appropriate forms of environmental remediation (Newman 2016). Challenges emerge, however, when the language of remediation is used to underscore and support deindustrialisation and class displacement. As Bryson and Wyckoff describe, a proposed tramway to the top of Mt. Haggin in Montana was halted after urban planners and city leaders cautioned that 'the tramway's spectacular views might alarm some visitors 'critical of such things as logging, mining, smelting ... that may not be pleasant to some tender eyes ... who think that the natural beauty is spoiled' (67).

The class dimensions of 'greening' rhetoric are even more visible in urban or inner-city working-class neighbourhoods. In Brightmoor – a post-industrial area of Detroit – Theodore Pride describes how incoming middle-class residents sought to proselytise green sustainability in the face of perceived industrial blight. He writes:

> These new eco-gentrifiers heighten the threat that urban agriculture becomes another vehicle for consumers to virtuously display their knowledge and adoption of the latest values while also perpetuating social distinction. . . . This creates a struggle over space and place between incoming higher class, green-minded consumers determined to own a 'farm in the inner city' and the original residents who either do not share their green/agriculturalist identity or cannot afford the green/agriculturalist lifestyle.
>
> *(2016: 38)*

In Brightmoor, the middle-class *perceptions* of the neighbourhood – if not the reality – was that it desperately required environmental remediation to become 'sustainable.' The same impulse is visible in other eco-gentrifying neighbourhoods, where the class relationships and cultures found within are rhetorically framed as component parts of the brownfield landscape that requires regeneration.

Conclusions

As public historians, the environmental dimension of deindustrialisation is difficult to represent. On its face, industrial work and the landscapes that it produces have clearly had a significant negative impact on both human health and the environment. In addition to the lives lost through industrial accidents, the environmental 'externalities' of centuries of industrial production has almost certainly cost millions of lives. This aspect of industrial work, the communities that it has sustained, and the contours of life in post-industrial areas must be included in our narratives – whichever form our interventions may take. But this requires a deft touch; after all, memories and meanings are not fixed in place. Environment and ecology are necessary for understanding working-class experiences of deindustrialisation, though we must be sure that concerns over these issues are not used to justify other forms of marginalisation or displacement.

Ultimately, deindustrialisation is a spatial process as much as an economic or cultural one. Recognising and representing how it has differentially impacted working-class residents in urban and rural areas is necessary work, and – as the memoryscapes described herein attest – scholars in Canada, the United States, and the United Kingdom are already taking up the challenge. Working-class communities have fought against their disruption through direct action, economic resistance, protests, strikes, and demands for political accountability. Cultural work, including commemoration, the development of memoryscapes, public history, community-based engagements, and public art are all, likewise, important methods of resisting these forms of displacement.

References

Apel, D. (2015). *Beautiful terrible ruins: Detroit and the anxiety of decline*. New Brunswick, NJ: Rutgers University Press.
Barlow, M. (2017). *Griffintown: Identity and memory in an Irish Diaspora neighbourhood*. Vancouver: UBC Press.
Berger, S., and Wicke, C. (2017). Introduction: Deindustrialisation, heritage, and representations of identity. *The Public Historian*, 39 (4): 10–20.
Butler, T. (2007). Memoryscape: How audio walks can deepen our sense of place by integrating art, oral history and cultural geography. *Geography Compass*, 1 (3): 360–372.
Bradley, S. (2016). *Archaeology of the voice: Exploring oral history, locative media, audio walks, and sound art as sitespecific displacement activities*. PhD thesis, University of Huddersfield.
Bryson, J., and Wyckoff, W. (2010). Rural gentrification and nature in the Old and New Wests. *Journal of Cultural Geography*, 27 (1): 53–75.

Chatterjee, P., and High, S. (2017). The deindustrialisation of our senses: Residual and dominant soundscapes in Montreal's Point Saint-Charles district. In: K. Holmes and H. Goodall, eds. *Telling environmental histories: Intersections of memory, narrative and environment*. Basingstoke: Palgrave Macmillan, pp. 179–212.

Clarke, J. (2011). Closing Moulinex: Thoughts on the visibility and invisibility of industrial labour in contemporary France. *Modern & contemporary France*, 19 (4): 443–458.

Dabakis, M. (1999). *Visualizing labour in American sculpture*. Cambridge: Cambridge University Press.

DeSilvey, C., and Edensor, T. (2013). Reckoning with ruins. *Progress in Human Geography*, 37 (4): 465–485.

Donatelli, C., Murray, C., Lionais, D., and Nicholson, M. (2017). Our practice has had to change because of this: Professional perceptions of long distance commuting in Atlantic Canada. *The Extractive Industries and Society*, 4 (3): 606–613.

Dooling, S. (2009). Ecological gentrification: A research agenda exploring justice in the city. *International Journal of Urban and Regional Research*, 33 (3): 621–639.

Edensor, T. (2005). *Industrial space, ruins, aesthetics, and materiality*. Oxford: Berg.

Gibbs, E., and Clark, A. (2017). Voices of social dislocation, lost work and economic restructuring: Narratives from marginalised localities in the 'New Scotland. *Memory Studies*, 1–21.

Harvey, D. (2012). *Rebel cities: From the right to the city to the urban revolution*. London: Verso.

Heron, C. (2000). The labour historian and public history. *Labour/Le Travail*, 45: 171–197.

Herstad, K. (2017). Reclaiming Detroit: Demolition and deconstruction in the Motor City. *The Public Historian*, 39 (4): 85–113.

High, S. (2007). *Corporate wasteland: The landscape and memory of deindustrialisation*, Ithaca, NY: Cornell University Press.

High, S. (2010). Placing the displaced worker: Narrating place in deindustrializing Sturgeon Falls, Ontario. In: J. Opp and J. Walsh, eds. *Placing memory and remembering lace in Canada*. Vancouver: University of British Columbia Press, pp. 159–186.

High, S. (2013). The wounds of class: A historiographical reflection on the study of deindustrialisation, 1973–2013. *History Compass*, 11 (11): 994–1007.

High, S. (2017). Deindustrialisation on the industrial frontier: The rise and fall of mill colonialism in Northern Ontario. In: S. High, L. MacKinnon and A. Perchard, eds. *The deindustrialised world: Confronting ruination in postindustrial places*. Vancouver: University of British Columbia Press, pp. 257–283.

High, S. (2018). *One job town: Work, memory and betrayal in Northern Ontario*. Toronto: University of Toronto Press.

High, S., MacKinnon, L., and Perchard, A. eds. (2017). *The deindustrialised world: Confronting ruination in postindustrial places*. Vancouver: University of British Columbia Press.

Johns, R., and Veral, L. (2000). Class geography and the consumerist turn: UNITE and the Stop Sweatshops Campaign. *Environment and Planning A*, 32 (7): 1193–1213.

Jones, O. (2011). *Chavs: The demonization of the working class*. London: Verso.

Lehrer, U., and Wieditz, T. (2009). Gentrification and the loss of employment lands: Toronto's Studio District. *Critical Planning*, 16: 138–160.

MacKinnon, L. (2016). *Deindustrialisation on the periphery: An oral history of Sydney steel, 1945–2001*. PhD thesis, Concordia University, Montreal.

Mah, A. (2012). *Industrial ruination, community, and place: Landscapes and legacies of urban decline*. Toronto: University of Toronto Press.

Massey, D. (1995). Places and their pasts. *History Workshop Journal*, 39: 182–192.

Massey, D. (2005). *For space*. London: Sage Publications.

Minchin, T. (2012). 'Us is spelled U.S.': The crafted with pride campaign and the fight against deindustrialisation in the textile and apparel industry. *Labor History*, 53 (1): 1–23.

Neumann, T. (2016). *Remaking the Rust Belt: The postindustrial transformation of North America*. Philadelphia: University of Pennsylvania Press.

Newman, R. (2016). *Love canal: A toxic history from colonial times to the present*. Oxford: Oxford University Press.

Nora, P. (1989). Between memory and history: Les lieux de mémoire. *Representations*, 26: 7–24.

Norkunas, M. (2002). *Monuments and memory: History and representations in Lowell, Massachusetts*. Washington, DC: Smithsonian Institution Press.

Nova Scotia Commission on Building Our New Economy. (2014). *Now or never: An urgent call to action for Nova Scotians*. Halifax: OneNS.ca, p. 84.

Paisley Peoples' Archive and Scottish Oral History Centre. (2018). *Paisley Thread Mills memoryscape*. Accessed 22 June 2018, www.paisleypeoplesarchive.org/memoryscapes.aspx

Peck, J., and Whiteside, H. (2016). Financializing Detroit. *Economic Geography*, 92 (3): 235–268.

Perchard, A. (2012). *Aluminiumville: Government, global business, and the Scottish Highlands*. Lancaster: Crucible Books.

Phillips, J. (2012). *Collieries, communities, and the miners' strike in Scotland, 1984–85.* Manchester: Manchester University Press.

Pride, T. (2016). *Resident-led urban agriculture and the hegemony of neoliberal community development: Eco-gentrification in a Detroit neighborhood.* PhD thesis, Wayne State University, Detroit.

Rankin, K., and McLean, H. (2015). Governing the commercial streets of the city: New terrains of disinvestment and gentrification in Toronto's inner suburbs. *Antipode*, 47: 216–239.

Ross, J. (2017). Interview by Edras Rodriguez-Torrez, 13 July 2017, https://detroit1967.detroithistorical.org/items/show/638

Schaffer, L. (2016). Interview by Hannah Sabal, 23 July 2016, https://detroit1967.detroithistorical.org/items/show/352

Slater, T. (2009). Missing Marcuse: On gentrification and displacement. *City*, 13 (2–3): 292–311.

Smith, L., and Campbell, G. (2011). Don't mourn, organize: Heritage, recognition and memory in Castleford, West Yorkshire. *In:* L. Smith, P. Shackel and G. Campbell, eds. *Heritage, labour, and the working classes.* London: Routledge, pp. 85–105.

Stoler, A. (2008). Imperial debris: Reflections on ruins and ruination. *Cultural Anthropology*, 23 (2): 191–219.

Sugrue, T. (2005). *The origins of the urban crisis: Race and inequality in postwar Detroit.* Princeton, NJ: Princeton University Press.

Vickers, S. (2013). *Making Co-opville: Layers of activism in Point St-Charles (1983-1992).* MA thesis, Concordia University, Montreal.

Walley, C. (2013). *Exit zero: Family and class in postindustrial Chicago.* Chicago: University of Chicago Press.

Zukin, S., Trujillo, V., Frase, P., Jackson, D., Recuber, T., and Walker, A. (2009). New retail capital and neighborhood change: Boutiques and gentrification in New York City. *City & Community*, 8 (1): 47–64.

17
REMEMBERING SPACES OF WORK

Emma Pleasant and Tim Strangleman

Introducing memory at work

What is the relationship between memory and the world of work? Taken to its most extreme, memory and work have a very problematic relationship. Marxists, for example, find the idea of attachment to work as antithetical to the process of alienation and estrangement. Here, remembering economic life in a positive way presents a series of real contradictions. To be attached through economic activity is at least in part a denial of the centrifugal forces at play in modern employment, pressures that progressively detach workers from their labour, their fellow workers, and ultimately their species being. However, in study after study we get glimpses of a sense of embeddedness, meaning, and value.[1] Arguably the strongest examples of this occur not when workers are in work, but after they lose it as part of shutdowns or layoffs (see Dudley 1994; High & Lewis 2007; Strangleman 2019).

In past research, Tim has used the idea of the 'breaching experiment' coined by Harold Garfinkel (see Strangleman 2019). This was the idea that underlying social structures were usually taken for granted in everyday life; it was only when social norms were breached that these hitherto disguised structures were revealed. The breach then exposed the meaning often attached in social interaction as well as a set of underlying rules. In the cases of work, redundancy, or wider deindustrialisation, these can equally act on a larger scale as a breaching experiment, forcing those caught up in the process of change to reflect on their working lives in the *absence* of work. Importantly, however, these memories of work almost always involve being rooted in particular times and spaces. Workers are not abstractly valuing a theoretical notion of work and attachment, but are rather remembering through people and place. They are socialising place and space through memory.

Economic geographers are critical of work sociology for failing to take space and place seriously. They suggest that the spatial, when invoked at all, is merely a device for containing action, a map reference where agency is exercised. Geographers, by contrast, claim they have a more sophisticated analysis where actors act on place and space, and equally the spatial makes up the actor (see for example Herod et al. 2007; Castree et al. 2003). There is a degree of co-production wherein the spatial is constantly live and contingent, not a passive receptacle. This chapter challenges these assumptions, suggesting that sociological accounts of work have often been alive to the potential of space. In particular, we suggest here that a complex and sophisticated understanding of place and memory emerges from writing about the loss of work through plant shutdown, or wider processes of deindustrialisation. Within this field, attachments to place are already being widely discussed whether it be through the economic dislocation of workers by political economists (Bluestone & Harrison 1982), the impact on cultural spaces described through cultural geography (Nayak 2006), or by employing literary terms to make sense of change (Linkon 2018). Through all of these accounts, the ideas of memory and space, articulated here through memoryscapes, helps us interrogate how workers remember their former labour, and how place-based memory often reveals an embedded attachment to work. This is often a relationship that was taken for granted when industry was open. In what follows we examine ideas of space and memory before focusing down on the workplace specifically as a site of memory. Finally, we draw on two short vignettes from our respective research showing differing forms of memory evoked by workplaces.

Before we take our discussion further, it is worth reflecting on a powerful counter narrative on memory and work, namely the suggestion that positive accounts of the past at work are 'smokestack nostalgia' (Cowie & Heathcott 2003). We take this to be the very real danger of romanticising an industrial past, one that edits out inequality around gender or race or glosses over pollution or industrial injury. As Jefferson Cowie and Joseph Heathcott say:

> we have to strip industrial work of its broad-shouldered, social-realist patina and see it for what it was: tough work that people did because it paid well and it was located in their communities.
>
> *(2003: 15)*

Equally, though, it is vital that the attachment felt by former workers to an industrial past is not simply dismissed as nostalgia. Whenever Tim has talked with retired or redundant workers they almost always leaven their positive accounts of the past with an acknowledgement of the negative aspects of labour (see Strangleman 2004, 2016, 2019). In other words, they engage in what Fred Davis (1979) labelled reflective or critical nostalgia, rather than simple nostalgia. 'Simple' nostalgia is a pejorative term which effectively diminishes claims to knowledge about the past. It usually implies an uncritical, romantic, rose-tinted view of events. Davis rejected the idea that this version of nostalgia is one that is ordinarily present. Instead he saw reflective and even critical forms of nostalgia as being more accurate interpretations of reminiscences. Here there is a questioning of the past alongside the present, even an evaluation of why particular memories are being triggered. These forms of nostalgia have the potential to be critical, analytical, oppositional and even radical (see Bonnett 2010; Boym 2001; Strangleman 1999, 2004).

Spaces of nostalgia

This nostalgia forms from affective attachments to place. Workers use this radical spatial remembering to compete for visibility by rupturing narratives of spatial and temporal progress. In *Theatres of Memory*, historian Raphael Samuel notes the temporal fracture present in sites of redevelopment, as he says; 'the past is seen not as a prelude to the present but as an alternative to it, "another country"' (2012: 221). In spaces where memories are shaped into a performance of history, there is a sense of a bygone past that is foreign to the now. Literatures that examine the relationship between space and memory will often echo this sentiment by articulating the different representations of a community's past, putting individuals at odds with official discourses (Linkon & Russo 2002; Dudley 1994; Misztal 2003). This most commonly occurs in spaces of 'official' remembering such as museums, ruins, or even entire cities and towns (Samuel 2012; Smith 2006; Edensor 2005a, 2005b). Within these spaces power and memory intersect. Indirectly, public spaces can also be enrolled into these controlled dialogues of remembering through what Samuel (2012: 39) called 'the historicisation of the built environment.' When these public spaces are 'rebranded' the evidence of destruction must be hidden, a process Tim Edensor (2005a) refers to as 'exorcising.' Non-elite memories are invariably marginalised, skipped over, or excluded altogether. The complexity of lived reality is flattened, simplified, and rendered linear.

When taken to former spaces of work, this relationship becomes even more complicated as nostalgia, and the affective attachment from which it forms, becomes a disruptive force. Edensor (2005a, 2005b) examines this in industrial ruins, spaces that are either forgotten or demolished to make way for regeneration. Agents of restoration attempt to strip particular areas of memory, washing over 'haunted' spaces. These practices aim to produce a single narrative of history, removing the possibility for contestation. Those with direct links to these sites of memory act as spectres, fluid entities who have the potential to disrupt the fixed narrative of modern capitalism. The dominant strategies of official and unofficial heritage spaces impose a certain identity on sites, but these get muddled by the spectral presence. Capitalism requires these spaces to perform a certain identity which becomes disturbed when those with competing discourses are present (Edensor 2005a). These ghosts, however, haunt all spaces of memory as they embody the contestation of power which characterise memoryscapes, acting as a transgressive form of remembering; they occupy a grey space between past and present as they draw in competing discourses of memory to these sites.

This theme is an elaboration of the ideas of American sociologist Avery Gordon and her book 2008 book *Ghostly Matters*. Gordon makes a powerful argument for the idea of haunting within the sociological imagination, and, in particular, the way ideas and memories linger long after the context which originally framed them disappeared. Long suppressed or forgotten memories and ideas bubble up in the present, reminding individuals, groups, or societies about an aspect of the past. Alongside Edensor's writing, Gordon's social haunting is suggestive of the revelatory moment, a surprise uncovering, and the keeping alive of memory of a suppressed past. Workers are drawn out of this past and into the present, fracturing the temporal shift of the redevelopment of space.

The affective attachment of workers to their spaces of employment makes this radical nostalgia place based. Ghosts reintroduce alternative memories and representations of loss, pain, and pride: the affective responses to change that are difficult to linearise or standardise. Affective responses are less successful in 'authorized heritage discourses' which cleanses history of emotion, reducing the past to observable and teachable facts (Smith 2006). Ghosts, by contrast, invoke empathy, making contact with the viewer and giving a more human face to change. Spaces of memory disrupt the clean break between past and present. They can challenge the linearity of capitalism and development, bringing forth the grey space in between now and then, highlighting the experience of change. So, spaces of memory reflect structural processes of capitalism, power imbalance, and control over discourse. They are microcosmic battle grounds for privilege and dissent where these dynamics become visible. This becomes most apparent when individuals remember workspaces and problematise the idea that industrial closure was inevitable and a part of the natural order of capitalist progress.

Remembering sites of work

Attachment to space is a temporal, spatial, and cultural negotiation of place and memory. Spaces of work physically and culturally imprint communities and communities imprint on the spaces in which they work. Memory is often used as a tool in workplaces to gain autonomy in labour and reclaim a form of communal control and solidarity. Workplaces act as 'circuits of memory' as routines and identities based from communities of memory make claim to the space (Edensor 2005a). These communities are what Barbara Misztal (2003) calls 'mnemonic' as they are formed from individuals adhering to a collective memory. As Misztal (2003) describes, remembering is a social process that forms a social identity. As new members at work engage in the same cultural knowledges, they ascribe to the same social identity that is transgressive by nature. Therefore, they are joining a wider communal identity which can be 'identified primarily with habit; its authority is derived from the felt need to reiterate the wisdom bequest by the past' (Hutton 1993: 17). Through these rituals, the workplace is renegotiated which alters the cultural space. Workplace cultures are reliant on memory as, through collective remembering, solidarity is formed. The destruction of these workspaces does not erode these communities entirely as Steven High and David Lewis (2007: 12) remind us, 'communities are lived in social networks, created in places, and imagined at a distance.' Therefore, they have the potential to exist beyond the walls of the workplace and imprint upon the physical site of labour.

When workplaces close, the community loses access to its physical focal point. However, this does not mean that they become entirely disassociated from it. Sherry Linkon and John Russo (2002) in *Steel Town* examine the role of communal and collective rememberings that are oppositional when trying to reconstruct the identity of a former industrial town. Here, the individual has a distinct form of remembering the workspace that contradicts the memories constructed by the official planners. Remembering workspaces here is too invested with dialogues of power as they try to understand who owns the collective identity of the community. These imbalances over power, however, can also work to restore the community of memory and refocus their collective solidarity into a singular memory that acts as a membrane to members. Valerie Walkerdine (2010) discusses this phenomenon in South Wales whereby memories of industrial culture allow members to draw strength in their remembering and use it to collate a singular sense of a communal self. This acts as a force to resist official discourses that would seek to reconstruct their identity in a way that they do not recognise. Within both cases, there is a struggle over memory and over how individuals and communities of workers attempt to renegotiate the forced terrain of detachment. They have been spatially disassociated from their sites of labour that allowed their solidarity and community to reproduce, and therefore, without ownership or direct relationship with the physical marker of their community, they must try to restore this focal point elsewhere.

What we have tried to pick out from the discussions so far is that recurrently the literature on sites of memory return to the notion that there is a power struggle in these places. When we began to consider what form remembering took at work and particularly within work, it appeared central to how work solidarities were formed through the renegotiation of cultural space. Of course, when the workplace closed, these cultural spaces became even more ephemeral as they lost their material attachment. Detachment and attachment are central components to what it is to remember a site of work because the community has been removed from its physical grounding. But, beyond that there seems to be two aspects of this that need to be examined a bit more. What happens to the workspace itself, the physical space of memory, and the workers who form the cultural space of memory.

In *Corporate Wasteland*, High and Lewis pose an important question: 'Industrial ruins are memory places, for they make us pause, reflect, and remember. But remember what, and to what end?' (2007: 9). We offer two vignettes from our respective research to illustrate some of the complexity of the relationship between memory and the place of work, and in the process hope to answer, at least in part, High and Lewis' question. Firstly, Tim will talk about a small aspect of one of his interviews from his Guinness research which highlights the bonds that exist between former workers, their memories and space. Secondly, Emma discusses an interview with former Chatham Dockyard worker, Stuart Pollitt, who further details the role of nostalgia and perspective in this triadic relationship. What both accounts taken together show is that memoryscapes are political. They incite dissent, and, particularly within the relationship between workspaces and workers, the complexity and fragmentation of rendering memories as a physical space is highlighted.

The brewery as a memoryscape

In the summer of 2015 Tim interviewed former brewery worker Terry Aldridge. Terry had worked at the Guinness brewery at Park Royal in west London for two decades from 1975. In the interview Terry had reflected warmly of his time at Guinness and those he had worked with there. Tim's abiding memory of this interview with Terry was of them both laughing as he recalled the pranks that he had been a victim or perpetrator of, these usually involved getting wet. Throughout his career Terry had been a keen and talented amateur photographer, taking pictures of the brewery from the time he began working there. Towards the end of their time together he showed Tim the images he had taken of the site while it was in the process of demolition. As he showed Tim the images he had taken of destruction, he told him about the impact of what he saw:

> I was very saddened by that. I was a bit angry in some ways because I thought they were going to [deep inhalation] use bit of it maybe the brew house as a museum. I went up there during the time that they were demolishing the place, and it's a bit of a lump in the old throat job. You're looking at this poor old building, the way you see it, and you can see all the floors – I got a picture in here – you can see all the floors and you say 'Oh, that's where I stored the bucket there, when I was wetting that' [laughs] God! It was sad. Yeah, very sad.[2]

Terry made several pilgrimages to his former place of work but, by the final visit, there was little to record:

> I went back again at a later date and there was absolutely nothing left, there was just these pyramids of rubble. And I'm standing there thinking, 'Wait a minute, where was it? Was it slightly that way or this way?' Because you've lost it [bearings], really odd, to see it like that. Yeah it was sad.[3]

Tim was fascinated to understand why Terry, who had not worked at the plant for over a decade, would make the long trip from his south coast home to record the fate of a former place of work. After a long sigh he told him:

> Curiosity I suppose. Can't help but go up back there. You work there all those years, and you just – it's funny, I wouldn't go back and see Lancia's [his former workplace] be demolished. And they did demolish that building. No it was something about the brewery, the people in it, the work, the camaraderie of all the people that you work with – you couldn't help but go back. It's almost the last tribute.[4]

A little later that summer Tim interviewed Henry Dawson in his north west London home, ten years after he had initially talked to him while the brewery was open. Tim asked him if he had returned to the site during its demolition. He said simply:

> No I stayed away. I didn't want to see it being knocked down. I've been back up the headquarters now and I look at it and think, "What used to be there was an iconic building. Now there's nothing, there's just a hole."[5]

Tim asked Henry if he had seen any of Terry's photographs of the demolition:

> Yeah I know Terry did. He's tried to get me to have a look at the pictures – oh no. Don't particularly want to – I've seen a couple of odd pictures but I don't particularly want to see it.[6]

As he explained to Tim, 'I'd rather remember it how it was working ... and people smiling while they were at work.'

There are some fascinating things going on in the material from Terry and Henry. In their different ways both shows a great love of their former site of work, seeing great value in it, and each was sad, and even angry, to witness the buildings destroyed. But while Terry felt compelled to take photographs of the ruination process, to record the events *because* of this attachment, Henry cannot bear to look at these same images, precisely because of a desire to remember attachment in a different form. Through images memories can be stored and regenerated but also perhaps eroded; they allow a complex range of emotions to surface and circulate.

It is clear that both Terry and Henry valued their work and the people they had worked with: looking back fondly, even nostalgically on the working lives at Park Royal. But their reflection on the past is not uncritical. Neither worker had given an unalloyed positive rendering of the past. Equally in looking at the present and the past, the place of labour is used to provide a critique of work now and its organisation. Complexity, then, is the marker of remembering through space. As Terry and Henry show, their presence and memories muddle any attempts to find coherence in the responses to memoryscapes whether they be physical or cultural. When revisiting these spaces as the two men show, their attachments render the spaces political.

The dockyard as a memoryscape

Similarly, Emma found this contestation and politics of attachment to take form in her discussions with former Chatham Dockyard worker, Stuart Pollitt, who had started at the Dockyard as a yard boy doing general maintenance duties in 1964. At 15, Stuart had been called into the Number 8 Machine shop and told by one of the senior engineers of his selection for the apprenticeship programme. The next ten years of Stuart's work life would happen within that machine shop until the day he left the yard. During Stuart's interview, the machine shop repeatedly emerged as a central theme in how he remembered his work, particularly, referring to the first time he visited the yard after its 1984 closure. Since then, the machine shop had been stripped of its cladding and kept as a skeleton frame which had been listed as a heritage site. The area surrounding it had been drastically transformed from a working industrial Dockyard to a retail outlet. The other machine shop that stood opposite Number 8 has been repurposed as a shopping centre, restaurants, and a cinema. The skeleton therefore seems odd against this landscape. Stuart revisited this preserved frame and stood amongst it:

> and I went in there and I could actually walk as if I was going to clock in through the big doors but the doors aren't there now 'cause it's just a skeleton structure but I was able to walk down the gang way that that chap I told you about swept in my imagination and I could see the bolts where the machines were bolted to the concrete floor . . . and also strangely enough think to myself, and I think this is old boy syndrome isn't it, standing there thinking all these things and then you think about it afterwards that when I was standing there, imagining what it was like and that I was gifted to have been part of it . . . that all these people around me now . . . they're not stopping to think of what was going on in there, in the days when it was thriving and had all these machinery sounds and the smells . . . all of the people who are milling about outside of the skeleton building as I'm standing there imagining what it was like and I'm thinking to myself, none of these people know . . . what I experienced or they're not . . . seeing what I saw.

Listening to Stuart shows that there is something important about remembering through workspaces, his own personal memoryscape. Stuart conjured up a mental image of the past rooted in that space, which gives him the perspective to see how invisible that past is to those passersby. He occupies a space much like Edensor's (2005a) ghosts, as an entity that complicates the narrative of the retail district that surrounds the skeleton by drawing upon his attachment to place. He draws the past into dialogue with the present and disrupts the notion of a clean break between the area's industrial past and its retail future. Stuart occupies the grey space between past and present and moves between these spaces to offer a reflection on what has changed.

Stuart was being nostalgic. He is looking back on the past and reopening the doors of what once was at the site to draw parallels with the future. Rather than being romantic and wistful, he is reflective, even critical about the changes to place and space. He does not have a blind desire to return to a better past; by standing in that frame and occupying the space between the then and the now, he has the privilege of perspective. He can cast a critical gaze on both the past and the present to see what has changed for the better and for the worse. He hazes the officially imposed normative discourse surrounding these sites of regeneration that the change is beneficial for the area as he reintroduces the critical voices of those who have lost out in these sites. He stands as a voice for the former workers who become ostracised from the spaces that were once central to their communities by the ongoing redevelopment projects. Stuart disrupts the linear narrative of change and progressiveness by being a ghostly figure of nostalgia.

Like Terry and Henry, Stuart is talking through a form of attachment; one that is evoked by a complex form of spatial remembering. He remembers his attachment to his community at work and the chap he remembered sweeping the floors – those that gave him the 'gift' of solidarity, communality, and all the other positive aspects of the workplace. Within the frame, he is reflecting on the memories shared in the workplace; he draws this community out of its past and reimagines its existence in that moment. Memory and spaces overlap in communities as groups imprint their collective conscious onto a physical site (High & Lewis 2007; Linkon & Russo 2002; Misztal 2003). Stuart acts as a vessel of this communal memory, finding the strength to reject their 'by-gone' status through being in the space of memory. However, being in the machine shop also forces him to consider the subsequent detachment felt to the passers-by. Those who remind him of his invisibility and cause his concern for the existence of his workplace in the present. Through these forms of attachment/detachment, he finds his critical viewpoint. His nostalgia therefore is a form of attachment as it is through his relationship with material and imagined spaces, that he draws the past into the present. In this respect, attachment to space is political as it is the catalyst for drawing out feelings of nostalgia, dissent, and complicated voices.

Conclusion

Overall, this chapter has thought through what happens when individuals and communities remember through spaces of work. Employing the concept of memoryscapes within the sociology of work allows us a new lens to examine a phenomenon that is already present within existing accounts. But this new articulation helps draw together the geographical, social, cultural, and political narratives that often imbue workers' relationships to place. How workers remember, imagine, and describe space becomes an essential part of the narrative of the place. Their agency is repositioned in the memoryscape and we understand more about their attachment to place and the places' attachment to them.

Workspaces therefore offer a unique form of spatial remembering. They are spaces where critical reflection is made easier and transgressive discourses flourish. Workers and workplace communities embed these sites with their memories; they are laced with spatial and temporal fixings that allow former workers to feel ownership over the space. Former factories, ruins, and workshops are imbued with meaning for those who have attachment to these communities of memory. The communal identities become fixed into the space which allows remembering outside of the official discourse. Particularly we have shown this to be brought into focus when workers engage with nostalgic remembering.

What Terry, Henry, and Stuart uncover through their presence and absence at their former workspaces is that memoryscapes are sites of politicised remembering. Within these spaces, attachment emerges as a form of nostalgia characterised by complexity and dissent. Through Stuart and his skeleton frame, we see this juxtaposition occur through the relationship between the past and the present. Henry and Terry show how this can occur within the attachment of the workers themselves as they negotiate their affective responses to their former workplace. Both examples show how conflict emerges both internally and externally for the workers.

Plant shutdowns and wider processes of deindustrialisation have acted as a breaching experiment: enabling, or rather forcing, former workers to reflect on their attachment to their now redundant trade. Memories are elicited through discussion, photography, and material object, and these reflections are often critical in nature rather than 'simple nostalgia,' or 'smokestack nostalgia.' While individual testament can be a powerful counter to sweeping industrial change, collectively the voices and memories of deindustrialised workers offer up a wider critique of capitalism and the decisions made about individuals, communities, and industry. Workers' memories therefore are powerful. Through engagement with their relationship to space, we learn more about how places are given meaning and the processes in which these are made and remade.

Notes

1 We are thinking here of both Marx himself discussing alienation across his work as well as the Marxist who developed ideas of alienation from the worker and the products of their labour.
2 Terry Aldridge, interviewed 2015: 26.
3 Ibid.
4 Ibid.: 26–27.
5 Henry Dawson, interviewed 2015: 22.
6 Ibid.

References

Bluestone, B., and Harrison, B. (1982). *The deindustrialisation of America*. New York: Basic Books, Inc Publishers.
Bonnett, A. (2010). *Left in the past: Radicalism and the politics of nostalgia*. New York: The Continuum International Publishing Group.
Boym, S. (2001). *The future of nostalgia*. New York: Basic Books.
Castree, N., Coe, N.M., Ward, K., and Samers, M. (2003). *Spaces of work: Global capitalism and geographies of labour*. London: Sage Publications.
Cowie, J., and Heathcott, J. (2003). *Beyond the ruins: The meanings of deindustrialisation*. Ithaca, NY: Cornell University Press.
Davis, F. (1979). *Yearning for yesterday*. New York: The Free Press.
Dudley, K. (1994). *The end of the line: Lost jobs, new lives in postindustrial America*. Chicago: University of Chicago Press.
Edensor, T. (2005a). The ghosts of industrial ruins: Ordering space and disordering memory in excessive space. *Environment and planning D Society and Space*, 23: 829–849.
Edensor, T. (2005b). *Industrial ruins: Space, aesthetics and materiality*. Oxford: Berg.
Gordon, A. (2008). *Ghostly matters: Haunting and the sociological imagination*. Minneapolis: University of Minnesota Press.
Herod, A., Rainnie, A., and McGrath-Champ, S. (2007). Working space: Why incorporating the geographical is central to theorizing work and employment practices. *Work, Employment and Society*, 21 (2): 247–264.
High, S., and Lewis, D. (2007). *Corporate wasteland: The landscape and memory of deindustrialisation*. Ithaca, NY and London: Cornell University Press.
Hutton, P. (1993). *History as an art of memory*. Hanover, VT: University Press of New England.
Linkon, S. (2018). *The half-life of deindustrialization. Working-class writing about economic restructuring*. Ann Arbor: University of Michigan Press.
Linkon, S., and Russo, J. (2002). *Steeltown USA: Work and memory in Youngstown*. Kansas: University Press of Kansas.
Misztal, B. (2003). *Theories of social remembering*. Berkshire: Open University Press.
Nayak, A. (2006). Displaced masculinities: Chavs, youth and class in the post-industrial city. *Sociology*, 40 (5): 813–831.
Samuel, R. (2012). *Theatres of memory. Past and present in contemporary culture*. London: Verso.
Smith, L. (2006). *Uses of heritage*. London: Routledge.
Strangleman, T. (1999). The nostalgia of organisations and the organisation of nostalgia: Past and present in the contemporary railway industry. *Sociology*, 33 (4), 725–746.
Strangleman, T. (2004). *Work identity at the end of the line: Privatisation and culture change in the UK rail industry*. Basingstoke: Palgrave Macmillan.
Strangleman, T. (2016). Deindustrialisation and the historical sociological imagination: Making sense of work and industrial change, *Sociology*, 51 (2): 466–482.
Strangleman, T. (2019). *Voices of guinness: An oral history of the Park Royal Brewery*. New York: Oxford University Press.
Walkerdine, V. (2010). Communal beingness and affect: An exploration of trauma in an ex-industrial community. *Body and Society*, 16 (1): 91–116.

18
MEMORY AND POST-INDUSTRIAL LANDSCAPES IN GOVAN (SCOTLAND)

Martin Conlon

Introduction

Deindustrialisation, defined here as the planned reduction of industrial activity in a region or economy, is a truly global phenomenon which has provoked focused study across multiple academic disciplines. Though initially understood as a primarily economic process, 'deindustrialisation studies' has expanded to acknowledge the deeper social and cultural implications of this shift. The study of the resultant 'post-industrial' condition has received considerable attention, with memory work integral to this. Works including *Beyond the Ruins* (Cowie & Heathcott 2003), *Heritage, Labour and the Working Classes* (Smith et al. 2011), and *The Deindustrialized World* (High et al. 2017) have sought to critique the complicated relationship between memory and the individualised and collective negotiation of industrial change in various locations.

One integral component of this emergent body of work has been the interrogation of the link between memory and the physical environment, which is seen in *Industrial Ruination, Community and Place* (Mah 2012) and Hilary Orange's diverse, edited collection *Reanimating Industrial Spaces* (Orange 2015). Another dominant strand has been the role of nostalgia in framing the perceptions and hopes of people and communities that have lived through processes of deindustrialisation. In *Beyond the Ruins*, Jefferson Cowie and Joseph Heathcott interrogated what they memorably labelled 'smokestack nostalgia' (2003: 15); a prominent and unhelpful force which they felt was preventing critical approaches to understanding industrial leftovers.

To combat this creeping form of nostalgia, the authors argued that it was necessary to 'strip industrial work of its broad-shouldered, social-realist patina and see it for what it was: tough work that people did because it paid well and it was located in their communities' (Cowie & Heathcott 2003: 15).

Since then, nostalgia has been reconsidered as a 'complex and nuanced emotion' (Smith & Campbell 2017: 609) which, though often dismissed, can lead to insightful contributions that broaden collective understandings of how people have come to comprehend the spaces around them. Similarly, Tim Strangleman (2013: 23) has convincingly argued that nostalgic narratives of deindustrialisation are useful as they show the 'desire to reflect back and find value in the industrial past.' This chapter builds on this work, connecting memory, nostalgia, and the physical environment by exploring the archaeological loss associated with the removal of shipbuilding cranes in Govan, Scotland.

Between 2007 and 2014, the Fairfield shipyard in Govan had its historic skyline (Figure 18.1) picked apart, with the yard's cranes being dismantled as part of a programme of modernisation. This chapter aims to show how cultural or social value has been ascribed to these structures, demonstrating the wider relationship between memory, monumentality, and post-industrial archaeology. Drawing on new oral history

Figure 18.1 The Fairfield cranes in 1999.
Source: Reproduced with the kind permission of Colin McPherson.

testimony and a critical analysis of material culture, this chapter will juxtapose the tangible and drastically altered physical landscape of Govan with the intangible landscapes of memory and wider industrial culture. The testimony was gathered in 2015 and 2016 as part of a wider doctoral research project on the last cranes along the River Clyde.

Studying deindustrialisation through industrial archaeology

One way of looking at industrial landscape is through Ingold's concept of the 'taskscape.' In 1993, Ingold published his classic paper 'The Temporality of the Landscape,' which broadly considered changing perceptions on landscape, and the benefits of unifying both archaeological and anthropological thought. Introducing the concept of the 'taskscape,' Ingold explained how a more socially constructed form of landscape can emerge through studying the activities of 'those who dwell therein.' The taskscape is useful in a variety in contexts. When applied to Industrial Archaeology, it can offer ongoing insights about place-making and the relationships people have with working landscapes. As Ingold comments, 'just as the landscape is an array of related features, so – by analogy – the taskscape is an array of related activities' (1993: 158). The taskscape is a therefore a continually unfolding, socially constructed space for human existence within a broader, spatially defined landscape. Ingold stresses that a taskscape can typically be assessed via five key factors: mobility, habitat, economy, nature, and public space. In the essay, Ingold demonstrates the usefulness of the taskscape as a methodological framework for analysing the temporality of the landscape in Peter Bruegel the Elder's painting *The Harvesters* (1565). Crucially, this concept can therefore provide a useful framework for approaching the connections that people have with changing industrial landscapes, the passing forms of industrial work associated with them, and the wider social or cultural reverberations involved. The conceptual framework of the taskscape allows us to go 'beyond the ruins' (Cowie & Heathcott 2003: 6) of industry, and look towards the more nuanced forms of temporal belonging.

In 1974, John Hume published the results of a ten-year survey project that produced an inventory of Glasgow's remaining industrial sites. When the research project began, even before many processes of deindustrialisation had been enacted, it had become apparent to Hume that 'most industrial buildings were no longer occupied by the firms for whom they had been built, and that the rate of demolition was likely to rise' (1974: 3). This prophecy became reality on a scale then unimaginable. Hume created a gazetteer of sites, supported by an enormous photographic record. In his introduction, Hume (1974: xviii) observed that

> it seems to me that the most vital task of the industrial archaeologist, particularly in urban areas, is to record extant buildings, machinery and structures before the scrap-man, the incendiary and the bulldozer come.

Over 40 years have passed since Hume's assessment, and the role of the industrial archaeologist has been largely redefined since, as the 'scrap-man, the incendiary and the bulldozer' have been and gone. In this light, the role of the industrial archaeologist should arguably be redefined. Hilary Orange (2014: 64) has stressed a potential refocus, writing;

> I am a post-industrial archaeologist. I study the materiality of closure and transition: the afterlives or 'reanimation' of industrial sites, including the continuing dynamics of industrial landscape, their future potential reuse and ongoing relationship with local communities.

Now dealing with the 'post-industrial' in many cases, academics have sought to examine the 'meaning' attached to passing industrial material, rather than merely survey and record change. These studies form part of a wider period of transition, described by Orange (2008: 83) as a 'post-abandonment' age, where industry's transience is a source of both 'acceptance and 'forgetfulness.' Crucial to this transition is the pace of change and its relationship with living memory, summed up by Bryne and Doyle (2004: 166):

> In mining parts of South Tyneside, an area which until the 1970s had four large collieries and where coal mining had been historically the largest single source of employment for men, there is actually more visual evidence of the Roman occupation which ended in the fourth century AD and has no historical connection to any contemporary experience than of an industry which at its peak in the 1920s directly employed more than 12,000 as miners, indirectly employed as many again as railway workers, dockers and collier seamen, and still employed more than 2,000 when it ended in 1993.

Since the turn of the millennium, industrial ruins have increasingly become the focus of academics, artists, activists, and urban explorers, leading to what Caitlin DeSilvey and Tim Edensor (2013: 465) describe as 'extraordinary intensification of academic and popular interest in the ruins of the recent past and associated realms of dereliction.' Despite this proliferation of interest and Don Mitchell's observation that 'the landscape is a significant site of social struggle' (2005: 84), academic literature that explicitly links memory to specific sites, or predefined spatial geographies, are still in short supply. Andrew Herod has also stressed the need to explore 'how workers' lives are structured and embedded spatially (2001: 4), whilst Raphael Samuel considered the importance of 'territorian attachments' (2012: 277) as the link between the ideological symbolism for work and the ways in which this can become imprinted on physical sites. This chapter will apply these holistic interpretations of place-connection to one case study in Govan, Scotland.

Govan

Govan is a former burgh, now part of the southwest of Glasgow, in the west of Scotland. It is situated on the River Clyde, where it meets the River Kelvin, approximately three miles from the city centre. Govan experienced a remarkable transformation in the nineteenth and twentieth centuries, transitioning from a small textiles and weaving town to arguably the centre of world shipbuilding in a relatively short space of time. T C F Brotchie (1905: 34) famously wrote that 'shipbuilding made Govan and Govan made shipbuilding,' a truism for the extent to which shipbuilding and its cultural by-products became inseparable from the area itself. A survey by the Glasgow Boundaries Commission in 1864 established the population to be around 9,500, with the census of 1901 suggesting this figure had rapidly risen to 82,174 (1888: 88). This rapid urbanisation brought with it a multitude of social issues, such as poor housing and ultimately overcrowding, ill-health, and material deprivation. As Chris Dalglish and Stephen Driscoll (2009: 99) point out, given its relatively compact geography, as well as making Govan now Scotland's seventh most populated town, made it 'one of the most congested communities in Scotland.' Vastly reduced from its population peak of 89,725 in 1911, Govan today has approximately 31,437 residents (ibid: 143). Many of the large tenement blocks have been gradually cleared to make room for low-rise housing schemes. As of 2009, the area had higher than average rates of unemployment: around 10% (ibid: 124).

Govan's cranes

From its construction in 1911, the Fairfield crane quickly became a monument to the dramatic change experienced in Govan, as evidenced in a postcard which proudly proclaims it to be the 'largest crane in the world,' standing at 170ft tall and said to be capable of lifting 250 tons (Glasgow Story 2004). Another photograph from 1915 shows children who appear to be in their best clothing, neatly assembled with what was – at the time – the world's largest and most powerful crane as a backdrop (Glasgow History 2010). The crane has also been immortalised in the famous folk song by Archie Fisher and Norman Buchan, 'The Shipyard Apprentice,' which uses 'being born in the shadow of the Fairfield crane' as a metaphor for the nature of local connections with the shipyards ('The Shipyard Apprentice' 1973). It serves as a lament to a Clydeside upbringing and the profound impact that industrialism had on the lives of many in Scotland, as well as an evocative introductory example of how this was expressed through popular culture of the time. The song chronicles the paradoxical reliance on industry for work and its role in the formation of local identity, despite its often-negative connotations, remarking 'I oft-times heard my mother say, it was tears that made the Clyde.'

The removal of the cranes

Figure 18.1 shows Govan's historic skyline that was picked apart between 2007 and 2014 as site occupiers BAE Systems redeveloped their yard. For BAE Systems, the cranes were now redundant; old technology that represented old working practices. Like the Clyde's other four giant cranes, the Fairfield crane had A-listed status, the highest designation of 'importance' by what was then named Historic Scotland, now Historic Environment Scotland. Despite this status, neither Historic Scotland nor Glasgow City Council wanted to be seen to be standing in the way of a modernisation programme that would secure long-term viability and ultimately jobs, which had been so historically sensitive to the region over the years. Since 1965, the shipyard has faced several moments of crisis, narrowly escaping threats of closure and mass job losses in 1972, 1988, and 1999. A project kick-started by several industrial heritage enthusiasts aimed to dismantle the crane and have it re-erected on a derelict shipbuilding site nearby. Yet these plans proved financially unviable, and the Fairfield crane was dismantled in 2007. In 2013, news broke in *The Herald* (2013) that the yard's remaining 80-ton berth cranes would be pulled down, leaving the yard with no cranes at all. The paper reported that 'fears are growing for the long-term future of 1500 workers at BAE Systems in Govan after it emerged the Glasgow shipyard's cranes are weeks away from being pulled down.' The article was accompanied with a cartoon that depicted the jib of the crane reimagined as a hangman's noose, suggestive of the peril that the workers, and the wider community of Govan, were facing. The changes to the yard's skyline came at a delicate time, with BAE pondering its future at Fairfield as well as its other site at Scotstoun across the river, against the backdrop of an impending independence referendum in Scotland and the associated speculation on future industrial development in Scotland, whether part of the United Kingdom or an independent country.

The speculation surrounding the cranes and jobs prompted the then Deputy First Minister and MSP for Glasgow Southside, Nicola Sturgeon, to comment:

> BAE has to come clean. The workforce has been through the mill too many times before. It would be outrageous if Govan doesn't have a future. . . . There is a duty on the company to make clear its intentions. I would hope the decision to take down the cranes would be put on hold until a decision on the future of the yards is made. I don't want to see Govan shipyard close.
>
> *(The Herald 2013).*

The newspaper article and the comments of the Deputy First Minister confirm the notion that area's cranes had become the physical embodiment of work: monuments to human toil. The cranes and shipbuilding had become inseparable to the point where, for many, the removal of the cranes automatically meant the removal of shipbuilding work. However, this was misguided. The departure of the cranes was evidence of the shipyard being prepared for future use by BAE. Removing the cranes showed that they were intent on making use of the space, building a modern shipyard onsite.

Cranes as monuments of work

The issue of sustained employment aside, the removal of the cranes was hugely symbolic, an almost ritualistic passing which pricked the consciousness of notions of cultural heritage, and the wider meaning of industrialism in the collective Glaswegian memory. The majority of the testimony that forms an integral part of this chapter comes from interviews conducted with former shipyard workers, as well as local people who have lived in Govan for long periods. Testimony is also drawn from a group interview with the Govan Reminiscence Group, a social history group that have been meeting in Govan for over 30 years.

The Govan skyline can be seen to represent the physical embodiment of work, but also act as a pyschogeographic locater of belonging. Stephen Farmer, who worked at Fairfield as an electrician, servicing and maintaining the yards cranes, expands upon this in a film made by Chris Leslie (2014);

> The cranes were everybody's idea of Glasgow and the Clyde, eh Kenneth McKellar singing *The Song of the Clyde*, you know, 'the hammers "ding dong" is the song of the Clyde', up and down the river, and the welding arcs and the burners, sparks come cascading out the side of the ship, that was all beautiful stuff, beautiful theatre when you sailed down the river.

In discussing the theatrical element, Farmer's evocative though romantic description seeks to stress the hyper-visual culture of shipbuilding of which cranes were an important part. Farmer (2014) adds:

> The place was a myriad of cranes, so the River Clyde without cranes is a bit alien, so it is a great loss to the skyline, and of the skyline of Govan because you would see them, it doesn't matter if you were in the north of the River, from the south, you'd see the cranes towering over the yards.

Farmer's job involved him spending time up the large Fairfield crane, servicing and maintaining it. He recalled the Fairfield crane as a unique place of work where routine, mundane, and often dangerous work was carried out at a great height. He remembered one incident where he narrowly avoided a fatal injury:

> I was up there fixing limit switches, now the crane has got hand rails all the way on the outside, but of course in the centre section is open because the bogey has to go along the jib, and I have gone to fix the wee box there, and its got a wee limit switch that tells me how far etc., and I am walking along there with my rucksack on, my satchel of tools, and I slipped and fell on my arse! Its greasy, it was horribly greasy on the top of this thing, so aye, bang – there but for fortune, you could have been right down the . . . that was . . . need a cup of tea after that one. I kept complaining to them, I says 'the only bit he greases is the stationary bits, if any rotating bit of machinery got any grease on it, it's by accident!'
>
> *(Farmer 2015)*

Another maintenance worker was Billy Dunn (2016), who reflected on the unique working environment in the sky:

> God it takes me back I must admit – I always remember the smell as well, either they were cooking – but it was hot oil cause there was a lot of grease on the ropes and that. Aye, they would have a wee kettle and wee hot stove, he [the driver] wisnae gonna come down, he would maybe come down at lunch time cause he had 45 minutes, but he is no going to come down for a cup of tea. . . . They used to have heaters in the cabin and they weren't very good, so they would smuggle an open bar fire up and we would disconnect it and take it away and the next day another one would appear!

Testimony of this nature is of importance, not only as it provides rich commentary on the nature of working lives at a distinct place in time, but it documents the cranes as working objects, rather than cultural artefacts that they became for many. The cranes' departure provoked a sense of absence that many had to negotiate. In confronting this new 'alien' existence, many people with connections to the cranes experienced a period of mourning at a point of dramatic lived change. 'That was shocking to see, that was' Farmer (2015) added. A *Herald* article from 2 November 2013 contained a quote from a yard employee, whose name was not given as a means of protecting his or her identity at work. Despite the paper stating the source fully acknowledged that the cranes had no working future, he or she commented:

> Their removal is hugely symbolic. Why remove them now when there is no need to do so? Why emasculate the shipyard? These iconic structures are part of the Glasgow skyline. These are berth cranes – the last on the Clyde actually used to build ships.
>
> *(The Herald 2013)*

Seen in this light, the removal of the cranes was an act of cultural vandalism designed to weaken the established connection between the shipyard and the wider community in Govan. Additionally, the suggestion of the yard's 'emasculation' demonstrates the implied masculinity of industrial space. The cranes gave what Strangleman, Rhodes, and Linkon referred to when discussing industrial culture as an 'illusion of permanence' (2013: 10); they often went unnoticed, their visual directness forming part of the rhythm of everyday life in the area. For life-long Govan resident Colin Quigley (2016), 'you walked out your close and that was the one of the first things you could see, it pointed the way hame!' Another Govan resident Bill Pritchard (2016) remarked:

> I could see it [the Fairfield crane], it was right across from my backyard, and it was always there, and wasn't until two months later to me it was pointed out to me that it was gone. I never even knew it was away. Yet, it was just part of the background. My wife says 'you don't take the rubbish out enough to notice!'

James McDonald (2016) began to develop an interest in industrial photography, prompted by the changing skyline:

> Living in Govan, you tend not to notice the cranes because you think they will always be there. Like most things, you take them for granted. It wasn't until I found out they were going to be demolished and that the river skyline was about to change that I decided to photograph them.

The contemporaneous desire to try to capture the lived change has been explored by Tim Strangleman (2013: 28) who described it as a 'mourning process,' where people 'reflect back' on industrialism and where it left them. For McDonald, this takes the form of a 'reawakening,' whereby he feels he took the crane's symbolism and its importance in shaping his memories 'for granted,' only realising this when it was to be removed. Life-long Govan resident Jean Melvin (2016) remarked on a failure to tap into heritage potentialities: 'I don't know why they took them away, they could have been a tourist attraction.' The significance of the passing of the area's last cranes was picked up by *The One Show*, a topical weeknight television programme broadcast live on BBC One. Glasgow poet Donny O'Rourke (2013) was commissioned to write a poem that marked the crane's passing, with the poem being recited by local residents as part of a short segment, interspersed with historical and contemporary footage of the structures:

> The last time I lay my eyes on
> Our city's steel horizon
> That the sun will never rise on–
> Til' the river drains; We'll mourn the cranes
>
> When there's a huge hole in the sky
> About a hundred meters high
> We'll ask the silent river why
> Glasgow maintains, only memorials to cranes.
>   (Donny O'Rourke 'The Cranes' 2013,
>   reprinted with permission)

In the poem, the cranes are an axis on which society and culture entwine around. For a topic of this nature to be broadcast on primetime television, across the United Kingdom, despite the considerable number of colloquial references in the poem and accompanying televised segment, is evidence of the broad appeal of the subject matter and its wide resonance with communities beyond the west of Scotland.

Class politics and representation in Glasgow

The demise of the cranes sparked wider questions about heritage and representation in the city amid a broader context of Glasgow's complex class politics. In the 1980s and 1990s, Glasgow embarked on a remarkable 'top-down' programme of investment in culture as a way of trying to officially move forward from the issues that arose from rapid industrialisation and later deindustrialisation. In 1983 Glasgow Corporation began the hugely successful *'Glasgow's Miles Better'* campaign to rebrand the city. The campaign aimed to alleviate the city's reputation for crime, poverty, and ill-health that sharp industrial decline had worsened. The corollary to this was an investment in culture, such as the inauguration of *Mayfest*, billed as the most ambitious public and visual arts festival in Britain, incorporating theatre, dance, music, and contemporary art. Similarly, Glasgow hosted the 1988 Glasgow Garden Festival, with over 4.3 million visitors descending on the derelict Princes Dock, which had been transformed as part of the event. Glasgow's cultural rebirth culminated in the city being awarded the title of European City of Culture in 1990, though this renaissance was not accepted uncritically. Damer (1990: 201) suggests the emergent Glasgow was a 'worker's city whose rulers resolutely pretend that it is something else' whilst Spring (1990: 192) remarked in this way on the 'myth of the new Glasgow': a city with 'its sooty-face scrubbed clean, like a recalcitrant schoolboy forced to visit a wedding or a funeral.' These quotes show the prevalent idea that working-class history has often been victim to erasure, and that the 'real' Glasgow has been lost. These comments came in 1990, though, in 2018, Glasgow's last fragments of industrial archaeology are still rife with contestations. Less than a mile down river from the site of Govan's last cranes lies the derelict Govan Graving Docks. Built between 1869 and 1898, the site is Category A listed by Historic Environment Scotland, described as being 'an outstanding graving dock complex without parallel in Scotland' (Historic Environment Scotland, 2018). Additionally, it is noted that 'the complex is of architectural/historic interest in an international context, of major significance in terms of the history of the world shipbuilding.' Despite being A-listed, the site has been derelict since the 1980s, becoming heavily fire-damaged, vandalised, and graffitied over time. More recently, the Graving Docks have become a battle ground as the owners of New City Vision seek to build 700 homes on the site, a plan which has been met with opposition from local people (McCall 2018). In response to similar plans unveiled in 2014, a group named the Clyde Dock Preservation Initiative was set up, launching a petition to transform the site into a 'shipbuilding heritage' park, garnering over 12,000 signatures (McGillivray 2014). The petition states that the site is

> one of the most important features of Glasgow's industrial heritage and represents a major opportunity to educate future generations about the city's past in a way that is more meaningful than looking at old photographs in a museum.

The quote demonstrates the idea that few remaining physical sites of Clydeside deindustrialisation are intrinsically important in terms of memorialising and locating people's experiences. This is in line with what Smith and Campbell (2017) call 'nostalgia for the future'; the idea that nostalgia can be innately positive in its hopes for lessons to have been learned from previous injustices, or powerful enough to lead to impactful societal future decision making and planning. Though this chapter focuses on shipyard cranes as structures where the contestations and reverberations of deindustrialisation play out, the case of the Graving Docks allows for useful comparisons, whilst also demonstrating wider industrial heritage narratives overall.

Conclusions

This chapter has collated the material culture relating to the historic skyline of Govan that was pulled apart between 2007 and 2014. It has added to the evolving body of literature that seeks to move beyond simply recording industrial apparatus, but rather assess its wider historic value within the context of the connections that people, through memory, have with the physical remnants of industrialism. Similarly, it has sought to further develop understandings of how spatial attachments are continuing to be dislocated by the long-term impacts of deindustrialisation. The case study of Govan shows that, over time, working objects can become cultural artefacts, with a resonance and meaning beyond their original purpose. Delores Hayden (1997: 9) argued that landscapes are 'storehouses of meaning,' a comment particularly true of Govan, where the cranes became both visual reminders of the lived experiences of work, and the fragility of work, as evidenced by the fears that their removal had a direct correlation with the future of the jobs within the yard. The cranes gave an impression of permanence, and their demise sparked the consciousness of notions of cultural heritage, and the wider meaning of industrialism in the collective Glaswegian memory. The case study of Govan also identifies the inherent problems in applying the term 'post-industrial' within certain spatial geographies in Scotland. Despite the unprecedented decline of industry in the second-half of the twentieth century, as measured quantitatively through reductions in employment figures and production levels, the remaining fragments of industry reveal a liminality that continues to evoke feelings of romanticism, attachment, and vulnerability that suggests the legacy of deindustrialisation has still not settled. By using methodological approaches like oral history, the intangible cultural or social landscape can be contrasted with the drastically evolving physical landscape, providing unique insights into wider temporal belonging.

References

Brotchie, T.C.F. (1905). *The history of Govan: Glasgow*. Govan: The Old Govan Club.
Bryne, D., and Doyle, A. (2004). The visual and the verbal: The interaction of images and discussion in exploring cultural change. In: C. Knowles and P. Sweetman, eds. *Picturing the social landscape: Visual methods and the sociological imagination*. London: Routledge, pp. 166–177.
Cowie, J., and Heathcott, J. (2003). *Beyond the ruins: The meanings of deindustrialization*. Ithaca, NY: Cornell University Press.
Dalglish, C., and Driscoll, S. (2009). *Historic Govan: Archaeology and development*. Edinburgh: Historic Scotland.
Damer, S. (1990). *Glasgow: Going for a song*. London: Lawrence and Wishart.
DeSilvey, C., and Edensor, T. (2013). Reckoning with ruins. *Progress in Human Geography*, 37 (4): 465–485.
Dunn, B. (2016). Interviewed by Martin Conlon, 19 February.
Farmer, S. (2015). Interviewed by Martin Conlon, 3 October.
Fisher, A., and Buchan, N. (1973). *The Shipyard Apprentice*. School of Scottish Studies Collection of recordings, Tobar an Dualchais/Kist o Riches. Accessed 15 August 2018, www.tobarandualchais.co.uk/en/fullrecord/100599/3
Glasgow History. (2010). *Sailing down the Clyde: "Doon the Watter."* Accessed 8 October 2018, www.glasgowhistory.com/wp-content/uploads/2010/07/Clyde-Fairfield-Crane-from-Elder-Park-Govan1.jpg
The Glasgow Story. (2004). *Fairfield crane*. Accessed 8 October 2018, www.theglasgowstory.com/image/?inum=TGSA00830
Hayden, D. (1997). *The power of place: Urban landscapes as public history*. Cambridge: MIT Press.
The Herald. (2013). *Fears over Govan yard as cranes removed*. 2 November. Accessed 2 February 2018, www.heraldscotland.com/news/13129894.Fears_over_Govan_yard_as_cranes_removed/
Herod, A. (2001). *Labour geographies: Workers and the landscapes of capitalism*. New York and London: The Guildford Press.
High, S., Mackinnon, L., and Perchard, A. (2017). *The deindustrialized world: Confronting ruination in postindustrial places*. Vancouver: University of British Columbia Press.
Historic Environment Scotland. (2018). *Buildings at risk register for Scotland: Govan Graving Docks*. Accessed 19 June 2018, www.buildingsatrisk.org.uk/details/909298
Hume, J. (1974). *The industrial archaeology of Glasgow*. Glasgow: Blackie and Son Limited.
Ingold, T. (1993). The temporality of the landscape. *World Archaeology*, 25 (2): 152–174.
Leslie, C. (2014). *Last of the Govan cranes*. Accessed 13 Jan. 2018, https://vimeo.com/101037589
Mah, A. (2012). *Industrial ruination, community and place: Landscapes and legacies of urban decline*. Toronto: University of Toronto Press.
McCall, C. (2018). Controversial housing plan for Govan Graving Docks moves forward. In: *The Scotsman*. Accessed 17 May 2018, www.scotsman.com/business/companies/controversial-housing-plan-for-govan-graving-docks-moves-forward-1-4621604
McDonald, J. (2016). Interviewed by Martin Conlon. 3 November.
McGillivray, I. (2014). *Restore Govan Graving Docks in Glasgow to create a shipbuilding heritage park #SaveGovanDocks*. Accessed 13 June 2018, www.change.org/p/glasgow-city-council-restore-govan-graving-docks-in-glasgow-to-create-a-shipbuilding-heritage-park-savegovandocks
Melvin, J. (2016). Interviewed by Martin Conlon, 3 November.
Mitchell, D. (2005). Working class geographies: Capital, space and place. In: J. Russo and S. Linkon, eds. *New working-class studies*. Ithaca, NY: Cornell University Press, pp. 78–97.
Orange, H. (2008). Industrial archaeology: Its place within the academic discipline, the public realm and the heritage industry. *Industrial Archaeology Review*, 30 (2): 83–95.

Orange, H. (2014). Changing technology, practice and values: What is the future of industrial archaeology? *Patrimonio: Arquelogia Industrial*, 6: 64–69.

Orange, H. (2015). *Reanimating industrial spaces: Conducting memory-work in post-industrial societies*. Walnut Creek, CA: Left Coast Press.

O'Rourke, D. (2013). *The cranes*, poem performed on The One Show, BBC One, 15 November.

Pritchard, B. (2016). Interviewed by Martin Conlon, 3 November.

Quigley, C. (2016). Interviewed by Martin Conlon, 3 November.

Samuel, R. (2012). *Theatres of memory: Past and present in contemporary cultures*. London: Verso.

Smith, L., and Campbell, G. (2017). "Nostalgia for the future": Memory, nostalgia and the politics of class. *International Journal of Heritage Studies*, 23 (7): 612–627.

Smith, L., Shackel, P.A., and Campbell, G. (2011). *Heritage, labour and the working classes*. London: Routledge.

Spring, I. (1990). *Phantom village: The myth of the new Glasgow*. Edinburgh: Polygon.

Strangleman, T. (2013). "Smokestack nostalgia", "ruin porn" or working-class obituary: The role and meaning of deindustrial representation. *International Labor and Working-Class History*, 84 (1): 23–37.

Strangleman, T., Rhodes, J., and Linkon, S. (2013). Introduction to crumbling cultures. deindustrialization, class, and memory. *International Labor and Working-Class History*, 84 (1): 7–22.

19
'HIDDEN IN PLAIN SIGHT'
Uncovering the gendered heritage of an industrial landscape

Lucy Taksa

Introduction

The impact of industrial heritage on men and masculinity continues to be a subject explored by sociologists and historians who note the paradoxical ways in which traditional industrial masculinity remains both visibly and invisibly influential, particularly in certain, de-industrialised locales (Roberts 2014; Ward 2015; Thurnell-Read 2016; Walker & Roberts 2018). As Salzinger noted, 'masculinity's historically accrued capacity to stand in for the general' has been enabled by the 'absence of explicit naming' (2004: 13). As a result, in her words, the 'unelaborated "worker" is always (already) male,' meaning 'that, socially, femininity is often specified, by name or by highly explicit conventions around embodiment, whereas masculinity literally 'goes without saying.' 'Hidden in plain sight, masculinity is easy to miss, whereas the embodied frill of femininity intrudes on the field of vision' (Salzinger 2004: 15).

A decade ago, Smith noted that 'gender tends to be overlooked in discussions of heritage' (2008: 159) and is often invisible in heritage practices (Smith 2008: 165). The promotion of a masculine 'vision of the past and present' through male-centred stories (Smith 2008: 159) is also reinforced by the marking of women and femininity. As Reading (2015: 401) pointed out more recently, gendered approaches to heritage have 'tended to use a simplistic framework in which gender is telescoped into a focus on women.' Accordingly, heeding Reading's advice, this chapter considers how 'constructions of masculinity and femininity interact with what is valued and included as heritage'. In order to uncover these relational aspects of gender, in time and in place, and 'whose identities are being "represented and reinforced"' (Reading 2015: 401), I consider past assessments and representations of the significance of one industrial place, the memories of those who once worked there and interpretations of their current heritage significance. In doing so I aim to provide one example of how the close association between industrial heritage, men, and masculine perspectives continues to dominate interpretations and management of industrial heritage (Labadi 2007: 162; Reading 2015: 404; Taksa 2003).

How has this masculine perspective been transmitted and reinforced over time? To address this question, I examine various 'vehicles of memory' (Confino 1997: 1386), including documentary and oral archival sources and more recently produced heritage assessments related to the past and present of the Eveleigh Railway Workshops in Sydney. Beginning with a consideration of the way the site was viewed and valued in newspapers and journals during its heyday, I then present memories of retired male workers who were employed at Eveleigh during the Second World War when the predominantly male workforce was augmented by women munition workers. In this way I contrast the unmarked masculinity associated with the site's industrial operations with the way the male workers remembered the women munition workers in terms of their femininity, marking them as outsiders. Finally, I reflect on the intersections between memory, place, and gender by reviewing how proposals for the site's heritage interpretation naturalise and legitimate the 'androcentric assumptions and messages' (Smith 2008: 167) identified in historic accounts of Eveleigh's industrial significance and the gendered assumptions and values identified in the male workers' memory narratives.

Gender and memory in place: the Eveleigh Railway Workshops case

The Eveleigh Railway Workshops encompass numerous buildings in which locomotives were maintained, repaired, assembled, and manufactured from the 1880s until the late 1980s. Although small numbers of women performed ancillary tasks there, Eveleigh's workforce was dominated by men throughout its century of rail operations. It is among Australia's most significant industrial heritage sites, yet one where the preservation of tangible material culture has dominated heritage management activities (Taksa 2003, 2005a, 2009). While there have been a number of proposals developed for heritage interpretation since Eveleigh's rail operations ended in the late 1980s (Godden Mackay 1996; 3-D Projects et al. 2012; Cooling & Vinton 2016), none have been implemented. The resulting marginalisation of Eveleigh's intangible cultural heritage has sustained the masculine bias identified by scholars in relation to the nature of paid work and technology, workplaces, work relations, and heritage (Cockburn 1983; Cockburn & Ormrod 1993; Collinson & Hearn 1997; Connell 2002; Devault 1990; Game & Pringle 1983; Shortliffe 2016; Smith 2008; Taillon 2001; Taksa 2005b; Wajcman 1991). By examining the men's memories of the women who worked at Eveleigh during the Second World War, I not only seek to challenge the ostensibly gender neutral representations of the site's significance throughout its operations but also the legitimacy of authorised gendered accounts of its value as industrial heritage.

The memory narratives used in this chapter draw on numerous different projects, some undertaken by trade union activists, some for heritage consultants funded by New South Wales (NSW) government bodies, and also for studies funded by the Australian Research Council (ARC). All the interviews referred to in this chapter were with Eveleigh's retired male workers. Among these are a number that were collected between 1983 and 1985 by sound recordist Russ Hermann (1995) for the Combined Railway Unions Cultural Committee's (1985a) NSW *Railway Historic Exhibition Project*. These recordings were combined into two cassettes entitled *Railway Voices*, together with reflections from labour historians (including me), and formed an important addition to the Exhibition, which was finalised in 1985 and subsequently called *Trains of Treasure* (1985b). The other interviews are drawn from various projects focused specifically on Eveleigh, such as the *Eveleigh Social History Project*, which I undertook with assistance from Joan Kent in 1996 for the government-funded *Eveleigh Workshops Management Plan for Moveable Items and Social History* produced by Godden Mackay heritage consultants (Godden Mackay 1996) for various government authorities; the NSW Department of Urban Affairs and Planning funded group interviews I conducted with retired Eveleigh unionists in 1996, which were recorded at Summer Hill Films and filmed at Eveleigh by Rod Freedman; and numerous interviews I conducted for the ARC-funded *Work, Technology, Gender and Citizenship at the Eveleigh Railway Workshops Precinct Project* between 1997 and 2000. Given my own involvement in all these projects, including the railway history exhibition, I claim no intellectual detachment from the records. On the contrary, I recognise my co-production of the memory narratives, although I acknowledge that the shared authority with the informants applies only to the creation of the records and not to their interpretation (Frisch 1990; Grele 2007; Pente et al. 2015: 33, 50). Other vehicles of memory referred to include newspapers and journal articles from Eveleigh's period of operation and heritage interpretation proposals produced since its operations ceased.

My analysis of these sources uses a gendered lens that aims to highlight what Confino (1997: 1393) referred to as the 'politics of memory.' In effect, I explore how these sources reflect institutionalised assumptions of masculinity that help to maintain the gender distinctions and unequal gendered power relations (Peretz 2016: 33) in individual memories and in the collective social memories that are transmitted through representations of heritage significance. Reading against the grain of the oral and documentary records to identify the unmarked masculine associations with industrial operations, and work and workers as 'male,' foregrounds gendered discourses and power relations. In this way, I investigate the 'symbolic and cultural acts, utterances and expressions' (Reading 2015: 407) that privilege masculine gendered roles, values, and norms and how they have been embedded and transmitted through cultural memory thereby creating '[a]n unequal gendered legacy' (Reading 2010: 11–12) in situ.

The Eveleigh Railway Workshops are treated here 'as a discursive space in which intangible industrial culture and the collective memories of "ordinary" people are also embedded' (Leung & Soyez 2009: 63). I view the workers' memories 'as an expression of collective experience' (Fentress & Wickham 1992: 25) and as a form of interpretation (Sturken 1997: 6). To the extent that the male workers' memories are 'imbued with cultural meaning' (Sturken 1997: 3), they can be said to form part of cultural memory. However, as Brockmeier pointed out, we not only need to recognise that cultural memory is a social process but also 'that this process itself is culturally mediated within a symbolic space laid out by a variety of semiotic vehicles and devices.' Such artefacts of cultural memory can include 'oral and written language, and other systems of communication,' as well as archives, 'memorials and other architectures and geographies in which memory is embodied and objectified' (2002: 25). According to Brockmeier, narrative provides 'a major integrating force' in this symbolic space (2002: 33). Two of the three narrative forms she identifies are pertinent here. The first is what she referred to as 'a semiotic order' in which a material structure or monument 'can be seen, or read, as a narrative text' (2002: 34). The second is 'a discursive or performative order,' which foregrounds 'the narrative event as a site where the social is articulated and its contradictory implications are struggled over' (2002: 35).

From this perspective, Eveleigh can be interpreted as a narrative text that encompasses not simply its buildings but also an authorised sexual division of labour and 'gender pattern of location,' which Cockburn and Ormrod (1993: 6) defined as 'the contemporary practices in which the sexes tend to be differentially positioned' in relation to different activities and technologies. In the context of this material and symbolic space, the workers' memory narratives can be read as a performative negotiation of the challenge posed to masculine culture and identity by the presence of women munitions workers in a traditionally masculine workplace. Here the metaphor of 'paradoxical space' elaborated by Rose (1993: 140–41 cited in Mahtani 2001: 299) is helpful because it focuses attention on women's confinement in space and their location in 'several social spaces simultaneously' at the centre and on the margins. A critical point made by Rose (1993: 152 cited in Mahtani 2001: 299) is that such 'simultaneous occupation of the centre and margins can critique the authority of masculinities.' In other words, I will suggest that in the paradoxical space of the Eveleigh Munitions Annexe, which was located in a mezzanine level of the main locomotive workshop building during 1942–1943, the women were simultaneously freed to undertake paid industrial work traditionally done by men and yet simultaneously contained (Desbiens 1999: 183) not only physically but also structurally and symbolically.

The Eveleigh Railway Workshops and its workforce

The Eveleigh Railway Workshops were constructed between 1880 and 1886 on a 51-hectare site four kilometers south of Sydney's Central Business District to maintain and repair steam locomotives. The locomotive workshops and large erecting shop occupied the southern side of the site and when railway operations ended in the late 1980s, the locomotive workshops were transformed into a technology park

(Brennan 1996). In 2015 they were sold by the NSW Government to Mirvac, one of Australia's largest property developers (Mirvac 2015).

During Eveleigh's century of continuous rail-related operation, its maintenance, repair, and manufacturing operations were performed by skilled and unskilled male blue-collar workers. By 1927, over 6,500 people worked there before the Great Depression caused a substantial reduction. After 1933 employee numbers increased to over 5,000 and remained around this figure until the outbreak of the Second World War (Guthrie c.1955: 7; Assistant Chief Mechanical Engineer 1955: 5). Although this workforce was predominantly composed of men, women were continuously employed there on a wide range of jobs, including as upholstresses, office cleaners and laundresses, clerks, and typists: 60 were employed as process workers involved with munitions work between 1942 and 1943 and seven were employed as industrial nurses after 1946 (NSW Government: State Archives & Records 1999). The latter two groups of women figure most in the memories of the male workers. Rarely do these women appear as part of the site's industrial heritage.

Eveleigh's significance: masculinity unmarked

Eveleigh's significance was based on its association with steam-powered technology, its capacity for technological innovation and the grand scale of its buildings, operations and machines. A few years after full operations began an article in *The Illustrated Sydney News* (1891: 11, 13) stressed that Eveleigh's machinery and appliances had no equal 'in the Southern Hemisphere,' nor 'out of England.' It was an undertaking that provided amazement for visitors who gazed at the 'magnitude of the operations,' its 'marvellous machinery and ingenious workmanship.' About 30 years later, Hyde (1922: 176, 179) wrote that '[t]he collection of machinery at Eveleigh' was 'magnificent' and extolled the 'excellence of workmanship and artistic craftsmanship.' As employee Stan Jones (1939) put it, Eveleigh was the 'heart of the NSW Transport System'; its 'Row upon row of drab smoke-grimed buildings' pulsed with 'a throbbing energy . . . to the accompaniment of the thump, thump, thump of giant presses torturing white-hot steel into servitude,' a 'steady drone of high-powered machinery, drilling, boring and turning in every possible fashion . . . and the staccato noise of the boilermakers' rattler.'

Decades later, in 1988, it was these buildings and machines and their contribution to the state's history of transport (Don Godden & Associates 1986) that justified the listing of the entire Eveleigh Railway Workshops complex on the Register of the National Estate as a site of national significance. In 1994, a heritage assessment for the NSW Government reiterated that the 'principal significance of the Eveleigh Precinct' was 'its association with and demonstration of railway history and technological development associated with that industry' (Thorp 1994: 17). Similarly, the Statement of Significance included on the NSW State Heritage Register in 1995 stated that the Eveleigh Railway Workshops were 'some of the finest historic railway engineering workshops in the world' whose value was 'increased by the fact that it is comprised of assemblages, collections and operational systems' (NSW Government: Office of Environment and Heritage n.d.).

These apparently gender neutral representations of Eveleigh's significance can be viewed as vehicles of memory that have cumulatively constituted an authorised cultural memory. However, if we acknowledge 'the ways in which gender gains expression in technology relations, and technology acquires its meaning in gender relations' (Cockburn & Ormrod 1993: 7), then we begin to recognise that the narrative surrounding Eveleigh's technological significance is not gender neutral. As Wajcman (1991: 19) pointed out, masculinity has been central to 'the language of technology [and] its symbolism.' Just as importantly, both men's physical capacity and their access to, familiarity with, and control over machinery has provided an important source of male power that has informed and relied on social practices, as well as on the definition of tasks and the selective design of tools and machines to match men's bodily strength (Cockburn 1986: 93, 97–98). This affinity between technology and men and also between skill and masculinity can, according to Wajcman (1991: 38), be 'seen as integral to the constitution of male gender identity.'

In railway workshops, where steam fitters, blacksmiths, lathe-operators, and boilermakers, among others, manufactured, assembled, repaired, and maintained steam locomotives, workers shared knowledge of and control over machine technology, and formed bonds with each other. Such bonds were reinforced by a range of social practices including practical jokes (Wajcman 1991; Klein 1994) that projected and affirmed 'powerful visions of manliness' (Drummond 1995: 79–80). As was the case in other traditional metal shops, 'dirt, noise, danger' and men's 'machine-related skills and physical strength' were all 'suffused with masculine qualities' that provided 'fundamental measures of masculine status and self-esteem' (Wajcman 1991: 143). According to Taillon (2001: 39), all these characteristics mediated the railway workers' 'sense of themselves as men.' Moreover, these hallmarks of hegemonic masculinity (Connell 1995: 63, 77) sustained a culture of machismo, which made these workplace cultures unfriendly to women. It relied on women being 'at home providing the backdrop' (Wajcman 1991: 141); that is, on the margins in traditionally feminine roles. Eveleigh's masculine culture during the steam era typified this gender order.

Barry Smith (as quoted in Burke 1995: 243) thought that during 'the steam days there was great comradeship between the men. We had to look after one another.' Similarly, Vaughan Givillian (1987) said that during the steam era 'the trades all got together' because they 'were all in the same sort of situation . . . more or less like a brotherhood . . . it was a challenge to see this big monstrous thing go by . . . and say with self-satisfaction well look I helped to put that thing out there.' For Givillian, the camaraderie was built on day-to-day interactions in the sheds. He therefore stressed that, during the steam era, the driver would 'come in and he'd prepare his engine and oil it and you'd have a yarn with him and you'd know how his wife and kids were' (1987).

Dealing with dangerous conditions was central to Eveleigh's workplace culture. In recalling the 'terrible job' his father had as a gland packer in the steam locomotive working in the smoke-boxes, John Willis (1996) said this was 'the dirtiest part you could work in.' In those days, 'safety wasn't thought of,' there were no gas masks 'no air flows, nothing.' Because 'it was a very dangerous place to work . . . you had to know what you were doing. . . . You took a lot of chances, a lot of risk.' For John Bruce (1996a) 'Noise was a great problem.' In the boiler makers' shop, it was 'absolutely deafening.'

These memory narratives highlight the 'patterns of talk and interaction' through which Eveleigh's male workers constituted 'a shared reality' (DeVault 1990: 97) about the nature of their work and workplace and through which they sustained their individual and collective identities as men.

Eveleigh at war

Eveleigh was drawn into the war effort in 1939, after Defence Department officials made a request to the Commissioner for Railways for space to be made available for the manufacture of shells (NSW Government Railways c.1947). On 3 March 1942, in response to labour shortages and the increased demands on the Railways Department for the transport of troops and equipment, the NSW Railways were proclaimed a 'Protected Undertaking' under National Security (Man Power) Regulations, and on 23 April previous permissions for rail workers to enlist were cancelled. At the very time when the War had extended to the Pacific and thus closer to Australia, and manliness was being construed in terms of bravery associated with war service and battle (Bolton 1990: 7–10, 15–16), Eveleigh workers had no choice but to stay at home. It was in this context that a mezzanine level was built above one of the bays in the locomotive workshop for the Munitions Annexe. This space consisted of a meal room, a change room with lockers, and a rest room exclusively for women who began there on work traditionally done exclusively by men in November 1942. In fact, the women's incursion into the workshops was short lived; the Annexe ceased production on 12 June 1943 (NSW Government Railways c1947: 57; NSW Department of Railways n.d.: 1–9, 42–43). Nevertheless, it did affect the wartime experiences of Eveleigh's male workers.

Men's memories and the paradoxical space of the Munitions Annexe

John Bruce (1996a), who joined the railways in January 1940 recalled, 'There was a gradual build up to a war footing' and 'the manufacture of the 25 pounder shells' in 'the wartime annexe.' It was all 'a bit hush hush,' he said, as the area was 'fenced and taboo to the general staff.' Similarly, Bob Matthews (1996), who also started in the locomotive workshop in January 1940, recalled that the influx of the women munitions workers 'dramatically' changed the nature of the 'organisation.' They had what he referred to as 'a separate identity.' 'They were all fenced off,' he said, 'and nobody was allowed in' the Annexe, 'because security was very tight.'

The Munitions Annexe can be read as a 'paradoxical space' in which the women process workers were located in 'several social spaces simultaneously' at the centre of pivotal war work and on the margins of Eveleigh's masculine industrial workplace. On the one hand, the Annexe challenged traditional gendered structures, roles, and contemporary practices by differentially positioning men and women in relation to various activities and technologies (Cockburn & Ormrod 1993: 6). On the other, it sustained the separation of men and women in the workplace since it provided a confined space within which the women were physically segregated from the male locomotive workshop employees. John Bruce (1996b) recalled that the women 'had a different stopping and starting time' to the men in 'the general workshop area . . . they worked around the clock with shifts. But when the mob of chappies were knocking off at 5 o'clock, the girls' had 'already left an hour earlier or an hour later of the changeover of shifts.' However, efforts to isolate the women from the men had mixed results. For instance, John Bruce (1996a) recalled, 'You could look through the wire to the girls working in there,' although 'You'd get moved on from perving on the girls' if caught in the act. This representation of the women munitions workers as objects of masculine desire was one way in which the women's challenge to traditional gendered assumptions and roles was negotiated. Other ways were also employed to limit the women's 'freedom' to traverse this masculine space. Jeff Aldridge (1999) remembered that:

> we used to see the afternoon shift coming in and they used to take a short cut through number 2 blacksmiths' workshop down to the middle road and along the middle road up into the annexe and it was mainly upstairs – and downstairs I think it was 6 and 7 bays. . . . But it used to be funny, most of the girls used to walk through to get on to the 3 o'clock afternoon shift. The smart alecs would put a ten-shilling note tied to a bit of string and they would have it laying on the centre of the floor and sit back in their fliers near the centre of the shop and as the girls come down they'd see the 10 dollar note see? And the blokes would give it a bit of a flick with the string and the girls used to go trying to catch it till their legs were spent and they'd fall down – and they done the splits to the ground and (laughs) this was done day after day – done for years.

This practice invoked what Reskin and Padavic (1994: 11) refer to as 'gender displays, which are language or rituals so characteristic of one sex that they mark the workplace as belonging to that sex.' It can also be read as a performative struggle to deal with the contradictions created by this paradoxical space.

In addition, the men downplayed the women's capacity to perform men's work by questioning their stamina for handling the lengthy shifts that accompanied the introduction of 24-hour operations (Bob Matthews 1996). As Keith Johnson (1996) elaborated:

> I used to see the Annexe . . . and you'd heard stories . . . because they were working so much that they put sand in one of the gear boxes of the lathe to slow it down so that they'd have a bit of time off. They probably got overworked, working so much overtime.

Such stories, as much as their retelling, are suggestive of a struggle to maintain the traditional representation of women as physically weaker than men and therefore not fit for industrial labour. The contrasting representations of how the men reacted to increased hours of work, during what Bob Matthews (1996) described as 'that bitter part of the war,' highlights the way masculine norms were upheld in line with the ideals of hegemonic masculinity, particularly those related to male bodily stamina for dirty and dangerous industrial employment. Matthews acknowledged that the men 'were very unhappy' about having to work 'for twelve hours a day, twelve days straight' followed by only 'two days off.' Nevertheless, he said 'we did it for quite a number of years until the war was finished.' In this narrative, the men were able to sustain the additional demands made on them without needing to resort to the sort of subversive evasion allegedly engaged in by the women.

The men's collective experience of the contradictory implications of the women's employment in the Annexe is also manifested in their narratives around the unsuitability of the material and social space of the locomotive workshops for women. For instance, Jeff Aldridge (1999) described Eveleigh as 'a complete workshop of men' and Keith Johnson (1996) explained that the workshops weren't appropriate for women because when 'You walked into the large erecting shop . . . you walked straight in past the urinal.' For Keith Johnson (1996) the women 'had to be protected in such a male dominated area,' even though he acknowledged that some of the women 'were quite capable of looking after themselves.' At the same time, he expressed concern for the men by commenting: 'fellows getting dressed at night, I don't know how they got on, when ladies were in the Annexe.'

These narratives illustrate the way the men sought to maintain the cultural ideals of masculinity and femininity and a gendered pattern of location that affirmed sexual differences and traditional gender stereotypes. The men also invoked traditional norms of femininity by depicting the women munitions workers as 'hard cases,' who had 'fun with the boys by telling them crude jokes . . . much to their embarrassment' (John Lee 1999). For John Lee, these women were 'a pretty tough lot,' 'and a unique pack of women, such as you wouldn't find at a lot of other social outings.' Hence, even though these women were confined in the Annexe most of the time, their presence in the masculine space of the locomotive workshops and their participation in the practises traditionally associated with masculine industrial workplace cultures transgressed acceptable gendered mores and challenged the gendered assumptions that construed the Eveleigh workshops as a site of male power. The men's memory narratives show how they deployed norms of femininity and 'highly explicit conventions around embodiment' (Salzinger 2004: 15) to negotiate the paradoxical space occupied by the women. To all intents and purposes these same norms have informed assessments of Eveleigh's heritage significance and proposals for its interpretation.

Interpreting Eveleigh's significance

In 1996, an Interpretative Concept was produced for the *Eveleigh Workshops Management Plan for Moveable Items and Social History* (Godden Mackay 1996: 103–115). Here the site was depicted as 'relentlessly masculine, in its scale, its design, its grittiness, its smells, its particular dirt, its muscular machinery.' The challenge, according to the author, was 'neither to sanitise the site, nor, pun intended, emasculate, it, but rather exploit its very nature to explore what it once meant to be a man working in such an environment . . . what it meant to be a man working in a heavy industrial site such as Eveleigh, where one's status was defined by one's capacity for and ability at physical work' (Godden Mackay 1996: 106–107). The masculine bias evident in this depiction was also reflected in the normative stereotypical gendered categories embedded in the proposed displays. Suggested mini-exhibitions for the 'Tools' display included one on 'Real Blokes' focused on machinery and engineering processes to highlight notions of physical strength, physical skill, and associated masculine values that were important for retaining employment in industrial workshops and 'in the daily battle for bread.' Another, on 'Hard Yakka' (Australian vernacular for work), centred on a 'piece of machinery' to examine 'industrial conditions, the process of work itself, the subdivision of labour within the plant.' In this display it was proposed to 'look at the way women were brought in for munitions work, and kicked out after the war ended' (Godden Mackay 1996: 114). The only other reference to women was as part of an exhibit of an 'Eveleigh worker's house – circa pre-War,' entitled, 'Hearth and Home' to allow 'the story of the women and the children of the Eveleigh workers' to be told (Godden Mackay 1996: 110–112). The assumption underpinning these proposals that Eveleigh was no place for women, reproduced the collective memory narratives provided by the men who had worked there. Subsequent Interpretation concepts similarly illustrate how normative representations of masculinity and femininity 'interact with what is valued and included as heritage' (Reading 2015: 401).

The *Eveleigh Railway Workshops (ERW) Interpretation Plan & Implementation Strategy* produced in 2012 for the Redfern Waterloo Authority noted that:

> the primary theme of the ERW site is the place itself – a nineteenth century industrial workshop for the construction and maintenance of NSW's locomotives and rolling stock. It is the preeminent site in NSW in which to explore the various aspects of heavy industrial work – the trades, the skills, the conditions, the products, the cultural life, workplace relations, the events of ERW, as well as the technology, the site organisation and its wider role in the maintenance and development of NSW railway system.
>
> *(3-D Projects et al. 2012: 7)*

Its proposal for interpretation of the social dimensions of the site centred on 'An installation of portraits of former ERW workers in one bay' that would draw on 'historic and contemporary photographs' in order to 'reflect something of the diversity of the trades, ethnicity and gender of former employees' (3-D Projects et al. 2012: 8). This was the only mention of gender contained in the report. More importantly, the treatment of the war period and the Munitions Annexe completely omitted any mention of the women who worked there, let alone their impact on the workshops' masculine culture (3-D Projects et al. 2012: 20). In short, the fact that gender rated only one mention in this proposal illustrates Salzinger's point that the 'unelaborated "worker" is always (already) male' (Salzinger 2004: 13).

Commentary of the site's significance replicated the Statement on the State Heritage Register mentioned earlier (3-D Projects et al. 2012: 46), while its social significance was related to the fact that Eveleigh was 'one of the largest employers in Sydney' and demonstrated 'the capacity of Australian industry' and the 'high level of craft skills,' which provided 'an important source of pride' (3-D Projects et al. 2012: 47). Although 'identity' and 'racism/sexism' were mentioned as possible subthemes (3-D Projects et al. 2012: 48–49), the major sample themes centred only on the following: 1) place, including 'buildings/operational divisions, structures, layout, logic'; 2) uses; 3) events, including the '1917 strike, union meetings, wartime production, social events, and ERW closure'; and 4) Workshops, including 'machines & workers in action, carriages & locomotives under construction' (3-D Projects et al. 2012, 51, 53). The centrepiece of this interpretation strategy was a '"floating" artefact and audio-visual installation to be located in Bay 2 of the ATP Locomotive Workshops' entitled 'Ghosts' to explore the 'personal stories and experiences of the place's past inhabitants' (3-D Projects et al. 2012: 8). This orientation was subsequently replicated in the Heritage Interpretation Strategy produced in 2016 (Cooling & Vinton 2016: 42, 57–62). In effect, all these proposals reinforce the 'androcentric assumptions and messages' (Smith 2008: 167) contained in representations of the site's significance both during its operations and since its transformation into a heritage place.

Conclusion

This chapter has considered Eveleigh as a narrative text that encompassed not simply buildings but also an authorised sexual division of labour and 'gender pattern of location' (Cockburn & Ormrod 1993: 6). In the context of this material and symbolic space, the workers' memory narratives have been read as a performative negotiation of the 'paradoxical space' created by the Munitions Annexe and the women's presence in their traditionally male-dominated workplace. Through this gendered perspective, I have shown how socially endorsed norms of masculinity and femininity were enacted at Eveleigh through a range of practises articulated in the memory narratives. Taken together with the ostensibly ungendered representations of Eveleigh's significance in historical articles and more recently produced heritage reports and assessments, we have seen how various vehicles of memory have contributed to a gendered 'politics of memory' (Confino 1997: 1393) that has been naturalised and legitimated in proposals for the site's heritage interpretation. As Smith (2008: 173) stressed 'the lack of extended discussion of gender issues in the heritage field helps to reinforce the continued legitimacy of authorised … accounts of history … and the uncritical maintenance of received gender identities', as well as hierarchical relations between them. As a result, Eveleigh and its ghosts stand out as a beacon of masculinity that reflects '[a]n unequal gendered legacy' (Reading 2010: 11–12).

References

3-D Projects, in collaboration with Artspace and Only Human. (2012). *Eveleigh Railway Workshops interpretation plan & implementation strategy*. Sydney: Redfern Waterloo Authority.

Assistant Chief Mechanical Engineer. (1955). Correspondence with F. P. H. Fewtrell, Works Manager, 14 April. *In*. State Rail Reference publications, monographs, pamphlets and papers, NSW State Archives and Records Series, NRS 17514 (originally A88/44 – Box 3, State Rail Authority Archives Sydney).

Brennan, W. (1996). On the track of new technology: Redfern railway shed hosts info superhighway. *Sunday Telegraph*, 30 June, p. 50.

Brockmeier, J. (2002). Remembering and forgetting: Narrative as cultural memory. *Culture Psychology*, 8 (1): 15–43.

Bolton, G. (1990). *The Oxford history of Australia, Volume 5, 1942–1988*. Melbourne: Oxford University Press.

Burke, D. (1995). *Making the railways*. Sydney: State Library of New South Wales Press.

Cockburn, C. (1983). *Brothers: Male dominance and technological change*. London: Pluto Press.

Cockburn, C. (1986). The material of male power. *In:* F. Review, ed. *Waged work: A reader*. London: Virago, pp. 93–113.

Cockburn, C., and Ormrod, S. (1993). *Gender and technology in the making*. London: Sage Publications.

Collinson, D., and Hearn, J. (1997). "Men" at "work": Multiple masculinities/multiple workplaces. *In:* M. Mac an Ghaill, ed. *Understanding masculinities: Social relations and cultural arenas*. Buckingham: Open University Press, pp. 61–76.

Confino, A. (1997). Collective memory and cultural history: Problems of method. *American Historical Review*, 102 (5): 1386–1403.

Connell, R. W. (1995). *Masculinities*. Sydney: Allen & Unwin.

Connell, R. (2002). *Gender*. Cambridge, UK: Polity Press.

Cooling, S., and Vinton, N. (2016). *Interpretation strategy for Australian Technology Park: Prepared for Mirvac*. Sydney: Curio Projects.

Desbiens, C. (1999). Feminism "in" geography. Elsewhere beyond and the politics of paradoxical space. *Gender, Place and Culture*, 6: 179–185.

Devault, M.L. (1990). Talking and listening from women's standpoint: Feminist strategies for interviewing and analysis. *Social Problems*, 37 (1): 82–101.

Don Godden & Associates. (1986). *A heritage study of Eveleigh Railway Workshops*. Sydney: Don Godden & Associates.

Drummond, D.K. (1995). *Crewe: Railway town, company and people, 1840–1914*. Aldershot: Scolar Press.

Fentress, J., and Wickham, C. (1992). *Social memory*. Oxford: Basil Blackwell.

Frisch, M. (1990). *A shared authority: Essays on the craft and meaning of oral and public history*. Albany, NY: State University of New York Press.

Game, A., and Pringle, R. (1983). *Gender at work*. North Sydney: Allen & Unwin.

Godden Mackay. (1996). *Eveleigh Workshops management plan for moveable items and social history*, Vols. 1–5. Sydney, https://eveleighstories.com.au/archive/eveleigh-workshops-management-plan-moveable-items-and-social-history

Grele, R.J. (2007). Oral history as evidence. *In:* T.L. Charlton, L.E. Myers and R. Sharpless, eds. *History of oral history: Foundations and methodology*. Lanham: AltaMira Press, pp. 33–91.

Guthrie, [no first name], Mr. (c.1955). *History of Eveleigh workshops*. (Unpublished notes). NSW State Archives and Records Series, NRS 17514 (originally A88/44 – Box 3, State Rail Authority Archives Sydney).

Hyde, H. (1922). The Australian engineer: Splendid work at Eveleigh Workshops. *Sea, Land and Air*, 1 June, pp. 176–179.

The Illustrated Sydney News. (1891). The NSW Railway Workshops at Eveleigh. *A State Enterprise*, 18 July, pp. 11–13.

Jones, S. (1939). Eveleigh: The heart of the transport system. *Daily News: Feature for Transport Workers*, 19 January, n.p.

Klein, M. (1994). *Unfinished business: The railroad in American life*. Hanover, NH: University Press of New England.

Labadi, S. (2007). Representations of the nation and cultural diversity in discourses on World Heritage. *Journal of Social Archaeology*, 7 (2): 47–170.

Leung, M.W.H., and Soyez, D. (2009). Industrial heritage: Valorising the spatial – temporal dynamics of another Hong Kong story. *International Journal of Heritage Studies*, 15 (1): 57–75.

Mahtani, M. (2001). Racial remappings: The potential of paradoxical space. *Gender, Place and Culture*, 8 (3): 299–305.

Mirvac. (2015). Mirvac successful tenderer for acquisition and renewal of iconic Australian technology park, 12 November. Accessed 6 January 2018, www.mirvac.com/About/News/Mirvac-Successful-Tenderer-for-Acquisition-and-Renewal-of-Iconic-Australian-Technology-Park/

NSW Government: Department of Railways. (n.d.). *History of the war effort, mechanical branch, September 1939–December 1942*. Typewritten manuscript, State Rail Authority Archives, M143.

NSW Government: Office of Environment and Heritage. (n.d.). *Eveleigh railway workshops*. Accessed 1 April 2018, www.environment.nsw.gov.au/heritageapp/ViewHeritageItemDetails.aspx?id=5045103

NSW Government: Railways. (1947c). *Railway at war: A record of the activities of the NSW government railways in the Second World War*. Sydney: NSW Government Railways held at NSW State Archives and Records Series, NRS 20017 item A92-4/11.

NSW Government: State Archives & Records. Gazette reels and railway personal history cards NRS 12922. Accessed in hard copy from 1999, www.records.nsw.gov.au/series/12922

Pente, E., Ward, P., and Brown, M. (2015). The co-production of historical knowledge: Implications for the history of Identities. *Identity Papers: A Journal of British and Irish Studies*, 1 (1): 31–52.

Peretz, T. (2016). Why study men and masculinities? A theorized research review. *Graduate Journal of Social Science*, 12 (3): 30–43.

Railways Unions Cultural Committee. (1985a). *Trains of treasure exhibition*. Accessed 7 April 2018, https://d1l35cu-3ko6uba.cloudfront.net/s3fs-public/H036%20Trains%20of%20Treasure%201985.pdf

Railways Unions Cultural Committee. (1985b) *Railway voices CD*. Accessed 7 April 2018, http://railwaystory.com/voices/

Reading, A. (2010). Gender and the right to memory. *Media Development*, 57 (2): 11–15.

Reading, A. (2015). Making feminist heritage work: Gender and heritage. In: E. Waterton and S. Watson, eds. *The Palgrave handbook of contemporary heritage research*. Basingstoke: Palgrave Macmillan, pp. 397–410.

Reskin, B., and Padavic, I. (1994). *Women and men at work*. Thousand Oaks, CA: Pine Forge Press.

Roberts, S. (2014). *Debating modern masculinities: Change, continuity, crisis?* Basingstoke: Palgrave Macmillan.

Rose, G. (1993). *Feminism and geography: The limits of geographical knowledge*. Minneapolis: University of Minnesota Press.

Salzinger, L. (2004). Revealing the unmarked: Finding masculinity in a global factory. *Ethnography*, 5 (1): 5–27.

Shortliffe, S.E. (2016). Gender and (world) heritage: The myth of a gender neutral heritage. In: L. Bourdeau and M. Gravari-Barbas, eds. *World heritage, tourism and identity: Inscription and co-production*. Abingdon: Routledge, pp. 107–120.

Smith, L. (2008). Heritage, gender and identity. In: B. Graham and P. Howard, eds. *The Ashgate research companion to heritage and identity*. Aldershot: Ashgate, 159–179.

Sturken, M. (1997). *Tangled memories: The Vietnam war, the AIDS epidemic, and the politics of remembering*. Berkeley, CA: University of California Press.

Taillon, P.M. (2001). To make men out of crude material: Work culture, manhood, and unionism in the railroad running trades, c.1870–1900. In: R. Horowitz, ed. *Boys and their toys? Masculinity, technology and class in America*. New York: Routledge, pp. 33–54.

Taksa, L. (2003). Machines and ghosts: Politics, industrial heritage and the history of working life at the New South Wales Eveleigh Railway Workshops. *Labour History*, 85: 65–88.

Taksa, L. (2005a). The material culture of an industrial artifact: Interpreting control, defiance and everyday resistance at the New South Wales Eveleigh Railway Workshops. *Historical Archaeology*, 34 (3): 8–27.

Taksa, L. (2005b). "About as popular as a dose of clap": Steam, diesel and masculinity at the New South Wales Eveleigh Railway Workshops. *Journal of Transport History*, 26 (2): 79–97.

Taksa, L. (2009). Labor history and public history in Australia: Allies or uneasy bedfellows? *International Labor and Working Class History*, 76: 1–23.

Thorp, W. (1994). *Heritage assessment: Archaeological resources, ATP master plan site Eveleigh*. Sydney: Report for City West Development Corporation.

Thurnell-Read, T. ed. (2016). *Drinking dilemmas: Space, culture and society*. Abingdon: Routledge.

Wajcman, J. (1991). *Feminism confronts technology*. North Sydney: Allen & Unwin.

Walker, C., and Roberts, S. eds. (2018). *Masculinity, labor, and neoliberalism: Working-class men in international perspective*. Basingstoke: Palgrave Macmillan.

Ward, M. (2015). *From labouring to learning, working-class masculinities, education and de-industrialization*. Basingstoke: Palgrave Macmillan.

Interviews and oral history

Aldridge, Jeff. (1999): Interview conducted by Lucy Taksa on 16 March for the *Work, Technology, Gender and Citizenship at the Eveleigh Railway Workshops Precinct Project*.

Bruce, John Robert (1996a): Interview conducted by Joan Kent on 25 March for the *Eveleigh Social History Project*.

Bruce, John Robert (1996b): Group Interview conducted by Lucy Taksa on 17 November for the NSW Department of Urban Affairs and Planning funded video on Eveleigh unionists.

Givillian, Vaughan (1987): Interview conducted by Russ Hermann for the *Combined Railway Unions Cultural Committee Oral History Project*.

Hermann, Russ (1995): Interviewed by Wendy Lowenstein, recorded on 9 June for the Communists and the Left in the arts and community oral history project. National Library of Australia, Sound Recording – ORAL TRC 3111/38. Accessed 15 September 2018, https://catalogue.nla.gov.au/Record/389499

Johnson, Keith (1996): Interview conducted by Joan Kent on 23 February for the *Eveleigh Social History Project*.

Lee, John (1999): Interview conducted by Lucy Taksa on 5 March for the *Work, Technology, Gender and Citizenship at the Eveleigh Railway Workshops Precinct Project*.

Matthews, Bob (1996): Interview conducted by Joan Kent on 20 February for the *Eveleigh Social History Project*.

Willis, John (1996): Interview conducted by Lucy Taksa on 5 February for the *Eveleigh Social History Project*.

20
REMEMBERED INTO PLACE

Jeff Benjamin

Introduction

Memory is deeply woven into considerations of place, and, akin to imagination (Ricouer 2004), it holds creative power. Archaeologists and others who study the past through the portal of memory enter into the very same realm where dreamers and artists linger, those who think about the possible: what could be, would be, will be. Likewise, the act of creation can be seen as *remembering in reverse*. To create is to *remember something into being*, to engage first and foremost with its absence as an inchoate constitutive substance: to piece together a presence out of this formless cloud. The following is a meditation on the upper Delaware river valley, which, I would like to suggest, is a place remembered into being.

The precise moment of encounter with any phenomena is decisive, for its perpetuation in memory forms the source of enchantment that propels all subsequent effort. Moreover, the *form of encounter* is important to acknowledge, especially when one considers the vast variety of ways of initiating a subject of research. My interest in the northern stretch of the Delaware River began by visiting it, repeatedly and compulsively, as an antidote to city life, spending hours and days canoeing and camping along its shores and small islands. Considering all of the different sensory and cognitive 'scapes' that emerge from a meditation on landscape, I think I approached the shores of the Delaware simply as an *e*-scape (Porteous 1990: 124): not from an impulse toward adventure but rather as a place to recover, 'collect myself,' recollect and simply be alone. For some, the notion of *escape* has a somewhat pejorative connotation – as a flight from responsibility or reality, but Yi-Fu Tuan suggests just the opposite:

> Now, consider wild nature. A sojourn in its midst may well be regarded as an escape into fantasy, far from the frustrations and shocks of social life. Yet nature lovers see otherwise. For them, the escape into nature is an escape into the real ... the real is the natural, the fundament that has not been disturbed or covered up by human excrescences.
>
> *(Tuan 1998: 24)*

This repeated ritual flight *into the real* that the openness of the Delaware river valley afforded was made possible by the absence of immediate anthropogenic intervention. It is an absence that, considering the surrounding urban growth, seems miraculous, although, as we will see, to characterise its presence as a miracle would be disrespectful to the hundreds of activists who fought to keep its shores undeveloped. Subsequent to these visits, and years later, my curiosity was piqued to try to understand *how* a location of industrial absence gains tenure.

The area of protected land and water known as the Delaware Water Gap National Recreation Area comprises only a small portion of the larger Delaware river watershed, and is situated along the northern portion of the Delaware river, forming the boundary between Pennsylvania and New Jersey. Its total area is 27,009 hectares spanning both sides of the 64 km length of the river. It is a steeply sloped, narrow strip of land, only 1.6 km wide (Thornberry-Erlich 2013). Because of its steep grade and relative absence of significant bordering wetlands (which act to modify and temper the effects of precipitation and thawing) the Delaware itself is susceptible to extreme flood events and drastic fluctuations in flow rates. Hydrological data from the United States Geological Survey (USGS 2015) taken at Port Jervis, New Jersey (situated at the northern extent of the river), for the period from 1904 to 1915 shows that the flow rate of the Delaware river at this particular location varied dramatically: from 300 cubic feet per second in the summer to 95,000 cubic feet per second in the spring. This extreme variation in flow rate suggests a dynamic, ever-changing environment. Combined with two sequential hurricanes, this volatility culminated in a catastrophic flood in 1955, which was ultimately used as the main reason for the proposal to build a dam on the main stem: flood control.

The geological feature known as a 'gap' occurs in other locations, such as the Cumberland Gap in East Tennessee. Although its formation is a complicated process, it can be explained simply as the result of a slow gradual erosion of a mountain ridge by water. The Delaware Water Gap began forming 500 million year ago, when a series of islands and small continents collided with the North American continent. In this series of collisions there was uplift (forming the Appalachians) as well as heat, which melted and fractured the quartzite conglomerate. This fracturing allowed for the future Delaware river to slowly erode through this fragmented stone. Glaciation, which occurred during the Pleistocene, from two million to 12,000 years ago, simply polished and gave the Gap its present form after the Wisconsin glacier retreated (Thornberry-Erlich 2013).

As it flows today, the Delaware river is the only major undammed river in the eastern United States. Because of this, it offers a unique opportunity to study the river and wider watershed area that has not been affected by this one particularly dramatic form of human intervention. Although archaeology has certainly demonstrated a concern with aquatic environments, water, rivers, meanders, flowscapes have only recently been considered as artefacts worthy of attention on their own (Edgeworth 2011; Pétursdóttir 2017). Perhaps this is due to the fact that – certainly as it pertains to rivers – an archaeologist can never step in, or touch, the same artefact twice. Recent research that fluidity itself is allowable for archaeological consideration also points to the opposite, that perhaps so-called static objects are more dynamic than we think.

Contending with the theoretical implications of flow within the concept of landscape, we can nonetheless state with confidence that the Delaware river is the defining 'feature' of the landscape in question in this present research.

My primary objective has been to investigate how the northern extent of the Delaware river has seen such sparse and faltering industrial development compared to other comparable rivers. Moreover, and following Fowles' exploration of absence in archaeology (2010), I am interested in how archaeology can approach industrial absence, which, in our highly industrialised age, is tantamount to human absence. The efficacy of archaeological attentiveness to repeated patterns comes not through a scrutiny of redundancy, but rather via an alertness to deviations therefrom. The proliferation of material culture under consideration by industrial archaeology is of a different magnitude than what we might find in pre-industrial societies. As noted by Olivier, the massive scale of industrialisation seems to warrant different methods of analysis, 'The world that was transformed by industrialisation can no longer be expressed in the same terms; its stability, and thereby its veracity, are no longer assured. In the wink of an eye, our physical world can be thrown into the ineffable' (2011: 82).

Encountering absence

One could argue that all things, events, and phenomena that are absent are equally so. Absence, like death, is a great leveler of all things. Absent entities are, at the very least, all 'elsewhere' although it could also be argued that they have a spatial distribution akin to presences (Fowles 2010). Through the act of material representation, we offer memories, other people, ourselves, and events as a kind of reprieve from absence, a solace of perpetuation of a second order. In the broad space of absence, that which *has been*, exists in a territory that is spatially and temporally contiguous with the infinite realm of potentialities and conditional possibilities: what might have been, what could be, what could have been, and perhaps even what will be. Ricoeur remarks on the peculiar quality of memory, the fundamental contradiction of memory, that it is the 'presence of absence' (2004: 7). In the broader space of absence, other events and phenomena occupy the same realm as the clearly or distantly remembered object.

In the act of remembering, past shapes and forms emerge just as landscape features coalesce or disperse on the far shore of a misty river, 'as seen through a veil' (Husserl in Casey 1987: 181). In this sense, memory itself is a *riparian* activity, for in its method and content it mirrors a riparian landscape, where forms of water and land merge and trade places. In ecology, riparian refers to the in-between zones of water and land, where ecosystems interact at the foothills bordering rivers and streams, where aquatic and terrestrial life forms intermingle. Although memory can possess an acute accuracy, its spectral and shifting quality as it coalesces and dissipates finds an apt metaphor within the experience of wetlands and rivers.

The fieldwork

A couple of years ago, I met with an archaeologist and cultural resource manager at the headquarters of the Delaware National Water Gap National Recreation Area in Milford, Pennsylvania. In the course of our discussions, we expressed amazement that a railroad was never built along the long flat stretch of the river which extends from the Catskill Mountains all the way to Philadelphia and the Atlantic Ocean. The river region did have several railroads that touched it in short sections, the most notable being the Delaware, Lackawanna, and Western Railroad, which carried passengers from New York City to the 'water gap' for recreation purposes as early as 1851. Nonetheless, for some reason this rail development did not propagate the otherwise predictable resource exploitation and industrial expansion. It is this very absence of a railroad – the device that changed the human sense of space and time in fundamental ways – that takes on the quality of an apparition. It *should* be there, but it is not. It would be easy at first to assume that this long narrow strip of land bordering the Delaware was somehow forgotten, a fluke, it avoided the railroad and its concomitant industries because it somehow slipped the gaze of the industrialists. However, this absence is, in fact, an example where 'absence can be aggressive; it can be cultivated; it can mark the overt rejection of that which is not present' (Fowles 2010: 37).

The absence of the railroad in the Delaware River Valley immediately conjures its omni-presence in many industrialised regions of the world as well as the omni-absence of a non-anthropogenically altered landscape. Throughout North America, the railroad and its myriad associated material and social forms displaced other forms, creating new absences, and a host of spatial and temporal adjustments ensued. The significance of the railroad on human beings' sense of place and time is not lost on contemporary philosophers. Foucault states: 'Europe was immediately sensitive to the changes that the railroads entailed' (Foucault 1997: 370). Thoreau was quick to observe a sudden change in human behavior: 'Do they not talk and think faster in the depot than they did in the stage office? . . . To do things "railroad fashion" is now the by-word' (Thoreau 2004: 202).

However, even with this striking presence of an absent railroad, it would be erroneous to characterise the Delaware River as a completely non-industrialised region. Although now a gentle, beautiful area of winding roads passing along meadows, forest, and occasional farmland, it would be false to state that the

Delaware River Valley was not connected to the process of 'carboniferous capitalism' as bleakly described by Mumford (1934). From the earliest days of European settlement, industrial activities were scattered along the river: the Pawahaquarry copper mine, active in the mid-1800s, is located within the confines of the park. Hundreds of small dams on the many tributaries are a testament to numerous mills in the area. In the early days of resource exploitation, the Delaware River remained protected *precisely because of* its usefulness as a thoroughfare of trade down to the Delaware Bay and the ports of Wilmington, Trenton, and Philadelphia. European settlers brought with them an acute awareness of the doctrine of riparian rights, and the primary objective for protecting the river, agreed upon by all, was transportation, primarily for floating logs down the Delaware to Philadelphia. In 1783 a treaty was signed by New Jersey and Pennsylvania (the two states bordering the main stem) prohibiting any full dams across the main channel. Even though the last log was floated down the Delaware in 1923, this treaty would remain in effect until 1953 (Reich 1976: 42).

There are hundreds of dams on the tributaries flowing into the Delaware, even though the Delaware itself was never dammed: the only major river east of the Mississippi to avoid this common fate (one can still drink its water without filtering or purification). Several of these dams are in disrepair and are in danger of failure, some even threatening small communities with flooding. The presence of these dams is quite in keeping with the larger pattern of industrialisation throughout the northeast United States. The geological configuration of the Delaware River channel and its tributaries, especially on the Pennsylvania side of the river, offered any aspiring millwright three necessities for mill construction: water, wood, and elevation differential. Very little is known about these early dams, their makers, or the specific purposes of their reservoirs.

Research into early European settlement of North America offers insights into how the Delaware River watershed may have appeared before settlement. The work of environmental historians is essential for the work of restoration ecology, for the particular forms and speciation of most pre-settlement North American wetlands is outside of living memory. Walter and Merritts offer a dramatic revision to our understanding of pre-settlement river landscapes in the eastern United States, arguing that rapid and expansive mill-dam construction on rivers and tributaries of the region from the 1600s onward (most likely numbering in the tens of thousands) resulted in a pattern of 'series of linked slackwater ponds' (2008: 303) along the tributaries which, in turn, collected sediment and eliminated a complex system of wetlands that bordered the streams themselves. The authors base their argument on historical documentation attesting to the existence of such wetlands and marshes before mass settlement. Early explorers related the presence of 'ubiquitous swampy meadows and marshes fed by springs at the base of valley side-slopes' (Walter & Merritts 2008: 302). In confirmation of this, the authors conducted sampling expeditions in several Pennsylvania streams and rivers demonstrating the presence of hydric (swampy) soils beneath a layer of sediment containing more recent anthropogenic matter within the matrix. The extent of the sampling is large enough to eliminate any doubt that this is a large-scale regional pattern which could be extended to the Delaware River itself.

However, the authors' assumption that the landscape of pre-European settlement constituted something 'natural' is to dismiss the effects of Native American habitation. If Native American intervention in the forested landscape of the northeast is any indication (thinning, burning), it quite possible that there was a similar 'human-hand' intervening in the configurations of rivers and streams. Also, the authors only briefly mention the once-ubiquitous presence of beaver dams along the river. The impact of the depletion of the beaver population by fur trapping on marshlands and river sedimentation remains less examined and is of huge importance (Vileisis 1997: 24).

The absence of a full dam across the main stem of the Delaware river is no accident. The Lenni Lenape, who arrived in the region 200 years before European settlement, wisely had no interest in obstructing their prime source of transportation and food. Kraft is careful to dismiss any romanticism or glorification of the Lenape in their relationship with the landscape and environment, stating that 'Contrary to popular belief, the Indians were not conservation-minded, as that term is understood today. . . . Indeed, some ten thousand years ago, the ancestors of the Lenape may have unwittingly contributed to the extinction of the slow-breeding mammoth and mastodon, and the native horse, sloth, and giant beaver, among other animals that formerly lived in prehistoric America' (Kraft 1986: xii). However, he does acknowledge that Lenape spirituality placed humanity in a fellowship with other living beings 'as an integral part of a natural world filled with almost infinite varieties of plants, animals, insects, clouds, and stones, each of which possessed spirits no less important than those of human beings' (Kraft 1986: 161). Soderlund also states that the Lenapes 'like other people of eastern North America, believed that the earth and sky formed a spiritual realm of which they were a part, not the masters' (Soderlund 2015: 21).

In a landscape such as the Delaware River Valley there are many seemingly incommensurate absences: histories, memories, 'real' pasts, and potentialities intermingle. One absence created and preserved the landscape as it now exists: the absence of a dam. The entire region that now comprises the Delaware River Water Gap was a planned reservoir for New York City, and the dam project was not fully scrapped until 1996. As early as 1949, the year of a serious drought for New York City, detailed plans were being made to submerge the entire area under the 'Tocks Island Dam Project.' In one proposal, the author alarmingly states that 'The region's productive capacity in terms of water . . . is currently brought into sharp focus by the need for "bathless Thursdays" in the largest city of the nation and in an age when American bath tubs are the envy of the world' (Allen 1950, 1). In the ensuing years after the decision to dam the Delaware, thousands of residents were displaced and their land was purchased. The area was transferred from the U.S. Army Corps of Engineers to the National Park Service which now administers it. Along with the absent railroad, the knowledge of this dam project which never came to pass (because it was forcefully resisted), imbues this landscape with a quality of serendipity – despite the knowledge that this absence was bought with months of hard work by countless individuals.

The main channel of the Delaware itself was seriously considered as a location for potential damming from the early 1900s onward. Ironically, it was its usefulness for transportation in other industrial processes (logging, anthracite coal mining) which prevented this from happening due to significant lobbying efforts. Finally, in 1962, the U.S. Congress authorised 11 separate dams in the Delaware watershed, including a dam on the main channel of the Delaware. This particular dam (known as the Tocks Island Dam because of its location at Tocks Island), would have created a reservoir 37 miles long upstream of the dam location: 12,000 acres of wooded lands would have been submerged had the dam been completed. In preparations for construction, the U.S. Army Corps of Engineers purchased land and moved *15,000 people*, leading to the abandonment or destruction of several entire towns, as well as extensive farmland (Duca-Sandberg 2011: 1). This act of mass displacement – augmented and accompanied by all of its concomitant mendacities, deceptions, and coercions – created an enduring social atmosphere of resentment and bitterness that has lasted, and perhaps increased, well into the present day (for an interesting comparative study, see Beisaw 2017). A local resident and activist Nancy Shukaitis lamented the plight of the residents removed from the Delaware Valley with Congressional assent:

> These persons (Delaware Valley homeowners) do not have a public relations group or lobbyists to front for them and are no competition for the interests of water for basin industry, proponents who want a fast return on their developer dollar. This brochure doesn't even mention that anyone lives in the project area. In the past few years land speculation has moved so rapidly, so lavishly and with such confidence of this bill's passage that it warrants a full-scale investigation. Of course it is difficult to uncover straw buyers and fictitious names, but the makeup of new corporations would be most revealing.
>
> (Shukaitis 1965, in Duca-Sandberg 2011: 51)

If the dam had been completed, a large portion of the region would now be under water, in fact, it is the boundaries of this phantom reservoir that now defines the basic contours of the preserved area. In an ironic twist of fate, the preparations for this massive industrial project (which was also intended to incorporate hydroelectric power) protected the entire region from subsequent anthropogenic incursion because the land was systematically purchased and cleared of its inhabitants in anticipation of the dam construction and has remained relatively vacant up to the present time. While some reports attribute the abandonment of the dam to improper geological formation, the most likely reason was a 'massive conservationist effort' (Thornberry-Erlich 2013: 11).

More often than dispassionately studying industrial phenomena, archaeologists find themselves caught up as players within the contingencies of its expansion. In the case of the Tocks Island Dam project, the Army Corps of Engineers conducted an extensive survey of the proposed reservoir location to remove significant artefacts that would otherwise be lost. Ironically, the results of this collection of cultural heritage played directly into the efforts of grass-roots organisers and conservationists to build an effective argument to scrap the entire project (Feiveson 1976: 78). As it now exists, the Delaware water gap has the quality of an outdoor museum. Abandoned houses remain standing and some are given perfunctory maintenance. Interpretive signage placed by historical societies and the National Park Service indicate the locations of significant features, such as the Old Mine Road, considered to the be one of the oldest roads in the country. The landscape is largely devoid of active human influence. Interviewed in a recent film on the region, *Ghost Waters* by Nick Patrick (2009), a former inhabitant of the valley, Mina Hamilton, returns to her house after over 30 years. While the sight of the house brings her to tears, she states, 'My heart is in the valley, not in this house.' The displacement from her childhood home led her to embrace another place as home, the valley itself. Adopting a philosophical tone while gazing at the structure, she states, 'Somehow I like the aspect that it's similar to an archaeological dig in terms of being a fragment from the past.' Hamilton was one of numerous individuals who, even though she lost her family home, worked to stop the damming of the Delaware. She notes that this was not a miracle from on high: 'It wasn't an accident. Some people think, oh, y'know ... there was going to be a dam and then there wasn't going to be a dam. A lot of people worked really hard to make that change.' Another resident, Sandy Hull, relates the serious effects of eviction: 'Once they bought their houses and moved these people out, the next thing you know, you're reading their obituaries in the newspaper. . . . A lot of these people who lost their homes were in their 70's' (Nick Patrick, *Ghost Waters* 2009).

When looking through the so-called grey literature of environmental planning that accompanied the resistance to the dam, one finds a language and passion that is anything but grey. This was a time when environmental concerns were at the forefront of many people's minds, leading some authors to question the imperatives of industrial control:

> Nature modulating society: is this something we could ever get used to? The thrust of most of industrial society has been in the opposite direction: to reduce man's vulnerability to nature's excesses and, by extension, to reduce man's subordination to nature's variability. . . . Yet the possibility of success in insulating ourselves from nature is a horror it is time to confront. Have we indeed instructed the engineers to produce a technology such that no natural event, however rare, would require us to react? Did we really mean to do this?
>
> (Socolow 1976: 20)

Memory and forgetting is woven through the pages of these reports, used by dam-proponents as well as anti-dam activists. Planners 'looked knowingly and without surprise at how the citizens' memory of the 1955 flood faded with time' (Reich 1976: 79). The 'Save the Delaware Coalition' was spearheaded by the Lenni-Lenape League, which was organised specifically to save Sunfish Pond, a nearby glacial lake that would have been transformed into a back-up reservoir for the main reservoir. While counting tribal members in their group, the choice of the name Lenni Lenape was a prudent one in helping the movement attain greater authority as far as memory is concerned. Finally, the project and the resistance around it attracted a great deal of press coverage, due to its propinquity to New York City. This is ultimately what coalesced into a wall of public opinion against the dam. The result is an other-worldly landscape that defies the 'thesis of inexorability' underlying mass industrialisation.

Archaeological possible worlds

> Landscapes can be deceptive. Sometimes a landscape seems to be less a setting for the life of its inhabitants than a curtain behind which their struggles, achievements and accidents take place. For those who, with the inhabitants, are behind the curtain, landmarks are no longer only geographic but also biographical and personal
>
> *(Berger 1967: 13)*

Perhaps the most interesting assertion within the philosophical scholarship of *possible worlds* is an insistence that these possible worlds are, in fact, real (Lewis 1986). While certainly debatable, one effect of this (logically tenable) claim is to offer philosophical support for the ontological tenure of projects of creative imagination, including projects of resistance or withholding of human intervention. While archaeology is an outward projection and expression of a form of curiosity, it is also the study of internal ruination, inner-remains, internal discovery, with the external object simply serving as a mirage. Possible (imagined) pasts – share equal ontological footing with actual pasts – thereby finding acceptance in the ever-expanding pathways of archaeological inquiry. I tend to agree with the following observation:

> Nostalgia, unlike mere sentimentality, describes the authentic emotion we feel in contemplating the objective fact that in many respects things were better in the past. Sentimentality sighs for a long-ago that never really existed, but nostalgia is a form of social realism . . . We have decided to pretend that 'another world is possible' – even if it isn't.
>
> *(Wilson et al. 2015)*

One of Coles' maxims in the employment of experimental archaeology is that 'Modern technology should not be allowed to interfere with the experimental results' (Coles 1984: 16), and I would submit that this should also be considered when doing the experimental work of thought. The creation of archaeological possible worlds can offer a great deal towards an expansion of phenomenological research, since it attends to the processes of consciousness itself. Ultimately, this is the goal of archaeological phenomenology: to understand through experience. Just as we are attempting to learn about 'who they were in the past,' we are engaging in a continual task of self-discovery and re-discovery. The archaeological subject emerges over time through learned, patterned embodied behavior, through the *habitus* of archaeological work, leading to the enviable lamentation: 'I cannot get rid of my technique' (Mauss 1992 [1934]: 364). Ranciere's 'distribution of the sensible' (2004), while perhaps a useful concept for those who maintain and handle spaces and objects of conventional value, ignores the removal and sequestration of the *insensible*. It is this recovery, this *retrieval of the insensible*, that falls upon the shoulders of the archaeologist, including the archaeologist's own discarded sensibilities. Through the creative act of imagining archaeological possible past worlds, we can locate abandoned pasts, the discarded sensibilities and wisdoms which we will need to draw upon as the carbon imaginary (Povinelli 2016) rapidly transitions into the carbon agony.

The trajectory of the *path not taken* carries its own momentum, continues in consciousness even though it did not happen, and it has the kind of vivid quality of a memory of an event which occurred. For instance, in the experience of narrowly avoiding death or catastrophe, a sense of amazement ensues. While Western philosophy has successfully wrested the power of absence from religion it seems that there is a transference between realms of absence. For an historian or an archaeologist, to write about what *has been* or what *was* creates an opening for what *might be*, or *what could have been* and there is often an overlap between the boundaries of these imaginaries. Absence embraces the past and potentiality in equal proportions.

With hard lessons learned from Europe, from the earliest days of American industrialisation there was caution, skepticism, and resistance to the kinds of changes that industrial expansion might bring to the landscape. Industrialisation was by no means a 'natural' occurrence: it was (and is) actively resisted in ever so many ways (Zerzan 1988; Sale 1995). For instance, the painters of the Hudson River School were literally painting industrial sites out of their landscapes, completely omitting them, and choosing viewpoints where industry was hidden or obscured. This seems understandable when one considers that the founder, Thomas Cole, was born and spent his early years in Lancashire, England, a major industrial location. The Hudson River and its tributaries were in the process of rapid industrialisation, and Cole and others, perhaps as a cautionary measure or simply to preserve what they sensed would be lost, painted the landscape as a harmonious assemblage of natural and agrarian structural forms. This attempt to somehow create a visual harmony between anthropogenic and natural forms culminated in what Leo Marx termed the 'machine in the garden' ideal in American thought (Marx 1964). But it should be pointed out that, from the outset, there was also a strident attempt to keep the machine *out* of the garden, and this resistance would sometimes come from unlikely places. It cannot be mere coincidence that Leo Marx served as a consultant during two symposia at the American Academy regarding the Tocks Island Dam controversy (Feiveson et al. 1976: xi).

A ubiquitous feature of critical discourse is its 'after-the-fact' nature. Once the dam is built and the villages are submerged, the after-the-facters casually appear out of the mist with infinite retrospective prowess. However, archaeological imagination holds significant potential for considerations of the future, and if we are to think of the future at all, isn't it logical to think about the best possible scenarios? The simplest act of representation is also a form of honoring, even critical or pejorative representation honors the object. Moreover, archaeological representations of the past are also forms of perpetuation, they are projections into the future. Archaeologists are well versed in these kinds of cautionary considerations.

Unless a new body of evidence emerges, it would seem that mass industrialisation has been a catastrophic failure for humanity as well as most other species. One can find the antithesis to any social dynamic not by looking for what is actively opposed by the status quo, but rather by paying close attention to what is casually dismissed out of hand. In our age this is the luddist temperament, expressed by those who question the fundamental premises of technological progress. While being dismissed as irrelevant is the greatest insult, it also offers a means of escape, to slip away unnoticed and attend to the more important projects of discovery, searching inward and outward for happy memories (is it any mistake that Ricouer explicitly based his magnum opus on 'happy memory'?): moments of archaeological exceptionalism, where the city or dam *wasn't* built, where the species *survived*, where the gentler, less exploitative temperaments *endured*. Is archaeological imagination strong enough to remember these memories into place?

References

Allen, J.H. (1950). *The Delaware river basin integrated water project*. Bethlehem, PA: Lehigh Valley Flood Control Council, Inc.
Beisaw, A. (2017). Ruined by the thirst for urban prosperity: Contemporary archaeology of city water systems. *In:* L. McAtackney and K. Ryzewski, eds. *Contemporary archaeology and the city: Creativity, ruination, and political action*. Oxford: Oxford University Press, pp. 132–148.
Berger, J. (1967). *A fortunate man*. New York: Pantheon.
Casey, E.S. (1987). *Remembering: A phenomenological study*. Bloomington: Indiana University Press.
Coles, J. (1984). *The archaeology of wetlands*. Edinburgh: Edinburgh University Press.
Duca-Sandberg, K. (2011). *The history and demise of the Tocks Island Dam Project: Environmental war or the war in Vietnam*. Seton Hall University Dissertations and Theses (ETDs). Paper 30.
Edgeworth, M. (2011). *Fluid pasts: Archaeology of flow*. Bristol: Bristol Classical Press.
Feiveson, H.A. (1976). Conflict and irresolution. *In:* H.A. Feiveson, F.W. Sinden and R.H. Socolow, eds. *Boundaries of analysis: An inquiry into the Tocks Island Dam controversy*. Cambridge: Ballinger, pp. 75–122.

Feiveson, H.A., Sinden, F.W., and Socolow, R.H. eds. (1976). *Boundaries of analysis: An inquiry into the Tocks Island Dam controversy*. Cambridge: Ballinger.
Foucault, M. (1997). Of other spaces: Utopias and heterotopias: Panopticism (extract): Space, knowledge and power (interview conducted with Paul Rabinow) *In:* N. Leach, ed. *Rethinking architecture: A reader in cultural theory*. New York: New York University Press, pp. 350–355.
Fowles, S. (2010). People without things. *In:* M. Bille, F. Hastrup and T.F. Sørensen, eds. *An anthropology of absence: Materializations of transcendence and loss*. New York: Springer, pp. 23–41.
Kraft, H.C. (1986). *The Lenape: Archaeology, history, and ethnography*. Newark: New Jersey Historical Society.
Lewis, D. (1986). *On the plurality of worlds*. Oxford: Blackwell.
Marx, L. (1964). *The machine in the garden: Technology and the pastoral ideal in America*. New York: Oxford University Press.
Mauss, M. (1992 [1934]). Techniques of the body. *In:* J. Crary and S. Kwinter, eds. *Incorporations*. New York: Zone Books, pp. 454–477.
Mumford, L. (1963 [1934]). *Technics and civilization*. New York: Harcourt, Brace and World.
Olivier, L. (2011). *The dark abyss of time: Archaeology and Memory*. Lanham, MD: AltaMira.
Patrick, N. (2009). *Ghost waters*. Accessed 10 August 2017, https://vimeo.com/142291370
Pétursdóttir, Þ. (2017). Drift. *In:* S.E. Pilaar Birch, ed. *Multispecies archaeology*. New York: Routledge, pp. 85–102.
Porteous, J.D. (1990). *Landscapes of the mind: Worlds of sense and metaphor*. Toronto: University of Toronto Press.
Povinelli, E. (2016). *Geontologies: A requiem for late liberalism*. Durham, NC: Duke University Press.
Ranciere, J. (2004). *The politics of aesthetics*. New York: Continuum.
Reich, M. (1976). Historical currents. *In:* H.A. Feiveson, F.W. Sinden and R.H. Socolow, eds. *Boundaries of analysis: An inquiry into the Tocks Island Dam controversy*. Cambridge: Ballinger, pp. 45–74.
Ricouer, P. (2004). *Memory, history, forgetting*. Chicago: The University of Chicago Press.
Sale, K. (1995). *Rebels against the future: The Luddites and their war on the Industrial Revolution. Lessons for the Computer Age*. Reading, MA: Addison-Wesley.
Socolow, R.H. (1976). Failures of discourse. *In:* H.A. Feiveson, F.W. Sinden and R.H. Socolow, eds. *Boundaries of analysis: An inquiry into the Tocks Island Dam controversy*. Cambridge: Ballinger, pp. 9–40.
Soderlund, J.R. (2015). *Lenape country: Delaware valley society before William Penn*. Philadelphia: University of Pennsylvania Press.
Thoreau, H.D. (2004 [1854]). *Walden and other writings*, ed. J. Wood Krutch. New York: Bantam.
Thornberry-Erlich, T.L. (2013). *Delaware water gap national recreation area: Geologic resources inventory report*. Fort Collins, CO: U.S. Department of the Interior.
Tuan, Y-F. (1998). *Escapism*. Baltimore: Johns Hopkins.
USGS River flow and hydrological data. Accessed 16 September 2015, http://waterdata.usgs.gov/nwis/discharge
Vileisis, A. (1997). *Discovering the unknown landscape: A History of America's Wetlands*. Washington, DC: Island Press.
Walter, R.C., and Merritts, D.J. (2008). Natural streams and the legacy of water-powered mills. *Science*, 319: 299–304.
Wilson, P.L., Kelly, R., Pollack, R., Stein, C., and Spurlock, K. (2015). *The old calendrist: Tracts for our time*. Pearl, CO: Enemy Combatant Publications.
Zerzan, J. (1988). *Elements of refusal*. Seattle: Left Bank Books.

21
THINKING VOLUMETRICALLY ABOUT URBAN MEMORY

The buried memories and networked remembrances of underground railways

Samuel Merrill

Introduction

To date, scholarship on urban memory has focused primarily on the city's surface with insufficient research conducted into the mnemonic qualities of the places and landscapes that lie below the world's metropolises. This chapter therefore argues for the greater investigation of urban memory in subterranean spaces in order to think through its vertical alignments and volumetric qualities. After recounting the groundwork studies on urban memory and noting their surficial bias, it turns to the research agendas of vertical urbanism and volumetric geography, and the setting of underground railways in order to inspire a more three-dimensional approach to urban memory. It then conceptualises this approach, with regard to underground railways, using the notions of 'buried memory' and 'networked remembrance.' Finally, by briefly presenting two case studies relating to the struggle to memorialise a victim of far-right violence in Berlin's Samariterstrasse U-Bahn station and the heritagisation of London's disused Aldwych Underground station, it illustrates the value of investigating questions of urban memory in the more taken for granted subterranean margins of the city.

The groundworks of urban memory

Cities and their constituent urban landscapes have long been acknowledged to play an important role in the production and consumption of shared memories. This was evident to Maurice Halbwachs – one of the first scholars to use the term 'mémoire collective' or 'collective memory' in the 1920s (1992[1925]) – but also to others like Marcel Proust (1981[1920]) and Walter Benjamin (1999[1932]), who each recognised the city's ability to concretise and trigger acts of remembrance. The later rediscovery of Halbwachs' ideas by, amongst others, Aldo Rossi (1982[1966]), in his architectural study of cities as loci of collective memory, helped activate a second wave of collective memory research across the social sciences and humanities in the 1980s. Within this wave, Pierre Nora's concept of 'lieux de mémoire' came to influentially denote the replacement of environments of immediate memory ('milieux de mémoire') and embodied forms of 'true memory' with objects and locales invested with historical significance (1989).

Variously translated to mean 'sites,' 'realms,' or 'places of memory,' Nora's concept has helped foreground the mnemonic function of physical space even though it also denotes non-spatial repositories of memory. Geographers, architects, archaeologists, sociologists, and many others have all used the concept to refer to a range of physical features created for, or retrospectively assigned to, the task of collective remembrance, including memorials, ruins, museums, street names, cemeteries, heritage attractions, archives, and public architecture. Cities in turn have been recognised to index collective memory in all of its social, political, and cultural complexity because they contain overlapping concentrations of these features (Boyer 1996; Crang & Travlou 2001; Crinson 2005; Gross 1990). By extension, and due partly to their population density and ability to retain traces of memory in their rapidly changing fabric, cities and their constituent urban landscapes have been productively conceived as layered palimpsests of memory which reflect generational turnovers and shifts between communicational and cultural forms of remembrance alongside broader social and political transformations (Assmann & Czaplicka 1995; Connerton 1989; Huyssen 2003).

However, the power to collectivise memory in the urban built fabric is not equally shared. Instead, it is unevenly distributed across different 'mnemonic communities' according to social and political hierarchies, which generally advantage groups in positions of authority (Zerubavel 1996). In the past, nation states and their central and municipal governments tended to monopolise the production of urban memory while other groups were disadvantaged by their lack of access to archives, professional historiography, and resources to build and maintain physical memorials (Legg 2005, 2007). These inequalities have often been conceived by distinguishing between intermingling forms of 'official' and 'counter' memory, with acts of collective remembrance not only sustaining hegemony from above but also subverting it from below (Confino 1997; Olick & Robbins 1998).

The mnemonic significance afforded to different spaces within the city is also not equally distributed and it is still common to consider that 'there are places for remembering and places where memories and the past are irrelevant' (Edensor 2005a: 833). That urban memory is subject to different spatial hierarchies is clearly illustrated by the allocation of commemorative street names and the accumulation of so-called memory districts (Azaryahu 1996; Till 2005). Urban symbolic structures, planning schemes, and policies all influence which parts of the city are deemed most suitable for the task of collective remembrance (Stevens & Sumartojo 2015). When it comes to memorials, central, better-connected urban places with pre-existing mnemonic associations and cultural value are often deemed more desirable than other areas for their placement, although such hierarchies can be complicated by concerns for locational authenticity in the case of the memorialisation of traumatic events (Till 2005). Likewise, different memorialisation and heritagisation practices are encouraged or constrained by the different vertices of the city even if such vertices have rarely been academically scrutinised in relation to the production and consumption of shared memory. The urban underground, for example, is often considered to be a less suitable setting for memorials and heritage sites not only due to the logistical limitations on physical access that it regularly poses but also because of the negative psychological connotations, not least those of claustrophobia, that are frequently attached to its enclosed spaces.

These issues have, however, seldom been studied 'in depth,' and to date the city's spatial hierarchies of remembrance have been conceived primarily in horizontal terms revealing the surficial biases that afflict the study of urban memory in general. This bias is highlighted by the aforementioned literature's heavy reliance on the investigation of urban mnemonic features that are mostly evident on or above the street level, to repeat: a city's memorials, ruins, museums, street names, cemeteries, heritage attractions, archives, and public architecture – in other words, its groundworks. The promise of a fuller understanding of urban memory and its associated spatial hierarchies of remembrance thus lies in the consideration of these phenomena within the vertical alignments and volumetric margins of the city – the same alignments and margins that have recently received increasing academic attention with regards to matters beyond memory.

The city's mnemonic volumes

Since the turn of the millennium there has been a marked growth in research dedicated to spaces beneath the city (Gandy 2002). Building on a handful of earlier works dating to the 1980s and 1990s (see Lesser 1987; Williams 2008[1990]), this research has addressed urban subterranean space in general (see Dobraszczyk et al. 2016; Pike 2005) and a whole host of more specific underground infrastructural networks including transport systems (see Ashford 2013; Gibas 2013), sewers (Gandy 1999), water and energy facilities (see Kaika & Swyngedouw 2000; Moss 2009), communication networks, and pedestrian walkways (see Graham & Marvin 2001; Bélanger 2007). According to broader relational and topological conceptions of place, and in ways that echo Lefebvrian understandings of the social production of space, these infrastructural networks have been reconceived as landscapes composed of material, representational, and experiential dimensions (see Gandy 2011; Lefebvre 1991; Massey 2005; Merrill 2017).

This research has also contributed to an incipient 'vertical turn' in urban geography that helps counter the continuing scholarly overreliance on two-dimensional ontological perspectives and has led to the increasing adoption of volumetric approaches to urban space (Graham 2004, 2016; Harris 2015). For example, Stephen Graham and Lucy Hewitt promulgate a 'fully *volumetric* urbanism' capable of addressing 'the ways in which horizontal and vertical extensions, imaginaries, materialities and lived practices intersect and mutually construct each other within and between subterranean, surficial and suprasurface domains' (2013: 74–75). In addition, Stuart Elden has called upon scholars to think volumetrically about the specific issues of territory, power, and security (2013); Andrew Harris has recommended a more diverse research agenda that is open to 'a wider world of more ordinary vertical urbanisms' (2015: 601–602); and Peter Adey has highlighted the need to investigate urban volumes within a broader variety of social and cultural registers (2013). Memory can serve as one such register.

Those who have studied memory in the urban underground have mostly focused on more exceptional spaces such as bunkers (see Bennett 2017) or the high-profile memorials and museums that have been dug into the subsoil of cities in order to commemorate past traumatic events (see Young 1992). In these cases, the mnemonic capacity attributed to different underground spaces usually stems from their locational authenticity and direct connection to remembered events or, failing that, the symbolic and metaphorical currency attached to the act of digging down, whether in order to excavate or to bury. Fewer scholars have considered issues of memory in more quotidian subterranean infrastructural landscapes like underground railways. This reflects the common conception of these railways primarily in terms of their functional and technical convenience, in the interest of which the past should be conquered and 'there can be nothing inherited' (Groys 2003: 117–118). For similar reasons, underground railways are often considered to be examples of what Marc Augé has called 'non-lieux' or 'non-places' even though Augé himself has stressed how they are deeply social places (2008; see also Gilloch 2014).

In fact, Augé is amongst only a handful of scholars to explicitly consider questions of memory in underground railways through his account of the autobiographical memories sparked by the use of the Paris Metro (2002). Petr Gibas has also explored collective memory in the Prague Metro, and especially that which has 'largely been suppressed in favour of the smooth efficiency of commuting, undisturbed by any undesired meaning and effect' (2013: 488). But neither Augé nor Gibas pursued their efforts within a geographical approach that appreciated the quintessentially volumetric nature of their sites of enquiry.

A volumetric approach is useful because the railways that pass beneath the surfaces of approximately 150 cities worldwide are rarely hermetically sealed subterranean networks. They emerge onto and above the city's surface to connect with other transport systems and urban landscapes, either physically in representations or through the pathways and experiences of those who use them. In other words, they are multi-levelled, sitting within and crossing the wider vertices and horizons of their cities, just one part of wider urban infrastructural landscapes that are simultaneously vertical and horizontal. In this sense, the memories and forms of remembrance that they play host to are both stratified and stretched, buried and networked.

Buried memory and networked remembrance

Many of the shared memories that are discernible across the material, representational and experiential dimensions of underground railways' infrastructural landscapes are 'buried' not only in terms of their subterranean spatial settings but also in terms of their status within the collective consciousness of their city's populations (Merrill 2017). These are the memories that at various times and places have been concealed within (or by) official and mainstream renderings of the past or, alternatively, in the words of Simon Schama, have lain 'hidden beneath layers of the commonplace' (1996: 14). Buried memories can thus also be considered to be counter memories especially given that both can relate to repressed forms of remembrance (Legg 2005). But buried memories need not be explicitly anti-hegemonic in the way that counter-memories are commonly assumed to be. They can also become occluded for more mundane reasons connected to the specifics of their spatial context and broader forces of cultural inertia.

Conceptualising these memories as buried does not mean to mask their potential dynamism. They can but need not remain buried. They can surface, face reburial, and resurface. In other words, they percolate throughout the stratified layers of underground railways (Dobraszczyk et al. 2016; Serres & Latour 1995). But not in the orderly manner often assumed by many of those who employ vertical and archaeological metaphors in order to conceive of different pasts neatly arranged in clean stratigraphic sequences beneath their feet (Holtorf 2005). They are subject to unruly flows as they travel to the surface – intentionally and unintentionally – and while they are in motion they interpenetrate with other memories meaning that they rarely travel alone (Merrill 2017). Their surfacing is influenced by but also influences the wider infrastructural networks to which they are connected. In this way, the various acts of remembrance that lead to the return of buried memories, and these memories themselves, can be thought of as networked.

The physically interconnected qualities of their infrastructural settings, across which particular jurisdictions that structure forms of remembrance spread, means that the consequences, repercussions, and issues associated with the resurfacing of buried memories in one place can radiate to many others. Within any underground railway's assemblage of connections, the emergence of buried memories thus creates what Michael Rothberg (2010) has called 'noeuds de mémoire,' 'knots,' or 'nodes of memory' lying within dynamic multidirectional networks of remembrance, as much as singular 'places of memory' in Nora's sense (1989). These networks of remembrance, in crossing physical, representational, and experiential domains, can also be productively conceived of as 'memoryscapes' – layered accretions of material, imaginative, and embodied memory that are stretched across geographical scales (Basu 2013). An underground railway's memoryscape and nodes of memory are not only spatially stretched through their physical networks: they are also stretched thanks to the mutual imbrication of computational code and urban space (Kitchin & Dodge 2011).

Whereas underground railways previously represented 'dead zones' in mobile internet coverage due to weak vertical connections, they now find themselves increasingly embedded in the information flows that a new, third wave of memory scholars have recently conceived of as creating digital and global networks of connective memory (see Hoskins 2009; Reading 2011; Zaporozhets 2016). Taking account of these rhizomatic digital and non-digital connections when studying their memories and associated forms of remembrance not only helps to avoid the ontological weaknesses of assuming a strict distinction between so-called 'real' and 'virtual' space and safeguards against the obfuscation of memory's spatial dimensions by a growing research agenda in memory studies that is dominated by a focus on digital media and technology. When combined with an interest in buried memories, it also helps ensure a volumetric approach to memory that emphasises not just its horizontal stretching but also its vertical stratification.

The value of approaching urban memory in this manner can be illustrated by reference to some examples from the Berlin U-Bahn and London Underground. Buried memories lay latent throughout the representational, material, and experiential dimensions of these transport networks' volumetric landscapes (see Merrill 2017). They can be identified in the U-Bahn and Underground's maps, station names, memorials, and disused spaces, but the rest of this chapter will restrict itself to two examples relating to the final pair of these mnemonic nodes.

Memorial struggles at Samariterstrasse U-Bahn Station

In one of the sub-surface entrance halls to the Samariterstrasse U-Bahn station in the Berlin district of Friedrichshain Kreuzberg there is a small plaque embedded in the wall dedicated to Silvio Meier. Meier was a member of the local alternative left and squatting scene who was killed in the station in November 1992 during a confrontation with a group of far-right youths. For the most part the plaque sits inconspicuously in the background of everyday life at the station but around the anniversary of Meier's death it is mnemonically reactivated during a memorial vigil and demonstration which have been organised every year for the last 25 years by changing constellations of Berlin's antifascist activist groups. During these events, which usually draw between hundreds and thousands of participants, the station's regular use is severely disrupted due to the enclosed nature of its space (Figure 21.1). The city's police force and transport authority – the Berliner Verkehrsbetriebe – however, have become well accustomed at operationalising these acts of public remembrance in order to minimise travel disruption.

The wider recognition of the station's mnemonic status as the location of Meier's death has been hard won by the communities that have carried forward memory of the event and its victim. In fact, the plaque that is on display in the station today is only the most recent material attempt by these communities to designate the memorial function of the station against the will, originally at least, of the transport authority and despite opposition from far-right groups. The first homemade plaque was installed in the station soon after Meier's death and remained in place until it disappeared just before the sixth anniversary. When it was discovered that station staff had removed the plaque local pressure forced its reinstallation in time for the anniversary but soon after it disappeared again. Thereafter the legitimacy of a second plaque, which was covertly installed in 1999, was rejected by the transport authority even though it agreed to temporarily tolerate it until the station's planned renovation.

Local politicians attempted to negotiate the plaque's retention; although, following a station fire elsewhere in the network, the transport authority argued against what it saw as the politicisation of the space and suggested that any future plaque should be relocated at street level. Such sentiments not only indicated the networked nature of public remembrance in underground railways but also how memory claims associated with subterranean spaces could be influenced by tacit mnemonic spatial hierarchies expressed in vertical as much as horizontal terms. In contrast, the views of local politicians at around this time suggested that Berlin's underground stations should be conceived as public spaces like those at street level and therefore reasonable places for memorialisation. With the station's renovation delayed, the transport authority finally agreed that the plaque would be retained; but when works were completed in 2005 it was revealed that the plaque had been accidentally given to a far-right group purporting to be associated with Meier.

Acknowledging this mistake, the transport authority installed a new plaque and thus fully accepted the station's status as a focal point for Meier's remembrance. However, this plaque was also removed, as was a temporary replacement, during the memory flashpoint of the 2006 anniversary. A replacement was installed in 2007 and remains in place today, firmly fastened to the wall by hidden metal anchors in order to prevent its easy removal. Still it continues to be attacked by far-right groups and is regularly graffitied with Nazi insignia. One such attack was described by a transport authority spokesperson as 'unacceptable vandalism against a democratically endorsed memorial' (Litschko 2010). Such condemnation reveals again that in this case local efforts to create an underground node of memory have been successful even if it simultaneously highlights the transport authority's selective amnesia of its earlier opposition to the memorial.

Meier's remembrance continues to branch out from its underground roots as indicated by the creation of an annual Silvio Meier prize for efforts against far-right extremism, racism, exclusion, and discrimination and the renaming of a local street after Meier on the occasion of the twentieth anniversary of his death. Since then it has also spread to become translocally networked thanks in part to its adherents' increasing use of digital

Figure 21.1 Attendees of Meier's annual memorial vigil lay offerings beneath his plaque on the twentieth anniversary of his death.

Source: Author 2012.

technology and internet and social media platforms like Indymedia, Wikipedia, YouTube, Twitter, and Facebook (see Merrill & Lindgren 2018).

The heritagisation of the disused Aldwych Underground station

Aldwych Underground station is amongst the best known of London's disused subterranean stations and is often lauded as an important 'site of memory' and part of the city's heritage. But it and the 15 or so other disused stations like it have not always commanded such public attention. The fragmented transport planning of late nineteenth-century London left Aldwych station at the end of a branch line, served only by a shuttle service, and rising costs meant that parts of it were abandoned even before its 1907 opening. After the service failed to generate sufficient profit the station's eastern tunnel was soon taken out of use. After its use as an air raid shelter during two world wars Aldwych faced recurrent calls for its permanent closure on a near decadal basis. These calls were eventually heeded in September 1994 and thus the station's convoluted closure conformed to one of the main causes of industrial ruination – 'capitalist development and the relentless search for profit' (Edensor 2005b: 5).

Partly because of its networked characteristics the closure of Aldwych served as a conduit for the wider public remembrance of the Underground's other disused stations even if some specialist groups like Sub-terranea Britannica had visited these places since the mid-1970s. This was evident in the media coverage that the stations received but the chance to encounter them first-hand was still limited mostly to Aldwych, which before 2000 had its doors opened to the public on irregular occasions during London's annual Open House weekend and National Museums Week. The closure of London's Post Office Railway in 2003 served similarly as a conduit for the remembrance of these places but the chance to visit them became even more restricted due to Transport for London's growing security, health, and safety concerns, especially following the 2005 terrorist attack of the network. The restricted accessibility of the disused stations in turn limited their opportunity to host the kind of unregulated and subjective mnemonic experiences often associated with ruinous places.

However, from 2009 onwards, the separate activities of an entrepreneur and a group of urban explorers helped improve the physical and virtual access to the stations. Ajit Chambers, founder of the Old London Underground Company (OLUC) generated considerable public interest in the stations by publicising his plans to develop profitable reuse strategies for them, which included their conversion into commercial heritage attractions (see Merrill 2017). His plans contrasted heavily with the efforts of the London Consolidation Crew (LCC), a group of urban explorers, to access all of London's disused stations in order to photograph them and experience their mnemonic capacity in an authentic and unregulated manner (see Garrett 2013; Merrill 2017). The photographs and written accounts thereafter distributed via the blogs of these urban explorers widened awareness of the mnemonic, experiential, and aesthetic potential of London's disused stations and, along with Chambers' plans, made the true extent of these places' public appeal evident to Transport for London.

Given this, the transport authority has increasingly adapted the strategy through which it previously secured its disused stations to rely partly on their heritagisation. Within this strategy Aldwych has been used both as a test case and a flagship subterranean heritage attraction partly because it is actually less networked than London's other disused subterranean stations due to its position at the end of an out-of-service line. Some of the earliest signs of this change in strategy were apparent when Aldwych was used by the London Transport Museum to host a living history experience marking the seventieth anniversary of the Blitz in 2010 and also when the station received UK's national heritage listing in 2011. Since then, Transport for London has run an increasing programme of public tours to the station via the London Transport Museum (Figure 21.2). Interestingly, however, rather than continuing to rely on the living history approach that was also featured in OLUC's plans, these tours have increasingly attempted to emulate the kinds of experiences sought by the LCC as indicated by their recent rebranding in 2015 under the name 'Hidden London.' Following the success of its Aldwych tours the Hidden London program of events has expanded to also include other disused parts of the Underground including forgotten tunnels at Euston station, the Clapham South air-raid shelter, and the disused Down Street station. Aldwych has thus once again served as a conduit for the networked remembrance of these places.

Figure 21.2 Visitors during an Aldwych open day in 2011.
Source: Author 2011.

Conclusion

This chapter has called for a more three-dimensional approach to the study of urban memory. It has done this by highlighting how the traditional surficial approaches to the investigation of collective memory in cities can be supplemented by the integration of perspectives offered by the emerging geographic subfields of vertical urbanism and volumetric geography and via the development of new concepts, such as buried memory and networked remembrance, suited for the examination of urban memory in taken-for-granted subterranean spaces such as underground railways. The need to develop new investigatory approaches to urban memory is clear not least because growing levels of urbanisation worldwide mean that cities will continue to serve as important settings in which to critically investigate how and when societies prioritise collective remembrance and which societal groups have the power and ability to collectivise memories. But also because the mounting urban population densities brought about by these urbanisation processes has stimulated the construction of an ever-growing number of new spaces beneath the city's streets and encouraged those already in existences to be reused in new ways. These developments provide good reason for scholars to turn their consideration of urban memory downwards beyond the exceptional towards more quotidian subterranean places – not just the underground railways foregrounded here but also the sewers, water and energy facilities, communication networks, pedestrian walkways, and pipelines that lie beneath the city's streets.

References

Adey, P. (2013). Securing the volume/volumen: Comments on Stuart Elden's plenary paper 'Secure the volume.' *Political Geography*, 34: 52–54.
Ashford, D. (2013). *London underground: A cultural geography*. Liverpool: Liverpool University Press.
Assmann, J., and Czaplicka, J. (1995). Collective memory and cultural identity. *New German Critique*, 102 (5): 1386–1403.
Augé, M. (2002). *In the metro*, trans. T. Conley. Minneapolis: University of Minnesota Press.
Augé, M. (2008). *Non-places: An introduction to supermodernity*. Second Edition. London: Verso.
Azaryahu, M. (1996). The power of commemorative street names. *Environment and Planning D: Society and Space*, 14 (3): 311–330.
Basu, P. (2013). Memoryscapes and multi-sited methods: Researching cultural memory in Sierra Leon. In: E. Keightley and M. Pickering, eds. *Research methods for memory studies*. Edinburgh: Edinburgh University Press, pp. 115–131.
Benjamin, W. (1999 [1932]). Berlin chronicle. In: M.W. Jennings, H. Eiland and H.G. Mith, eds. *Walter Benjamin: Selected writings, volume two, part two, 1931–1934*. Cambridge, MA: Harvard University Press, pp. 595–637.
Bennett, L. ed. (2017). *In the ruins of the cold war bunker: Affect, materiality and meaning making*. London: Rowman and Littlefield International.
Bélanger, P. (2007). Underground landscape: The urbanism and infrastructure of Toronto's downtown pedestrian network. *Tunnelling and Underground Space Technology*, 22: 272–292.
Boyer, M.C. (1996). *The city of collective memory: Its historical imagery and architectural entertainments*. Cambridge, MA: MIT Press.
Confino, A. (1997). Collective memory and cultural history: Problems of method. *The American Historical Review*, 102 (5): 1386–1403.
Connerton, P. (1989). *How societies remember*. Cambridge: Cambridge University Press.
Crang, M., and Travlou, S. (2001). The city and topologies of memory. *Environment and Planning D: Society and Space*, 19 (2): 161–177.
Crinson, M. ed. (2005). *Urban memory: History and amnesia in the modern city*. New York: Routledge.
Dobraszczyk, P., Garrett, B.L., and Galviz, C.L. eds. (2016). *Global undergrounds: Exploring cities within*. London: Reaktion.
Edensor, T. (2005a). The ghosts of industrial ruins: Ordering and disordering memory in excessive space. *Environment and Planning D: Society and Space*, 23 (6): 829–849.
Edensor, T. (2005b). *Industrial ruins: Spaces, aesthetics and materiality*. Oxford: Berg.
Elden, S. (2013). Secure the volume: Vertical geopolitics and the depth of power. *Political Geography*, 34: 35–51.
Gandy, M. (1999). The Paris sewers and the rationalization of urban space. *Transactions of the Institute of British Geographers*, 24 (1): 23–44.
Gandy, M. (2002). Hidden cities. *International Journal of Urban and Regional Research*, 26 (1): 183–190.
Gandy, M. (2011). Landscape and infrastructure in the late-modern metropolis. In: G. Bridge and S. Watson, eds. *The new Blackwell companion to the city*. Oxford: Wiley-Blackwell, pp. 57–65.
Garrett, B.L. (2013). *Explore everything: Place-hacking the city*. London: Verso.
Gibas, P. (2013). Uncanny underground: Absences, ghosts and the rhythmed everyday of the Prague Metro. *Cultural Geographies*, 20 (4): 485–500.
Gilloch, G. (2014). Eurydice at Euston? Walter Benjamin and Marc Augé go underground. *Societies*, 4: 16–29.
Graham, S. (2004). Vertical geopolitics: Baghdad and after. *Antipode*, 36 (1): 12–23.
Graham, S. (2016). *Vertical: The city from satellites to bunkers*. London: Verso.
Graham, S., and Hewitt, L. (2013). Getting off the ground: On the politics of urban verticality. *Progress in Human Geography*, 37: 72–92.
Graham, S., and Marvin, S. (2001). *Splintering urbanism*. London: Routledge.
Gross, D. (1990). Critical synthesis on urban knowledge: Remembering and forgetting in the modern city. *Social Epistemology: A Journal of Knowledge, Culture and Policy*, 4 (1): 3–22.
Groys, B. (2003). The art of totality. In: E. Naiman and E. Dobrenko, eds. *Landscapes of Stalinism*. London: University of Washington Press, pp. 96–122.
Halbwachs, M. (1992 [1925]). *On collective memory*, trans. L.A. Coser. Chicago: University of Chicago Press.
Harris, A. (2015). Vertical urbanisms: Opening up geographies of the three-dimensional city. *Progress in Human Geography*, 39 (5): 601–620.
Holtorf, C. (2005). *From Stonehenge to Las Vegas: Archaeology as popular culture*. Oxford: Altamira Press.
Hoskins, A. (2009). Digital network memory. In: A. Erll and A. Rigney, eds. *Mediation and remediation, and the dynamics of cultural memory*. Berlin and New York: De Gruyter, pp. 91–106.
Huyssen, A. (2003). *Present pasts: Urban palimpsests and the politics of memory*. Stanford, CA: Stanford University Press.

Kaika, M., and Swyngedouw, E. (2000). Fetishizing the modern city: The phantasmagoria of urban technological networks. *International Journal of Urban and Regional Research,* 24 (1): 120–138.

Kitchin, R., and Dodge, M. (2011). *Code/space: Software and everyday life.* Cambridge, MA: MIT Press.

Lefebvre, H. (1991). *The production of space,* trans. D. Nicholson-Smith. Cornwall: Blackwell.

Legg, S. (2005). Contesting and surviving memory: Space, nation, and nostalgia in Les Lieux de Mémoire. *Environment and Planning D: Society and Space,* 23 (4): 481–504.

Legg, S. (2007). Reviewing geographies of memory/forgetting. *Environment and Planning A: Economy and Space,* 39 (2): 456–466.

Lesser, W. (1987). *The life below the ground: A study of the subterranean in literature and history.* Boston: Faber and Faber.

Litschko, K. (2010). Gedenktafel verunstaltet: Silvio wird nicht vergessen. *Die Tageszeitung.* Accessed 24 April 2016, www.taz.de/!5132143/

Massey, D. (2005). *For space.* London: Sage Publications.

Merrill, S. (2017). *Networked remembrance: Excavating buried memories in the railways beneath London and Berlin.* Oxford: Peter Lang.

Merrill, S. and Lindgren, S. (2018). The rhythms of social movement memories: The mobilization of Silvio Meier's activist remembrance across platforms, Social Movement Studies, OnlineFirst, 1–18. Available from: https://www.tandfonline.com/doi/full/10.1080/14742837.2018.1534680

Moss, T. (2009). Divided city, divided infrastructure: Securing energy and water service in postwar Berlin. *Journal of Urban History,* 35 (7): 923–942.

Nora, P. (1989). Between memory and history: Les Liex de Mémoire. *Representations,* 26: 7–24.

Olick, J.K., and Robbins, J. (1998). Social memory studies: From 'collective memory' to the historical sociology of mnemonic practices. *Annual Review of Sociology,* 24: 105–140.

Pike, D.L. (2005). *Subterranean cities: The world beneath Paris and London, 1800–1945.* Ithaca, NY and London: Cornell University Press.

Proust, M., (1981 [1920]). *Remembrance of things past,* Vol. 3, trans. C.K.S. Moncrieff, T. Kilmartin and A. Mayor. London: Penguin.

Reading, A. (2011). The London bombings: Mobile witnessing, mortal bodies and globital time. *Memory Studies,* 4 (3): 298–311.

Rossi, A. (1982 [1966]). *The architecture of the city.* Cambridge, MA: MIT Press.

Rothberg, M. (2010). Introduction: Between memory and memory – from Lieux de mémoire to Noeuds de mémoire. *Yale French Studies,* 118/119: 3–12.

Schama, S. (1996). *Landscape and memory.* Bath: Fontana Press.

Serres, M., and Latour, B. (1995). *Conversations on science, culture and time,* trans. R. Lapidus. Ann Arbor, MI: University of Michigan Press.

Stevens, Q., and Sumartojo, S. (2015). Memorial planning in London. *Journal of Urban Design,* 20 (5): 615–635.

Till, K.E. (2005). *The new Berlin: Memory, politics, place.* Minneapolis: University of Minnesota Press.

Williams, R. (2008 [1990]). *Notes on the underground: An essay on technology, society, and the imagination.* New Edition. Cambridge, MA and London: MIT Press.

Young, J.E. (1992). The counter-monument: Memory against itself in Germany today. *Critical Inquiry,* 18 (2): 267–296.

Zaporozhets, O. (2016). Subway and digital porosity of the city. *National Research University Higher School of Economics Basic Research Program Working Papers.*

Zerubavel, E. (1996). Social memories: Steps to a sociology of the past. *Qualitative Sociology,* 19 (3): 283–299.

PART V

The body

Sarah De Nardi and Hilary Orange

Introduction

The body is a site of knowing. An emplaced ethnography attends to the issue of experience by paying close attention to the relationships between bodies and the materiality and sensoriality of place, which often exceed the representational canon. Embodiment can be considered as a process that is integral to the relationship between humans and their environment (Pink 2009: 27; Birth 2006a). Our encounters with memoryscapes are frequently non-verbal. Smells, touch, sounds, and taste can transport us to places in the past; they can aid in the recall of events, people, and places. Moreover, they act both as prompts to memory and inseparable parts of the memory landscape.

The body is also a site of political memory, and serves as a filter to guide individuals in the social worlds in which they are enmeshed. Whether prompted by identity or body politics, we make sense of our social and perceptual worlds in tandem. Judith Butler advocates the primacy of the body and interpersonal stance in the performance of identity and social and cultural norms, which she calls 'performativity,' or rather 'not a singular 'act,' for it is always the reiteration of a norm or set of norms' (1997: xxi). Resistance to the norms and conventions of a social group are concepts very close to Butler's work, which finds purchase in this context.

Thus, the focus of this section is on the dynamics between place, space, the body and emotion, starting from the premise that collecting and sharing memories means dealing with tales of the body and remembered corporeal experience, as well as through the gestures of the body in recollection which may be filtered and shaped by the socialised body. Memory can also be felt, enacted, and experienced through the body.

The body as a place of memory can be simultaneously a vehicle of mobility through memory spaces and a site of memory situated in the places and stories of memory. The tethering or unsettling of the body to/from place varies. Experience is key, as is memory, and the autobiographical experience we carry is skin deep. Kevin Birth (2006b: 176) asserted that 'remembering can use far more than the written word . . . it can rely on buildings, spaces, monuments, bodies and patterns of representing self and others.'

Contributors in this section make a powerful case for the body as part of an experiential and epistemological framework with which to better understand how memory and remembrance work.

The section begins with two chapters, by Emma Waterton and Shanti Sumartojo, that override prescribed narratives and modes of consumption at memorial sites though the lens of affectual re-imagining and disclosure. Emma Waterton leads the reader through a sensory experience of visiting the heritage that is connected to the attack on Pearl Harbor in Hawai'i. This chapter articulates the layers of intensity that define Pearl Harbor as a place of memory, together with the meanings and identities that this iconic site evokes and attracts. Through the optics of 'bright objects,' Warterton constructs the idea of the 'encounter' at Pearl Harbor and other wartime memory sites as being linked to various 'registers of affect' that can do unbidden work when we visit sites with a traumatic past. Thus, this chapter could be read alongside the contributions in the *Handbook's* section on 'Difficult Memories (Part 2).'

In a similar vein, Shanti Sumartojo also advocates for sensory experience in guiding and building intellectual and imaginative understandings of memory sites, drawing on her previous work on commemoration as *lieux de mémoire*. Sumartojo links the imagined and the affectual to the historical memory of historic sites, positing that the material is fundamental in framing the ways that memory 'not so much pierces the present, but ongoingly comprises it through our sensory experiences.' The chapter develops around the idea of spontaneous and unruly impact of the senses and the body during these encounters. Experiencing a *lieu de mémoire*, she argues, functions as a phenomenological encounter with the remembered past that percolates visitors' own logics, identities, and expectations.

Dan Hicks then examines the relationship between the body as represented in photographs of his younger self, found in an archaeological archive, and the embodied practice of archaeological excavation and performance: photology as a visual knowledge of the past. Hicks frames this disclosure in Proustian and Barthesian terms of inquiry, developing the intriguing notion that archaeological depictions and the human body are compromised in the same assemblage. Timelines and materialities intersect through the uncanny yet pervasive absence-presence of the photographic trace and the phantomic qualities of the body-as-archive.

The body that moves through this section, in its various guises, then goes walking. Ceri Morgan explores the fundamental human practice of walking, set within a burgeoning cross disciplinary interest in walking as a process and method, but here mainly addressed from psychogeographic perspectives. In this foray into alternative ways of navigating space, Morgan discloses the potent links between industrial spaces, abandonment, and presence. We become engulfed in a dreamlike world of movement and spectral encounters, framed as a compelling commentary that considers absence in relation to industry while articulating creative walking methods at the core of a series of student workshops. This chapter could be read alongside the other contributions in Part 4 'Industry' and in Part 3 on 'Memoryscapes' it also finds resonance with Toby Butler's contribution.

Walking then segues into the relationship between maps, mapping, and memory. In Patrick Laviolette, Anu Printsmann, and Hannes Palang's chapter, 'anthropography' is articulated as one of the ways in which people find meaning in maps, mapping experiences, and map-like images. The authors explore some of the ways that individuals may attach cartographic significance to places through the optics of tourist brochures, counter mapping and other 'charting' practices. Here, a comparative review of the many forms of mnemonic mapping serves to entrench the central role of the body in how life experiences are qualified, made sense of, and represented. This chapter also assesses popular items of visual/material and digital culture depicting such a relationship (such as tourist brochures, guidebooks, postcards, Points of Interest [POI], and Global Positioning System [GPS]). Two themes come to the fore: map production and mapping practices and globalisation, as expressed through new technological production and the promotion of international tourism. This contribution could be read alongside the chapter by Sebastien Caquard, Emory Shaw, José Alavez, and Stefanie Dimitrovas (Chapter 5, 'Mapping memories') on migration and diaspora story mapping.

Finally, Luis C. Sotelo's chapter foregrounds 'voicing' and listening as a mechanism of bridging past, present, and future in Colombia. We follow his core inquiry about the meaning of effective 'voicing' and listening in a post-conflict context. The author tells us how performing artists in Colombia started working with both victims and offenders of abuses of human rights in an attempt to stage collaborative performances informed by or based on the real-life stories of those directly affected by the armed conflict. Here, we can glean the performativity of memory trumping political and social divisions in a shared rehearsal of memory through carefully timed and choreographed strategies and practices of uttering memories and listening.

The contributions in this section of the volume have more in common than just an attention to the body and the senses as forces of memory-making: they highlight memory experience as something which occurs at the embodied level as it unfolds in the political and shapes the social.

References

Birth, K. (2006a). Time and the biological consequences of globalization. *Current Anthropology*, 48 (2): 215–236.
Birth, K. (2006b). The immanent past: Culture and psyche at the juncture of Memory and history. Introduction to the special issue "The Immanent Past." *Ethos*, 34 (2): 169–191.
Butler, J. (1997). *Excitable speech: A politics of the performative*. New York and London: Routledge.
Pink, S. (2009). *Doing sensory ethnography*. London: Routledge.

22
MEMORIALISING WAR
Rethinking heritage and affect in the context of Pearl Harbor

Emma Waterton

Bright objects and the affective power of heritage

Places of war, along with their intangible meanings and the intensities of feeling they foster, often hold sway in the public imagination. Part of their evocative power lies in their ability to help visitors make conscious links between the physical spaces in which they stand and what is known to have happened there. The way such sites are officially framed and represented plays a significant role in the overall memorialisation process, hinting at the cultural, economic, and political agendas that sit behind them and signal at least part of the range of responses a visitor may have when reflecting on the horrors of war. But, in addition to those more cognitive tones, there are also other forces at work: registers of affect. Those registers are not always immediately expressible but are deeply felt, and bear down on us as we react to atmosphere, eye-witness accounts, memories, and, on occasion, a sense of haunting. In such moments, we, as visitors, are asked to grapple with both the anticipated meanings and narratives that places of war absorb and accumulate, as well as the involuntary and unexpected personal responses that are often physically felt there, and expressed as fear, sadness, foreboding, despair, anger, or pride.

This chapter attunes to the range of affective and emotional responses registered by people as they visit heritage that relates to, or represents, war. In using the term 'heritage,' I refer to those places, objects, and memories that have acquired *shared* meanings, beliefs, and values to the extent that people alter their behaviour in relation to them (Waterton & Watson 2014). In keeping with my interest in places of war, the example of 'heritage' I take as my focus is that associated with the attack on Pearl Harbor during the Second World War. Prior to 7 December 1941, Pearl Harbor might have been known for its pearl-producing oysters or its associations with the Hawaiian Shark God, and certainly as a naval base, at least from 1908. On 7 December 1941 all that changed. But it was not until 1962 that the first elements of the site – the USS *Arizona* – were opened to the public as a site of memory, thereby transforming the area, through post-conflict preservation and reconstruction, into a place of heritage.

Today, the heritage of Pearl Harbor encompasses a visitor centre, the USS *Arizona* Memorial, the Bowfin Submarine Museum and Park, the USS *Missouri* Memorial (affectionately named 'the Mighty Mo'), and the Pacific Aviation Museum. In addition, the complex also includes the mooring quays that formed part of Battleship Row, plus six historic chief petty officer bungalows, the USS *Utah* Memorial and the USS *Oklahoma* Memorial, all of which are located on or near Ford Island, or Moku'ume'ume, on Oahu, Hawai'i[1] (Figure 22.1).

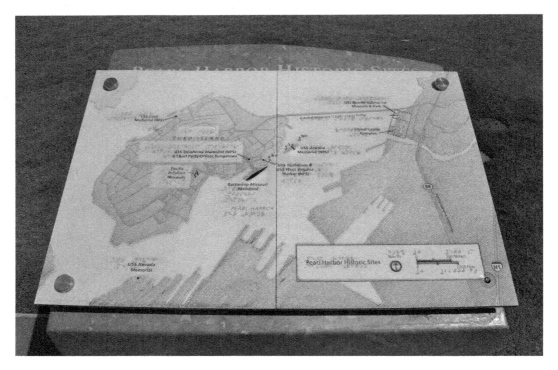

Figure 22.1 Layout of the Pearl Harbor Memorial Complex.

In what follows, I explore the contours of intensity that define Pearl Harbor as a place of memory, as well as the meanings and identities that it evokes and attracts. My purpose is to point to the way that memories accumulate around it, which then act as vectors of intensities that pull in a range of emotions, identities, and feelings. Pearl Harbor thus contains strong peaks of intensity, though I have often struggled to name and describe them. Indeed, while some memories associated with Pearl Harbor have been packaged and displayed to the public for many years in literature, films, posters, TV shows, conspiracy theories, social media posts, and newspaper articles, others have been far more difficult to capture.

To understand the way such memories become legible in situ, I adopt the idea of 'encounter.' This is a theoretical construct I have developed and elaborated on elsewhere in collaboration with Steve Watson (see Waterton & Watson 2014, 2015; Watson and Waterton 2019), and it draws on the eloquent proposition put forward by Bull and Leyshon (2010: 126) that 'individuals are always encountering their own lives, in places and in moments.' Our sense of encounter, or that 'momentary alignment between person and place' as DeSilvey (2012: 47) describes it, homes in on the representational qualities that accrete around nations, cultures, and public memories *as well as* the felt, embodied, and emotional experiences that they engender (Waterton & Watson 2014; see also Waterton and Watson 2015; Waterton 2014; Watson & Waterton 2019). It is philosophically linked to the notion of 'affect' and the concomitant idea that bodily experiences, registered in feelings and emotions, are key to understanding the body's power to act when interacting with other bodies, events, and places. This understanding emerges from the work of Spinoza, who defines such intensities as the capacity 'to affect and be affected' in a manner that is pre-discursive (see Massumi 2015: ix; see also Anderson 2006). Drawing from Spinoza, Massumi (1987: xvi) has also argued that affect is 'a prepersonal intensity corresponding to the passage from one experiential state of the body to another and implying an augmentation or diminution in that body's capacity to act.' A key feature of both Spinoza's and

Massumi's work is thus its prepersonal proposition, in which affect is understood as 'something' that happens. That 'something' is registered in the body, but it is not necessarily represented or expressed discursively, either to oneself or to others, nor is it necessarily confined to any one individual alone.

While theories of affect have clear potential for both Memory Studies and Heritage Studies, particularly given both fields' interests in embodied performances and sense registers, I have space here only to skim the surface of this theoretical terrain (but see Waterton & Watson 2013 and Waterton 2014 for a more fulsome discussion). Suffice to say that my interest lies with how this approach can tailor our focus towards not only what heritage places, objects, and experiences *mean* but what they *do*. Here, Ahmed's (2004) reflections on the cultural politics of emotions are instructive. Ahmed sees emotions and affects as 'things' that circulate among and between individuals, objects, and place, sticking to one or other and leaving traces in the present. The turn to affect thus opens a line of questioning about what a heritage object or place means, what feelings it evokes, and which emotions 'stick' to it over time as people interact with it. As a place of heritage, Pearl Harbor – as with many, if not all, heritage sites – works differently on different visitors; its capacity to 'stick' to people and their memories, and pull them into its orbit, depends on the affective power of the residues of war that are found there, and *how* they are encountered.

At Pearl Harbor, much of that power is traced to one moment in America's past: the surprise Japanese attack on the US Pacific fleet at their moorings in 1941. Using the work of Gordillo (2014), who in turn draws on Bryant (2012), I understand this power as being at least in part a consequence of the way objects, things, and bodies relate to, or are networked with, other objects, things, and bodies. This notion of networking is important, as while Pearl Harbor is itself a significant site of memory, evidenced by the declaration of a National Pearl Harbor Remembrance Day observed annually, it is also explicitly networked to September 11 (9/11) as the only other site of a surprise attack by outside forces on American soil. Both may be considered 'bright objects,' to use Byrant's (2014) terms, by their many visitors, particularly those who come from far away. By adopting the phrase 'bright objects,' I mean to imply that both events – Pearl Harbor and 9/11 – are key hubs or nodes 'in a network, exercising gravity that influences and defines the paths of most other objects in [their] vicinity' (Byrant 2012: n.p.; see also Byrant 2014). But their brightness is not assured; instead, as Gordillo argues, it is 'the result of the networks [they are] part of' (2014: 22). For international visitors, Pearl Harbor might be more of a satellite object in Bryant's scheme, in that the site is known to exist but perhaps only on the periphery of the gravitational network defined by the attraction of Hawai'i as a tourist destination.

To explore the relationships between heritage, memorialisation, and affect, examined through the lens of Pearl Harbor as a place of war, this chapter details the findings from a qualitative project undertaken in 2012. The study revolved around a discourse analysis of the site's interpretive material, in-depth semi-structured interviews with key staff working across the site ($n=6$), semi-structured interviews with visitors ($n=105$), and a performative autoethnography. Over the course of a fortnight, I talked to 114 visitors and staff: 58 women and 66 men. Only 30% of them were international tourists; the rest self-identified as American citizens. However, before discussing material that emerged from the semi-structured interviews with visitors – the focus of this chapter – I will first introduce my own encounters with Pearl Harbor as well as the emotions that surfaced as I sat down to write this chapter. The inclusion of this immediacy of experience will, I hope, allow for a more fulsome understanding of what happens when people encounter heritage through corporeal proximity, and give insight into the experiences of heritage as places of memory.

Getting there

I first came upon Pearl Harbor on my second trip to Hawai'i, in July 2012. I was there for research, tasked with interviewing people visiting the memorial complex over the course of a fortnight. Each morning I would board Honolulu's Route 20 bus as it inched its way from Waikiki to Pearl Harbor, slowed by the city's infamous traffic. On the first morning of my fieldwork, I arrived onsite at 7.30am, when the air was still, fresh and dewy. A handful of cars were dotted around the car park and a small group of visitors loitered at the memorial's entrance. Clutching my ticket for the 8am USS *Arizona* Memorial tour, I moved into the site – watchful and a little apprehensive. After a short wait I was ushered into the Pearl Harbor Memorial Theatre where I watched a National Park Service (NPS) film introducing the attack. From there, I was quietly herded onto a small, US Navy-operated shuttle boat and taken out to the Memorial (Figure 22.2).

Composed of an entry room, the Assembly Hall, and a Shrine, the Memorial straddles the bruised and sunken hull of the USS *Arizona*. I remember vividly the sound of water lapping against its concrete footings as we approached, so silent was our group. The water's surface shimmered with the flighty, rainbow-slick of oil still leaking from the ship below, which was covered by hundreds of barnacles fused to its rust-crusted surface. With the introductory film still fresh in my mind I could just about make out the dark outline of 183 fighters and torpedo bombers filling the sky, followed by strafing, shooting, fire, and panic: 185 vessels of the US Navy were moored at Pearl Harbor that Sunday morning, including eight battleships clustered together in an area that has since become known as 'Battleship Row.' Two of those battleships never returned to service: the Oklahoma capsized and the Arizona exploded, sinking in just nine minutes.

Pearl Harbor, like many places of war, rubs up against a range of other presences. It drifts into politics, the media, the imagination, and everyday life, so much so that when we visit such places we find ourselves on the cusp of what Stewart (2003: n.p.) calls the 'charged border between public and private.' I have been to similar places, of course: Port Arthur, Hiroshima, Auschwitz, Derry. These are all places where forces gather to a point of impact (after Stewart 2003: n.p.). The benches, trees, restrooms, flowers, fences, voices, signs, and posters housed within them – perfectly ordinary things – resonate with something profoundly

Figure 22.2 The USS *Arizona* Memorial.
Source: Photo taken by the author.

*un*ordinary. They become potent, unsettling, laying claim to a sort of agency, and they are alive not so much in our minds but in our bodies. Unsurprisingly, over the course of my time at Pearl Harbor I watched people eventually settle into prosaic tourist practices: strolling, eating, reading, talking, and posing. But those practices were enlivened not so much by grins and chatter but by a watchful, precarious tension. Posing for a photo, for example, seemed to matter more at Pearl Harbor: was it insensitive, inappropriate, inadequate? Such questions composed and recomposed the memorial; as visitors, we recognised its sombreness because we had already felt it and let it influence our behaviour. Now, thinking back, it is hard to say which one was complicit to the other.

Being there

As Wachtel (1986: 216) pointed out some 30 years ago, 'the preservation of recollections rests on their anchorage in space.' Thus, before engaging with my data it is useful to provide a quick overview and chronology of the memorial site itself. The USS *Arizona* Memorial was first opened to the public in 1962. Initially operated by the US Navy, this early iteration of the memorial site revolved around water transportation to and from the shoreline and the *Arizona* Memorial. The first visitor centre came in 1980, at which time the Navy handed over management and organisation of the site to the NPS, though it retained control over transportation to and from the *Arizona* (and still does). Shortly thereafter, in 1981, the USS *Bowfin* submarine arrived, effectively developing into the Bowfin Submarine Museum and Park by 1987. Just over a decade later, in 1998, the USS *Missouri*, recently decommissioned and struck from the Naval Vessel Register, was donated to the museum complex. In 2006, a final historic site was added with the Pacific Aviation Museum, which includes two of the original aeroplane hangars and the Ford Island Control Tower, all three of which were in operation in 1941. At Pearl Harbor, then, there is no shortage of material remains left in place that might serve as reminders of trauma. Collectively, the sites attract over 1.8 million visitors a year. The boat tour alone, which operates out of the visitor centre and visits the *Arizona* Memorial, is almost always sold out. Indeed, it is not unusual for all tickets to sell out by 8.00am on any given day, especially in summer.

But 'the constitutive relationship between memory and place,' as Hoelsher and Alderman (2004: 350) remind us, 'is also, and no less, performative.' I therefore turned my attention to the residual affects that lingered in the landscapes of Pearl Harbor, and to the way visitors 'performed' or attuned themselves to those affects during my visit. From the outset, I should acknowledge that my responses were undoubtedly hypersensitive – I was there as a researcher, after all. But I was not the only one making *something* out my visit, nor was I the only one ready to *feel* something. It is nonetheless important to note that some visitors seemed barely to feel anything at all. If they did, it was more akin to boredom or a burst of irritation at some element of the site's organisation: the expensive cafes, the waiting, or a lack of shade, perhaps. These were in the minority (4.76%), however, so in this chapter I will leave aside such responses and focus instead on those indicative of a willingness to *feel*. Quite often, those feelings were named by respondents. A steady list grew across the fortnight, a list which eventually settled into a fairly contained collection of feelings named as 'humbled,' 'sad,' 'solemn,' 'proud,' 'in awe,' 'heartbroken,' or 'tearful,' with well over 50% of participants nominating at least one (if not more) of those words when prompted to think about how they felt (Table 22.1).

While an acknowledgement of feeling was frequent, I was struck by their cajoling into fairly distinct affective registers, mobilised by attempts to attend to processes of memorialisation. Two such registers will be the focus of this chapter: 1) attuning to unimaginable atmospheres; and 2) responding to the affective intensities of being in place.

Unimaginable atmospheres

In starting with the theme, 'unimaginable atmospheres,', I point to some of the more obvious configurations between memory and affect in my data, configurations that take us beyond (or add to) the myriad

Table 22.1 How does it make you feel?

How does it make you feel as you are making your way around the site?				
Response		Frequency	Total	Percentage
1	Moved, emotionally but a general sensation	7	105	6.67%
2	Moved, specifically naming feelings of pride	5	105	4.76%
3	Moved, specifically naming feelings of patriotism	9	105	8.57%
4	Moved, specifically naming a feeling of being humbled	2	105	1.90%
5	Honoured/privileged to be able to visit	6	105	5.71%
6	Nominates sad, sombre, or solemn feelings at a general level	29	105	27.61%
7	Nominates sad, sombre, or solemn feelings due to family connections and/or links	10	105	9.52%
8	Nominates feelings of awe	17	105	16.19%
9	Nominates a feeling of nostalgia	3	105	2.86%
10	Mixed feelings of sadness and pride	5	105	4.76%
11	Impressed by the materiality of the site (e.g. the ship, submarine)	4	105	3.81%
12	Indicates a critical engagement with the site's history/contemporary times	3	105	2.86%
13	No strong feelings	5	105	4.76%
	Total	105		99.98%

representational qualities of place. In other words, while Pearl Harbor is clearly presented and represented as a place of war, my merging of memory and affect is an attempt to underscore the fact that *it is also felt, remembered, and lived* in the multi-dimensional complexities of the experiences it creates (Waterton & Watson 2015). To understand these experiences, I draw on Sumartojo's (2016) work on atmosphere (see Sumartojo's contribution to this volume, Chapter 23; see also Waterton & Dittmer 2014; Edensor & Sumartojo 2015), which she argues are 'accounted for as aspects of space that are more-than-representational, part of the "feel" and experience of events, encounters and places' (2016: 544). They surround and envelope, she goes on to suggest, and emanate from 'a dynamic combination of the built environment, place and people' (p. 544). But while they 'exist' – in that they are perceived and apprehended by subjects as they rise out of events, encounters, and places – they are also always approximate, precarious, and indistinct (Anderson 2014). Atmospheres are thus akin to things like tones, moods, or ambiance, all of which heritage places like Pearl Harbor exude or generate in ways that change individual and collective experiences of it: they are part of the material and more-than-representational qualities of place that prompt feelings, actions, and emotions. They are therefore at once personal yet collective; contained yet diffuse; conditioned yet conditioning; material yet ethereal (Lorimer et al. 2017).

There is a burgeoning of work on atmosphere in the literature that deals with theories of affect, each taking up a particular element such as light, air, weather, architecture, sound, and so on. The focus of Sumartojo's work centres on her alignment of atmosphere with memory, the latter of which works to trigger, tint, and sustain atmospheres in heritage settings. The nexus between the two is thus a central dimension that shapes a visitor's receptivity to a site, its narratives, and its intensities. At Pearl Harbor, it is easy to see this in the way that the site has been deliberately harnessed to produce atmospheric feelings of presence and unfathomable absence. Such apprehensions were rife at Pearl Harbor, and worked to intensify (and render personal) the extent to which visitors were 'touched by the past,' to borrow from Trigg (2012: xviii):

> The feeling you get here ... it was worth coming.
>
> *(male, plant operator, aged 45–54)*

When I was walking around the submarines, the list of people that were lost at sea, I felt a little something. I can't really explain it. I felt a little something.

(male, student, aged 25–34)

I . . . got goose-bumps going on the Arizona . . . it even makes me a little angry or makes me glad that I'm doing something about it . . . the Arizona's very powerful, definitely. For me, it was a big deal. I just really . . . I want to honour the people that were here. We should never forget the guys that were here and the men and women that died.

(male, US Army, aged 18–24)

For some visitors, this translation from 'absence' to 'atmosphere' to 'affect' culminated in a complex mix of feelings that were transitory yet vivid:

Mixed feelings. . . . Very impressed with the sailors of the time that had to go through this, but also very remorseful. I mean, there were a lot of tragic events obviously involved. But a lot of pride, a lot of pride comes through it. I know now where my forefathers walked in order for me to get where I am.

(male, US Navy, aged 35–44)

Captivating, because in some sense you can't really feel the pain that was felt here. There was pain, there was joy, there's a whole mix of emotions that I feel.

(female, research engineer, aged 25–34)

Thus, while Pearl Harbor is in many ways overdetermined by its representational qualities, it remains at the same time strangely unfathomable. Unfathomable, yet personal. It is this sense of the unfathomable or unimaginable, as a distinct atmosphere, that seemed to fill at least part, if not all, of the memorial complex, acting as a catalyst for personal reflections for many visitors. To borrow from Landsberg's (2018) argument, it was the atmosphere that helped visitors bridge the chasm between themselves and past events:

God knows I don't know how I'd feel if I was going through it too . . . I couldn't imagine myself doing that . . . and being under attack here as well, not knowing what's going on. Crazy.

(male, US military, aged 25–34)

It's unbelievable that something like that could have happened and it's hard to imagine that that can happen and what those people have gone through, families and survivors. Devastating.

(female, retired, over 65)

This atmospheric catalyst was particularly active near Battleship Row, where visitors, already articulate about their perceptions of atmosphere, were prompted to connect with a specific array of terrifying conditions described at the time of the attack. In the following quote, for example, the experiences of one visitor were instigated by an atmospheric quality conjured by prosthetic memories of sailors trapped in the capsized hull of the USS *Oklahoma*. The visitor, in trying to imagine the unimaginable or assemble that which is absent, registers first disbelief, which soon flickers into an expression of empathy before flittering into something akin to helplessness:

Yeah, to think about the loss that some people have gone through, to hear your friend that you're trying to rescue out of the USS *Oklahoma* and they're beating on the hull and you can't get to them and you, as a sailor, know that that means . . . So, the man on the outside trying to rescue and the man on the inside trying to get rescued . . . both of them knowing that drowning is imminent . . . it's pretty touching.

(male, teacher, aged 35–44)

Using Landsberg's terminology I would term this a prosthetic memory, which she describes as memories that are 'worn on the body' and enable a 'sensuous engagement with the past' (2018: 149–150). This is so even though they are not 'natural' or possessed by a single individual but are borne from engagements with mediated representations (Landsberg 2018). Prosthetic memories emerged at other times and in other parts of the site, too, where visitors seemed to be enveloped by unimaginable atmospheres. Sometimes, such atmospheres were created simply by the blissful sunny weather, natural light conditions, or the beauty of the harbour in the present day, which acted as a vector into an affective experience so contrary to that which must have been part of the place in 1941. Others, still, were moved to make profound and emotive reflections on the nature of humankind today, expressing regret that with the passing of time so little had changed.

In her discussions of affective atmospheres, Sumartojo (2016) is particularly keen to explore the way some atmospheres give way to the more cognitive categories of identity, especially national identity. At Pearl Harbor, a 'flickering presence of nation-ness' (Merriman & Jones 2017: 600), visible within the folds of memory, people, and place that the site brings together, points to different scales of intensity: personal, local, regional, and national. I am therefore interested not only in the ability of its atmospheres to gather and attract people and affects, but to gather and attract intensities. This draws back to my introductory comments about 'bright objects' and their ability to attract because of their relations to other places and events. Pearl Harbor offers a clear illustration of the way a place of war can, and does, become a 'bright object,' not only because there is a certain resilience to its own gravitational pull but because of its atmospheric affects, which seem to pulse and intensify when encountered in relation to objects and events visible at the level of the nation. Indeed, there is no doubt that Pearl Harbor's brightness as a prosthetic memory would have diminished or reduced in intensity were it not for the occurrence of 9/11, which brought renewed focus to American wartime history and memories. Significantly, these two events represent the only times America has been attacked on her own soil, with some 25% of visitors referring to correlations between Pearl Harbor and 9/11:

I keep thinking about 9/11 because that's something that I experienced in my life and I'm sure somebody will come here in 50 years – or there – and kind of feel the same thing that we feel today.

(female, market researcher, aged 18–24)

I was reading a book and the way we felt about 9/11, they felt here, only it was a much larger attack here. Then it just makes . . . I don't know . . . I guess I just wondered, too, why did they have them all so concentrated together? My God, why would you put everything on this little, tiny island?

(female, retired, aged 55–64)

And so, this is no neutral space: indeed, there were undeniably times when the memorial spaces gathered together a world of intense ideology, into which some visitors placed themselves in relation to a strongly felt patriotism that they perceived to be signified by the site and its narratives. It is a theme that undoubtedly undergirds Pearl Harbor, where patriotism glides over every surface, though space precludes a more fulsome exploration here.

Memorialising war

The affective intensities of being in place

The earlier discussion around unimaginable atmospheres foregrounds a capacity to be affected by place. Relatedly, it also points to the way that atmospheres, when encountered, allow heritage places to spill beyond the here-and-now. We might therefore argue that atmospheres are neither trans-historical (they are fleeting, after all) nor a-historical but something in between. At memorial sites like Pearl Harbor, visitors see and feel history in context, onsite, via stark material remains and artefacts that, once curated, become an important component of the affect/memory dyad (Figure 22.3).

Yet, as the previous section illustrated, even the curatorial choreography that occurs in such tightly managed places cannot smooth over all of the 'cracks in the surface of the present where time can be otherwise,' to borrow from Crang and Travlou (2001: 175). Thus, I want to draw attention to the ability of these memorial sites – situated as they are on the site where *something* happened – to conjure up feelings of 'being there' or, put another way, produce authentic feelings of 'being in *that* moment.' This is because of the affective intensities that lie behind the knowledge that elements of the site, and indeed the site itself, is *authentic* to that moment, as some respondents remarked:

> Wow. Just to know what people really went through and experienced, and the – you get the emotion of it, the emotional side, instead of just a black and white reading about it. To actually walk on it is much more personal to me.
>
> (female, student, aged 18–24)

Figure 22.3 Looking back at the Pearl Harbor memorial complex from the USS *Bowfin*.
Source: Photograph by the author.

Very sad. The stuff that you read and the stuff that you watch, and you actually get to see and touch too is two different things. It's more real. When you hear, what is it, the voice, the dialogue, it makes it more real.

(male, engineer, aged 35–44)

Pearl Harbor, as a bright object, is contrived from its network of affects, meanings, and memories; their lodgement in place presents a further opportunity to intensify it as an experience. Indeed, at Pearl Harbor, the ability to enter *the* USS *Bowfin* submarine is vitally important, as is standing above *the* USS *Arizona* or gazing down on the Surrender Deck of *the* USS *Missouri*. All three give a sense of confinement, touch, silence, claustrophobia, and helplessness, and have been curated in ways that conjure up atmospheres that are contingent on historical context and that uncanny sense of presence yet absence (see Turner & Peters 2015). It is this sense that produces *place* at Pearl Harbor. Already we can see it as multi-temporal in its references to 9/11. But there are traces (a presence yet absence) of so many other times and versions of the past in play there, too, as many participants confirmed in their articulations of their movement between history and individual experience:

The first time I was at Pearl Harbor in 1984, my uncle was a chief of . . . a destroyer, and he took me out to the *Arizona*. We were all alone. Very powerful experience, that, just to watch the black tears, the diesel fuel coming out, it's quiet, it's early in the morning, and just reflect on what that means and what you're really standing on.

(male, teacher, aged 35–44)

This site makes me feel sad. My grandfather was in World War II and it makes me think of him and his past, so it makes me feel sad.

(female, accountant, aged 25–34)

Tearful, kind of . . . I'm not sure how to phrase it. My father would have been here so it's kind of hard to take.

(female, teacher, aged 45–54)

In addition to the visceral eruption of the past evident in these quotes, there is also a sense of being transported. What I mean here is that Pearl Harbor is made in multiple times, and regularly poaches the bodies and memories of others from other times and brings them into the present. There is never, then, a single point of contact between *a* past, *a* present, and *a* place; rather, there are multiple connections with multiple pasts in a place that is made over and again, 'activated and energised by the affective and emotional motilities that the non-human site and its human event together engender' (Waterton & Watson 2015: 100). Pearl Harbor, as a place, is thus produced out of 'discordant moments,' to borrow from Crang and Travlou (2001: 173), which do not 'form a singular time flowing forwards, but instead offer multiple temporalities.'

The notion of multiple temporalities was evident in many participant responses, in some of which we can see the past erupting into the present in ways so unpredictable they must surely exceed the expectations of curatorial design. As Drozdzewski et al. (2016) argue, it was often the deeply personal, sensorial cues that triggered such moments of meaning-making:

Yeah, the fact I'm going to be a marine, so it really screams out when I saw one of the big old monuments over there that was talking about all marines and stuff and how many survived and how many were killed. That really screamed out to me and was like, you've got to do something.

(male, student, aged 18–24)

I think, for me, [it speaks to] my nursing side. I wonder about what it was for those nurses that were here. They were probably friends with all those guys and dating them. These were horrific wounds too. I mean swimming in water with oil on fire. I mean there's being shot, loss of limbs, oh gosh. It would have been very hard to – I mean how did they control their pain?

(female, nurses, aged 45–54)

Well, for work, being in the Navy in a site like this. . . . it kind of revitalises my pride in my job. After a while, you kind of get the day-to-day routine and things become mundane and you go out to see sites like this and it kind of refuels that vigour that you had. Then, for home life, it reminds me of sacrifice. Being a submariner, we go out for months at a time and we come home – heck, I had a time when my daughter didn't even recognise me. But it reminds me of why I do that.

(male, US Navy, aged 35–44)

In these intimate reflections, certain elements of the site are being subtly worked into how participants define not only themselves, but their relationships to others. In responding to such sensorial prompts, affective intensities leach between public spaces and our everyday lives, shaping how identities *feel* and how they are *understood* (see Drozdzewski et al. in press, my emphasis). In this, we can see a mingling of shock and wonder – the unfathomable – in a process of meaning-making that is emergent from the experience of the encounter itself. This is an affective encounter in which an engagement with the past, in place, is not fleeting, as some theories of affect might suggest. Rather, it is laced with histories and memories – personal or otherwise – that are folded together. Indeed, as Bergson (2007[1912]: 24) has so powerfully argued, '[t]here is no perception which is not full of memories. With the immediate and present data of our senses, we mingle a thousand details out of our past experience.'

Leaving there

On my third trip to Hawai'i, I did not make it to Pearl Harbor. I spent only a few hours in Honolulu, in transit from one island to another, but caught myself on several occasions watching the Route 20 bus rattle by, on route to Pearlridge. As I close this chapter now, sitting at my desk on a cool, winter's day, it would be fair to say that I am still haunted by the place, just as I was as I stood on the Memorial in 2012. I remember feeling thoroughly drawn in by the silence – the atmosphere – emanating from the stark, white structure as it held together all that was left of the lives of over 1,000 crew members. I can almost feel the heat of the sun bearing down on my skin, the occasional stirring of air, and I struggle, as I did then, to decipher the bewitching juxtaposition between the beauty of the harbour and the trauma of its past. In some ways, I participated in the same personal processes of memory-making that I have tried to detail here, and found myself trapped within the folds of an impossibly large event. As a heritage site, and one that articulates the horrors of war, Pearl Harbor oscillates between the large and the personal, the representational and the more-than-representational, the extraordinary and the everyday, and memory plays a powerful role in drawing everything together to produce an understanding of place. Visitors, as I have tried to illustrate, are affected *by* and *in* Pearl Harbor in a process we might term the affective experience of place. They trace themselves through the memorial complex, with atmospheres and affects discharged by the confined spaces of the submarine, the kamikaze deck on the Mighty Mo, the bullet holes at the Pacific Aviation Museum, and the *Arizona* Memorial, which seems to hold onto atmospheres like it holds onto the remains of those who died there. Together, they form a bright object, intense and durable, energised by human engagements yet returning that energy in their relations with the people who visit them. This is the 'work' of heritage: constructing and maintaining memories, meanings, and identities though the interplay of the material and the affective. It is in these interactions, between bodies and place, and made up not only of representational practices but performative moments reconstituted in every visit, every encounter, that memories are accumulated and re/constituted.

Note

1 At the time of research, the NPS's Pearl Harbor Virtual Reality Center had not yet opened.

References

Ahmed, S. (2004). *The cultural politics of emotion*. Edinburgh: Edinburgh University Press.
Anderson, B. (2006). Becoming and being hopeful: Towards a theory of affect. *Environment and Planning D: Society and Space*, 24: 733–752.
Anderson, B. (2014). *Encountering affect: Capacities, apparatuses, conditions*. Farnham: Ashgate.
Bergson, H. (2007 [1912]). *Matter and memory*. New York, NY: Cosimo.
Bull, J., and Leyshon, M. (2010). Writing the moment: Landscape and memory-image. *Spatial Practices*, 10: 124–148.
Bryant, L.R. (2012). Five types of objects: Gravity and onto-cartography. Available online. Accessed 16 July 2018, Https://larvalsubjects.wordpress.com/2012/06/17/five-types-of-objects-gravity-and-onto-cartography/
Byrant, L.R. (2014). *Onto-cartography: An ontology of machines and media*. Edinburgh: Edinburgh University Press.
Crang, M., and Travlou, P. (2001). The city and topologies of memory. *Environment and Planning D: Society and Space*, 19: 161–177.
DeSilvey, C. (2012). Copper places: Affective circuitries. In: O. Jones and J. Garde-Harrison, eds. *Geography and memory: Explorations in identity, place and becoming*. Basingstoke: Palgrave Macmillan, pp. 45–57.
Drozdzewski, D., De Nardi, S., and Waterton, E. (2016). Geographies of memory, place and identity: Intersections in remembering war and conflict. *Geography Compass*, 10 (11): 447–456.
Drozdzewski, D., Sumartojo, S., and Waterton, E. (in press). Cultural memory and identity in the context of war: Experiential, place-based and political concerns. *International Review of the Red Cross*.
Edensor, T., and Sumartojo, S. (2015). Designing atmospheres. *Visual Communication*, 14 (3): 251–265.
Gordillo, G.R. (2014). *Rubble: The afterlife of destruction*. Durham, NC: Duke University Press.
Hoelsher, S., and Alderman, D. (2004). Memory and place: Geographies of a critical relationship. *Social and Cultural Geography*, 5 (3): 347–356.
Landsberg, A. (2018). Prosthetic memory: The ethics and politics of memory in an age of mass culture. In: P. Grainge, ed. *Memory and popular film*. Manchester: Manchester University Press, pp. 144–161.
Lorimer, J., Hodgetts, T., and Barua, M. (2017). Animals' atmospheres. *Progress in Human Geography*, Online First. Doi:10.1177/0309132517731254.
Massumi, B. (1987). *A thousand plateaus: Capitalism and schizophrenia*. Minnesota: University of Minnesota Press.
Massumi, B. (2015). *Politics of affect*. Cambridge: Policy Press.
Merriman, P., and Jones, R. (2017). Nations, materialities and affect. *Progress in Human Geography*, 41 (5): 600–617.
Stewart, K. (2003). The perfectly ordinary life. *The Scholar and Feminist Online*, 2 (1). Accessed 17 July 2018, http://sfonline.barnard.edu/ps/stewart.htm
Sumartojo, S. (2016). Commemorative atmospheres: Memorial sites, collective events and the experience of national identity. *Transactions of the Institute of British Geographers*, 41: 541–553.
Trigg, D. (2012). *The memory of place: A phenomenology of the uncanny*. Athens: Ohio University Press.
Turner, J., and Peters, K. (2015). Unlocking carceral atmospheres: Designing visual/material encounters at the prison museum. *Visual Communication*, 14 (3): 309–330.
Wachtel, N. (1986). Remember and never forget. *History and Anthropology*, 2 (2): 307–335.
Waterton, E. (2014). A more-than-representational understanding of heritage? The 'past' and the politics of affect. *Geography Compass*, 8 (11): 823–833.
Waterton, E., and Dittmer, J. (2014). The museum as assemblage: Bringing forth affect at the Australian war memorial. *Museum Management and Curatorship*, 29 (2): 122–139.
Waterton, E., and Watson, S. (2013). Framing theory: Towards a critical imagination in heritage studies. *International Journal of Heritage Studies*, 19 (6): 546–561.
Waterton, E., and Watson, S. (2015). A war long forgotten: Feeling the past in an English country village. *Angelaki: Journal of the Theoretical Humanities*, 20 (3): 89–103.
Waterton, E., and Watson, S. (2014). *The semiotics of heritage tourism*. Bristol: Channel View Publications.
Watson, S., and Waterton, E. (2019). The Spanish imaginary: A trilogy of frontiers. In: H. Saul and E. Waterton, eds. *Affective geographies of transformation, exploration and adventure*. Abingdon: Routledge, pp. 31–48.

23
LIEUX DE MÉMOIRE THROUGH THE SENSES

Memory, state-sponsored history, and sensory experience

Shanti Sumartojo

At a moment of widespread digitalisation of archival collections, the creation of virtual reality heritage 'experiences,' and the presentation of sounds, objects, and testimonies from the past in official memory sites through multi-media displays, this chapter insists in the irreplaceability of the material in understanding how memory not so much pierces the present, but ongoingly comprises it through our sensory experiences. It advocates an approach to official and state-sponsored memory sites that interrogates how sensory experience is central for shaping visitors' understandings of site-specific historical narrative and its capacity to heighten and nuance empathetic connections with memory sites.

As such, it treats memory sites as locations of emergent experience that we encounter through our senses as well as our intellectual and imaginative understandings. This allows us to think of sensory experience in terms of its potential to make new ways of understanding the past possible, but also what those understandings might do to cohere (or fracture) collective identity. In making this argument, I build on the notion that

> the embodied experiences that visitors have in such sites remain vitally important in communicating about the past – that matter matters, and should not be abandoned or neglected in favour of virtual or digital replacements. In this sense, the actual places and objects of memory are irreplaceable in forging the affective encounters that not only allow memory to perdure, but also to connect feeling to narrative.
>
> *(Sumartojo & Graves 2018: 341)*

These arguments, however, must be prefaced with three important points. Firstly, it is not sensible to categorise people's surrounding environments into discrete sites of memory, because all places can carry the echoes of the past in various ways. Our own knowledge of our homes, towns, and cities or workplaces obviously evoke spatially specific memories that condition how we feel about these locations. Having said that, this chapter is focused on state-sponsored memory sites, those that have been identified, refurbished, designed, and narrated by official bodies, and that usually have pedagogical aims for visitor audiences that seek to frame the state and its past in particular terms. But this is not to suggest that these meanings are fixed or comprehensive, or that they end when visitors leave the site – indeed, such places often resonate in our memories long after our visits have ended.

Secondly, it follows that place as a spatially bounded category that somehow has memories stuck to it is in itself problematic. This is because where place begins and ends has been thoroughly problematised, not least in significant works in anthropology (Ingold 2015) and human geography (Massey 2005). However, state-sponsored memory sites are often treated as internally coherent, not least because they are spatially delineated in their larger spatial contexts and subject to planning or design regulations. They are resourced and designed in ways that often define them in limited spatial terms. This means, however, that when we think of them (and do research about them) as spatially bounded places, we can lose sight of their larger significance.

Finally, I do not seek to dispute the importance of representation, narrative, or discourse in providing the contours for specific forms of collective memory which may take the form of rituals at commemorative events, the evocative testimonies at memorial sites or the oral histories that are powerful in directing post-conflict restorative justice. However, memory as a set of specifically configured stories about the past – as particular representations of official forms of history – here should not be conflated with memory as a part of how we understand the world as a process of ongoing and emerging encounter (Muzaini 2015), or how it is very much a part of an anticipatory mode of engaging with and making sense of our surroundings (Sumartojo 2016). In this chapter, while I am interested in sensory experience as a path to understanding the significance and impact of official memory sites, the representations of the past in such places remains crucially important to their meanings.

Memory, place, and the senses

There is now a robust body of work that attends to the affective entanglements of memory sites at all scales, and that draws together how people emotionally and sensorially feel such places, and how this makes memory piquant in particular ways. Sather-Wagstaff (2017: 18), for example, highlights the emergent and unpredictable nature of memory when she asks us to attend to 'the dynamic relationship between the senses, feeling, emotion, cognition and memory as continually in process.' Others highlight how the senses provide a route for memory to interrupt our daily lives, diverting our attention to the past as it suddenly springs into notice. Drozdzewski et al. (2016: 447), for example, insist that 'sensory cues provoke remembrance; they install pauses and digressions in our normative thought processes; and they transport us, however momentarily, to different times and different places.' This shows how memory accessed through the senses is part of the ebb and flow of our everyday lives as we move through urban landscapes (Muzaini 2015), handle particular objects (Zhang & Crang 2016; Freeman et al. 2016), or visit official sites such as museums or memorials (Waterton & Dittmer 2014; Turner & Peters 2015; Sumartojo 2016).

This last category of official site carries particular connections to national narratives, often related to war or violence. These are important to attend to because they can give rise to intense encounters where affective and emotional engagement become part of the narratives themselves; or, put differently, affect becomes part of how discourse 'sticks' (Ahmed 2010), with flow-on effects for the work that these narratives might do in terms of cohering national groups, defining who belongs in them and who does not, and crafting a version of the national group with an eye towards preserving aspects of it (and forgetting others). Apprehending such places in affective terms is closely entwined with sensory experience as we have bodily encounters that also have emotional impact.

For example, in research on Anzac Day (Sumartojo 2015: 283), Australia's national day for war commemoration, I have argued that, during the annual Dawn Service, 'darkness retells and reinforces the narrative of the Anzac attack at Gallipoli, working this into the bodies of commemorants through their sympathetic experience of dimness, stillness and anticipation.' Here, participants' sensory experience of dark, cool pre-dawn conditions; the gradually increasing illumination of sunrise that changes how they perceive their surroundings; and the repetition of well-known ritual phrases alongside thousands of other visitors create powerful emotional and atmospheric experiences that are anchored in the senses.

Indeed, attending to the sensory allows us to move beyond the poles of memory and history (Nora 1989) that have dominated memory studies, and engage with official *lieux de mémoire* in fruitful, more-than-representational ways. As I have argued elsewhere, foregrounding the sensory carries with it an approach that finds an alternative way into official memory sites: 'a better understanding of the political heft and potential of memory sites can come into focus by attending to the minor, by rehabilitating the emergent, and from an understanding of the concomitant subtle and unpredictable slippage of multiple pasts into the perception and experience of the present' (Sumartojo & Graves 2018: 340). By centring sensory experience, we might be able to notice what usually goes unremarked, which creates possibility for reimagined ways of doing things. Attuning to experience makes potential visible – that something new could be made possible that escapes control, in this case perhaps the control of the state, or of established and conventional ways of designing, promulgating, or encountering official memory. Here Massumi (2015: 57–58) offers an opening via affect:

> Even in the most controlled political situation, there's a surplus on unacted-out potential that is collectively felt. . . . No situation simply translates ideological inculcations into action. There's always an event and the event always includes dimensions that aren't completely actualised, so it's always open to a degree, it's always dynamic and in re-formation.

This has an echo in my point above that place should not be treated as bounded or constrained. Instead, using feelings – both sensory and emotional – as a way to understand our surroundings offers productive possibilities to make sense of state-sponsored sites that can move us beyond what we might already know about them.

Implications and possibilities

Such a reorientation offers, I suggest, a range of implications for how we might understand official memory sites, what they do, and what they make possible. These begin with the vital role that materiality plays in terms of how we come to know the past. As I said at the beginning of this chapter, while this has been thoroughly discussed by others, it bears repeating in an era of digitalisation where changes to the mode of sensory encounter with historical material often goes unremarked and unchallenged. Virtual encounters with historical sites, archives, or testimonies certainly can make aspects of them accessible to a wider audience, but the material and sensory aspects of this encounter are inevitably transformed. The implications of this change need to be understood given the established importance of sensory and material encounters at official sites. The manual handling of archival documents, for example, affords very different material encounters and bodily sensations that viewing the same documents on a screen, from the physical environment of the archival space, to the smell and feel of the paper, to the small thrill transmitted through the pencil marks made by another hand years before – all these open the imagination to the lives and bodies of others and enrich understanding overall.

It follows that objects, sites, buildings, landscapes, and the range of other physical and spatial substance of *lieux de mémoire* are alive with affects, both mundane and intense, that work to lodge them in visitors' bodies and minds, by way of the spatial encounter and its evocation of personal memory, empathetic imagination, and bodily resonance (Waterton & Dittmer 2014; Sumartojo & Graves 2018). Ahmed (2010) reminds us that affect slides between and sticks together objects, representations, and our understandings of them, making things meaningful and significant. At the same time, she insists on the 'messiness of the experiential, the unfolding of bodies into worlds' (Ahmed 2010: 30), which reminds us of the complexity of how people come at official narrative – or how we all 'take away' official discourse based on our own terms of engagement with it. This might be shared with larger groups, but is also intimate to our individual experiences, values, and aspirations.

This complexity of encounter, response, and assimilation of state-sponsored memory sites extends to the tangled chronologies that emerge in such places. Or, put differently, the paradoxically relative unimportance of linear chronology in places dedicated to telling stories in a particular order. Indeed, Crouch (2015: 178) describes heritage as a set of entanglements in the present that unfold for each visitor in unique ways: 'Memory is not simply "placed" in time in a linear "ordering" of being but tumbles among the memories of others, or exists in a net with others, open to being regrasped anew in other moments.' In my own research (Sumartojo & Graves 2018: 341), participants described how personal memories were evoked by official memory sites, and how these blended together with and tinted their understanding of the site in the present. It also shaped how they understood the experiences of people who had previously dwelt in the site:

> Prompted by the senses, and by the material environments that nudge them into being, intimate memories return, the past becomes imaginable through the perceptions and feelings (including those of invisible others), and empathy and other affective intensities can be both fostered and turned to political and official ends.

Indeed, such understandings, made possible by affective encounters as much as by material ones, rely on the empathetic imagination of the visitor (Sumartojo & Pink 2017). Sensory experience provides a route to such imaginative engagement, drawing visitors closer to others, potentially including groups or individuals who may have been excluded from previous official histories. Here we begin to see how attending to sensory experience can open up, complicate, or recast who belongs and whose stories might be told and recognised in official sites (Sumartojo 2019).

This begins to show how state-sponsored memory sites – which are so often subject to relatively fixed narratives about the state, its histories, and how these are meaningful in the present – might have the potential to offer something that exceeds the limits of official discourse. In other words, attending to sensory experience – and the associated unfolding and emergent encounters, affective engagements, and connections with personal memory they pull into being – allows us to move beyond what the state intends in such sites and probe a richer vein of meaning based in experience. Memory is thus revealed as a part of the ongoing present and potential future as much as a version of the past.

References

Ahmed, S. (2010). Happy objects. *In:* M. Gregg and G. Seigorth, eds. *The affect theory reader*. Durham, NC: Duke University Press, pp. 29–51.

Crouch, D. (2015). Affect, heritage, feeling. *In:* E. Waterton and S. Watson, eds. *The Palgrave handbook of contemporary heritage research*. London: Palgrave Macmillan, pp. 177–190.

Drozdzewski, D., De Nardi, S., and Waterton, E. (2016). Geographies of memory, place and identity: Intersections in remembering war and conflict. *Geography Compass*, 10: 447–456.

Freeman, L., Nienass, B., and Daniell, R. (2016). Memory | Materiality | Sensuality. *Memory Studies*, 9 (1): 3–12.

Ingold, T. (2015). *The life of lines*. London: Routledge.

Massey, D. (2005). *For space*. London: Sage Publications.

Massumi, B. (2015). *Politics of affect*. Cambridge: Polity Press.

Muzaini, H. (2015). On the matter of forgetting and "memory returns". *Transactions of the Institute of British Geographers*, 40: 102–112.

Nora, P. (1989). Between memory and history: Les lieux de mémoire. *Representations*, 26: 7–24.

Sather-Wagstaff, J. (2017). Making polysense of the world: Affect, memory, heritage. *In:* D.P. Tolia-Kelly, E. Waterton and S. Watson, eds. *Heritage, affect and emotion: Politics, practices and infrastructures*. London: Routledge, pp. 12–30.

Sumartojo, S. (2015). On atmosphere and darkness at Australia's Anzac Day Dawn Service. *Visual Communication*, 14 (3): 267–288.

Sumartojo, S. (2016). Commemorative atmospheres: Memorial sites, collective events and the experience of national identity. *Transactions of the Institute of British Geographers*, 41 (4): 541–553.

Sumartojo, S., and Graves, M. (2018). Rust and dust: Materiality and the feel of memory at *Camp des Milles*. *Journal of Material Culture*, 23 (3): 328–343.

Sumartojo, S., and Pink, S. (2017). Empathetic visuality: Go-Pros and the video trace. *In:* E. Gómez-Cruz, S. Sumartojo, and S. Pink, eds. *Refiguring techniques in digital-visual research*. London: Palgrave Pivot, pp. 39–50.

Sumartojo, S. (2019) Sensory impact: Memory, affect and photo-elicitation at official memory sites. *In:* D. Drozdzewski and C. Birdsall, eds. *Doing memory research: New methods and approaches*. London: Routledge, pp. 21–37.

Turner, J., and Peters, K. (2015). Unlocking carceral atmospheres: Designing visual/material encounters at the prison museum. *Visual Communication*, 14 (3): 309–330.

Waterton, E., and Dittmer, J. (2014). The museum as assemblage: Bringing forth affect at the Australian war memorial. *Museum Management and Curatorship*, 29: 122–139.

Zhang, J.J., and Crang, M. (2016). Making material memories: Kinmen's bridging objects and fractured places between China and Taiwan. *Cultural Geographies*, 23 (3): 421–439.

24
MEMORY AND THE PHOTOLOGICAL LANDSCAPE

Dan Hicks

I

Any archaeologist can call a diversity of categories of landscape to mind. We can conjure up the city and the country, the Indigenous and the colonial, the monumental and the ruinous or abandoned, the industrial and the primordial (a myth), the desert and the wetland, and so on, with little effort. The history of landscape archaeology is a punctuated oscillation between the experientially nuanced and the spatially patterned. The practice of field archaeology requires a movement from outdoors to the indoor, artificial, darkened secondary 'landscape' of the museum storeroom. Laurent Olivier suggested that we might understand these archaeological landscapes not as fragments of material history but forms of human memory (Olivier 2008). Jacquetta Hawkes was clearly on to something when she wrote that archaeologists are 'instruments of consciousness who are engaged in reawakening the memory of the world' (Hawkes 1951: 26). Archaeological knowledge must always begin with some kind of return (Hicks 2016). We defer writing-up until after back-filling the trench. As we come indoors from outdoors, we project time across space, which is to make memory through a dislocation from one place and one state of consciousness to another. We come to know past lives through such deferrals. If were to think of them as a genre, they would come close to the form of autobiography.

Thinking about a variety of categories of memory comes less naturally to us. Perhaps it is easier for other disciplines. From psychology there's *episodic, procedural, semantic, declarative, subliminal*, and so on; and from psychanalysis there's *repressed, recovered, false*. From literature there's *Proustian*. From geography, *situated*. From 'critical heritage studies', that most hyper-Foucauldian of sub-disciplines, there is the sense of memory as simply bogus – the past as one great hegemonic stitch-up. But what form of memory does archaeology offer? It traces the role of objects as mnemonic technologies – a practice that we might trace to a long history of the technologies of bureaucracy and commerce, stretching from the development of writing and even money as the form of clay tablets, to silver coins, to plastic VISA cards (Malafouris 2013, Hart 2000). Perhaps any archaeological gesture, in that it is an act of erasure, creates a memory of debts and obligation in material form. These forms of knowledge – artefacts, drawings, photographs, texts, etc. – contain the stress that follows any trauma. In these forms knowledge mixes freely with melancholia, just like the changing nature and depth of any single reminiscence over any human lifetime. For too long, we have imagined that archaeological knowledge is made through the performance of excavation. Perhaps we might put the lasting attraction of the reflexive vision of the fieldworker constructing the past, that dominant trope of the representational archaeology of the past 40 years, to a misplaced authenticity – an impossible origin myth for the moment in which knowledge of the past is constructed in the field. The mistake has been to miss the fact that archaeological knowledge of the past is a kind of memory of places wrought through the archive, dislocating both time and space in the museum. Let us be clear that archaeological knowledge of the past is a kind of memory of places wrought through the archive as a technology of dislocation across both time and space in the museum (part memory bank, part mausoleum).

Photography is one form of archaeology's dislocated knowledge of the past. Modern archaeology emerged hand in hand with photography from the 1830s. We might think about it as visual archaeology – making the past visible. What kind of memory does archaeological photography constitute, what kind of landscape, what kind of knowledge? Chiaroscuro wrought through available light and time taken, the duration of exposure extended through the curatorial gesture, transparencies mounted in plastic, contact prints folded in acid-free paper. A deferral that salvages endurance. Any photograph is 'absolutely, irrefutably present, and yet already deferred' suggested Roland Barthes.[1] In archaeology, as in anthropology, this deferral involves some kind of 'deep hesitation' that enables us 'to *not* make connections (start comparing) before the moment is right' (Holbraad & Pedersen 2010: 17). As Marilyn Strathern put it,

> Not to know what one is going to discover is self-evidently true of discovery. But, in addition, one also does not know what is going to prove in retrospect to be significant by the very fact that significance is acquired through the subsequent writing, through composing the ethnography as an account after the event. The fieldwork exercise is an anticipatory one, then, being open to what is to come later.
>
> *(1999: 9)*

The archaeologist's photographic archive is not a mimetic aftereffect of fieldwork. It is not an epiphenomenal simulacrum or a mere by-product or remnant of practice. It is an anticipated postmodifier, in that 'significance is acquired through the subsequent writing' (Strathern 1999: 9). It is a visual practice that excavates and thus casts light, discovers landscape, builds memory, generates self-knowledge, and reveals appearance. We need a way of understanding archaeology not as interpretive and representational (trapped in an illusory present moment receiving the past) but speculative, durational, and transformational. We might call this visual knowledge of the past 'photology.' In these terms, the movement between the museum and the field is a photological landscape that is made up of places made visible after an interval of time: places that are rediscovered.

II

I have a photological story about landscape and memory. Some months ago, while researching some sites that I excavated early in my career, I came across some unexpected photographs. Catalogued alongside the pot sherds and bone fragments in the archive of Warwickshire Museum, among images of Iron Age ditches and medieval post-holes, there were photographs of me, in the field, 25 years ago. An extreme instance of involuntary memory or optical unconsciousness, this is no straightforward autobiographical trace simply enduring through the archive in a life parallel to my own. The confusing reunion is more *autopoiesis* (self-renewal) than flashback or reconciling denouement, an apparition through *mathesis* (institutional ordering) in which we might identify the archive with myths and music in the account of Lévi-Strauss, and thus with the trowel and the bulldozer alike – as 'machines for the deletion of time' (Lévi-Strauss 1964: 23–34).

III

In Figure 24.1, it is, or at least it was, May 1993 and my 20-year-old body is half way along a trench cut across a ploughed field in south Warwickshire, part of an archaeological evaluation in advance of the

Figure 24.1 Evaluation trench for the A435 Alcester-Evesham Bypass, Warwickshire, May 1993. The excavator nearest to the camera is the author.

Source: Photograph by John Thomas.

construction of the Norton-Lenchwick Bypass (Phase 2 Field Evaluations). Behind me are, or at least were, Kev and Sarah, and behind Sarah you can just see the edge of Tim who's wearing a hard hat. Our names aren't mentioned in the excavation report (Warwickshire County Council 1993). Steel surveying arrows secure the 50-metre tape, held taut with a bulldog clip, which runs along the centre of the trench from which the ploughsoil has been removed. The trial trench intersected a crop mark. A mattock leans in the far right-hand corner, the spoil heap along the other side. The crop is perhaps spring onions. Trees and hedges. Geologically we are on the second Terrace Gravel of the River Arrow. We are sampling the Iron Age and Romano-British site at Marsh Farm, Salford Priors. The trowel in my hand is tracing the edge of a Romano-British ditch, distinguishing the different shades of layers and fills of different dates of deposition in anticipation of the box graders and bulldozers that would construct the dual carriageway (opened August 1995). The exercise in which our group, museum workers in the field, is engaged is one of 'preservation by record.' It involves the destruction of in situ remains carried under archaeological conditions, sampling and archiving what will be erased.

Figure 24.2 shows the following week, another trench, the same field project, the adjacent field. Tim is wearing his hard hat again, and Kev has one knee up awkwardly on the baulk as he shovels the loose from the trench, a straw hat on his head, a bottle of water, jackets, another hard hat along the side of the trench. My right hand is flat on the soil next to Tim's boot. My face is a blur. The blur is an effect of the time spent looking, not so much a lack of clarity of form as a particular quality of movement made through accumulated durations (Knight & McFadyen 2019). My body is moving and the variation in speed changing form – 'an effect of action not representation' (ibid: xx).

The photograph, either photograph, is hardly a still. A photological perspective reveals that the photograph not as remnant or trace but as an unfinished trick of the light, as a visual mode of knowledge.

Memory and the photological landscape

Figure 24.2   Evaluation trench for the A435 Alcester-Evesham Bypass, Warwickshire, May 1993. The blurred excavator to the right is the author.

Source: Photograph by John Thomas.

It is moving. To discover in the archive one's own anticipatory performance of the human past is to encounter, unexpectedly, some form of a group's prefigurative thought, of which you were a part, which now takes the form of visual knowledge. A photological landscape.

IV

What has my body become? A memory in a landscape? 'A ghost that will not stand still' (Knight and McFadyen ibid.)? Not just an artefact present in an archive, nor just a concept, nor just a multiplication. That phrase, a 'multi-temporality,' is inadequate, like always.

No, this has to be some form of nonmimetic distribution. That's what I mean by transformation – a prediction of the past. As Barthes has it, 'Photography every now and then makes what you never perceived in a real face (or reflected in a mirror) appear.'[2] These photographs emerge in (Barthes again):

> that very subtle moment when, to tell the truth, I am neither a subject nor an object but rather a subject who feels himself becoming an object: I am then living a micro-experience of death (of parenthesis): I really am becoming a ghost.[3]

In this image system the language of absence or immateriality is inadequate too for these ongoing presences: the blue jeans, that Throwing Muses T-shirt (a US rock band), Dr. Martens boots, a sunburnt arm, the hands/my hands on the soil. Flattened prosthetics, subtle decompositions. Hypotexts for our knowledge of this Romano-British landscape.

These archaeological images are human. Technological and public documents that show what I did with my body. My 1993 trowel is tracing, and thus repeats, and so retraces the outline of a cut made perhaps 1600 or 1700 years earlier – but we are by no means in the moment. The photographs are much less traces or remnants or ruins than they are media and frames for sustained regression. We start to retrace the scene: the probability that there is rolling tobacco in my back pocket; the likelihood that the van is parked out of shot; the certainty we had called out the answers to the Guardian quick crossword over lunch, filling it out in biro or permanent marker; that if it is perhaps 2.30pm then one of us will leave to light the gas for the kettle for tea quite soon.

And of course uncertainties: where I had woken that morning; how I imagined this work as environmentalism, as counter-cultural, in this field where we were witnessing that there was not yet a road, while our friends were still in Hampshire, at the Twyford Down road protest; what I wanted from the future; what I wanted from the past; who I was sleeping with, and so on. This storm of regression neither awakens the dead or deadens the living, and it's hardly nostalgic, this inalienable salvage, this palimpsest of durations. The archaeologist's body is 'revealed, riven, exposed, in the photographic sense of the term.'[4] It is an effigy that does not represent the past but calibrates time (cf. Ginzberg 2001). A human scale.

Has the body has aged twice, once outdoors in flesh and once indoors in paper? Has the landscape aged twice, those trees, some erased by the road (now a quarter of a century old, a four-lane highway from the ploughed fields of Alcester to the orchards of Evesham). And what new blurs of life fill these protended years of exposure? Artist Gerhard Richter suggests that 'What the photograph mourns is both death and survival, disappearance and living-on, erasure from and inscription in the archive of its technically mediated memory' (2010: xxxii). (He says nothing of objectification, but these photographs certainly inscribe my body.)

These images are human, environmental, Roman, 90s, unexpectedly anticipatory, surprisingly prefigurative, unconfessional but not insincere, alive and approaching us unrepressed, intervening at once from the future and the past. The body does not re-encounter itself in this once future moment, comparing one right hand to another. We *identify* with each other, like Roland Barthes 'identified with' Marcel Proust: 'I identify with him (*je m'identifie à lui*'): a confusion of practice (*pratique*), not of value (*valeur*).'[5]

This is not quite Lacan's notion of *aplianisis* (the fading of the subject) that Barthes had employed the previous autumn in his James Lecture at NYU, on the theme of *Proust et Moi*,[6] but it is a kind of prefigurative distribution of personhood. It operates in the only register through which artifice can transform into reality: a register of temporal imagination. Sontag's distinction between archaeology (along with weather forecasting and espionage) and fiction breaks down.[7] Confusions of contemporaneities reveal the untimely and speculative nature of photological knowledge. 'The translation from idea or effort into material object and back again marks the boundaries of the person' (Strathern 2005: 154).

The collapse of immediacy and distance that the archaeological image brings is captured by W.G. Sebald's commentary (1995) on Thomas Browne's *Hydriotaphia* (1658). The eyepiece of an astronomer's telescope is lit up by the most remote past refracted across light years. But this is doubled in the antiquary's melancholic account of the excavation of particles of cremated human bone in earthenware vessels buried shallow in the Norfolk soil, which reveal that in the practice of archaeology 'the more the distance grows, the clearer the view becomes':

> You glimpse the tiniest details with the utmost clarity. It is as if you were looking through a reversed telescope and through a microscope at the same time.[8]

Or, as Walter Benjamin put it, 'in the optics of history – in this respect the counterimage of the spatial – movement into the distance means growing larger.'[9] And so the experience of this archaeographic encounter seems like an intervention in the passage of time that casts a light on my skin, like the projections of the magic lantern described by Proust. Through the 'lyrical resurrection of past bodies' we 'make History into an enormous anthropology' (Barthes again) (Barthes 1978). And reveal 'the illusory logic of biography' if it follows a 'purely mathematical order of years,' suggesting how 'systems of moments succeed each other, but also respond to each other' (Barthes 1984 [1978]: 318).[10]

V

A transformational archaeology. An intimate distance. My body distributed across time. Involuntary memories and surrogate landscapes. In white gloves my hands returned what had lain on the museum lightbox to the embalming grip of its plastic sleeve, where the photographs remain, suspended from a metal strip under lock and key in a fireproof cabinet, objects dated as if we were numismatists, where the currency is also the witnessing of ongoing obligation across time.

Mutual knowledge made through two types of body living apart to make recollection a possibility. It is both my own body and is not. I *identify with* it. This working body, hired labour, hands eyes and brain, still working away in the archive. We emerge not as analogies for each other (a confusion of value) or as schizophrenic multiples, but through some confusions of practice (we appear to each other).

I look through this screen, fingers on keys, at the archaeologist with eyes downcast, focused on the ground, and in a blur of ongoing movement. We are not a trace or a ruin but an untimely ghost returning from the future to recall and to haunt the landscape and its proxy, the archive. We resist any variety of reflexivity that would offer to this two-dimensional form a second biography. I/we resist being reduced to a construction or a metaphor. is by no means just some sense of the uncanny. Barthes evoked 'that small terror that is there in all photography: the return of the dead.'[11] But as a method, both archaeology and photography have functioned here as some form of speculative intervention in time, in which my body is implicated. The embodied future archive. A reverse phenomenology. What is method after all if it is not 'a premeditated decision' (Barthes 2002: 3)?

Landscape, memory, photology. Human life and human memory would surely be strange yardsticks to apply to the nonhuman past – the trace, the remnant, the ruin with which so many believe we are confronted in the archaeological past. But the truth is more human than that, as we see in photology a visual knowledge of the past. The photograph and the body operate in the same register, such that memory and landscape constitute a visual mode of thinking.

What archaeologists write, draw, collect, and photograph in the landscape is a loose genre of memoir; what they recall are places; we conduct temporal experiments with our own bodies, our excavations are always re-enactments; the time passing during fieldwork and after calibrates the time-depth of the archaeological sequence.

In Barthes' terms, the 'essence' (*noeme*) of the archaeological photograph is not 'what has been' (*ça-a-été*), but 'as if' (*comme si*), not fixed on the remnant, but at once both scientific and imaginary (compare Barthes 1980: 120, 1984 [1978]: 325).

As if a photograph were not a still. As if the archival box were a photographic aperture. As if the museum were a dark room. As if you could ever look for long enough. As if human life were a blur. As if archaeology could be more than interpretive, and less than representational. As if the past were more than ruins.

As if this landscape were here. As if our memories could be experienced. As if it were me in these photographs. As if archaeology could hold forth and transform the world. A photological landscape made through rediscovery.

Notes

1 'Il a été absolument, irrécusablement présent, et cependant déjà différé' (Barthes 1980: 121).
2 'La Photographie, parfois, fait apparaître ce qu'on ne perçoit jamais d'un visage réel (ou réfléchi dans un miroir): un trait génétique, le morceau de soi-même ou d'un parent qui vient d'un ascendant' (Barthes 1980: 160–161).
3 'ce moment très subtil où, à vrai dire, je ne suis ni un sujet ni un objet, mais plutôt un sujet qui se sent devenir objet: je vis alors une micro-expérience de la mort (de la parenthèse): je deviens vraiment spectre' (Barthes 1980: 30).
4 'se dévoile, se déchire, se révèle, au sens photographique du terme' (Barthes 1977: 34).

5 'By putting Proust and myself on the same level, I in no way mean that I am comparing myself to this great writer: but, in a quite different way, that I identify myself with him: confusion of practice, not of value' (Barthes 1984 [1978]: 314, my translation). Barthes expands on this distinction between comparing with and identifying with in the first lecture of the *The Preparation of the Novel* series (Barthes 2011: 3).
6 See discussion by Blonsky 1985: xiii–xiv. The lecture was originally envisioned as *Proust et Moi* but eventually titled '*Longtemps, je me suis couché de bonne heure*'.
7 'Photographs are valued because they give information. They tell one what there is; they make an inventory. To spies, meteorologists, coroners, archaeologists, and other information professionals, their value is inestimable. But in the situations in which most people use photographs, their value as information is of the same order as fiction' (Sontag 1973: 16).
8 'Je mehr die Entfernung wächst, desto klarer wird die Sicht. Mit der größtmöglichen Deutlichkeit erblickt man die winzigen Details. Es ist, als schaute man zugleich durch ein umgekehrtes Fernrohr und durch ein Mikroskop' (Sebald 1995: 29–30).
9 'Jedoch bedeutet in der Optik der Geschichte – darin das Gegenbild der räumlichen – Bewegung in die Ferne Größerwerden' (Benjamin 1972 [1927]: 348).
10 'Des systems d'instants . . . se succédent, mais aussi se répondent. Car ce que le principe de vacillation désorganise, ce n'est pas l'intelligible du Temps, mais la logique illusoire de la biographie, en tant qu'elle suit traditionnellement l'ordre purement mathématique des années' (Barthes 1984 [1978]: 318).
11 'un peu terrible qu'il y a dans toute photographie: le retour du mort' (Barthes 1980: 23).

References

Barthes, R. (2011). *The preparation of the novel: Lecture courses and seminars at the Collège de France, 1978–1979 and 1979–1980*, trans. K. Briggs. New York: Columbia University Press (translation of Barthes 2003).
Barthes, R. (1977). *Fragments d'un discours amoureux*. Paris: Seuil.
Barthes, R. (1978). *Leçon. Texte de la leçon inaugurale prononcée le 7 janvier 1977 au Collège de France*. Paris: Editions du Seuil (published in English, translated by R. Howard, in 1979 as Lecture in Inauguration of the Chair of Literary Semiology, Collège de France, 7 January 1977, 8 October, pp. 3–16).
Barthes, R. (1980). *La chambre claire*. Paris: Seuil/Gallimard.
Barthes, R. (1984 [1978]). Longtemps, je me suis couché de bonne heure. In: *Le bruissement de la langue (Essais critiques IV)*. Paris: Editions du Seuil, pp. 313–325.
Barthes, R. (2002). *Comment vivre ensemble: Simulations romanesques de quelques espaces quotidiens: Notes de cours et de séminaires au Collège de France, 1976–1977*, ed. C. Coste. Paris: Editions du Seuil.
Benjamin, W. (1972 [1927]). Moskau. In: *Gesammelte Schriften*, Vol. 4, No. 1. Frankfurt: Suhrkamp, pp. 316–352.
Blonsky, M. (1985). Introduction: The agony of semiotics: Reassessing the discipline. In: M. Blonsky, ed. *On signs*. Baltimore, MD: Johns Hopkins University Press, pp. xiii–li.
Ginzberg, C. (2001). *Wooden eyes: Nine reflections on distance*, trans. M. Ryle and K. Soper. New York: Columbia University Press.
Hart, K. (2000). *The memory bank: Money in an unequal world*. London: Profile Books.
Hawkes, J. (1951). *A land*. London: Cresset Press.
Hicks, D. (2016). The temporality of the landscape revisited. *Norwegian Archaeological Review*, 49 (1): 5–22.
Holbraad, M., and Pedersen, M.A. (2010). Planet M The intense abstraction of Marilyn Strathern. *Anthropological Theory*, 10 (1): 1–24.
Knight, M., and L. McFadyen. (2019). 'At any given moment': Duration in archaeology and photography. In: L. McFadyen and D. Hicks, eds. *Archaeology and photography*. London: Bloomsbury.
Lévi-Strauss, C. (1964). *Mythologiques: Le Cru et le Cuit*. Paris: Librarie Plon.
Malafouris, L. (2013). *How things shape the mind a theory of material engagement*. Cambridge, MA: MIT Press.
Olivier, L. (2008). *Le Sombre Abîme du temps*. Paris: Éditions de Seuil.
Richter, G. (2010). Introduction. Between translation and invention: The photograph in deconstruction. In: J. Derrida, ed. *Copy, archive, signature: A conversation on photography*, trans. J. Fort. Stanford, CA: Stanford University Press, pp. ix–xxxviii.
Sebald, W.G. (1995). *Die Ringe des Saturn: Eine englische Wallfahrt*. Frankfurt: Eichborn Verlag.
Sontag, S. (1973). *On photography*. New York: Farrar, Straus and Giroux.
Strathern, M. (1999). *Property, substance and effect: Anthropological essays on persons and things*. London: Athlone Press.
Strathern, M. (2005). *Kinship, law and the unexpected: Relatives are always a surprise*. Cambridge: Cambridge University Press.
Warwickshire County Council. (1993). *A435 Norton Lenchwick bypass. Phase 2: Archaeological field evaluations. Part II.* Warwick: Warwickshire County Council.

25
WALKING, WRITING, READING PLACE AND MEMORY

Ceri Morgan

The last 15 years have seen an interest in walking in many disciplines, including geography, history, sociology, literary studies, performance, dance, and creative writing (Middleton 2011; Mock 2009; Sinclair, e.g. 2002). Indeed, 'walking studies' (Lorimer 2011: 19) is emerging as a sub-field or subdiscipline in its own right. In what have come to be some of the founding texts on the subject, walking is represented as a liberating and subversive – even revolutionary – act. Consequently, in 'Theory of the dérive,' Guy Debord (2006 [1956]) defines the '*dérive*' or 'drift' as a dropping out of the regular routines of work and domesticity in order to 'be drawn by the attractions of the terrain and the encounters [found] there' (2006 [1956]: 62). Best undertaken as a collective, the '*dérive*' aims to 'study a terrain' and/or 'to emotionally disorient oneself.' This is achieved via a 'rapid passage through varied ambiances' (Debord 2006 [1956]: 62, 64) within the city – Debord considers the '*dérive*' or 'drift' as an essentially urban practice. The 'drift' challenges what he and other members of the Situationists International saw as the alienating numbness of modern capitalist society. The utopian aspects of Debord's theory are echoed, to a degree, in Michel de Certeau's less collectivist but similarly urban-centric, Walking in the City (1984 [1980]). In this 'ur-text' of cultural studies (Morris 2004: 675) – and other disciplines – Certeau describes walking as one of several 'everyday' practices with the potential to challenge the workings of power. This is because it escapes full representation, and hence observation by all-seeing authority regimes: 'escaping the imaginary totalizations produced by the eye, the everyday has a certain strangeness that does not surface, or whose surface is only its upper limit, outlining itself against the visible' (1984 [1980]: 93). Famously, Certeau contrasts the 'path' made by walking, with the panoptical 'map' which, he argues, seeks to fix space in a readable way. Eluding representation, since the 'path' can only trace the route undertaken and not the act itself, walking gives rise to 'a *migrational*, or metaphorical city' (1984 [1980]: 103). It is connected with legend and remembering – the second another form of potential micro-resistance – with Certeau claiming, 'haunted places are the only ones people can live in' (1984 [1980]: 108).

Certeau has been critiqued by Meghan Morris for mobilising a binary model of power (1992: 13). *Walking in the City* has also been revisited by Nigel Thrift, who argues that it fails to acknowledge the modern city's symbiosis with the car, and driving as a complex, embodied practice (2008: 75–88). More generally, walking has been recognised as something which is not always done for pleasure, easily, or on feet. In this way, John Urry warns, 'there are . . . many ways to walk, sometimes walking is mundane (to shop), sometimes the basis of unutterable suffering (to go on a forced march)' (2007: 65). Phil Smith challenges the romanticism of some of the recent writing on walking, highlighting that 'an overriding quality of walking . . . can be its connectedness. But not at the expense of disruption, of tripping up and over, stumbling and righting, of falling' (2014: 27). For their part, Richard Keating and Sue Porter remind us that walking is not necessarily a bipedal activity (2015). Nevertheless, walking is seen as 'the most "egalitarian" of mobility systems' (Urry 2007: 88). It continues to be taken up as a research method, with one example being Maggie O'Neill's project, *Methods on the Move: Experiencing and Imagining Borders, Risk and Belonging*. It still informs artistic practice, as in Rosana Cade's performance piece, *Walking: Holding* (2011–). In this, participants are invited to walk a route holding a number of different individuals' hands at certain points along the way; prompting reflections on sexual identity, intimacy, and abled, disabled, or un-able bodies. Walking is taken up, too, in works which combine method and practice, as in the 'histories' by Rebecca Solnit (2002) and Robert Macfarlane (2013 [2012]). Both these writers use creative nonfiction to bring together walking with the personal and the collective, the present and the past. Consequently, the introduction to *Wanderlust: A History of Walking*, sees Solnit reflect on her experiences as an activist as part of a broader investigation of this 'most obvious and most obscure' practice (2002: 3). In *The Old Ways: A Journey on Foot*, Macfarlane describes (largely non-urban) paths as 'ghost-lines,' which are collectively made and remade over time and across generations (2013 [2012]: 21). In so doing, he not only recalls his own walks with his late friend, the nature writer and environmental activist, Roger Deakin, but also acknowledges the ways in which literary texts – and, by implication, other forms of cultural production – can haunt us as we move through a landscape (Macfarlane 2013 [2012]: 32).

The combination of mobilities, personal and collective memories, and broader connections with places and their representations is found in psychogeography, with Will Self and Iain Sinclair being two of the best-known contemporary writers working in this mode. In his characteristically mock-arch style, Self identifies psychogeography as the pursuit of the (implicitly heterosexual) middle-class male, as he sets out on a series of walks which underpin his creative non-fiction book, *Psychogeography* (2007). Self humorously associates himself with what his friend, Nick Papadimitriou, describes as a 'psychogeographic fraternity': presenting this as a group of 'middle-aged men in Gore-Tex, armed with notebooks and cameras . . . querying the destinations of rural buses' (Self 2007: 12). Sinclair's fiction, creative non-fiction, and film have drawn on the author's walks around London, including around the M25 motorway in *London Orbital* (2002), and along the Thames in *Downriver, or, The Vessels of Wrath: a narrative in twelve tales* (2004 [1991]). The author's writing blurs temporal layers, so that the past bleeds into the present, as in the visions of one of *Downriver's* characters which occur during the present of the novel in Thatcher's London. These are of the victims of the *Princess Alice*, a pleasure steamer sunk in 1878 following a collision with a colliery boat on the Thames:

> From beyond the curve of the power station, Bobby saw them coming up on the tide. . . . They were all dead. They swam to fetch him. . . . There were women in hats, holding their children above the waterline.
>
> *(Sinclair 2004 [1991]: 10)*

Sinclair's novel mobilises a dreamy or trippy atmosphere found in the work of other psychogeographic practitioners, such as the artist, Laura Oldfield Ford, who, in working on *Chthonic Reverb* (2016), walked repeatedly around Birmingham's Digbeth and Eastside – formerly industrial neighbourhoods targeted for regeneration. In so doing, she recalled her younger, anti-capitalist, and rave-frequenting self as she attempted to tune into the past. The exhibition featured photographs, written texts, and visual symbols such as the smiley face associated with 1990s rave culture on plywood and sitex (perforated steel): materials used to board up disused buildings. It suggested the ignored, neglected, and marginalised spaces lying alongside or beneath the shiny new-builds of redevelopment. *Chthonic Reverb* also featured an audio piece of a '*dérive*' undertaken by the artist and a friend from her youth, which Pawas Bisht describes as 'resisting . . . the newly regenerated (sterile/sepulchural/unhaunted/uninhabitable) Eastside Park' (Bisht & Morgan 2017: n.p.). At least in its popular perception, psychogeography has tended to be identified with cis-, straight-, and often middle-class and white masculinity. However, as Oldfield Ford illustrates, there have been, and continue to be, creative practitioners and critics who do not identify with this model. A collection edited by Tina Richardson (2015a) features essays from a range of contributors, including Victoria Henshaw, who has a chapter on sensory walks (2015), Morag Rose, who includes a contribution on the LRM (the Loiterers Resistance Movement) she founded in Manchester (2015), and Richardson herself, who writes on her adaptation of 'schizocartography' (2015b: 164–175).

Although psychogeography is a genre which lends itself particularly well to creative productions which move between the individual and the collective as well as the past and the present, there are other walking research practices which engage with memory. One example is the *The Walking Library* which Deirdre Heddon and Misha Myers originally created for the 2012 Sideways Arts Festival. This featured volunteer librarians carrying a selection of purchased and donated books across Belgium (2014). The project included some public readings at festivals, but also discussion, song, and conversations, as well as reading, on the move. Heddon and Myers argue that reading and walking are altered by doing both at the same time (2014: 651–652). They also suggest that reading-walking/walking-reading add further layers to the strata of material and imaginary geographies from the past and present which make up any landscape, revealing that 'the books we read *in* place' function as 'mnemonics of place' (2014: 649, their italics). Another engagement with memory is found in Myers's article on pre-recorded and 'live' audio walks, with the latter being defined as those in which a performer reads aloud at points on the route (2010: 59). Myers's critical-creative piece gives a sense of the interplay of her own memories with the collective recollections she hears whilst undertaking Graeme Miller's *Linked* (2003). *Linked* is a pre-recorded soundscape featuring former residents of an area of East London whose homes were expropriated for the M11 Link Road. It plays continuously along a three-mile route via radio transmitters. Myers writes of her experience of it, 'I look through windows into front rooms just a few feet away, the memories I hear interweaving with a displaced memory-house of my own' (2010: 62).

Myers's participation in *Linked* is a kind of staged haunting, a deliberate listening out for voices from the past. As such, it differs from the many 'micro-hauntings' (Morgan 2018: 49) we likely encounter every day without noticing. In an article structured around his commute to work through suburban Manchester, Tim Edensor highlights how hauntings are a regular feature of daily life, arguing that 'traces of the past linger in mundane spaces by the side of the road to renewal, haunting the idealistic visions of planners, promoters and entrepreneurs' (2008: 314). In his contribution to a special journal issue on 'spectro-geographies' (Maddern & Adey 2008: 291), Edensor suggests that the repetition across generations of everyday tasks and routines ensures these 'become sedimented in neighbourhoods and are often physically inscribed on place' (2008: 325). These 'mundane hauntings' (Edensor: 2008: 313) have the capacity to disrupt chronological time like Derrida's ghosts (1993: 72). In *Spectres de Marx: L'État de la dette, le travail du deuil et la nouvelle Internationale*, Derrida points to the possibility of positive hauntings in the form of ghosts who remind us of the necessity for social change by warning us, like Hamlet's father, that '*the time is out of joint*' (1993: 47, Derrida's italics). These spectres can come from the future as well as from the past (Derrida 1993: 71). Walking and writing (where 'writing' is defined in broader terms than purely literary) can be combined to layer presents, pasts, and imagined futures, and offer possibilities for potential exchanges between various temporalities. As a long-term member of the geopoetics research group, la Traversée, at l'Université du Québec à Montréal, in recent years, I have drawn on methodologies shared by Rachel Bouvet, André Carpentier, and others, notably the '*ateliers nomades*,' or mobile workshops (Bouvet 2015: 239–240). Planned and organised by a small sub-group in advance, these take the form of trips to particular places, discussion, reading, and creative practice. The resulting outputs are published in the series, *Carnets de navigation*. Geopoetics has tended to be largely associated with the work of its founder, Kenneth White. For White, geopoetics is a world writing which 'applies not only to poetry . . . but also to art and music, and can be extended . . . into science and even social practice' (2004: 241). Seeking to break with post-Romantic traditions which position the world as the object of study of the observer, geopoetics proposes a relationship between humans and the world which has similarities with deep ecology in its recognition of the interdependence of the human and non-human.

Both practice and method, geopoetics can be undertaken using the most basic technologies – pen, pencil, paper, and legs (composed of flesh or other materials). In recent years, I have adopted the collective approach practised by Bouvet, Carpentier, and other members of La Traversée (Bouvet 2015: 239–240), combining this with creative methods drawn from psychogeography, literary, and cultural geographies, and oral history. I currently work with a group of graduate students who are using walking methodologies in their creative writing. Nicknamed 'the Dawdlers,' after Carpentier's nonfiction 'flâneries' (e.g. 2005), we draw variously on a theoretical mix of geopoetics, psychogeography, geocriticism, ecocriticism, and related methods in our place-writing on Manchester, Montreal, Stoke, Kent, Somerset, and South Wales. To date, these methods have produced two main projects: 'Tapping Ware' and 'Memories of Mining.' 'Tapping Ware' took the form of two mobile workshops and an exhibition. The first workshop was at the Spode Works in Stoke-on-Trent, where tableware, teaware, and decorative ceramics were made from 1776 until 2008. There, scattered design transfers, paperwork, and health and safety signs left behind following the pottery factory's closure gave the impression that the workers had only just gone home. A testament to Stoke-on-Trent's industrial past as the world-leading producer of ceramics, the factory functioned as what Caitlin DeSilvey and Tim Edensor term a 'new rui[n]' (2012: 466). In their overview of the extensive scholarship on ruins, DeSilvey and Edensor point to the ways in which spaces deemed no longer useful by capitalism 'may become forces for mobilizing and materializing collective anger and resistance; but … may also simply be painful reminders of loss' (or 2012: 468). The hauntedness of the Spode site was intensified by traces of previous visits on the part of other artistic and educational groups. These traces included the occasional artfully arranged artefact, such as an open drawer containing a teacup, and a piano which had been installed for an arts or performance project and forgotten or abandoned.

At the time of the workshop, the Spode Works comprised what Karen E. Till terms a 'wounded plac[e].' For her, such places 'are understood to be present to the pain of others and to embody difficult social pasts' (2008: 108). Unquestionably, working conditions in Stoke's potteries factories were often exploitative. However, it is not so much the memories of these but the loss of the majority of the jobs associated with ceramics and other industries which opens a 'wound' in the city-region. Unemployment, job and housing precarity, poverty, and ill health affect a high proportion of Stoke-on-Trent's residents, and the city is ranked 14th out of 326 local authorities in England with respect to deprivation levels (CHAD 2018: n.p.). There are many local activist and community groups working to improve people's lives, and the last couple of years have seen a certain degree of renewed civic pride in response to increased awareness of, and growth in, many forms of cultural activity. This has been especially the case in the lead-up to, and following, the bid for UK City of Culture 2021, which saw Stoke-on-Trent make the five-long shortlist but ultimately lose out to Coventry. Some of the recent artistic activity takes place at the Spode Works, which were renovated and refurbished in 2016 to include artists' studios. Nevertheless, economic and social forms of suffering persist, as do emotional 'ruins' wrought by multi-generational marginalisation. These give rise to a kind of identity dysmorphia by which some residents see themselves and their city as worthless – a phenomenon highlighted by Mark Featherstone in a piece on the Bransholme housing estate in Hull. Featherstone argues that the 'banal horror, the despair of wasted lives in a wasted world' fosters a 'being-on-Bransholme,' whereby residents excluded from participation in a social and economic life shaped by neoliberalism define themselves as nothing (2013: 195).

The 'Tapping Ware' project aimed to avoid focusing solely on the city's historic successes, contrasting these with the failures of the moment. For, as DeSilvey and Edensor point out, 'new ruins' (2012: 466) can become sites of a socially conservative nostalgia (2012: 469–70). A second workshop was held at the highly successful Emma Bridgewater Factory, which produces vernacular ceramics using traditional techniques. The factory occupies buildings originally opened in 1883 for Charles Meakin of the Meakin brothers, a company best known for its ironstone tableware. As with the Spode workshop, participants had an hour's visit to the site, during which they used their own preferred technologies to make notes, take photographs, or record ambient sounds. As the majority were undergraduate students in their twenties, the smartphone was the most commonly selected technology. Photographs, recordings, and notes from the site visit informed poetry and fiction produced during a creative writing session and later refined following peer critique. Influenced by the lyricised depictions of manual labour in Michael Ondaatje's novel on the building of modern-day Toronto, *In the Skin of a Lion* (1987), my own stories evoke working-class labour practices and industrialised pasts in Stoke and South Wales (2016a, 2016b). Excerpts from some of the poems and stories formed part of the *Back to the Drawing Board* and related exhibitions on the Bridgewater-Rice family at Keele University (Autumn to Winter 2016–2017). The creative writing was displayed alongside photographs and excerpts from interviews with former potteries workers on their everyday work routines.

A third workshop, entitled 'Memories of Mining' was held at Silverdale Country Park in 2016 to mark the fiftieth anniversary of the Aberfan Disaster. This saw a spoil tip collapse on Pantglas Junior School in the South Wales village, killing most of the pupils and teachers inside. I have a personal connection to the event, in that my aunt is a survivor of the Disaster. Silverdale Country Park is situated on what was the site of the village's coal mine. The workshop was attended by a mix of academics, undergraduate and graduate students, parish councillors, an archivist, and a former miner who had worked at Silverdale Colliery and his wife. Drawing on research by James Evans and Phil Jones on the walking interview, which suggests that 'it seems intuitively sensible for researchers to ask interviewees to talk about the places that they are interested in while they are in that place' (2011: 849), I carried out an informal interview with the last two participants. Not surprisingly, given the theme, the discussion centred on striking as well as working practices, with local memories of the Miners' Strike of 1984–1985 contrasting with my own from South Wales during the same period. Poems and short stories prompted by the workshop were displayed alongside quotations from the interviews and photographs at the community library in Silverdale during Autumn 2017. An exhibition launch event featuring readings and a performance of a song was attended by colleagues, villagers, and former miners. In June 2018, an opportunity to work with local participatory performance company Restoke led to a collaborative performance called *Seams*. This featured readings by several Dawdlers, dance, music, and sound. A small number of professional dancers and singers were joined by community volunteers from a range of backgrounds and ages. Several of the volunteers had performed with Restoke before, and others had an interest in, or connection to, mining. One of the most memorable moments of the show featured an extract from the recorded interview from 2016, in which the former miner describes some of the fun and camaraderie of mining. It was accompanied by music composed by Restoke Co-artistic Director, Paul Rogerson, and an athletic and playful solo by male dancer, Frankie Hickman. The music and solo articulated some of pleasures found in the dangerous and difficult work of mining and celebrated miners as a collective. They also paid tribute to the individual miner whose voice the audience heard, with the dancer's performance a sort of ventriloquising or translation of the youth of the man now in his 60s, who was sitting in the audience.

Till argues that 'places are embodied contexts of experience, but also porous and mobile, connected to other places, times and peoples' (2008: 105). *Seams* suggests this interplay of bodies, temporalities, and places. It was performed at Keele Chapel, a modernist building by architect George Pace. The choice of venue represented something of a change of practice for Restoke, whose work is usually site-responsive. However, under the direction of Co-artistic Director, Clare Reynolds, performers succeeded in producing contemporary dance sequences which responded at one and the same time to the (read) written texts, the landscapes of Silverdale and South Wales represented within the poems and stories, and the chapel itself. *Seams* and 'Memories of Mining' did not claim to aim to heal any past or present wounds – as if such a thing were possible. Rather, they acknowledged the wounds' presence whilst offering a moment of hope: a 'beautiful gesture', as White describes geopoetics (2004: 230). Combining walking and writing methods enables a (fictionalised, aestheticised) tracing of social histories to produce textual memoryscapes which can speak with the present. Given that geopoetics refers to cultural or artistic production in general rather than creative writing specifically, it has the potential to be used with a variety of publics. My own future adaptations of the method include a project with screendance theorist and practitioner, Anna Macdonald on walking, screendance, cyclical processes, and chronic pain (forthcoming, 2018–2019), as well as further development of the 'Memories of Mining' project. Other possibilities, of course, are as wide as the horizon.

References

Bisht, P., and Morgan, C. (2017). *"Rescuing" place: Haunting, memory & mobility in Birmingham's Digbeth*. (Unpublished conference paper), Nordic Geographers' Meeting, Stockholm, 18–21 June.
Bouvet, R. (2015). *Vers une approche géopoétique: Lectures de Kenneth White, Victor Segalen, J-M.G. Le Clézio*. Montreal: Presses universitaires de Montréal.
Carpentier, A. (2005). *Ruelles, jours ouvrables*. Montreal: Boréal.
CHAD (Centre for Health and Development). (2018). *Health inequalities*. Accessed 17 July 2018, www.chadresearch.co.uk/health-inequalities/
Debord, G. (2006 [1956]). Theory of the derive. In: K. Knabb, ed. *Situationist International Anthology*. Berkeley, CA: Bureau of Public Secrets, pp. 62–66.
de Certeau, M. (1984 [1980]). Walking in the city. In: S. Randall, trans. *The practice of everyday life*. Berkeley, CA: University of California Press, pp. xi–xxiv.
Derrida, J. (1993). *Spectres de Marx: L'État de la dette, le travail du deuil et la nouvelle Internationale*. Paris: Galilée.
DeSilvey, C., and Edensor, T. (2012). Reckoning with ruins. *Progress in Human Geography*, 37 (4): 465–485.
Edensor, T. (2008). Mundane hauntings: Commuting through the phantasmagoric working-class spaces of Manchester, England. *Cultural Geographies*, 15 (3): 313–333.
Evans, J., and Jones, P. (2011). The walking interview: Methodology, mobility and place. *Applied Geography*, 31: 849–858.
Featherstone, M. (2013). Being-in-Hull, being-on-Bransholme: Socie-economic decline, regeneration and working-class experience on a peri-urban council estate. *City*, 17 (2): 179–196.
Heddon, D., and Myers, M. (2014). Stories from the walking library. *Cultural Geographies*, 21 (4): 639–655.
Henshaw, V. (2015). Route planning a sensory walk. In: T. Richardson, ed. *Walking inside out: Contemporary British psychogeography*. London: Rowman and Littlefield, pp. 220–234.
Keating, R., and Porter, S. (2015). In-between places: Envisioning and accessing new landscapes. In: C. Berberich, N. Campbell and R. Hudson, eds. *Affective landscapes in literature, art and everyday life: Memory, place and the senses*. Farnham: Ashgate, pp. 223–246.
Lorimer, H. (2011). Walking: New forms and spaces for studies of pedestrianism. In: T. Cresswell and P. Merriman, eds. *Geographies of mobilities: Practices, spaces, subjects*. Farnham: Ashgate, pp. 19–33.
Macfarlane, R.E. (2013 [2012]). *The old ways: A journey on foot*. London: Penguin.
Maddern, J., and Adey, P. (2008). Editorial: Spectro-geographies. *Cultural Geographies*, 15 (3): 291–295.
Middleton, J. (2011). Walking the city: The geographies of everyday pedestrian practices. *Geography Compass*, 5 (2): 90–105.
Mock, R. ed. (2009). *Walking, writing & performance: Autobiographical texts by Dierdre Heddon, Carl Lavery and Phil Smith*. Bristol: Intellect Books.
Morgan, C. (2016a). San Francisco sigh. *Coordinates Society*. Accessed 11 January 2018, www.coordinatessociety.org/single-post/2016/04/21/San-Francisco-sigh
Morgan, C. (2016b). Weak spot. *Coordinates Society*. Accessed 11 January 2018, www.coordinatessociety.org/single-post/2016/11/04/Weak-Spot
Morgan, C. (2018). Sonic spectres: Word ghosts in Madeleine Thien's *Dogs at the Perimeter* and the digital map project, fictional Montreal/Montréal fictif. *London Journal of Canadian Studies*, 33 (4): 40–57.
Morris, B. (2004). What we talk about when we talk about "Walking in the City". *Cultural Studies*, 18 (55): 675–697.
Morris, M. (1992). Great moments in social climbing: King Kong and the human fly. In: B. Colomina, ed. *Sexuality and Space*. Princeton, NJ: Princeton University Press, pp. 1–52.
Myers, M. (2010). "Walk with me, talk with me": The art of conversive wayfinding. *Visual Studies*, 25 (1): 59–68.
Ondaatje, M. (1987). *In the skin of a lion*. Toronto: McClelland & Stewart.
Richardson, T. ed. (2015a). *Walking inside out: Contemporary British Psychogeography*. London: Rowman and Littlefield.
Richardson, T. ed. (2015b). Developing schizocartography. In: T. Richardson, ed. *Walking inside out: Contemporary British psychogeography*. London: Rowman and Littlefield, pp. 205–219.
Rose, M. (2015). Confessions of an anarcho-flâneuse, or psychogeography the Mancunian way. In: T. Richardson, ed. *Walking inside out: Contemporary British psychogeography*. London: Rowman and Littlefield, pp. 169–186.
Self, W. (2007). *Psychogeography*. London: Bloomsbury.

Sinclair, I. (2002). *London Orbital: A walk around the M25*. London: Granta.
Sinclair, I. (2004 [1991]). *Downriver, or, The Vessels of Wrath: A narrative in twelve tales*. London: Penguin.
Smith, P. (2014). *On walking . . . and stalking Sebald: A guide to going beyond wandering around looking at stuff*. Axminster, Devon: Triarchy Press.
Solnit, R. (2002). *Wanderlust: A history of walking*. New York: Viking.
Thrift, N. (2008). *Non-representational theory: Space, politics, affect*. London: Routledge.
Till, K. (2008). Artistic and activist memory-work: Approaching place-based practice. *Memory Studies*, 1 (1): 99–113.
Urry, J. (2007). *Mobilities*. Cambridge: Polity Press.
White, K. (2004). *The wanderer and his charts: Essays on cultural renewal*. Edinburgh: Polygon.

26
MNEMONIC MAPPING PRACTICES

Patrick Laviolette, Anu Printsmann, and Hannes Palang

Introduction

> What, a map? What are you talking about?
> Don't you know what a map is? There, there,
> never mind, don't explain, I hate explanations;
> they fog a thing up so that you can't tell anything about it.
> Run along, dear; good-day; show her the way, Clarence.[1]
> (Mark Twain 1889)

Samuel Langhorne Clemens chose the pen name Mark Twain because it was an old Mississippi River term for navigating through shallow waterways. As the boatman's call 'mark number two,' it signified the second marking on a line that measured a depth of two fathoms – or 12 feet, that is, a safe depth for a steamboat to travel. In a sense then, this name choice is itself about mapping a mnemonic spatial identity, in this case by reference to a vocational shipping idiom linked with the Mississippi delta.[2]

The quotation from Twain is all the more relevant here because we want to use it as a means of introducing certain methodological and epistemological tensions. In particular, those of being lost in translation, as well as the ones that exist between a person's spoken justification on the one hand (as revealed by the dismissal of listening to the explanation, 'there, there, never mind, don't explain') and the more phenomenologically informed descriptive knowledges, which arise from observational or embodied experiences on the other. In this chapter, we navigate through these tensions by relying on evidence that comes from discursive explanations about maps and mapping, as found in a wide body of interdisciplinary literature about the historical relationship between maps and memory.[3]

Maps are quintessential tools and symbols for geographers that often make up an important component of their research results. Images in general are of interest in this field because of their common association with various cognitive, spatial, or representational forms of mapping (Cosgrove 1999; Dorling & Fairbairn 1997). Anthropologists, for their part, have largely overlooked the cultural significances of the geographic map. Crick provides one exception by explicitly conceptualising map usages in semantic anthropology and social psychology. The value that he ascribes to a map, which we will play with here, is that it is 'something that is itself a representative device [and] can be employed as a means of representation' (1976: 129). He divides mapping metaphors into two categories: those that fit either into 'mirror theory' where they are iconic reflections of spatial reality, or those that are a part of a 'semantic field' theory where they generate a figurative spatial language.

Though this dichotomy might be limiting (and this chapter attempts to open it up), Crick nevertheless makes the astute claim that anthropologists should appropriate the task of devising methods for reading maps that chart out the world-views and life-worlds of different social groups. Despite his initial concern, the broader cultural use, interpretation, and understanding of cartographic images has not been of particular interest in anthropology. There are, however, significant exceptions. Indeed, important parallels have been put forward, not only in terms of mnemonic mappings, but also in relation to deciphering the artistic, nationalistic, ritualistic, and navigational/way-finding properties of cultural landscapes (Bender 1992; della Dora 2011; Gaffin 1995; Gell 1985; Ingold 2000; Jones 2004; Küchler 1993; McDonald 1990; Morphy 1991; Nash 2011; De Nardi 2014).

For instance, Gell's work (1985) on how to read a map draws on ethnographic material on the navigational skills of Micronesian seafarers to show how maps can be used to generate images indexed to specific locations, defined in terms of an imagined set of spatial coordinates, that can then be matched to the perceptual image at any particular place, allowing the navigator to know precisely where he or she is. But what of the non-navigational purposes of these maps? Maps can indeed stand outside or at the edge of most of our relations to place. In this respect they are both iconic of national ideologies and express feelings of difference. Cultural distinction here gets played out in the realm of territory rather than on more topographically two-dimensional playing fields (Turnbull 1989).

This line of questioning paves the way for a different level of understanding in the study of more conventional topographic representations – what we will call 'anthropography' – the written ethnographic account about how people give meaning to maps, mapping experiences and map-like images. That is, how they inscribe place with cartographic significance and, through these loaded representations, shape their own identities as well as the identities of these objectified artefacts and the areas they represent.

This chapter also draws on various analyses of popular items of visual/material and digital culture depicting such a relationship (such as tourist brochures, guide books, postcards, Points of Interest [POI], and Global Positioning System [GPS]). The conceptual frame for much of the material presented here centres upon two main themes that are of particular relevance to the study of visual culture: 1) map production and mapping practices, and 2) globalisation, as expressed through new technological production and the promotion of international tourism. For instance, Elizabeth Edwards reveals that the artefacts linked with holidaymakers 'are an important facet in the ongoing consideration of the politics of representation' (1996: 216). Travel is relevant to this framework since Harley (1992) and Anderson (2004) remind us that cartographic processes are themselves inextricably linked with the development of colonial and national ideologies, as well as with the formulation of imperial subjectivities. Their research emphasises that if we seek to better our understanding of hegemonic and counter-hegemonic social structures, we need to be more attentive to the manipulations of symbolic power occurring in the business of mapping places, identities, and histories. Further still, according to Black (1997), Harmon and Clemans (2010) and Hieslmair and Zinganel (2013), social analysts are showing a growing interest in the representations of identity which occur in artistic, promotional, and other tourist-related memorabilia. Such research stresses the importance of semiotic ethnographies of media-construed images of the travel market.

Brown and Turley (1997) amplify this call by warning that Baudrillardian views of hyper-reality – where simulacra or reproductions take precedence over authentic images – often do not do justice to the diversity of real-life holiday experiences.[4] Consequently, they suggest that scholars should consider the materiality of travel in its habitual contexts. Hence, it is to everyday life-worlds that we turn in the first part of this chapter in order to explore how the notions of locality, identity, and memory work off each other visually (Fernandez 1986; Holm 2005). The second looks at whether there is any value in exploring how 'anthropography' could help define certain links between mapping depictions, the appropriation or countering of difference, and the addition of landscape diversity through popular visual/digital media (Crampton 2009; Monmonier 2007; Wilson 2017).

The parallel fields of visual and material culture studies have burgeoned in the social sciences and humanities for at least two decades now. One reason for this is that vision in particular is probably the least subjective sense in Occidental science, in part because it allows us to encounter material reality in large sweeps (Ingersoll & Bronitsky 1987; Said 1978; McEvedy 1998). In much the same way, landscapes and maps also deal with large expanses of people and places. And maps are indeed highly visual objects, sometimes even considered as fleeting works of art (in addition to *technê*). As a result, Western understandings of topography often take shape through sight and are abstracted from the sensual contact of the other senses, which are usually experienced at a much smaller scale.

Many scholars interested in visual culture argue that it is important to challenge these 'visualist' assumptions by conceptualising the visual in terms of more phenomenologically diverse forms of experience (Banks & Morphy 1999; Selwyn 1996). A metaphorical framework helps us in this respect because it permits us to employ the visual representations of one domain to highlight relationships in another (Tilley 1991, 1999). In these terms, the visual realm becomes something that we can consider as an objectified other. It acquires an added dimension of materiality, one in which the position of visual objects – their locality and the situation in which we exist as 'observers' – are gestured back to one another so that the emphasis on the visual setting has transformed itself 'into a generalised other and into something extremely close to the concept of the artefact as congealed labour' (Richardson 1987: 389).

From this vantage point, the visual setting defines itself as a 'collapsed act.' It becomes a social vehicle for the actions occurring and the materials being shaped – a microcosm of form, image, and representation. This process provides a telling social record of the thoughts, symbols and embodied experiences that give meaning to social behaviour. Hence, as Tuan points out: 'valued artifacts must be maintained by human discourse' (1980: 466), a discourse that, as this chapter demonstrates, refers to a complex set of ephemeral relationships between visuality, mapping, and the mnemonic.

Regarding the realm of 'anthropography,' one could immediately think of the *Pianta Grande di Roma* (Great Plan of Rome) – usually referred to as the Nolli Plan. Giambattista Nolli's 1748 map is one of the most significant physical representations of a specific location. This post-Renaissance Enlightenment map set an historical and anthropological precedent for an urban mapping of what can be defined as 'interpretative mapping,' a form of mapping social convention, more than a mere sensorial survey. Many heuristic projects use the Nolli Plan to locate hidden secrets within it, such as the way spaces are defined as public or private. The grey-scale colour is an effective medium in reflecting cultural and national identities. Voids and figures of the city are easily revealed and are narrative symbols of the most basic questions on the ties between individuals and cities (Monmonier 1996). One could easily go into depth on the means in which the Nolli Plan is a substantial artefact. The plan has its own material narrative in a sense that is 'anthropographic.'

Forming and following function

Tourist brochures

Onto the landscapes that we see and live in, we constantly broadcast images – cultural crests, emblems of our identities and memories – and vice versa. By examining attitudes to signifiers in the paraphernalia of the popular media, this section highlights how visual cultures propagate numerous opposing and complementary images. The section thus provides an iconological and semiotic look at what are essentially visual items of popular material culture. The map images that we rely on most heavily occur in tourist brochures. As images, these socio-cultural mediators are not simply methods of recording data but become data in and of themselves (Banks & Morphy 1999).

Map images continue to be of great significance although a shift of emphasis has occurred in late modernity regarding some of the search for scientific rigour in relation to mapping representations (Casey 2002). The need for traditional scientific cartography has required a different type of description – one that facilitates accounts of walking, hiking, and sightseeing, or how far (planar length, in absolute and relative height and calories) is the next parking lot or POI.

In his classic volume *Mythologies*, the critical theorist Roland Barthes (1957) analyses the role that world holiday guides play in France by examining Hachette's series of global 'blue guides' (*Guides Bleus*). Overall, he argued against the idea that these items were essentially products destined to enhance one's appreciation of landscape. Nor, he claimed, did they act as educational devices in the service of increasing one's cultural capital in terms of perception and geographical awareness. Instead, Barthes claimed that travel guides are blinding agents, directing the user's attention away from the everyday. They mask what is 'real' in the mundane history of human experience. Consequently, by this profusion of sensationalism, these guides advance an ideology of individualism that considers the ethos of travel to be 'effort and solitude.' Through introducing the concept of sightseeing, such guides have come to allow for the purchase of effort and it is in this indirect sense that they are able to serve the bourgeoisie. Duncan and Duncan (1992) thus suggest that these travel guides form part of an ideological strategy of leisure:

> originating in a nineteenth-century 'Helvetico-Protestant morality' that promotes the aesthetic appreciation of uneven ground, mountains, gorges, torrents, and defiles. This ideology is described as a 'hybrid compound of the cult of nature and of puritanism' which espouses regeneration through clean air, moral ideas at the sight of mountain tops, summit climbing as civic virtue, etc.
>
> *(1992: 20)*

In recognising the conflating of such notions as scenery and the picturesque in the 'blue guides,' Barthes has uncovered the leading, although not altogether uncontested, ideological process that underscores the bourgeois gaze. He concludes that these guides mystify socio-political realities via an almost totalising interest in monuments. Further, for all intents and purposes, their imagery depopulates the landscape. We can compare this with Mels (2002) on how National Parks are presented without humans (especially places such as Lapland). These factors contribute to a sickness or even a 'disease' of thinking in essentialist terms. A populist and decadent order, he argues, governs the bourgeois gaze and lies behind its mythologies (Anon Institut Francais Estonie et al. 2015).

Mapping imagery acts on many metaphorical levels as a visual expression and mnemonic device. Visual icons and landscape features shape a place's material historicity or biography. In selling tourism, these representations advertise the local; by talking of leisure they use vernacular labour; through promoting accessibility, they advocate escapism or even parochiality (see Kaur et al. 2004). Many of the images themselves are actually quite ordinary – but paradoxically their role is to mystify. The more mundane, the more enigmatic. Clearly then, brochures with maps are materially embodied visual artefacts. They are good post-modern examples of what Tilley (1991) calls an 'art of ambiguity.'

Counter mapping

There is a growing literature on so-called counter-mapping, much of it the result of anthropologists' interests in the land rights claims of indigenous peoples. Far from being neutral instruments, maps have the power to include and exclude peoples and places. So-called community-mapping or counter-mapping is increasingly employed by indigenous peoples, often with the assistance of anthropologists, to defend and protest their homelands, artefacts, and knowledges or to set boundaries and stake claims to customary lands.

In the Malaysian state of Sarawak, for instance, community maps have featured in court cases mounted by local communities against corporations whose activities threaten to degrade customary lands and forests

(Peluso 1995). Combining detailed knowledge of their homelands with the techniques and manner of representation of dominant groups, 'maps drawn by communities utilize memory (oral histories) and markers (fruit trees, sites of old settlements) as tools for claiming territory and customary rights' (Cooke 2003: 266). Similarly, Käyhkö et al. (2013) give an example for Zanzibar, Tanzania whereby the voices of farmers influence government interventions on the transition of forest cover.

Often remarkably detailed and including a great many named places and other spatial associations that go unrecorded in conventional topographical maps, the everyday mapping and map making of local communities can mount challenges to state hegemony, act as a medium of empowerment and resistance (De Nardi 2014), help bolster claims to land and other resources, and (re)establish community bonds (Cross 2012), as well as project, reflect, and reinforce cultural identities.

The idea that map depictions can sometimes act as artefacts of dissidence parallels Anderson's (1983) and McDonald's (1990) research. Anderson demonstrates that the 'map-as-logo' has come to underscore the significant delimitation of political domains. Though the rationale might have been benign at the outset, the practice of cartographically isolating and colouring colonies with particular dyes on the part of Imperial states transformed maps into pure signs (Etherington 2007; Olwig 2007). No longer did they need to have an actual bearing on the world. Under this new guise, maps entered a political economy of commodification, befitting for the market of books, magazines, posters, tea towels, and so forth. Anderson (1983) gives as an example of the way in which the new form of territorial map brought quarrelling young West Papuan nationalists together against the Indonesian state. The maps, he demonstrates, came to form an ideological iconography that was instantly recognisable at once. Visible everywhere, logo-maps profoundly entered the popular imagination, establishing a compelling emblem for the arising anti-colonial sentiments of nationalism.

For her part, McDonald reveals how Breton militants (i.e. those involved in the nationalist movement) use 'historical' maps to emphasise a long tradition of Celtic defiance in the face of persecution. She suggests that the way in which such militants frequently map the modern Celtic countries disregards contemporary reality and cultural history. Through such cartography, France is largely ignored while Brittany stands alone and is portrayed as a Celtic homeland west of a line that extends from Mt St. Michel to the Loire.

McDonald claims that, although these map images might propose that the Breton/French language duality is comparable with the duality between Brittany and France, this is not and has never been the case either socially or geographically. Furthermore, she briefly traces the shifting boundaries of Celtic identity in Eurasia:

> In a book of which the title translates as 'Breton, Celtic language' there is a map entitled 'Celtia, 25 centuries ago' in which Celts are shaded in from Asia minor to Hibernia. Some pages later comes a map called 'Celtia now', where the shading is limited to Brittany, Cornwall, Ireland, the Isle of Man, Scotland and Wales, each given the name in its respective Celtic language.
>
> *(McDonald 1990: 117)*

From McDonald's research on this discourse of Celtic identity, we are made aware of another cartographic category: nationalist propaganda maps. To a degree, such maps are becoming increasingly controversial given that they are responsible for outlining highly charged contours which appropriate as well as convey loaded notions of spatial 'ethnicity' (Harley 1992; Pickles 1992).

As we see, national shapes are themselves intrinsic to the formulation of many types of identities. This is so in a material and not functional or deterministic way however. In other words, through the repeated uses of the actual contour we encounter a self-perpetuating visual metaphor about belonging. The shape acquires an agency and an identity that become in themselves sources of identification.

In this instance, maps have many meaningful layers – they are good to travel with, good to think with and good to affiliate identity with. Put differently, maps help harmonise and integrate a more coherent vision of identity. They are not simply tools used to control the masses through some esoteric power struggle, but instead can be used to subvert authority or at least signal certain forms of social identity. In these terms, maps are part of a material process for generating the significance of spatial difference. Despite their seeming reflections of it, they actually inaugurate meaning (Harley & Woodward 1987). Consequently, maps can potentially give structure to ethnicity. That is, they have a ubiquitous presence that metaphorically reconstitutes a shape for belonging.

In tandem with the increasing perception that modern tourism threatens the very fabric of certain identities, we are witnessing a rise in the use of emblems meant to mitigate the contradictions in promoting ethnicity. But it is a point of contention as to whether what outlines identity locally and to the outside world is becoming increasingly colourful and complex or more simplistic – more black and white as it were. Such a debate is of course healthy if it problematises perceptions of ethnicity. It should not, however, reach a point where it becomes obsessed with the controversy regarding who should contribute to, or have control over, the (re)production of indicators of belonging, for some authors have warned us of the problems of this slippery slope (see Thomas 1999 for a parallel discussion about aboriginal art).

Counter-mapping has gained a momentum beyond indigenous power struggle to engage larger sweeps of populace aided by technological savviness. While Cantwell and Adams (2003) focused on determining the aboriginal sacred sites' connection to ecological suitability, the availability of crowdsourcing can map other issues of concern like environmental governance. A community mapping tool was also used for example after hurricanes in Puerto Rico to designate areas without power or where it was safe to land with helicopters.

Conclusion

Behind individual mnemonic perceptions are more systematically structured discourses about such notions as landscape, cartography, scenery, and sight. These validate specific activities and ways of seeing, especially those related to walking, tranquillity, and the search for open-air leisure. The evolution of these historically and socially variable memories have been transformed so that sight stands as a spatial icon. We thus see an attempt to generate a sense of longing to be in place from what are essentially two dimensional representations, until recently when three-dimentional virtual reality and augmented reality have become more accessible. These have a ubiquitous presence in everyday life.

Indeed, maps are indoctrinating of a *habitus* that is constantly attempting to reconcile the rupture between home and destination. Such items of visual material culture are especially relevant because they inform us on how the landscape is a socio-political medium that possesses exchange value and status. Consequently, such articles of popular culture, which also include oral history mapping portals, are part and parcel of a global currency of images. They are amongst the most widespread means that societies have of publicising their foreign elements to the modern world. And they are strategic tools in keeping the memory industry afloat (Selwyn 1996).

In this light, the visual realm is not strictly speaking a vernacular one since it enters many commodified fields that the tourist industry and leisure consumers can control. Landscape maps are examples of both exported and imported signifiers of belonging. But, nonetheless, sight acquires many levels of agency through alternative forms that occur in maps and other similar popular media images. Modernity's visual cultures thus come to form part of a metaphorical realm in which people dabble when creating their social identities.

We should not fail to recognise the centrality of the visual realm which exists outside the art world in confusing and solidifying those identities. To a certain degree, then, it seems unavoidable to conceive of memory in terms of fostering a strongly visual *habitus*. Consequently, one of the things to come out here is that belonging is part of a process of appropriating locality and affiliating with place whereby the media as symbols of memory are used to propagate many visual cultures which shape and shatter traditional images of modernity.

Acknowledgements

This chapter has been supported by the Estonian Research Agency, Project IUT 3–2.

Notes

1 From Chapter XI, *The Yankee in Search of Adventures* (pp. 127–140). Extract from a conversation between the main protagonist Clarence and a young lady from the land of Moder, when their cultural worlds collide due to a fanciful shift in time and space. To this gallant new arrival, she purports to being 'the Demoiselle Alisande la Carteloise.' The absurdity of the claim infuriates his rational mind to no end.
2 A different yet no less interesting literary example is found in Lewis Carroll's *The Hunting of the Snark* (1876), when the Bellman, the leader of a crew of ten, uses an empty map of the ocean to find the island home of a mysterious creature.
3 Strictly speaking of course, Ordnance Survey maps record what meets the eye and property boundaries for taxation purposes. These can be distinguished from mental maps and experiential maps as *Tabula Peutingerianas* (see Gould and White 1975).
4 Landscapes inspire people to represent it 'in a variety of materials and on many surfaces – in paint on canvas, in writing on paper, in earth, stone, water and vegetation on the ground' (Daniels and Cosgrove 1988: 1), at the same time the representation thus created influences people to make the 'real' landscape match with the representation, this creates a sort of circular reference (Olwig 2004).

References

Anderson, B. (1983). Census, map, museum. *In:* B. Anderson, ed. *Imagined communities: Reflections on the origins and spread of nationalism*. London: Verso, pp. 163–186.
Anderson, J. (2004). Talking whilst walking: A geographical archaeology of knowledge. *Area*, 36 (3): 254–261.
Anon. Institut Francais Estonie and Dept. of Semiotics, Univ. of Tartu. (2015). Semiotics conference "Paris/Tartu – Barthes/Lotman" (June 8). See www.flfi.ut.ee/en/news/semiotics-conference-paristartu-bartheslotman-june-8-2015
Banks, M., and Morphy, H. eds. (1999). *Rethinking visual anthropology*. New Haven, CT: Yale University Press.
Barthes, R. (1957). *Mythologies*. Paris: Editions du Seuil.
Bender, B. (1992). Theorizing landscapes, and the prehistoric landscapes of Stonehenge. *Man (N.S.)*, 27: 735–755.
Black, J. (1997). *Maps and politics*. London: Reaktion Books.
Brown, S., and Turley, D. (1997). *Consumer research: Postcards from the edge*. London: Routledge.
Cantwell, M., and Adams, C.W. (2003). An aboriginal planning initiative: Sacred knowledge and landscape suitability analysis. *In:* H. Palang and G. Fry, eds. *Landscape interfaces: Cultural heritage in changing landscapes*. Dordrecht: Kluwer Academic Publishers, pp. 163–184.
Carroll, L. (1876). *The hunting of the Snark: An agony in eight fits*. London: Macmillan & Co.
Casey, E.S. (2002). *Representing place: Landscape painting and maps*. Minneapolis: University of Minnesota Press.
Cooke, F.M. (2003). Maps and counter-maps: Globalised imaginings and local realities of Sarawak's plantation agriculture. *Journal of Southeast Asian Studies*, 34 (2): 265–284.
Cosgrove, D. ed. (1999). *Mappings*. London: Reaktion Books.
Crampton, J.W. (2009). Cartography: Performative, participatory, political. *Progress in Human Geography*, 33 (6): 840–848.
Crick, M. (1976). *Explorations in language and meaning: Towards a semantic anthropology*. London: Malaby Press.
Cross, S. (2012). *Look, listen and learn. Back in Kilkenny: Hurling and community heritage mapping*. Accessed 22 August 2018, http://susancrosstelltale.com/2012/09/15/look-listen-and-learn-community-heritage-mapping/
Daniels, S., and Cosgrove, D. eds. (1988). Introduction: Iconography and landscape. *In: The Iconography of landscape: Essays on the symbolic representation, design and use of past environments*. Cambridge: Cambridge University Press, pp. 1–10.
De Nardi, S. (2014). Senses of place, senses of the past: Making experiential maps as part of community heritage fieldwork. *Journal of Community Archaeology and Heritage*, 1 (1): 5–23.
Dora della, V. (2011). *Imagining mount athos: Visions of a holy place from homer to world war II*. Charlottesville, VA: University of Virginia Press.
Dorling, D., and Fairbairn, D. (1997). *Mapping: Ways of representing the world*. Harlow: A.W. Longman.
Duncan, J., and Duncan, N. eds. (1992). Ideology and bliss: Roland Barthes and the secret histories of landscape. *In:* T.J. Barnes and J.S. Duncan, eds. *Writing worlds: Discourse, text & metaphor in the representation of landscape*. London: Routledge, pp. 18–37.

Edwards, E. (1996). Postcards – greetings from another world. *In:* T. Selwyn, ed. *The tourist image: Myth and myth making in tourism.* London: John Wiley & Sons Ltd., 197–221.

Etherington, N. ed. (2007). *Mapping colonial conquest: Australia and Southern Africa.* Crawley: University of Western Australia Press.

Fernandez, J. (1986). *Persuasions and performances: The play of tropes in culture.* Bloomington: Indiana University Press.

Gaffin, D. (1995). *In place: Spatial and social order in a Faeroe Islands community.* Prospect Heights, IL: Waveland Press Inc.

Gell, A. (1985). How to read a map: Remarks on the practical logic of navigation. *Man (N.S.),* 20 (2): 271–286.

Gould, P., and White, R. (1975). *Mental maps.* London: Routledge.

Harley, J.B. (1992). Deconstructing the map. *In:* T.J. Barnes and J.S. Duncan, eds. *Writing worlds: Discourse, text, and metaphor in the representation of landscape.* London: Routledge, pp. 231–247.

Harley, J.B., and Woodward, D. eds. (1987). *Cartography in prehistoric, ancient, and medieval Europe and the Mediterranean.* Chicago: Chicago of University Press.

Harmon, K., and Clemans, G. (2010). *The map as art: Contemporary artists explore cartography.* Hudson, NY: Princeton Architectural Press.

Hieslmair, M., and Zinganel, M. (2013). Stopover: An excerpt from the network of actor-oriented mobility movements. *In:* S. Kesselring, G. Vogl and S. Witzgall, eds. *New mobilities regimes in art and social sciences.* London: Routledge, pp. 115–134.

Holm, J. (2005). *Caught mapping: The life and times of New Zealand's early surveyors.* Christchurch, NZ: Hazard Press Publishers.

Ingersoll, D.W., and Bronitsky, G. eds. (1987). *Mirror and metaphor: Material and social constructions of reality.* Boston: University Press of America, Inc.

Ingold, T. (2000). To journey along a way of life: Maps, wayfinding and navigation. *In:* T. Ingold, ed. *The perception of the environment: Essays on livelihood, dwelling and skill.* London: Routledge, pp. 219–242.

Jones, M. (2004). Tycho Brahe, cartography and landscape in 16th Century Scandinavia. *In:* H. Palang, H. Sooväli, M. Antrop and G. Setten, eds. *European rural landscapes: Persistence and change in a globalising environment.* Kluwer: Dordrecht, pp. 209–226.

Kaur, E., Palang, H., and Sooväli, H. (2004). Landscapes in change – opposing attitudes in Saaremaa, Estonia. *Landscape and Urban Planning,* 67: 109–120.

Käyhkö, N., Fagerholm, N., and Mzee, A.J. (2013). Local farmers' place-based forest benefits and government interventions behind land cover and forest transitions in Zanzibar, Tanzania. *Journal of Land Use Science,* 10 (2): 150–173.

Küchler, S. (1993). Landscape as memory: The mapping of process and its representation in a Melanesian society. *In:* B. Bender, ed. *Landscape – politics and perspectives.* Oxford: Berg, pp. 85–106.

McDonald, M. (1990). *We are not French! language, culture and identity in Brittany.* London: Routledge.

McEvedy, C. (1998). *The Penguin historical atlas of the Pacific.* London: Penguin Books.

Mels, T. (2002). Nature, home, and scenery: The official spatialities of Swedish National Parks. *Environment & Planning. D, Society and Space,* 20 (2): 135–154.

Monmonier, M.S. (1996). *How to lie with maps.* Chicago: Chicago of University Press.

Monmonier, M.S. (2007). Cartography: The multidisciplinary pluralism of cartographic art, geospatial technology, and empirical scholarship. *Progress in Human Geography,* 31 (3): 371–379.

Morphy, H. (1991). *Ancestral connections: Art and an Aboriginal system of knowledge.* Chicago: Chicago of University Press.

Nash, J. (2011). *Insular topologies.* PhD thesis, Department of Social Science, Adelaide, https://digital.library.adelaide.edu.au/dspace/bitstream/2440/71015/8/02whole.pdf

Olwig, K.R. (2004). "This is not a landscape". Circulating reference and land shaping. *In:* H. Palang, H. Sooväli, M. Antrop and G. Setten, eds. *European rural landscapes: Persistence and change in a globalising environment.* Kluwer: Dordrecht, pp. 41–66.

Olwig, K.R. (2007). The landscape of 'customary' law versus that of 'natural' law. *Landscape Research,* 30 (3): 299–320.

Peluso, N.L. (1995). Whose woods are these? Counter-mapping forest territories in Kalimantan, Indonesia. *Antipode,* 27 (4): 383–406.

Pickles, J. (1992). Text, hermeneutics and propaganda maps. *In:* T.J. Barnes and J.S. Duncan, eds. *Writing worlds: Discourse, text and metaphor in the representation of landscape.* London: Routledge, pp. 193–230.

Richardson, M. (1987). A social (ideational-behavioural) interpretation of material culture and its application to archaeology. *In:* D.W. Ingersoll and G. Bronitsky, eds. *Mirror and metaphor: Material and social constructions of reality.* Boston: University Press of America, Inc, pp. 381–403.

Said, E. (1978). *Orientalism.* London: Penguin Books.

Selwyn, T. (1996). *The tourist image: Myth and myth making in tourism.* London: John Wiley & Sons Ltd.

Thomas, N. (1999). *Possessions: Indigenous art/colonial culture.* London: Thames & Hudson.

Tilley, C. (1991). *Material culture as text: The art of ambiguity*. Cambridge: Cambridge University Press.
Tilley, C. (1999). *Metaphor and material culture*. Oxford: Blackwell.
Tuan, Y-F. (1980). The significance of the artefact. *Geographical Review*, 70 (4): 462–472.
Turnbull, D. (1989). *Maps are territories: Science is an atlas*. Victoria: Deakin University Press.
Twain, M. (1889). The Yankee in search of adventures (Chap XI). *A Connecticut Yankee in King Arthur's Court*. New York: Charles L. Webster & Co., pp. 127–140.
Wilson, M.W. (2017). *New lines: Critical GIS and the trouble of the map*. Minneapolis, MN: University of Minnesota Press.

27
FACILITATING VOICING AND LISTENING IN THE CONTEXT OF POST-CONFLICT PERFORMANCES OF MEMORY
The Colombian scenario

Luis C. Sotelo

Introduction

Performing artists in Colombia began some decades ago to work with both victims and offenders of abuses of human rights in order to produce collaborative performances informed by or even fully based on the real-life stories of those whose lives have been directly affected by the armed conflict. The overall aims of these practices are to use the power of their narratives and/or their presence on stage to contribute to ongoing peace-building, post-conflict efforts to claim for the rights of the victims, and for restorative justice purposes. Some examples include Carlos Satizabal's *Antigonas Tribunal de Mujeres* [Antigonas, a Women's Tribunal] a play in which four women share testimonies related with the killing or disappearance of their beloved sons and husbands and with persecution by state agents (Satizabal 2015); Constanza Ramírez Molano's *Vivificar* [Vivifying] (2015), a flash mob in a shopping centre in which she stages the forced disappearance of musicians and the conductor of an orchestra as a means to engage the public in direct conversations with the relatives of disappeared people, who appear on stage at the end of the flash mob; choreographer Alvaro Restrepo's commemorative ceremony *Inxilio: el Sendero de Lágrimas* [Inxilio: The Trail of Tears] (2010–2013), in which more than 150 victims of forced internal displacement appear on stage along with trained contemporary dancers, a symphonic orchestra, an actress, and a soprano singer. In the context of that performance, a selection of the performers/survivors presents the self; they act as representatives of other victims and voice their thoughts on aspects relevant to the post-conflict moment that Colombia was going through at that time; Luis C. Sotelo's audio-guide *La Salida Más Conveniente* [The Most Convenient Way Out] (2014–2018), in which one person at a time is guided for 25 minutes by a young man, in silence, about a building in a city (the project is adaptable to any city). As they walk, they listen together via headphones to a fragment of a life story by a young ex-combatant of rebel group Revolutionary Armed Forces of Colombia (FARC). The life story that they get to listen to presents the ex-combatant's early childhood, and refers to what made him join the guerrilla, how he was trained, why he decided to escape from it, and what life-threatening hurdles he faced in trying to do that.

Artists doing these kinds of work in a post-conflict context are faced with a specific set of aesthetic, ethical, and political challenges, and are often not aware of each other's work. This practice has received little academic attention in Latin America.[1] It is worth, thus, to put this practice in dialogue with debates that have emerged in other contexts. In *Performance Affects: Applied Theatre and the End of Effect*, James Thompson (2011) discusses applied theatre practice in the Sri Lankan post-conflict context. His discussion may be seen as seminal for the reflections that need to happen within this field. In that book, he reflects on an 'incident' in which 27 young ex-combatants were killed while being 'held as surrendered child soldiers' in a rehabilitation centre. Three months before the massacre, he and a team of practitioners had conducted applied theatre workshops with them, which resulted in a public performance to all camp residents including staff. On being asked whether he thought there was a link between the massacre and the applied theatre project, one of Thompson's Sri Lankan colleagues said, 'of course there was.' Thompson sets off to explore that response, and, more specifically, 'how the problems revealed in this example can illustrate the broader limits of applied theatre in conflict or crisis situations' (2011: 16). His argument is that in such a charged context, the claims made by practitioners about the transformative power of their practice need to be carefully rediscovered in the light of the ways that the performances might be read by some of the armed actors or their supporters. More concretely, he acknowledges that the presence of the project and the artists involved in such a context may strengthen, shape, or even challenge agendas by either the government or any of its armed enemies. In consequence, the participants in the project are at risk of being perceived as symbols of a political agenda and to be targeted as such by some of the armed groups involved in the conflict. Such a re-positioning of the participants and their narratives poses a huge ethical risk for the participants. A performance of memory in such a context ends up being completely intertwined with the mechanisms for either constructing or challenging (sometimes violently) public political post-conflict agendas.

Just as Thompson focuses on a single case in his book to illustrate the risks of working with real-life stories and people in a post-conflict performance of memory, I focus in this chapter on Restrepo's *Inxilio: el Sendero de Lágrimas* (2013) to illustrate how aesthetic, ethical, and political decisions are intertwined and embodied in the craft of facilitating voicing and listening in the context of such performances. I have chosen this project for various reasons. Firstly, it reaches an unprecedentedly large audience. It has been presented on three occasions, one in 2010, and twice in 2013. In one of its 2013 versions alone, the one that I discuss here, it reached 4,000 audience members who attended the sports centre in which it was staged (entry was free of charge), plus a national television audience via a special live broadcast edition of a national television program. More importantly for the purposes of this chapter, there is one section of the 2013 version on which I focus here, that deserves critical attention. In that section, the Colombian president Juan Manuel Santos (in office 2010–2018), who was leading at that time a peace process with guerrilla group FARC in Havana (Cuba), appears on stage within a circle as a person who listens to the 11 victims' representatives. The analysis made here of the decisions that were made to stage the president as a figure of authority who listens to what the victims of forced internal displacement have to say, will illustrate how performance aesthetics, ethics, and politics are intertwined in this practice.

Theoretical framework – literature review

Drawing on Brown, Langer, and Stewart's (2011) typology of post-conflict environments, I conceptualise a post-conflict scenario 'as a process that involves the achievement of a range of peace milestones.' Those milestones include, for instance, the cessation of hostilities and violence, the signing of a peace agreement, demobilisation, displaced people and refugee repatriation, reforms at institutional level, societal reconciliation and integration, and economic recovery (2011: 4). Taking a process-oriented approach is productive in that it allows one to see that countries such as Colombia lie along a transition continuum which sometimes moves backwards and in which a conflict persists with some armed groups, while peace milestones are achieved with others. In that sense, there are multiple processes taking place in Colombia's post-conflict context. For instance, the transition from war to peace with paramilitary in 2005 or the peace agreement with guerrilla group Revolutionary Armed Forces of Colombia (FARC) in 2016, while there is still an ongoing struggle both for the victims to render accountable perpetrators on all sides, including state actors, and, more broadly, the struggle to achieve justice.

Colombia has been torn by a series of non-stop internal armed conflicts since 1948. The armed groups of the last five decades have included various guerrilla groups and armed paramilitary fronts, state security forces, and drug dealers. The most significant registered violent acts include forced displacement, killings, threats, kidnapping, terrorist attacks, torture, sexual crime, and child recruitment.[2]

Victims of the conflict and human rights organisations have not stopped vindicating the victims' rights, searching for the disappeared relatives, fighting for their kidnapped beloved ones to be freed (Sotelo 2018), and fighting for acknowledgement as victims, even amidst the conflict.[3] To do so, a plethora of activist memory work initiatives have emerged.[4] Socially engaged artists have also contributed innovative work in which they often collaborate with victims. In that sense, acts in which painful memories have been voiced in Colombia have not been limited to moments within the transitional, post-conflict legal scenarios that have been implemented over the last two decades. As director of the National Historic Memory Centre, Gonzalo Sánchez has said, in Colombia, memory work is established as a 'militant response' to the ongoing and 'multiple silencing projects that are at place' in the country: 'memory is an expression of rebellion against violence and impunity' (Grupo de Memoria Histórica [GHM]. 2013: 13).

Literature across a range of disciplines has made reference to listening as a concept to be considered in discussions of post-conflict performances of memory, but only a few sources address it in detail. Within oral history (Field 2006; Cave & Sloan 2014; Greenspan 1998), for example, listening is often addressed as an aspect of an interview process, but its social dimension as part of a public history event is rarely discussed.[5] Practices such as the one that will be discussed as a case study in the next section have not been the subject of critial inquiry in terms of notions of listening. However, previous studies within the fields of memory studies (Jelin 2002, 2007), social psychology (Aranguren 2012), sociology (Payne 2000), anthropology (Borneman 2002; Riaño-Alcalá & Uribe 2016; Castillejo-Cuellar 2007), and performance studies (Phelan 2016; Thompson 2011) have identified listening as a keyword in the context of post-conflict memory projects. Although they have not discussed it as a distinct craft, their reflections make useful contributions towards the establishment of a critical framework for discussing how this craft may be exercised effectively, ethically, and with political sensibility.

With reference to the South African context, but from a theoretical standpoint, anthropologist John Borneman's (2002) article *Reconciliation after Ethnic Cleansing: Listening, Retribution, Affiliation* argues that cultivating 'practices of listening' after a violent conflict is key for advancing reconciliation goals. In his own words:

> My focus on listening here is not meant to discount the importance of the voice, which has received much recent attention in the human and social sciences. Voicing projects, however, tend to emphasize the authority of the location of the speaker and to be concerned with the constructedness and autonomy of discourse – over and against the truth-value of the message and its relation to listeners. . . . And it fails to direct us to the ineffectiveness of speaking if no one listens, and to the question of how and on what basis one should act upon listening, should it occur.
>
> *(Borneman 2002: 295)*

To act upon listening, I propose, is precisely to turn both listening and listening facilitation into a craft of sorts. The listener can be trained as a listener and a facilitator of listening can develop the skills and knowledge needed to facilitate listening acts by others. However, a key underlying question is what does effective voicing and listening mean in a post-conflict context? This question is connected with the call for facilitators of voicing and listening in such contexts to develop a clear *vision* as to what is intended with their intervention: what is the social process that a post-conflict performance of memory facilitates working towards?

In Borneman's (2002) words (drawing on Gadamer's (1975) *Truth and Method*), voicing in a post-conflict context is about 'telling the truth,' and it is about making the value of lived experience matter in the public sphere. Truth-telling is focused on the victims (not on the perpetrators), he adds, and it is about taking the risk of contradicting what is said or believed by others (or willingly confessed by others, mainly by the perpetrators). Truth-telling makes a difference (is effective) when the disclosed experience elicits exposure or an orientation to a new experience, that is, when it runs counter to personal and collective expectations (Borneman 2002: 291). In a transitional context we need to listen for truth of concrete lived experience, that is, for statements grounded in experience that contradict power dynamics. In that sense, truth only becomes truth 'when plugged into practices and systems of power' (2002: 296).

For a listening act to be effective in a post-conflict scenario one needs to ask *who* listens? Borneman answers: there needs to be 'third-party' listeners, that is, people external to the conflict that would be the subject of narration. Examples of, third-party listeners are United Nations' observers, anthropologists, peacekeepers and other professionals who could be trained 'to listen for departures from violence' (2002: 301). An individual, or a group starts to depart from violence, he explains, when they tell the truth, acknowledge their wrongdoings, when they depart from abstract political visions of the future and move toward concrete human beings and ways of defending them (2002: 292). Ultimately, he concludes, third-party listening is required because who listens intervenes 'simultaneously on the side of accountability, trust, and care for the Other, including the enemy' (2002: 302). He ends his contribution on a controversial note: effective voicing and listening in such a context needs to be linked with the purpose of allocating responsibilities and, more specifically, with decisions of retributive justice. This is controversial because transitional justice, for instance in the current model that is being implemeted in Colombia in the wake of the Juan Manuel Santos – FARC peace agreement, promotes restorative rather than retributive justice. Previous models in Colombia have even resulted in amnesties to the perpetrators (Jaramillo Marin 2014 in Riaño-Alcalá & Uribe 2016: 10).

Borneman's (2002) idea of 'third-party' listening resonates with ideas by perhaps the most influential author within Latin America's field of memory studies, Argentinian sociologist Elizabeth Jelin. In her influential book *Los Trabajos de la Memoria* (2002) and essay *Testimonios Personales, Memorias y Verdades frente a Situaciones Límite* (2007) Jelin addresses the relationships between the act of giving testimony (voicing) and listening in the transitional, post-dictatorial Latin American contexts of the 1970s onwards. To Borneman's question for the effectiveness of voicing memories of a violent past, she adds the question of what makes listening possible in such contexts (Jelin 2007: 375). Jelin argues that such voicing and listening is 'social' and coincides with Borneman, but rather than third-party listening, she lists 'alterity' as a key element of social, post-conflict listening.

Social listening is possible if it is performed.[6] by an-Other person, that is, by someone who is not part of the inner circle of the one who gives testimony, Jelin (2007) explains. Alterity, as experienced in the course of an empathic, dialogic encounter, drives voicing and social listening acts, it creates the 'social conditions' in which the interlinked acts of voicing and listening may take place (Jelin 2007: 382).[7] Further, alterity is important, Jelin clarifies, for what is needed in a post-conflict context is different from autobiographical recounting. While autobiographical performance is normally a representation of lived experience in the first person singular 'I,' post-conflict testimony is mediated, dialogic, and speaks on behalf of a social condition first person plural, 'we' (Jelin 2007: 385). This statement suggests, importantly, that the post-conflict testimony relates to a site of conflict, not to an isolated individual and not even to single communities. Rather, following Colombian anthropologist Alejandro Castillejo's line of argument in a 2007 research report on testimonies in the context of the Peruvian Truth Commission, the post-conflict testimonies of victims that he interviewed during his field work draw 'a social cartography and a political economy of experience and of exclusion' (Castillejo-Cuellar 2007: 83).

In Jelin's approach, the *how* to perform listening becomes to an extent more important than the *who* listens. To an extent, one may say, that even alterity can be performed, it is not of the essence of social positions. What matters is that listening is *performed* in a specific way, that is, with engagement, empathy, dialogically and with a level of emotional and cognitive distance. A distance that allows the listener to remain curious, ask sensible, relevant questions, show compassion, and even acknowledge his or her own limitations to understand what is voiced to him or her (2007: 382). Now Jelin claims explicitly that social projects of listening and of 'rescuing' (rescatar) testimonies need to meet some special requirements. However, her discussion falls short of describing in full detail such requirements.

In a book which I consider to be a major point of reference on this emergent field of inquiry – *La Gestión del Testimonio y la Administración de las Víctimas: El Escenario Transicional en Colombia durante la Ley de Justicia y Paz* – Colombian psychologist Juan Pablo Aranguren furthers Jelin's questions and investigates social listening in the context of a transitional justice mechanism called *versiones libres*/voluntary depositions.[8]

The voluntary depositions are a step within a formal legal transitional justice procedure by which members of illegal armed groups (mainly paramilitary) were encouraged to confess the criminal acts that they had comitted as members of their group prior to demobilisation. By confessing and contributing to a re-construction of what had happened, and to the localisation of bodies of disappeared victims and the further identification of networks of support of their activities, those who made a voluntary deposition would receive a significantly lower criminal sanction (maximum eight years of prison – Article 29, Law 975 of 2005). The mechanism was implemented in 2005 by former President Alvaro Uribe Vélez (in office 2002–2010).

Aranguren argues that this transitional justice procedure implements a 'differential' approach to managing (rather than facilitating) voicing and listening acts. (Aranguren 2012: 50) This is, he argues, because behind the entire procedure lurks a war strategy against the guerrilla. While it creates a 'social disposition' for the confessions of paramilitary to be listened to, it creates legal and logistical barriers for both the victims and the wider public to listen directly to those confessions and, thus, to question them. In part, Aranguren claims, this is so because the war strategy that Uribe Vélez' government, the army, and the paramilitary deployed included to spread the view that those who had fallen victims of the paramilitary were usually part of the network of support of the guerrilla or, in any case, people inclined towards the left and thus sympathetic with the guerrilla's cause (Aranguren 2012: 49).

Aranguren (2012) also highlights the following:

> The desire of the victim actually goes beyond making his case or that of his relative known: in his testimony there is an attempt to fight for justice and not only for recognition.
>
> (Aranguren 2012: 73)

The actual problem of facilitating voicing in a post-conflict context, Aranguren claims, is connected with how to facilitate effective listening (2012: 73). 'To give voice' to those who apparently do not have one in the public sphere and to disseminate their message is not precisely what some would consider a high landmark of solidarity, he adds. Rather, drawing on the previously mentioned contribution by anthropologist Alejandro Castillejo (2007), Aranguren argues that effective listening has to do less with giving voice to previously unheard voices and more with adjusting the ability of those who listen so that they do listen with historical depth. In Aranguren's argument to adjust the ability of those who listen means to revise the context in which the voicing and the listening acts take place. A post-conflict context, in particular a transitional legal mechanism, may end up being experienced by the victims as a continuation of the abuses against the civilian population by other means. These other means, more specifically, involve disguising a lack of ability to listen to their lived experience – their truth – behind a theatrical curtain that, while it claims to give a space for their voices to be heard, does so under very specific and defined conditions of enunciation. When such highly regulated (and ideological) conditions of enunciation operate, it is no longer appropriate to use the verb 'facilitate' to refer to the action of dealing with voicing and listening acts. Rather, in such a situation the flow between testimonies and listening acts gets managed. While it may be said that there is quite a bit of management in the craft of a facilitator and that a good management involves good facilitation, management in Aranguren's account refers to bureaucratic, legal hurdles that stop both voicing and listening from being spontaneous, dialogic, interdependent live acts.

The literature offers a set of critical tools for the discussion both of performances of memory mediated by transitional legal mechanisms and by 'memory-work entrepreneurs' (Jelin 2002) such as socially engaged artists. The study of the following case illustrates the usefulness of these theoretical tools, but also the need to do further, more detailed research on the craft of facilitating voicing and listening in the context of post-conflict performances of memory.

Case study: *Inxilio, el Sendero de las Lágrimas* [*Inxilio: the Trail of Tears*] (2010, 2013)

This project has been described by dance and literary critic Iván Jiménez García as a 'choreographic symphony' (Jiménez García 2013). The term refers to the fact that in the piece simple everyday life actions such as marching in a procession, attending a ritual, and sitting in a circle to present the self and listen to pre-recorded testimonies by a diverse range of internally displaced people are carefully choreographed and performed to the Symphony of Lamentations Songs by Polish composer Henryk Górecki. American soprano Sarah Cullins sings the songs, while the local philarmonic orchestra performs the symphony. As this happens, the aforementioned actions take place, 25 trained contemporary dancers perform a dance and, at one point in the 2013 version which I focus on here, the president of Colombia Juan Manuel Santos joins the performers on stage and is seen performing listening (sitting on a chair and paying attention) to 11 pre-recorded testimonies in the presence of those who voiced them originally. This live act of appearing on stage in front of the president and of the wider national audience and being listened to was described by many members of the public and by the artistic team as 'very moving' (*muy conmovedor*), as a genuine act of symbolic reparation.[9] The victims themselves, according to the available project documentation,[10] consider it a powerful moment in the sense that they felt supported and acknowledged. This 'voicing and listening scene,' however, navigates difficult dilemmas and tensions that are worth discussing. What makes both voicing and listening effective in the context of a post-conflict performance of memory as illustrated by this case, and based on what criteria should a sensible answer to that question be explored?

Inxilio was created by Colombian dance artist Alvaro Restrepo in collaboration with French-Colombian choreographer Marie-France Delieuvin. It was commissioned in 2010 by the City of Bogotá to commemorate the bicentenary of the Independence from Spain. In that sense, it is an innovative piece of public, political art; in fact, because of its large scale (featuring some 300 people on stage including the musicians of the orchestra and the performers), it is a magnificent, unprecedented, living, commemorative, state-supported monument. Drawing on Thompson (2011) one may say that this project strengthens the government's post-conflict agenda. Restrepo would agree with this in the sense that he is a public supporter of the peace process, as evidenced in his multiple publications and statements for the press (Restrepo 2013). In saying this, I am not questioning the artistic independence of the project. The point that Thompson (2011) makes, and that I apply to the discussion of this example, is that by strengthening the government's post-conflict agenda, participants may be seen by enemies of the peace process, as political symbols of what the government aims at achieving. What does the artistic team do to manage that risk?

Alvaro Restrepo's motivation for doing the project is to do 'a poetic manifesto that calls for the spectators' sensibility so that we understand that the drama of being internally displaced, the drama of the war is a drama that implicates us all' (Restrepo in Jiménez García 2013: 56). In other words, his intention is to use the presence of the 150 'oficiantes' (participants) and the selection of testimonies that he makes public to 'expand the sphere of [personal] grief so that we cry together with the victims as a single, collective body, a nation in grief, and so that we acknowledge that parts of our social body are still in pain' (Sotelo & Restrepo 2017) Such a space for collective grief, he argues, is where we can not only grieve for those who are now in pain but also for those who have suffered historically, in particular Indigenous communities and peoples of African descent. Rather than an act of activist memory or for victims to voice legal claims, *Inxilio* becomes a space for a danced and sung funeral of sorts in which 150 voices and bodies, including those of Indigenous communities in their own language, are offered a safe performance space for the city and the nation to listen to them.

The content of what the internally displaced people voice at one point during the commemoration is facilitated, curated, and carefully produced. Working in groups, they were asked to write thoughts in response to a series of themes such as 'territory,' 'forgiveness,' 'reconciliation,' or 'the role of art.' Thus, they were not asked to share narratives of their painful past in public, but rather thoughts about the future based on their position in life as victims of the armed conflict. Then, the participants themselves selected 11 responses as representatives of the 150 participants. The artistic team had no input on that selection process. These 11 testimonies or rather thoughts were pre-recorded. The participants themselves decided not to deliver them live. They were not trained actors. In fact, the performing arts world is completely new to them as inhabitants of rural Colombia. The intention with presenting a pre-recorded version of their voices was to help them feel more confident on stage. These recordings were played back for the president and all to hear during the second section of the ritual titled 'The Circle of Memory and The Word' (*El Círculo de la Memoria y la Palabra*). As their pre-recorded statements were played back through speakers, those who had voiced them originally stood silent and appeared in public at the centre of the circle, facing the president and slowly turning to also face the audience. Rather than voicing their thoughts directly, what the participants do in this scene is to perform voicing: they stand still, in silence, as their pre-recorded voices are heard in public. Alvaro Restrepo explains how this procedure came about:

> We [both the artistic team and the victims-participants] thought it was more prudent and more interesting that the participants were not going to exceed in time or emotion or in what knows what kind of harangues.
>
> *(Sotelo & Restrepo 2017)*

It is important to note that, at the beginning of the project in Medellín, the artistic team did not know that the president would want to appear on stage with the victims. This was a last-minute request by the president that Alvaro Restrepo had to deal with. Restrepo's immediate reaction was to ask for some time to think about it.[11] His fear was that the work could be 'instrumentalised' for political purposes. After consultations with the participants and with his artistic collaborators, Restrepo agreed for the president to appear on stage. To minimise that risk, he asked the president to appear barefoot as everybody else. Restrepo asked the president to appear on equal terms with the victims, and to sit with them within the 'Circle of Memory and the Word.' He also requested that the president would speak within that circle, just as everybody else (even if via pre-recorded testimonies). The president agreed on all but speaking. As a result, he appears within the circle as a figure of authority who listens but does not speak.[12]

Final discussion

According to both Borneman (2002) and Jelin (2007), in a post-conflict context there must be either third-party listeners, or there must be the sense of alterity in the encounter between the one who voices memories and the one who listens to them. The idea is that the one who listens does so for 'departures from violence' (Borneman 2002: 301), and that the one who performs listening does so with engagement, empathy, dialogically, and with a level of emotional and cognitive distance (Jelin 2007: 382). The figure of the president, however, can hardly be seen as a 'third-party listener' or as someone who can show the kind of alterity that Jelin expects. As the supreme leader of the Army, he represents one of the main parties involved in the armed conflict. It is also known that many human rights abuses were committed by members of the army. The questions of who listens and how to perform listening become all the more relevant when it is the president himself who listens. His appearance on stage as the president who listens to the victims' voices may be seen either as a genuine listening act or as a symbolic perfomance of listening, that is, as a political performance. The key question then is how does he perform listening in the context of such an act? Further research is needed to answer this question.

The political and symbolic power of the act in which the president performs listening to the victims for all to see may be read in connection with the fact that he is enacting a social disposition for victims to be listened to. He is showing that under his presidency it is possible to listen to what the victims have to say. However, to what extent is such (symbolic) listening effective? In *Inxilio* (2013), from a live, spontenous, dialogic act, voicing became a choreographed representation of voicing (a mechanical reproduction in the presence of the original narrator, a performance of an archived voice). What would the 11 representatives have said to the president if they had *authorised* themseves to speak freely? The decisions made both by the artists and the victims themselves as their voicing acts were facilitated also managed their ability to voice their deepest concerns. If post-conflict voicing and listening is about truth-telling (Borneman 2002), and truth-telling is about challenging established narratives, how can the presidents' narratives be challenged by the victims if he listens but does not speak? Now, the fact that the president got to listen to a range of different pre-recorded voices, including those of Indigenous representatives, may be seen as a strategy to adjust the ability of the president to listen to the victims in the light of their ethnic and cultural diversity, as well as in the light of historic, colonial roots of exclusion (Jiménez García 2013; Castillejo-Cuellar 2007).

What the 11 representatives say in the pre-recorded files does reveal where they were displaced from, what their name is, what they stand for, and, broadly, what community they are a part of. Thus, their voices do allow for a kind of social cartography (Castillejo-Cuellar 2007) of the internal displacement to be drawn by any attentive listener. They do not speak as individuals but as the living face of a site of conflict, a 'we.' However, the pre-recorded voices also hide what they might have to say in relation to the peace process or to any other contentious issue led by the president and potentially affecting their interests. The choreography not only determines body movement but also timing, and *Inxilio* is not the moment in time for victims to express uncomfortable concerns ('*harengues*') in public. In that sense, similar to what Aranguren (2012) writes in relation to transitional justice institutions, this example shows that artistic decisions may lead (at times unintentionally) to regulating what the victims get to voice and thus what their interlocutors get to listen to.

This example shows that artistic, ethical, and political criteria come together in the craft of facilitating voicing and listening in the context of a post-conflict performance of memory. While *Inxilio* did allow for the 11 representatives who took the centre stage to present the self via the pre-recorded statements and by being present on stage, it shifted both voicing and listening from a dialogic interactive affair, to a choreographed or sculpted object of attention. Further research is needed to test this statement by interviewing the 11 representatives who were selected to take the centrestage, the president and the audience. The case shows the need for positioning the craft of 'facilitating voicing and listening' in the context of post-conflict performances of memory as a subject of critical inquiry.

Notes

1 A 2016 special issue of Argentinian memory studies journal *Clepsidra* focuses on theatricality and uses of the body in the context of recent Latin American performance dealing with memories of the recent violent past in the region. There are useful contributions in that dossier from authors across the region. However, the problem of how to facilitate voicing and listening acts in the context of post-conflict performance of memory is not addressed (See Verzero et al. 2016).
2 See Registro Unico de Víctimas [Accessed 2 March 2018] https://rni.unidadvictimas.gov.co/RUV
3 See for instance Gonzalo Sanchez' introduction to the report *Basta Ya. Memorias de Guerra y Dignidad* (Grupo de Memoria Histórica [GMH] 2013), in which he states that what is unique about Colombia's bloom of memory

projects of the last decades is that it has happened amidst the conflict rather than within a more typical post-conflict environment.

4 Two useful sources for gaining a panoramic insight on the type of memory initiatives that have emerged in Colombia are *Memorias en Tiempo de Guerra. Repertorio de Iniciativas*, a report by the GMH (2009) and Chapter 5 of the Report *Basta Ya! Memorias de Guerra y Dignidad* by the Historic Memory Group (GMH 2013)
5 An interesting exception is Steven High's (2013) 'Embodied ways of listening: oral history, genocide and the audio tour.' However, it is not concerned with post-conflict performance of memory.
6 This is not a word that she uses in Spanish, I am adding it in my interpretation of her statements. See next note.
7 Elisabeth Jelin states that 'Se requieren "otros" con capacidad de interrogar y expresar curiosidad por un pasado doloroso, con capacidad de compasión y empatía. Sugiero que la 'alteridad' en diálogo, más que la identificación, ayuda en esa construcción' (Jelin 2007: 382)
8 I borrow this translation from Riaño-Alcalá and Uribe (2016).
9 Martha Cecilia González Avalos, a representative from the Unidad de Víctimas (Victims' Unit) in Medellín, says at minute 2:40 of the documentary *mentios supra* (see note 2) that: 'Nosotros invitamos a 150 víctimas del conflicto a que nos acompañen en este proceso en Inxilio para que sea un proceso de reparación simbólica para ellas y para la ciudad y para el país.' [Accessed 18 March 2018] https://youtu.be/XrIvCTRqfzE
10 See note 8.
11 See Restrepo (2013).
12 A recording of the entire event. [Accessed 23 March 2018]. www.youtube.com/watch?v=cltNJ-mqIkA.

References

Aranguren Romero, J.P. (2012). *La gestion del testimonio y la administracion de las victimas: El escenario transicional en Colombia durante la Ley de Justicia y Paz*. Bogota, Colombia: Siglo del Hombre Editores.
Borneman, J. (2002). Reconciliation after ethnic cleansing: Listening, retribution, affiliation. *Public Culture*, 14 (2): 281–304.
Brown, G., Langer, A., and Stewart, F. (2011). *A typology of post-conflict environments*. CPRD working paper 1. Centre for Research on Peace and Development University of Leuven, Belgium.
Castillejo-Cuellar, A. (2007). La globalizacion del testimonio: Historia, silencio endemico y usos de la palabra. *Antipoda*, 4: 76–99.
Cave, M., and Sloan, S.M. (2014). *Listening on the edge: Oral history in the aftermath of crisis*. Oxford: Oxford University Press.
Field, S. (2006). Beyond healing: Trauma, oral history and regeneration. *Oral History*, 34 (1): 31–42.
Gadamer, H-G. (1975). *Truth and method*. London: Continuum.
Grupo de Memoria Histórica [GMH]. (2009). *Memorias en Tiempo de Guerra Repertorio de iniciativas*. Bogotá: Imprenta Nacional, Bogotá: Puntoaparte Editores.
Grupo de Memoria Histórica [GMH]. (2013). *Basta Ya. Colombia: Memoria de Guerra y Dignidad*. Bogotá: Imprenta Nacional.
Greenspan, H. (1998). *On listening to Holocaust survivors: Recounting and life history*. Westport, CT: Praeger.
High, S. (2013). Embodied ways of listening: Oral history, genocide and the audio tour. *Anthropologica*, 55 (1): 73–85.
Jaramillo Marín, J. (2014). *Pasados y presentes de la violencia en Colombia estudios sobre las comisiones de investigación (1958–2011)*. Bogotá: Editorial Pontificia Universidad Javeriana.
Jelin, E. (2002). *Los trabajos de la memoria*. Madrid: Siglo XXI de España.
Jelin, E. (2007). Testimonios personales, memorias y verdades frente a situaciones limite. In: S. Gayol and M. Madero, eds. *Formas de historia cultural*. Buenos Aires: Prometeo Libros, pp. 373–392.
Jiménez García, I. (2013). Exclusión y figuras de la comunidad en Inxilio. *Desde el jardín de Freud*, 13: 55–70.
Payne, L.A. (2009). *Testimonios perturbadores: Ni verdad ni reconciliación en las confesiones de violencia de estado*; Traducción: Julio Paredes. Bogotá: Editorial Uniandes.
Phelan, M. (2016). Lost lives: Performance, remembrance, Belfast. In: D. O'Rawe and M. Phelan, eds. *Post-conflict performance, film and visual arts: Cities of memory*. London: Palgrave Macmillan, pp. 171–207.
Restrepo, A. (2013). Santos caminó descalzo por las víctimas. In. Periódico *El Tiempo*, 21 May. Accessed 10 April 2018, www.eltiempo.com/archivo/documento/CMS-12814877
Riaño Alcalá, P., and Uribe, M.V. (2016). Constructing memory amidst war: The historical memory group of Colombia. *International Journal of Transitional Justice*, 10 (1): 6–24.
Satizabal, C. (2015). Memoria poetica y conflicto en Colombia – a proposito de Antigonas Tribunal de Mujeres, de Tramaluna Teatro. *Revista Colombiana de las Artes Escénicas*, 9: 250–268.
Sotelo Castro, L.C., and Restrepo, A. (2017). Interview on Inxilio. Montreal: Concordia University Press.

Sotelo Castro, L.C. (2018). "Mr president: Open the door please, I want to be free": Participatory walking as aesthetic strategy for transforming a hostage space. *In:* A. Breed and T. Prentki, eds. *Performance and civic engagement*. Cham, Switzerland: Palgrave Macmillan, pp. 243–267.

Thompson, J. (2011). *Performance affects: Applied theatre and the end of effect*. Basingstoke: Palgrave Macmillan.

Verzero, L., La Rocca, M., and Diz, M.L. (2016). Dossier: Teatralidades y Cuerpos en Escena en la Historia Reciente del Cono Sur. In. *Revista del Núcleo de Estudios de Memoria*, 3 (5): 6–10.

PART VI

Shared traditions

Sarah De Nardi and Hilary Orange

Introduction

The concept of place is embroiled in and constitutes a constant process of becoming (Belcher et al. 2008: 501). There exists, then, a dynamic tension between the 'fixedness' of the past and the forces pulling the past into the present through time's connection to place; this creates a flux of shifting meanings that may or may not be tethered to place, to a community, or to a traditional worldview. Agencies are at stake in this dynamic assemblage, too. Who can, or should, tell stories about the past? At what scale does it make sense to investigate notions of tradition, custom, local colour, ancestral practices, and performance? Is the past, is memory performed, or narrated?

Sociologist of memory, Barbara Misztal, has written extensively on the configurations and grammars of social remembrance, but imagination does not appear as central to her conceptions. She does explore the idea of invented traditions, but the two 'ideas,' while they intersect and mutually inform each other, are not that closely aligned or mutually packed. For rationalist social historians, the past can be known 'not through imaginative stories but through the rationalization and the conventionalization of experience' (2003: 118). However, Misztal herself maintains that the powers of memory and the imagination are interconnected and inseparable.

In this section, contributors explore some of the themes that lie firmly at the core of what socially and culturally shared traditions do. The collective rehearsal and crafting of narratives often leads to the 'shared' production of knowledge about the past that makes sense to a social group or cultural context. By going beyond the confines of cultural heritage or ethnographic knowledge, an organic attention to 'memoryscapes,' with their coterminous stories and affects can channel the otherwise elusive imaginaries and the mnemonic spaces encountered during 'cultural' fieldwork. Here, the authors explore the materiality of our relationship with 'the past' through the voice of heritage interventions, and the performative logics and politics of traditions. The emplacing of rituals, re-enactments, relics, rewilded areas, heritage sites, and remembering bodies interweave narratives of memory and nostalgia, along with senses of national and local identities, in ways that exceed the traditional focus on *lieu de mémoire* as constructed monuments despoiled of an often ephemeral, fragile and shifting materiality.

In the first chapter in this section, Ray Cashman addresses the processes of place-making in the context of commemoration of political movements and their violent history in both urban and rural Ireland.

While the city of Derry is famous for its murals commemorating the clashes between police forces and catholic nationalists, he notes that one rural village lacks such visible murals and graffiti. Listening to local people, however, reveals a rich body of stories about un-marked scenes and absent places that have instead materialised in prison art and posters found in private homes. Remarks about outward international links in the urban materialised commemoration and the inwardness of rural folklore's logic is highly interesting and draws on a body of significant research in anthropology and folklore research, highlighting the importance of studying the (local) non-visible historical narratives that play a significant role in the processes of place-making.

Nadia Bartolini and Caitlin DeSilvey also explore a rural locale, this time in Portugal. They consider the ways in which rewilding projects and the development of 'rewilded' place identities and culture has encouraged global connections and intergenerational communication in local communities, bringing different community and external agents together in ways that has changed, and is changing, community discourse, traditions, and self-narratives. Re-wilding is framed from the perspective of heritage studies, as the creation of new natural heritage has prompted people to reconsider their own, and their communities', shared traditions in relation to memories and place. The seeds and plants have triggered more than ecological habitats and cover, the 'transformation of the landscape,' they argue, 'triggers a reworking of heritage in place.' Theirs is, therefore, a case study and exploration of place and heritage in transition.

The growth evoked in Nadia Bartolini and Caitlin DeSilvey's chapter, segues neatly into the next two contributions by Nadia C. Seremetakis and Cathy Stanton on food cultures and memory. Reflecting on the contemporary Greek and American contexts, respectively, these two chapters provide a fascinating foray in a world of taste and remembrance, in which sensory and cultural cues fill our social lives with longing or nostalgia for less globalised pasts and simpler ways of life, but also contending with the complexities that are brought into the equation of the body politics and social justice. The social acts of food preparation and its shared, convivial consumption are, in both cases, in stark contrast to the commercial fast food cultures of a zeitgeist in which instant gratification is imperative.

In Greece, Seremetakis critiques the globalisation-related erasure of food cultures through the articulation of Taste and Memory, a public educational project in which the participants rediscover the joys of traditional slow cooking and the sensual, ancient Hellenic cuisines. Seremetakis makes a case for the unhurried, hands-on handling of ingredient quantities, textures, and patience as the core ingredients of the traditional food cultures in Greece. In New England, Cathy Stanton highlights the deep roots of food cultures in agrarian culture and regional identity and their entanglements with the food reforms and in particular, the foodshed movement. Tension between the traditional and the new create a poignant framework with which Stanton analyses how food and agricultural remembrance practices 'make' New England. She argues that one of the ways that New England memoryscapes have been crafted is with a nod to 'the edible heritage' of the region, a framing, in other words, of New England's imaginary and the commemorative infrastructures geared towards tourist attractiveness.

Rounding off this section on Shared Traditions is a chapter by Tanja Vahtikari that looks at historical pageants as a form of historic re-enactment. Pageants have, she demonstrates, been a popular form of public entertainment since the first half of the twentieth century and provide a vehicle through which people can engage with the past in a multiple and imaginative ways. Making sense of the performative as well as societal aspect of these events helps us to contextualise pageants as more than edutainment events, but as social crafting of narratives about the past. Thus, Vahtikari argues, that pageants are a heritage phenomenon that can contribute to discussions regarding place-making and senses of belonging.

The chapters in this section, then, explore the social and material lives of shared traditions, positing their lenses on how communities variously produce and assemblages of things, people, places, affects, atmospheres, and stories, adding up to myriad storyscapes. They unpack the many meanings and affectual-mnemonic processes at work in renewing, preserving, and celebrating the past across cultural forms and media. We feel that the attention the contributors pay to their respective frames of references and topic makes for a compelling review of the implications of place, and memory practices, in shaping culturally defining traditions and customs.

References

Belcher, O., Martin, E., Secor, A., Simon, S., and Wilson, T. (2008). Everywhere and nowhere: The exception and the topological challenge to geography. *Antipode*, 40 (4): 499–503.

Misztal, B. (2003). *Theories of social remembering*. Maidenhead: Open University Press.

28
FOLKLORE, POLITICS, AND PLACE-MAKING IN NORTHERN IRELAND

Ray Cashman

Place and place-making

According to Yi-Fu Tuan, place is undifferentiated space that has been made meaningful through 'the steady accretion of sentiment over the years' (1977: 32). Tim Cresswell would add 'in the context of power' (2004). Folklorists would be justified in adding 'quite often through vernacular oral traditions, customs and rituals, public performances, and material culture' (2004: 19ff.e). In Tuan's formulation and its elaborations by himself and others, place is not a pre-existing fact in the world but an achievement. More process than essence, place is a way of knowing and a tradition of meaning-making.

What many, following Tuan, refer to as 'place' in contrast to abstract and undifferentiated 'space,' Arjun Appadurai calls 'locality' in contrast to 'location.' In a move similar to Tuan's, Appadurai asserts that locality is fundamentally a 'phenomenological quality,' a 'property of social life,' and an 'inherently fragile social achievement.' Furthermore, locality or place as achievement must be continually reproduced and maintained as 'structured feeling' (1996: 178–182). Indeed, contemporary methods of ethnographic fieldwork, especially long-term participant observation, reveal that place as structured feeling is quite often achieved through conversation and storytelling, and much of what ethnographers write is essentially a record of people's discursive efforts to produce and reproduce place or locality under conditions of social, political, economic, and ecological uncertainty and volatility (Appadurai 1996: 181).

Investigating how Western Apaches construct a sense of place through storytelling and abbreviated references to shared narratives set in known places, Keith Basso (1996) uses the term 'place-world' for storied sites of manifold personal and collective significance. Drawing from Bakhtin's notion of chronotopes as 'points in the geography of a community where time and space intersect and fuse' (1981: 7), Basso characterises a place-world as a site through which elements of the past may be animated as the needs of the present demand. 'Place-making,' then, is a strategy – observable around the world and through time – for constructing emplaced, rhetorically significant historical narratives, for fashioning novel, usable versions of 'what happened here.' Every place-world – developed through an accretion of association and narrative – manifests itself as a possible state of affairs. Whenever these constructions are accepted and circulated as credible and convincing, plausible and provocative, they enrich the common stock on which everyone may draw to contemplate past events, interpret their present significance, and imagine them anew as changing wants and needs dictate.

Surveying the oral traditions of a rural community in County Fermanagh, Northern Ireland, Henry Glassie contends that history is the essence of the idea of place (1982, especially chapter 31). Moreover, his work demonstrates the centrality in Irish oral traditions of a commemorative genre traditionally known as *dinnseanchas* or 'place lore' (cf. Ó Catháin & O'Flanagan 1975). Through storytelling ordinary people engage in place-making, inscribing the landscape with meaning. In the process, history – the amalgam of past precedents relevant to the present – becomes organised in the mind largely in terms of where rather than when. Likewise, through narrative, places become sites of memory that invite the initiated to use the past as a removed but relevant realm through which to comprehend the present and undertake informed action in the future. This notion of landscape as a vast mnemonic device for narratives treating the past, commenting on the present, and informing the future is also born out in the works of Maurice Halbwachs (1992 [1941]), Pierre Nora (1989), and Tim Robinson (1986, 1995), and especially in the work of folklorists Mary Hufford (1992), Terry Gunnell (2009a, 2009b), and the present author (2002, 2008, 2016).

Paying attention to places, place-making, and place lore continues the tradition of what Kent Ryden (drawing on E. V. Walter, drawing on Ptolemy) calls *chorography*: mapping the invisible landscape of the local imagination, demonstrating how people transform arbitrary space into meaningful place (1993). The promise of chorography is to offer a more emic or insider perspective on people's senses of place and to reveal aesthetics, values, orientations, ideologies, and other aspects of *habitus* or worldviews that might not be recognised or comprehended otherwise.

Focusing on the invisible landscape should not, of course, preclude attention to that which is visible. The visible and invisible – or the tangible and intangible, the material and the verbal – contextualise and in some cases mutually constitute each other. Whereas oral traditions and verbal art may have dominated early folklore studies, folklorists have long since expanded productively and insightfully into areas such as popular art, material culture, vernacular custom, public display, embodiment, and performance. Indeed many works by folklorists on material, embodied, performed, or at any rate primarily nonverbal aspects of traditional culture have been important additions to research on sense of place and place-making (e.g. Gabbert 2011, Gabbert & Jordan-Smith 2007, Glassie 2000, Noyes 2003).

That being said, in some contexts it is easy to overlook the centrality of narratives that may be ephemeral – in the sense that storytelling performances end – but that are no less conceptually resonant in people's everyday memories and senses of place, themselves, and others. Here I would like to investigate one case study in which the full range of vernacular expressive culture, including storytelling, is crucial to understanding both local political orientations and motivations and the broader dynamics of place-making as a form of social achievement with real-world consequences.

Northern Ireland

For decades, folklorists and anthropologists have been engaged productively in what amounts to an ethnography of what there is to look at in Northern Ireland. As Jack Santino notes in his *Signs of War and Peace* (2001), one thing that sets Northern Ireland apart is a preponderance of visual displays and public performances of collective identities (cf. Buckley 1998, Buckley & Kenney 1995, Bryan & Stevenson 2009). Political murals and sectarian graffiti (Goalwin 2013; Jarman 1997; Jarman 1998; Rolston 1991, 1992, 1995, 1999, and 2003; Sluka 1992) and annual commemorative parades (Bryan 1996, 2000a, 2000b; Jarman 1997; Racioppi & O'Sullivan 2000; Walker 1996) fuel ongoing conflict by reinforcing and opposing ethnic/sectarian/political compound identities – Irish Catholic nationalist versus British or Ulster Protestant unionist. Locals tout such displays as the very bedrock of their respective nationalist or unionist traditions. Such self-conscious displays of differential identity (Bauman 1971) provide insight into social and political realities in Northern Ireland's urban areas where these displays are common. But what of rural Northern Ireland where these displays are not as common? Can we take this dearth to mean that rural Catholic or Protestant enclaves are somehow less committed to nationalist or unionist agendas and identities?

Whereas deadlines and the desire for provocative visuals may impede the journalist, the fieldworker – particularly the folkloristic fieldworker – is well-positioned to answer these questions. For ethnographers in rural areas it is all the more important to investigate how locals invest landscape and everyday material culture with meaning through commemorative narrative. If political messages are more overtly expressed through urban displays, the very different visual scene in rural areas intimates no less forceful reminders of collective memory and identity that inform contemporary politics. Contextualising the visual scene with oral traditions, we may also discover significant differences between urban and rural conceptions of such fundamental political issues as belonging, birthright, and authority.

The visual scene of one rural nationalist community in Northern Ireland makes explicit the sort of political messages people associate with a range of objects, from domestic material culture to ostensibly unmarked localities in the wider landscape. To set this in relief, however, consider first the self-conscious visual displays of Irish nationalism found in urban areas.

Expressions of Irish nationalism in urban Derry

In majority Catholic areas of Derry, Belfast, and smaller cities such as Strabane and Newry, colorful murals on the gable ends of working-class public housing are painted to convey nationalist messages through text and visual symbolism (see Conflict Archive on the Internet (CAIN) website (2018a) for a wide range of visual examples). They are ubiquitous, striking, and important. In fact, the murals seem to anchor other visual displays such as flags and painted curbstones, joining in a self-conscious *assemblage* (Santino 1986, 2001) that signifies nationalist turf and serves as the backdrop for visually rich, annual parades. Decoding this urban visual scene, punctuated and anchored by murals, benefits from local knowledge and from an appreciation of visual semiotics, but it may not require the extent of contextualisation required in rural areas. Northern Ireland's second largest city, Derry, provides good examples.

From the seventeenth-century walls defending the core of Protestant Derry, one looks out over the solidly nationalist area known as the Bogside. In the 1960s the Bogside saw several civil rights demonstrations protesting Catholic unemployment and political disenfranchisement. The Bogside was also the site of several skirmishes between the nationalist community and the local Protestant police force. Most famous of these was the 1969 Battle of Bogside, a three-day conflict in which nationalists successfully repelled the police from what became known as 'Free Derry.' Introduction of the British army precipitated the tragedy of Bloody Sunday in 1972 when 13 Catholic civil rights protesters were shot dead in the Bogside by British paratroopers. Today these events and others from the 1960s through the 1980s are commemorated by several murals painted on terraced houses and apartment buildings that are clearly visible from the walls of Derry (Figures 28.1–5). Eleven were painted by a community group calling themselves 'The Bogside Artists,' whose work can be seen on the CAIN website (2018b).

Most Bogside murals replicate famous newspaper photos such as the image of a nationalist boy with gas mask and Molotov cocktail (Figure 28.2). Drawing from well-known news photos or from often-replayed newsreels lends these murals both the authenticity of historical documents and the personal familiarity of images that have come to stand in for these events over the last thirty years and more. Most murals of the photo-realistic variety create a pastiche of images relevant to nationalist memory and identity. In Figure 28.3 we see activist and British MP (1969–1974) Bernadette Devlin with a bullhorn

Figure 28.1 Murals in the Bogside, Derry, as seen from the city walls.

Figure 28.2 *The Petrol Bomber,* Battle of the Bogside. Mural by the Bogside Artists.

Figure 28.3 Civil Rights Leader, Bernadette Devlin. Mural by the Bogside Artists.

organising protesters before what became the Battle of the Bogside. In the middle ground Catholic women bang trash can lids on the pavement; this was a common method during the Troubles for warning fellow nationalists of the presence of British military patrols. In the near background we see nationalist youths at a barricade throwing stones amidst teargas. Dominating the far background and anchoring the image as a whole is the now widely broadcast message of resistance painted in 1969 on a Lecky Road gable. After the original building was demolished Bogsiders stepped in to preserve this now freestanding gable (seen in profile on the right in Figure 28.1, depicted in the mural in Figure 28.3, and viewable from the perspective of the Bogside in Figure 28.10) that marks entry to what is today referred to as 'Free Derry Corner.'

Another mural nearby offers a photo-realistic pastiche depicting Father (later Bishop) Edward Daly trying to secure safe passage for Jackie Duddy who died of his wounds on Bloody Sunday. The mural combines three famous news photos for maximum effect and stands only yards away from the site of the shooting, evoking a particularly affective evocative place-world.

Taken together, these and several other murals consistently portray the complementary themes of subjugation by British and local Protestants and resistance by Irish Catholic nationalists. On a denotative level, the mural in Figure 28.4 depicts a British soldier knocking down a door to search a nationalist household. On a connotative level, the theme is subjugation. On a denotative level, the mural in Figure 28.5 depicts a nationalist youth standing up to an armored troop carrier. On a connotative level, the theme is resistance. Both themes of subjugation and resistance are intertwined in representations of imprisoned members of the Irish Republican Army on a hunger strike. Although not a new tactic, these hunger strikes were used in the early 1980s in the attempt to secure political status for republican prisoners and to register republican violence as a military campaign of liberation rather than as acts of crime or terrorism.

In addition to subjugation and resistance, a third theme worth mentioning – a theme we will not see expressed in rural areas – is a comparison of the Irish nationalist cause with anti-colonial and liberation movements all over the world. In Figure 28.6 we see a certain international consciousness on the wall of a Derry betting shop. On the right is the standard portrait of Bobby Sands, the first Irish Republican Army prisoner to die in the 1981 Hunger Strike. On the left is a heraldic image conflating the cause of the Irish Republican Army and the wider republican movement with that of the Palestine Liberation Organization.

This international consciousness is in part a clever adaptation. It has become *de rigeur* for foreign (and domestic) television journalists covering Northern Ireland to use political murals as back-drops for their reports. Taking advantage of this exposure, urban nationalists have attempted to legitimate their cause to the rest of the world in the same way a musician might define himself or herself by saying who his or her influences are. Nationalist murals in Derry and Belfast have made tributes to the American Indian Movement, dispossessed Australian Aborigines, and Basque separatists. Che Guevara, Nelson Mandela, Emiliano Zapata, Malcolm X, and George Washington have also made appearances. The murals testify through analogy and implication, as if to say, 'If you like the ANC or the PLO, you'll love the IRA.'

Still, what we have seen so far does not require too much on the viewer's part to access intended messages. Sometimes the image is familiar enough for everyone in Northern Ireland to appreciate – 'I see, this is Bloody Sunday, we're talking about subjugation' or 'I see, this is the Battle of the Bogside, we're talking about resistance.' In other cases, images are compounded by other images to help amplify the intended message. In yet other cases, the image is anchored by text, telling us explicitly that Nelson Mandela, for example, is a 'comrade' and 'fellow freedom fighter.'

Yet, again, what of Derry's rural hinterland in County Tyrone, a scene almost devoid of public monuments and political art? Is this predominately Catholic area any less nationalistic? Is it somehow shielded from the divisive politics of Derry and Belfast?

Figure 28.4 British soldier. Mural by the Bogside Artists.

Figure 28.5 *The Rioter.* Mural by the Bogside Artists.

Figure 28.6 Mural of Bobby Sands.

Expressions of Irish nationalism in rural Aghyaran

If we are seeking the sort of nationalist displays common in Derry, the rural Tyrone parish of Aghyaran does offer a few, modest examples. Occasionally a bus stop or abandoned house is graffitied with '*Tiocfaidh ar lá*' – a common nationalist slogan translating from the Irish Gaelic 'Our day will come.' The retaining wall of one bridge reads, 'Ireland Unfree Shall Never Be at Peace.' The quote comes from Patrick Pearse, one of the leaders of the 1916 Easter Rising, the insurrection that acted as a catalyst for independence in the south of Ireland. Still, in Aghyaran there is no tradition of painting murals on the gable ends of houses.

Each year at Easter, Irish tricolors are flown from telephone polls and painted on roads in commemoration of the 1916 Rising. But graffiti and flags are not often found in the mixed Catholic-Protestant areas of the parish. In fact, the only permanent, semi-public visual displays of militant nationalist resistance are found on a handful of grave markers in the all-Catholic cemetery of St Patrick's chapel. These graves of past IRA members and republican activists are also decorated with wreaths by republican organisations at Easter.

So the nationalist themes of subjugation and resistance are keyed by some displays, but only in solidly nationalist areas and mostly at certain times of year. These displays, then, are much more restrained than those in Derry. In fact, both Catholics and Protestants in Aghyaran are much more restrained in publicly displaying any overt political symbols than their co-religionists in urban areas.

One way to interpret this pattern is to claim that rural society is more integrated, less confrontational. In places where Catholics and Protestants are more interdependent as neighbours, the argument goes, local identity and affiliations matter more than sectarian ones. This argument applies to certain but not all periods and rural locations in Northern Ireland. Glassie persuasively makes a similar case for the rural community of Ballymenone (1975, 1982, 2005), as do Anthony Buckley (1982), Mary Bufwack (1982), Rosemary Harris (1972), and Elliott Leyton (1975) for rural communities across the rest of Northern Ireland. Note that all these works are set during the 1970s. At least in Aghyaran from the mid-1970s through the mid-1990s, paramilitary and state violence exacerbated and entrenched sectarian divisions at the expense of the traditional non-sectarian common ground afforded by shared local identity and allegiances. There have been many recent efforts to regain stronger cross community integration in Aghyaran (see Cashman 2006a,

2006b, 2007, 2008). The fact remains, however, that rural border areas such as Aghyaran have witnessed nearly as many politically motivated attacks and murders, per capita, as less integrated urban areas.

If we turn to the more intimate spaces of Catholic homes, there is no lack of political artwork and material reminders of nationalist subjugation and resistance. Quite often in these homes, you will find prominently displayed collections of nationalist themed artwork made by friends and relatives while imprisoned for IRA involvement. Calling to mind romantic visions of a sometimes quaint, sometimes glorious Irish past, hand-carved Celtic crosses, Irish harps, and traditional thatched cottages serve as symbols of nationalist memory and identity, and they are displayed with pride to those who have access to the more intimate spaces of nationalist kitchens and parlors. That is, they are displayed to fellow nationalists.

The graffiti, the grave decorations, and the prison art are all part of the same phenomenon as the Derry murals, though when and where these objects are displayed is more restricted in rural Aghyaran. But for all the seemingly unmarked appearance of the wider landscape – for all the lack of intentional, artistic representation of subjugation and resistance out in the open – Aghyaran nationalists bring local knowledge to bear on many sites.

What appears to the outsider as a picturesque roadside setting is in fact, to local nationalists, the spot where a Protestant paramilitary, in collusion with the security forces, gunned down local republican political activist Patrick Shanaghan. What appears to the outsider as a welcoming pub in the one-shop village of Killeter is in fact, to local nationalists, the spot where a Protestant paramilitary detonated a car bomb that killed a young Catholic woman crossing the street to mail her wedding invitations.

Of course, other sites stand in for nationalists as symbols of resistance, rather than subjugation. Danny Gallen in Figure 28.7 is calling our attention to more than an unusual rock formation at a place called Carrickanaltar. Like his neighbours, Danny knows that, during the Penal Era, Catholics were forbidden to practice their religion, and churches were either destroyed or confiscated by the Anglican Church. Aghyaran Catholics then turned to the outlaw priest Fr. Cornelius O'Mongan who constructed a roofed altar between these two rock slabs and said mass in the open air. Worshiping at eighteenth- and nineteenth-century mass rocks is a common exemplum in collective nationalist memory of resistance to British rule. The imagined visualisation of this idea in Figure 28.8 has been mass-produced as a color poster found in nationalist homes throughout Northern Ireland; it has also appeared in mural form in Belfast.

Other sites that recall nationalist resistance are not as easy to see. For example, the Lough House, a hunting lodge built in 1891 by the local landlord, Lord Caledon, was seen by many local nationalists as a symbol of Protestant domination. In 1971 the IRA blew up the Lough House in a symbolic gesture, and eventually many of the leftover stones were quarried by locals building field walls, sheds, and farm paths. Today little remains. In fact it never occurred to me to take a picture of the *lack* of a building. But for Aghyaran nationalists the resonance of the Lough House depends on its significant absence, an absence that would be neglected without the benefit of local knowledge. As one nationalist declared, he often drives the back route to Donegal Town – a route that passes by this site – in order to enjoy *not* seeing the landlord's lodge.

It may take more patience, more questioning to discover what rural nationalists bring to bear on their ostensibly unmarked visual scene, but from what we have seen so far, nationalist memory and identity in both Derry and Aghyaran are founded on the same dual themes of subjugation and resistance. However, a third theme – not as apparent in Derry – is central to Aghyaran nationalists' understandings of their politically charged landscape. We have seen that the urban murals often gesture outward to an international consciousness, making connections between the Irish nationalist cause and that of American Indians, Basque separatists, or the Palestine Liberation Organisation. In rural Aghyaran, by way of contrast, mental connections between place and narrative often focus radically inward, clutching this spot, this land, as 'our birthplace,' 'our birthright.' At issue is nativism but also the prospect that the Irish Catholic side – autochthonous, ineluctably tied to the land – has greater access to divine power and sanction. A few examples will make this clear.

Figure 28.7 Carrickanaltar mass rock.

Figure 28.8 Nationalist poster depicting an open-air mass.

Folklore, politics, and place-making

In Magherakeel townland low foundations are all that is left of St Caireall's, the original parish church that fell into ruin after colonisation, Penal Laws restricting Catholic worship, and the rise of the Protestant Ascendancy starting in the 1600s. The field below is the site of an earlier monastery and graveyard, but no physical trace of this earlier site remains. For local nationalists, the ruins of St Caireall's are an obvious physical reminder of subjugation at the hands of the British, but the otherwise unremarkable lower field offers other ideas for contemplation.

According to local legend, sometime after the fall of St Caireall's, the new Protestant owner of the lower field prepared it to sow oats. Catholic neighbours warned him against this, saying it would desecrate holy ground, but the man waved off Romanist superstition and pressed ahead. In one version of the legend, the man was seen plowing up bones and flinging them to the ditches. In three of four versions I recorded, a Catholic priest told the Protestant that he, the Protestant, would never eat a meal off the field. Whether this was a prediction or a curse, the Protestant does indeed die in a freak accident unhitching his plough horse.

The story is rich for imagining the power of primeval, autochthonous status. Moreover, the story is rich for *contriving* the power of primeval, autochthonous status. The land itself is made of the bones of Irish Catholics, we are told. We know who the rightful stewards are. We know whose side God is on. Legitimacy of the nationalist cause comes from the very soil; its sanction is from the very highest power. Here in Aghyaran, in Magherakeel, in this lower field, Nelson Mandela and Che Guevara have precious little to do with nationalist birthright. Far-flung international comparisons pale in significance to the radically emplaced, sacred authority of the nationalist cause.

Consider one more example. Just over the border in Co. Donegal is the pilgrimage site of Lough Derg, which has drawn Catholic penitents for more than a millennium. In the recent past several medieval stone crosses still guided the way. The one shown in Figure 28.9 is a replica erected and dedicated in 1999

Figure 28.9 Replica of the Drumawark cross in Co. Donegal.

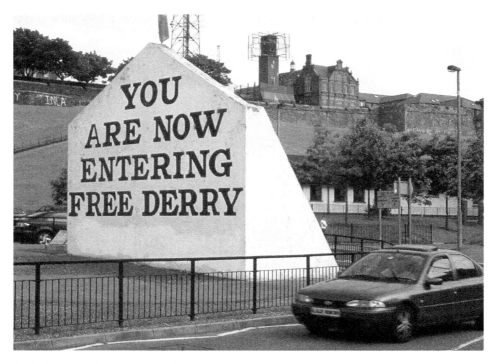

Figure 28.10 The Free Derry gable, Bogside.

on Drumawark Hill. In the early 1890s, a group of local Protestants destroyed the original Drumawark cross with sledgehammers, and, according to local legend, every one of the perpetrators suffered terrible misfortune afterward. One was unable to sleep from that day forward and died an early death. Another's livestock died, and his crops repeatedly failed. The chief instigator was afflicted with a stoop and stiffened limbs. In particular, his arms were frozen in the gesture of wielding a sledgehammer – a sign of his guilt till his last day.

For a century, a proposition about autochthonous power was keyed by the significant absence of a cross, like the significant absence of a graveyard, like the significant absence of a hunting lodge. Today the same proposition is keyed by a physical presence purposefully displayed, just like the Free Derry gable (Figure 28.10), just like the Bogside murals. But the Drumawark cross, like the lower field in Magherakeel, has no anchoring text, no compounded images to capture our attention and guide interpretation. Nothing on the surface tips us off to the fact that for many rural nationalists the categories of politics, ethnicity, locality, and sacred power are utterly collapsed.

Conclusions

If we choose what there is to look at as our entry into local culture in rural areas, we are liable to conclude that the lack of politically charged displays reiterates another long ethnographic tradition – representing rural society in Northern Ireland as relatively integrated, cooperative, and peaceful *despite* urban violence. However, once we appreciate physical presences and absences from the perspectives of local nationalists, we understand that they invest an ostensibly unmarked landscape with meaning through shared commemorative narrative. These stories can be just as politically charged as urban murals, monuments, and parades.

These stories also challenge an ideal *Gemeinschaft* vision of rural Northern Ireland. In particular, these stories call into question the notion that continuing sectarian tension and occasional violence in rural areas is merely a spillover from urban areas.

Moreover, paying attention to how oral tradition contextualises the visual scene points to significant differences between urban and rural ways of imagining and internalising the Irish nationalist cause. Both urban and rural nationalists are mindful of and animated by past subjugation and resistance. Yet, focusing outward, some Derry murals gesture to a secular, cosmopolitan, and international consciousness, while the Tyrone landscape – as contextualised by oral tradition – focuses radically inward on the local, autochthonous, and sacred. Even as the Northern Ireland peace process edges forward and civic issues of equal rights and opportunities are addressed, we may find that the rural, quasi-mythic charter for the nationalist cause has significant staying power, out in the margins, beyond the pale.

Material culture, custom, ritual, and public display and performance are all well within the wheelhouse of contemporary folklorists, but folklorists and non-folklorists alike would all do well not to ignore the lasting importance of intangible yet resonant narrative – historically the first concerns of folklorists – when seeking out those associations that transform neutral, undifferentiated space into a meaningful, even partisan place. Such a chorographic approach offers immeasurable ethnographic insight into the conceptual, aesthetic, and ideological worlds of others.

References

Appadurai, A. (1996). *Modernity at large: Cultural dimensions of globalization*. Minneapolis: University of Minnesota Press.
Bakhtin, M.M. (1981). *The dialogic imagination*, trans. C. Emerson and M. Holquist. Austin: University of Texas Press.
Basso, K. (1996). *Wisdom sits in places: Landscape and language among the western Apache*. Albuquerque: University of New Mexico Press.
Bauman, R. (1971). Differential identity and the social base of folklore. In. *Towards New Perspectives in Folklore*. Special issue of *Journal of American Folklore*, 84: 31–41.
Bryan, D. (1996). The right to march parading a loyal Protestant identity in Northern Ireland. *International Journal on Minority & Group Rights*, 4 (3/4): 373–396.
Bryan, D. (2000a). Drumcree and "the right to march": Orangeism, ritual and politics in Northern Ireland. In: T.G. Fraser, ed. *We'll Follow the drum: The Irish parading tradition*. Basingstoke: Macmillan, 191–207.
Bryan, D. (2000b). *Orange parades: The politics of ritual, tradition and control*. London: Pluto Press.
Bryan, D., and Stevenson, C. (2009). Flagging peace: Struggles over symbolic landscape in new Northern Ireland. In: M.H. Ross, ed. *Culture and belonging in divided societies: Contestation and symbolic landscapes*. Philadelphia: University of Pennsylvania Press, pp. 68–84.
Buckley, A.D. (1982). *A gentle people: A study of a peaceful community in Ulster*. Cultra: Ulster Folklore Transport Museum.
Buckley, A.D. ed. (1998). *Symbols in Northern Ireland*. Belfast: The Institute of Irish Studies, Queen's University of Belfast.
Buckley, Anthony and Mary Catherine Kenney. (1995). Negotiating Identity: Rhetoric, Metaphor and Social Drama in Northern Ireland. Washington, DC: Smithsonian Institution Press.
Bufwack, M. (1982). *Village without violence: An examination of a Northern Irish community*. Cambridge, MA: Schenkman.
Cashman, R. (2002). Politics and the sense of place in Northern Ireland. *Folklore Forum*, 33 (1–2): 113–130.
Cashman, R. (2006a). Critical nostalgia and material culture in Northern Ireland. *Journal of American Folklore*, 119 (472): 137–160.
Cashman, R. (2006b). Dying the good death: Wake and funeral customs in County Tyrone. *New Hibernia Review*, 10 (2): 9–25.
Cashman, R. (2007). Mumming on the Northern Irish border: Social and political implications. In: A. Buckley, C. Mac Cárthaigh, S. Mac Mathúna and S. Ó Catháin, eds. *Border-crossing: Mumming in cross-border and cross-community contexts*. Dundalk: Dundalgan Press, pp. 39–56.
Cashman, R. (2008). *Storytelling on the Northern Irish border: Characters and community*. Bloomington: Indiana University Press.
Cashman, R. (2016). *Packy Jim: Folklore and worldview on the Irish border*. Madison: University of Wisconsin Press.
Conflict Archive on the Internet. (2018a). *Political wall murals in Northern Ireland*. Accessed 11 December 2018, http://cain.ulst.ac.uk/murals/index.html
Conflict Archive on the Internet. (2018b). *The Bogside artists*. Accessed 11 December 2018, www.cain.ulst.ac.uk/bogsideartists/menu.htm

Cresswell, T. (2004). *Place: A short introduction*. Oxford: Wiley-Blackwell.
Gabbert, L. (2011). *Winter carnival in a western town: Identity, change, and the good of the community*. Logan: Utah State University Press.
Gabbert, L., and Jordan-Smith, P. eds. (2007). Space, place, emergence. Special issue of *Western Folklore*, 66, 3–4.
Glassie, H. (1975). *All silver and no brass: An Irish Christmas mumming*. Philadelphia: University of Pennsylvania Press.
Glassie, H. (1982). *Passing the time in Ballymenone: Culture and history in an Ulster community*. Philadelphia: University of Pennsylvania Press.
Glassie, H. (2000). *Vernacular architecture*. Bloomington: Indiana University Press.
Goalwin, G. (2013). The art of war: Instability, insecurity, and ideological imagery in Northern Ireland's political murals, 1979–1998. *International Journal of Politics, Culture, and Society*, 26 (3): 189–215.
Gunnell, T. (2009a). Legends and landscape in the Nordic Countries. *Cultural and Social History*, 6 (3): 305–322.
Gunnell, T. (2009b). Introduction. In: T. Gunnell, ed. *Legends and landscape: Plenary papers from the 5th Celtic-Nordic-Baltic folklore symposium, Reykjavik 2005*. Reykjavik: Iceland University Press.
Halbwachs, M. (1992 [1941]). *On collective memory*, ed. and trans. L.A. Coser. Chicago: University of Chicago Press.
Harris, R. (1972). *Prejudice and tolerance in Ulster*. Manchester: Manchester University Press.
Hufford, M. (1992). *Chaseworld: Foxhunting and storytelling in New Jersey's Pine Barrens*. Philadelphia: University of Pennsylvania Press.
Jarman, N. (1997). *Material conflicts: Parades and visual displays in Northern Ireland*. New York: Berg.
Jarman, N. (1998). Painting landscapes: The place of murals in the symbolic construction of urban space. In: A.D. Buckley, ed. *Symbols in Northern Ireland*. Belfast: Institute of Irish Studies, Queen's University of Belfast, pp. 81–98.
Leyton, E. (1975). *The one blood*. St. John's: Institute of Social and Economic Research, Memorial University of Newfoundland.
Nora, P. (1989). Between memory and history: Les lieux de mémoire. *Representations*, 26: 7–25.
Noyes, D. (2003). *Fire in the placa: Catalan festival politics after Franco*. Philadelphia: University of Pennsylvania Press.
Ó Catháin, S., and O'Flanagan, P. (1975). *The living landscape*. Dublin: Comhairle Bhéaloideas Éireann.
Racioppi, L., and O'Sullivan, K. (2000). Ulstermen and Loyalist ladies on parade: Gendering Unionism in Northern Ireland. *International Feminist Journal of Politics*, 2 (1): 1–29.
Robinson, T. (1986). *Stones of Aran: Pilgrimage*. London: Penguin.
Robinson, T. (1995). *Stones of Aran: Labyrinth*. London: Penguin.
Rolston, B. (1991). *Politics and painting*. London: Associated University Presses.
Rolston, B. (1992). *Drawing support: Murals in the North of Ireland*. Dublin: Colour Books.
Rolston, B. (1995). *Drawing support 2: Murals of war and peace*. Dublin: Colour Books.
Rolston, B. (1999). From king Billy to Cúchullain: Loyalist and Republican murals, past, present and future. *Éire-Ireland*, 33 (2): 6–28.
Rolston, B. (2003). *Drawing support 3: Murals and transition in the North of Ireland*. Belfast: Beyond the Pale Publications Ltd.
Ryden, K. (1993). *Mapping the invisible landscape: Folklore, writing, and the sense of place*. Iowa City: University of Iowa Press.
Santino, J. (1986). The folk assemblage of autumn: Tradition and creativity in Halloween folk art. In: J. Vlach and S. Bronner, eds. *Folk art and art worlds*. Ann Arbor, MI: UMI Research Press, pp. 151–169.
Santino, J. (2001). *Signs of war and peace: Social conflict and the use of public symbols in Northern Ireland*. New York: Palgrave Macmillan.
Sluka, J.A. (1992). The politics of painting: Political murals in Northern Ireland. In: C. Nordstrom and J. Martin, eds. *The paths to domination, resistance, and terror*. Berkley: University of California Press, pp. 190–216.
Tuan, Y-F. (1977). *Space and place: The perspective of experience*. Minneapolis: University of Minnesota Press.
Walker, B. (1996). *Dancing to history's tune*. Belfast: The Institute of Irish Studies, Queens University.

29
REWILDING AS HERITAGE-MAKING
New natural heritage and renewed memories in Portugal

Nadia Bartolini and Caitlin DeSilvey

Introduction

Since the early 2000s, scholars have examined rewilding through scientific, ecological and environmental perspectives. Attention has focused mainly on debating rewilding's different meanings, assessing the impact of species re-introductions and understanding the tensions involved in policy frameworks at different scales (see, for example, Svenning et al. 2016; Lorimer et al. 2015; Monbiot 2014; Rubenstein et al. 2006). In this chapter, we propose instead to look at rewilding from the perspective of heritage studies.

Heritage studies is most commonly considered as a field that focuses itself on the past, and the uses of the past in the present. However, there has been growing scholarly interest in understanding the future-orientation of heritage, and how its production is directed towards future generations and future social and material configurations (Harrison et al. 2016; Harrison 2013; Holtorf & Piccini 2009). So, while the practice of heritage appreciates historical features, memories, and traditions as they are found and subsequently conserved, we are interested in critically engaging with the ways through which heritage is made in the present, for the future. This heritage-making practice is, for us, the key to appreciating how heritage is activated in place. We focus on heritage as it is being deployed through engagement with vernacular memory and place attachment; not as a fixed and defined reference to an event in time, but as an ongoing mechanism that is produced relationally by the human and nonhuman entities in a given location.

To explore heritage-making in action, this chapter will hone in on a rewilding pilot initiative in the Côa Valley in Portugal. We will first consider rewilding as the creation of new natural heritage and follow with examples of how the rewilding initiative prompts local people to reassess and reengage with landscape-based memories and cultural heritage. We argue that rewilding, as an ongoing process of the transformation of the landscape, triggers a reworking of heritage in place.

Rewilding in the Côa Valley

As part of the Heritage Futures programme (a four-year interdisciplinary project funded by the UK Arts and Humanities Research Council), we spent four 2–3 week periods between 2015–17 doing ethnographic fieldwork in the Côa Valley in northeastern Portugal. Antony Lyons, Heritage Futures Senior

Creative Fellow, was part of the fieldwork team for two of these visits, in the context of his long-term involvement with the research enquiry. The Côa Valley, a UNESCO World Heritage Site, protected for its unique concentration of open-air Paleolithic rock art, shares the same landscape as a Rewilding Europe pilot initiative. Managed locally by the nature conservation organisation Associação Transumância e Natureza (ATN), the core Faia Brava rewilding reserve is made up of approximately 1,000 ha of land (Figure 29.1).

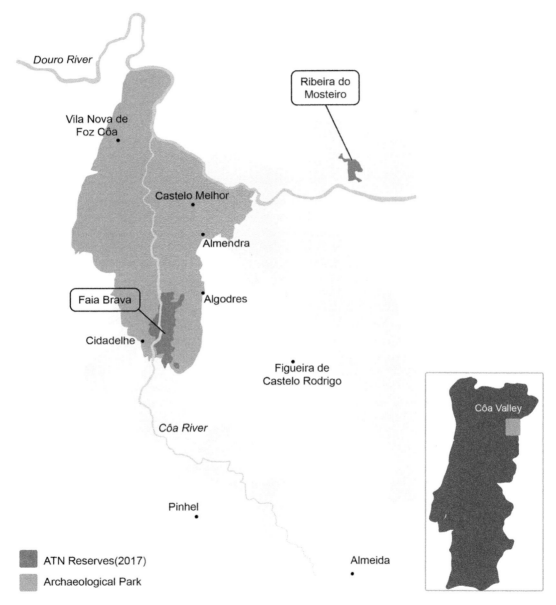

Figure 29.1 Map of the Côa Valley in Portugal.
Source: University of Exeter Design Studio.

Since the organisation was founded in 2000, ATN has been purchasing or leasing a series of individual plots previously used for farming. Following a period of emigration and depopulation in the 1960s and 1970s, much of the agricultural land in the Côa Valley was abandoned – and is still being abandoned, with younger generations opting to re-establish themselves in urban areas. This trend is evident across Europe, particularly in the Mediterranean countries, and is a focal point for rewilding and nature conservation initiatives, which seek to use abandoned lands to promote biodiversity and ecosystem restoration (Agnoletti 2014; Navarro & Pereira 2012).

There is increasing recognition by international organisations such as UNESCO and International Union for Conservation of Nature (IUCN) of the importance of conserving natural habitats for people's wellbeing. Indeed, the latest recommendations stemming from the IUCN's 2016 Congress suggest further insights are needed into the development, promotion, and implementation of 'nature-based solutions' to meet the needs of both human and nonhuman populations (IUCN 2016). European rewilding, as a process that seeks to re-balance ecosystems on lands that previously underwent extensive cultivation, follows similar reasoning. The working definition used by Rewilding Europe states:

> Rewilding ensures natural processes and wild species play a much more prominent role in the land and seascapes, meaning that after initial support, nature is allowed to take more care of itself. Rewilding helps landscapes become wilder, whilst also providing opportunities for modern society to reconnect with such wilder places for the benefit of all life.
>
> *(Rewilding Europe)*

From this perspective, Rewilding Europe alludes to there being a relationship between 'wilder places' and 'modern society' that benefits all life, human and nonhuman. To ensure that natural processes take hold effectively in rewilding initiatives, Rewilding Europe acknowledges the need for 'initial support.' This support suggests that re-balancing ecosystems is a process of gradual development, creating something new in the natural landscape. This assumption is corroborated in a Rewilding Europe policy brief, whereby rewilding is described as:

> Taking inspiration from the past but not replicating it by *developing new natural heritage* and values that evoke the past but shape the future – with the point of reference in the future, not in the past.
>
> *(Jepson & Schepers 2016: 2, emphasis added)*

Some scholars have investigated the philosophical underpinnings of rewilding as an initiative that involves natural heritage-making (Barraud & Périgord 2013). For Barraud and Périgord, the terminology used in rewilding initiatives is essential in developing a value system based on 'the wild.' However, the statement from Rewilding Europe clearly points to the value of 'developing new natural heritage.' Certain natural processes are being supported or encouraged through species re-introductions; because these new ecosystems are allowed to evolve through the gradual withdrawal of management and control, the result is ultimately open-ended and uncertain (Deary & Warren 2017). The unpredictable and emergent outcomes of this 'managed yet experimental' aspect of rewilding seem, on the face of things, incompatible with the prevailing understanding of heritage as a concept associated with the preservation of tangible and intangible elements of the past (Vidal & Dias 2015). In the case of rewilding, as 'new natural heritage' is being developed, there is a shift from preserving what was there to conserving what will become – although it remains unclear what nature(s) and culture(s) will emerge in these transitional landscapes.

While UNESCO World Heritage Sites are primarily framed through reference to either their cultural or natural heritage value, the addition of Criterion 6 (defined as 'cultural landscapes') in 1992 recognises a link between the natural environment and cultural traditions, as cultural landscapes are defined as being 'directly or tangibly associated with events or living traditions, with ideas, or with beliefs, with artistic and literary works' (UNESCO 2018). Rather than seeing these 'living' traditions as fixed entities, instead, it is useful to appreciate them as fluid, particularly, as Navarro and Pereira suggest:

> that landscapes result from the dynamic interaction of natural and cultural drivers (Antrop 2005), they cannot be perceived as anchored in time and we should anticipate occasional changes that will force us to reevaluate their definition.
>
> *(Navarro & Pereira 2012: 909)*

Navarro and Pereira highlight the dynamic interaction of both the natural environment and cultural traditions in shaping landscapes. As such, 'landscape' is not a fixed concept. Deary and Warren go further by suggesting that rewilding produces hybrid landscapes:

> the growing recognition that a large majority of the world's landscapes have been co-created by human and nonhuman processes over millennial timescales, meaning that cultural landscapes are the norm not the exception (Marris 2011; Marris et al. 2013; Pearce 2015). In most places, pristine is history – and ancient history at that – making the rewilding of hybrid landscapes a question of broad international relevance.
>
> *(Deary & Warren 2017: 213)*

While nature conservation organisations and environmentalists acknowledge the importance of re-establishing balanced ecosystems, there is room for more work on how these initiatives relate to an expanded concept of heritage and a need for better understanding the ways that human communities interact with these hybrid landscapes.

For the aging populations that make up the isolated villages around the Faia Brava reserve, their lives revolved around activities such as hunting and farming – whether they stayed in the villages or moved abroad and returned after the fall of the Salazar regime (1932–1968). The landscape they worked on and hunted in during their lifetimes is embedded in their memories and their attachment to the land. As such, the values they associate with the landscape differ from those embodied by the rewilding initiative: village residents speak of their experience working the land and tending their domesticated animals, and contrast this with the desire to let nature 'take more care of itself,' as mentioned in the rewilding working definition (Rewilding Europe). Their response suggests that more work, thought, and experimentation is required in order to develop 'ideas of how to re-engage with nature' so that both human and nonhuman populations can relate to landscape changes (Tokarski & Gammon 2016: 153).

In order to explore the human communities affected by the transformation of their landscape, the next two sections focus on how the development and gradual expansion of the Faia Brava rewilding reserve has influenced local populations in the Côa Valley, specifically in the villages of Cidadelhe and Algodres (Figure 29.1). Our aim in providing these site-specific examples is to expose the ways through which rewilding, when seen through the perspective of heritage-making, triggers a re-assessment (and sometimes a renewal) of cultural traditions and memories alongside the creation of 'new natural heritage.'

Reworking and sharing traditions

When we first visited Cidadelhe in November 2015, we were following The Côa Valley Grand Route, a 200 km trail that crosses the Côa Valley from its river spring (at Fóios, Sabugal) to the river mouth (at Vila Nova de Foz Côa) where it connects with the Douro river. The Grand Route enables visitors to explore the various heritages along the valley, from the rock art to the built environment, while hiking, cycling, or horseback riding through the landscape (Rota Vale do Côa 2014a, 2014b). As we walked around the village on that visit, we noted the abundance of ruined homes and outbuildings, but also saw the first signs that Cidadelhe was being renovated and restored. We learned that a young couple from a nearby town had started acquiring the abandoned houses and renovating them as hostel accommodations.

When we returned in August 2016, ATN and Rewilding Europe were about to release 16 Garrano horses into a newly expanded area of the Faia Brava reserve located at the edge of Cidadelhe. The Garrano horses they released into the reserve are understood to be a 'wilder' breed that can acclimatise themselves to rugged terrains, and contribute to the rewilding initiative's aim to re-establish natural grazing habitats in the area (ATN 2015). A few months later, we returned to see how the newly expanded rewilding area was affecting village life and changing people's relationship to the landscape.

The expanded rewilding area borders the oldest part of Cidadelhe, where many of the old stone houses lay derelict. We noted that more houses appeared to have been renovated, and a crane's ongoing presence suggested that more building work was to take place. The older villagers we spoke to shared their memories of growing up in the houses that lay in ruins, and expressed their desire to see them restored (Informal interviews, 4 October 2016; Interview, 11 October 2017; Leuvenink 2013). The old houses are also associated with what the young couple who own the hostel describe as the 'historical value of the place': this part of the village dates back to the medieval period, and as such, it is the heart of the historic community (Interview, 7 October 2016).

The young couple mentioned that, initially, the drive to renovate some of the old houses was to meet the needs of residents' relatives: villagers could not host under one roof all their relatives coming to visit them during the summer (Interview, 7 October 2016). However, the young couple has seen an increase in foreign tourists who wished to obtain accommodation in the vicinity of the Grand Route and Faia Brava. They are now renting out a number of suites through online travel sites, such as Booking.com. They believe that a renewed appreciation for what they describe as 'village tourism' (rather than eco-tourism) is developing, and they are keen to promote the village by highlighting its proximity to the rewilding reserve:

> The most important thing, and why Cidadelhe is different from the other villages, is how it allows contact with nature.... [I]t brings specificity and value. There are birds and animal species that are unique. All of this has created those business opportunities.
>
> *(Interview, 7 October 2016)*

In addition to their hostel accommodations, the young couple is also planning to build a restaurant and an ethnographic museum (Interview, 7 October 2016). While on the surface, these 'business opportunities' may appear to point to purely commercial interests, the young couple highlight that it is important for them to promote the village and its traditions, and the memories associated with local features and vernacular buildings. When we visited in August 2016, we stayed in one of their restored hostel buildings. In the courtyard outside the one-room structure, a group of village men were repairing the ancient granite step well ('*fonte*') which had fallen into disrepair. They explained that the well had an important place in local memory because it was where young men and women used to meet on their trips to collect water.

When asked to define their understanding of heritage in the area, the young couple responded:

> Nature, landscape, people, memory, traditions, traditional rock art, architecture, monuments: it's the whole. But to define heritage, you need to consider that heritage has to be shared with people.
>
> *(Interview, 7 October 2016)*

There are two aspects here that deserve critical attention. The first is the concept of seeing heritage as a whole, which incorporates both natural and cultural elements. As Harrison highlights (2013, 2015), recent critiques of the nature-culture divide have demonstrated that many communities around the world make no distinction between nature and culture in their understanding of tradition, heritage, and wellbeing. The young couple's statement contributes to the view that, in the case of Cidadelhe, nature and culture go hand in hand, and appreciation of place involves both in combination. The renovation of the medieval houses is associated with the village's memory and identity. Nonetheless, the tangible, visual reminder of a different time, apparent through the village's architecture, also relates to the appreciation of nature. Intergenerational views on heritage – the older generation of villagers who wish to have their childhood houses restored and remembered, and a younger generation that sees nature as having value in itself – are blended together and become mutually reinforcing. Cultural heritage is conserved and created even as 'new natural heritage' is produced through rewilding. This brings us to the second aspect that the previous statement raises: the idea that heritage needs to be shared in order to have value. This will be discussed by elaborating on another example from Cidadelhe.

Upon returning to Cidadelhe in October 2016, our visit was first mediated by a local resident who gave us a tour of the medieval village, and showed us the stone house that he grew up in. Antony Lyons had been doing ongoing research on a particular Portuguese song tradition reputed to be still extant in Cidadelhe (Lyons, in press). While the traditional mode – a 'song-duel' called *cantigas ao desafio*, which is performed by two 'antagonistic' singers – has now died out in the village, the local resident suggested, as an alternative, that we return at another time when he and another resident would be available to play us the traditional melody from the village (Informal interview, 4 October 2016). The next day, we found the two village residents with their musical instruments at the central café and playground where a number of other residents were waiting for their performance.

We were told that this traditional song from Cidadelhe – its melody, timbre, and form – was distinct from other villages in the region (Informal interview, 5 October 2016). However, the lyrics that were being sung included our names, and other terms, like 'Facebook' and 'internet.' The singer mentioned that he was inspired to compose these lyrics after he met us the previous day (Informal interview, 5 October 2016).

For us, this anecdote is useful to identify how locals in Cidadelhe wish to renew and share their traditions by communicating and performing them with visitors. Although we had never previously met, we demonstrated interest in their local customs and their everyday lives, and the villagers opened up about who they are and what they do. The incorporation of 'new' lyrics that reflected our encounter aligns with an understanding of heritage as fluid and open, not frozen in time or fixed in particular material forms. As this example demonstrates, the traditional can be adapted to incorporate present-day, modern influences. The song can further be modified as new people visit. In this example, heritage is a process woven in the present (Harrison 2013) that emerges through social encounters in a place that is transitional and still becoming (Massey 2005).

What is taking place in the vicinity of the Faia Brava reserve highlights 'the dynamic interaction of natural and cultural drivers' (Navarro & Pereira 2012: 909) in a rewilding landscape. The interaction engages with an understanding that creating 'new natural heritage' is a process that is produced in situ and in proximity to rewilding initiatives, where animals are re-introduced and are expected to acclimatise to their new environments alongside human communities who must contend with these changes. In that sense, heritage can be appreciated as a 'collaborative, dialogical and interactive, a material-discursive process in which past and future arise out of dialogue and encounter between multiple embodied subjects in (and with) the present' (Harrison 2015: 27). The next section focuses more specifically on ATN as nourishing intergenerational exchange, renewing memories and enhancing distributed place attachment.

Renewed memory and place attachment

The village of Algodres lies on the other side of the river from Cidadelhe, on the level plains above the river valley at the eastern edge of the Faia Brava rewilding reserve. In 1950, Algodres' population numbered 962 (Statistics Portugal 2014). The latest census figures from 2011 indicate a population of 294: the number of people in the village has dropped to less than a third of what it had been 60 years earlier (Statistics Portugal 2014). The decline in population can be explained by various factors, such as the emigration of families abroad to escape the authoritarian regime, the departure of many Portuguese men to fight for the Portuguese colonies in Africa (such as Guinea-Bissau and Mozambique), and the abandonment of subsistence farming to find employment in urban areas. Some people returned to their rural childhood lands after the 1974 revolution (Informal interviews, 15 October 2017). What remains, however, is a sense that rural populations are dwindling, with current aging populations wondering if their village will be the next one to disappear (Informal interviews, 10 October 2017).

Although the steep drop in population would suggest that Algodres faces a bleak future, the introduction of ATN to the area in the early 2000s has opened up alternative paths and possibilities. A small café still remains at the centre of the village. There, we met the owner, a septuagenarian who, upon his return from Mozambique in 1978, took over the café from his father. He told us that his son does not want the café, so he continues to serve the community:

> I have the age to retire, but I don't because I'm concerned that the village won't have a meeting place. . . . It's the meeting point: this is the value of the café.
>
> *(Interview, 10 October 2016)*

In Algodres, the café is valued for its capacity to gather the community. While this observation may appear banal, in a village that has experienced an exodus of population, meeting fellow residents ensures consistency, continuity, and familiarity in village life. As we got to know the café owner, we visited him at various times during our field trips, and we noticed that the café was particularly busy during lunch and after work: men, women, and children came by to talk about their day and enquire about family members and relatives. News travels fast in local communities, and usually the café owner and his wife are the first to know what is going on in the area. The café is a symbolic yet tangible focal point of village life. It is part of the everyday life of the village, and, in its own way, it assists in keeping the village alive.

When we first met him, the café owner proudly told us that Algodres was the first village to have embraced the rewilding initiative (Interview, 10 October 2016). The first landowners to sell lands to ATN were from Algodres, and those lands were the poorer-quality farmlands located at the edge of the village. One of those parcels of lands belonged to his wife. She remembers tending a vegetable garden with her grandfather on the plot (Informal interview, 10 October 2017). Yet, while she has good memories associated with the land, she also recounted a tragic incident. We were told that once she sold the land, she never returned; not because of the traumatic memory, but simply because the land was not productive anymore (Informal interview, 10 October 2017). The café owner told us that people's relationship with the land was to work it, to provide for their families. When land was sold, there was no reason and no need to return (Informal interview, 10 October 2017).

Nonetheless, the fact that these lands became the seed of the Faia Brava reserve has meaning to local people. The café owner specified that the ancestral lands of Algodres were the place where the rewilding project began, and the lands continue to be considered part of Algodres' identity. For him, keeping ancestral place names intact is one factor that links the rewilding reserve with the village: Faia Brava, for instance, means 'wild cliffs' locally (Interview, 10 October 2016). While people who have sold their lands to ATN may have relinquished their personal working connection to the land, a new conception of collective memory in place is being fashioned. Indeed, Algodres now has a new relationship with the landowner, ATN – an association that goes beyond ownership.

ATN's executive coordinator as well as its interns have made Algodres their home. During our interview, the café owner left the room and returned to show us a series of postcards and photographs from interns whom he had met over the years. He explained that he attempts to promote ATN to locals, and one way of doing this is by making interns feel at home at the café by introducing them to other residents in the village.

This is important to him because he believes that people make communities (Interview, 10 October 2016). For him, ATN now contributes to Algodres' spirit and vitality as a place:

> ATN brings people to Algodres. . . . It's good for business, but also for the village.
> *(Interview, 10 October 2016)*

The addition of ATN interns to the community instills new life through intergenerational exchange, as well as enabling the isolated residents to make global connections. When considering issues around heritage and globalisation, discussions tend to focus on the impact of international frameworks on local communities, economic development initiatives, and engagements with the developing world (see, for instance, Labadi & Long 2010). The notion of living in a globalised world may often be taken for granted in cosmopolitan urban environments. In villages that are virtually absent from regional place-promotion (or even cartographic representation), however, the impact of communicating with individuals from different cultures cannot be underestimated. When it does occur, it may have a lasting effect, as the café owner evidenced through his collection of postcards, letters, and photographs. Today, people from various parts of the world know this tiny village in northeastern Portugal, its café owner, and the everyday life of Algodres.

Moreover, the integration of the ATN interns into village life also contributes to revaluing and reframing local attachments to place. If place attachment refers to 'the bonds people form with places and the meanings they ascribe to them' (Gurney et al. 2017: 10078), then one way to understand what is occurring in Algodres consists in acknowledging that the influx of people from abroad has the effect of locals reassessing what matters to them, and appreciating, through the eye of outsiders, the value of their local landscape and traditions. This process is not a one-way street. As much as locals shape their memories to fit a changing landscape, so do the interns, as their attachment to place is also being reevaluated. In October 2017, we encountered a previous ATN intern from abroad who had come on holiday to revisit the Côa Valley and see the villagers (Field notes, 14 October 2017). In Algodres, the café owner mentioned that the intern had stopped by to greet him and local residents.

Rewilding, therefore, not only creates new natural heritage, but it can also be a catalyst in bringing together cultures that are otherwise located at a distance. Discussions with interns from abroad contribute to the reworking of memories which will have an effect on future generations; those generations who will not have known the landscape without rewilding. Place attachment, therefore, requires more than memories and tradition: it is also inclusive of change (Manzo & Devine-Wright 2014). As Lewicka notes, the challenge in a globalised world is 'to reconcile the need for close emotional ties to specific places with the fluidity of the contemporary world' (2011: 226). In this case, place attachment concerns the natural world, both through the circulation of ideas about rewilding, as well as an appreciation for cultural connections. And while emotional ties often reflect intangible considerations, the example from Algodres reminds us that place attachment can also be located, specific, and tangible. The café in Algodres is not only the central meeting point of the village, but also where visitors are made to feel welcome, and a place they remember and return to. While the landscape is transforming, certain locations continue to function as familiar reference points.

Conclusion

This chapter set out to think through rewilding as a heritage-making practice. Rewilding in ecological conservation contexts is associated with the re-introduction of species into a landscape in order to re-establish natural processes, with the long-term goal of reducing direct management and control. However, while rewilding is a planned and managed process, it also includes uncertainty in relation to the future character of the 'new nature' being generated (Deary & Warren 2017).

Our study showed that the active presence of rewilding activities has an effect on cultural practices and features in the rural villages adjacent to the Faia Brava rewilding reserve. In Cidadelhe, rewilding is shaping a new landscape while medieval stone houses are being renovated to cater to visitors' demands and to local residents who wish to remember their childhood homes. Both natural and cultural referents are in transition, and through this transition village tradition and history are blended with contemporary appreciation of the emerging landscape. The combination of past memories with modern influences is also expressed through the traditional Cidadelhe song. In both cases, traditional houses and music are meant to be shared. The sharing of heritage with people, as locals develop a sense of pride and visitors uncover new customs, exists *because* of the adjacent rewilding initiative. So, while 'new natural heritage' is being created through rewilding, the case of Cidadelhe shows how local communities adapt to a changing landscape. As people visit the rewilding area – and Cidadelhe – heritage will continue to be made and evolve along with the changes in the landscape.

The proximity of the Faia Brava rewilding reserve also encourages global connections and intergenerational communication in local communities. In Algodres, the café owner proudly reflected on how the village was the first to contribute to the establishment of the Faia Brava rewilding reserve. While personal connection to private lands sold off to ATN is slowly being lost, a renewed collective memory is emerging with the new use of its ancestral lands forming part of Algodres' identity. The connection between the village and Faia Brava is also evidenced through ATN and its interns making Algodres their home. Intergenerational exchanges with the rewilding interns enable locals to re-evaluate their heritage through everyday encounters with mobile youths. This increases the circulation of rewilding ideas as well as having the effect of creating new attachments to isolated villages. In this sense, place attachment is not restricted to long-term residents, as emotional ties are fostered locally and globally through place-based referents, such as the café in Algodres.

The site-specific examples in Portugal demonstrate how heritage is an ongoing process activated in place. Rewilding as an initiative shaping 'new natural heritage' runs alongside local communities' adaptation to their changing landscape, as they mediate their place-memories through new global connections and emerging landscape values.

Acknowledgements

The research presented here draws on field visits, interviews, and ongoing collaborations with Associação Transumância e Natureza, The Memory Archive and the local communities along the Côa Valley undertaken by the authors and Antony Lyons as part of a broader comparative study of transformation in heritage practice, one of four major areas of thematic foci for the Heritage Futures research program. We would especially like to acknowledge the invaluable assistance of Bárbara Carvalho during the October 2016 and 2017 visits. Heritage Futures is funded by an Arts and Humanities Research Council (AHRC) 'Care for the Future: Thinking Forward through the Past' Theme Large Grant (AH/M004376/1), awarded to Rodney Harrison (principal investigator), Caitlin DeSilvey, Cornelius Holtorf, Sharon Macdonald (co-investigators), Antony Lyons (senior creative fellow), and Nadia Bartolini, Sarah May, Jennie Morgan, and Sefryn Penrose (postdoctoral researchers), and assisted by Esther Breithoff, Harald Fredheim (postdoctoral researchers), Hannah Williams and Kyle Lee-Crossett. It receives generous additional support from its host universities and partner organisations. See www.heritage-futures.org for further information.

References

Agnoletti, M. (2014). Rural landscape, nature conservation and culture: Some notes on research trends and management approaches from a (Southern) European perspective. *Landscape and Urban Planning*, 126: 66–73.
ATN. (2015). *Nature conservation strategic plan*. Figueira de Castelo Rodrigo: ATN.

Barraud, R., and Périgord, M. (2013). L'Europe ensauvagée: Émergence d'une nouvelle forme de patrimonialisation de la nature ? [Rewilding Europe: A renewal of natural heritage-making?]. *L'Espace Géographique*, 42 (3): 254–269.

Deary, H., and Warren, C.R. (2017). Divergent visions of wildness and naturalness in a storied landscape: Practices and discourses of rewilding in Scotland's wild places. *Journal of Rural Studies*, 54: 211–222.

Gurney, G.G., Blythe, J., Adams, H., Adger, W.N., Curnock, M., Faulkner, L., James, T., and Marshall, N.A. (2017). Redefining community based on place attachment in a connected world. *Proceedings of the National Academy of Sciences*, 114 (38): 10077–10082.

Harrison, R. (2013). *Heritage: Critical approaches*. Abingdon: Routledge.

Harrison, R. (2015). Beyond "natural" and "cultural" heritage: Toward an ontological politics of heritage in the age of Anthropocene. *Heritage & Society*, 8 (1): 24–42.

Harrison, R., Bartolini, N., DeSilvey, C., Holtorf, C., Lyons, A., Macdonald, S., May, S., Morgan, J., and Penrose, S. (2016). Heritage futures. *Archaeology International*, 19: 68–72.

Holtorf, C., and Piccini, A. eds. (2009). *Contemporary archaeologies: Excavating now*. Frankfurt: Peter Lang.

IUCN. (2016). *IUCN resolutions, recommendations and other decisions*. Gland, Switzerland: IUCN.

Jepson, P., and Schepers, F. (2016). *Making space for rewilding: Creating an enabling policy environment. Policy Brief*. Oxford and Nijmegen: Rewilding Europe.

Labadi, S., and Long, C. eds. (2010). *Heritage and globalisation*. Abingdon and New York: Routledge.

Leuvenink, A. (2013). *Facilitating social learning to increase levels of local involvement: The case of Associação Transumância e Natureza in Portugal*. MSc thesis, Wageningen University & Research, Netherlands.

Lewicka, M. (2011). Place attachment: How far have we come in the last 40 years? *Journal of Environmental Psychology*, 31: 207–230.

Lorimer, J., Sandom, C., Jepson, P., Doughty, C., Barua, M., and Kirby, K.J. (2015). Rewilding: Science, practice, and politics. *Annual Review of Environment and Resources*, 40 (1): 39–62.

Lyons, A. (in press). Sensitive chaos (a geopoetic 'river movie'). In: R. Harrison, N. Bartolini, E. Breithoff, C. DeSilvey, C. Holtorf, A. Lyons, S. Macdonald, S. May, J. Morgan and S. Penrose, Co-authored monograph, *Heritage Futures: Comparatives approaches to natural and cultural heritage practices*. London: UCL Press.

Manzo, L.C., and Devine-Wright, P. eds. (2014). *Place attachment: Advances in theory, methods and applications*. Abingdon: Routledge.

Massey, D. (2005). *For space*. London: Sage Publications.

Monbiot, G. (2014). *Feral: Rewilding the land, sea and human life*. London: Penguin.

Navarro, L., and Pereira, H. (2012). Rewilding abandoned landscapes in Europe. *Ecosystems*, 15 (6): 900–912.

Rewilding Europe. (2015). Rewilding Europe presents working definition of 'rewilding'. Accessed 15/07/2019, https://rewildingeurope.com/news/rewilding-europe-presents-working-definition-of-rewilding/

Rota Vale do Côa. (2014a). *The Route*. Accessed 26 February 2018, www.granderotadocoa.pt/en/the-route

Rota Vale do Côa. (2014b). *Valley of Heritages*. Accessed 26 February 2018, www.granderotadocoa.pt/en/valley-of-heritages

Rubenstein, D.R., Rubenstein, D.I., Sherman, P.W., and Gavin, T.A. (2006). Pleistocene Park: Does re-wilding North America represent sound conservation for the 21st century? *Biological Conservation*, 132 (2): 232–238.

Statistics Portugal. (2014). *(Instituto Nacional de Estatística)*. Accessed 24 October 2017, http://censos.ine.pt/xportal/xmain?xpid=CENSOS&xpgid=censos_historia_pt

Svenning, J-C., Pedersen, P.B.M., Donlan, C.J., Ejrnæs, R., Faurby, S., Galetti, M., Hansen, D.M., Sandel, B., Sandom, C.J., Terborgh, J.W., and Vera, F.W.M. (2016). Science for a wilder anthropocene: Synthesis and future directions for trophic rewilding research. *Proceedings of the National Academy of Sciences*, 113 (4): 898–906.

Tokarski, M., and Gammon, A.R. (2016). Cultivating a dialogue: Rewilding, heritage landscapes, and belonging. *The Trumpeter: Journal of Ecosophy*, 32 (2): 147–154.

UNESCO. (2018). *The criteria for selection*. Accessed 14 February 2018, http://whc.unesco.org/en/criteria/

Vidal, F., and Dias, N. eds. (2015). *Endangerment, biodiversity and culture*. London and New York: Routledge.

30
TASTE AND MEMORY IN ACTION
Translating academic knowledge to public knowledge

C. Nadia Seremetakis

Introduction: globalisation as memory erosion

In the era of globalisation that totalises and homogenises human society and culture, a turn to the 'imperceptible' everyday can problematise the perceptual givens of our lives, and provide alternative perceptions of reality. A turn to the micrological is a necessary political task. It is also an ethnographic project.

Ethnography, as both research method and text or 'beyond text,' has always focused on everyday life experiences and specifically on the traumatic dimension of historical experience. Because ethnography does not focus on events per se, like the declaration of war, but on the first casualty of social crisis, that is, everyday life as a stable predictable structure. Traumatic history gives rise to new forms of everyday life, documents people making meaning and identity during and in the aftermath of cultural attrition, displacement, violence, separation, and exile.

Anthropology thus is the discipline that theorises the traumatic as everyday life.

The Greek understanding of 'trauma' or wound is far from a mere symptom of an underlying pathology or a visible injury caused by an external force, which demand a therapeutic catharsis. It speaks about long historical experiences, open wounds that travel through time and via bodies, natural and human, individual and social, national and global, and congeal into a collective memory. Greeks have historically been a migrating population to the 'developed' world. The recalling of this memory in action was what we witnessed recently in the streets of Greece where locals and invading others, immigrants or refugees, met in affective exchanges. These outsiders, like ghosts from the past, (re)emerged from the borders, awakening unconscious history into traumatic consciousness.

Salvaging alternative memories is thus important. But the role of anthropology is not simply to salvage. Ethnography must provide alternative affectivities. In my first ethnography on death rituals (Seremetakis 1991), for example, I presented an alternative and historically actual model of space-claiming in the context of modern Greece, and proposed that we attend to these historical models which teach the history of public space and truth-claiming in public space, but also women's history. Greek tragedy and contemporary Greek mourning rituals are about dialogue from below based on performative media that allow for communal participation, not just passive reception.

This is particularly salient in globalising modernity where media – one of modernity's principal constituent forces – displace and monopolise all possible dialogue that a society can have with itself. The media simulate dialogue in monological forms and leave the audience as passive spectator of the process, or dialogue is sold as a commodity to be consumed by those who are barred from it. This is in contrast to the dialogical performance of Greek tragedy or the Maniat mourning rituals.

Modernity does not engage with the tension between media-technological domination of information and lived experience, which is where the dialogical emerges. The gap between official commodified knowledge and lived experience widens.

Any exploration of cross-cultural and historical difference, instead, leads inevitably to a reflection on our present. This is politically important, for it points to an alternative mediology of everyday life that is now experienced as colonised by public media.

Context

At the advent of globalisation and before the so-called Greek crisis, the modern globalised Greek urban scape was characterised by a schizoid oscillation between the inflation of consumptive acts versus the utopian desire of the disappearing consumer. Greeks were inundated with fatter and fatter newspapers – inflated media artefacts offering more and more information for readers to consume in competition with the increase in fashion magazines and private television channels and programs. Coincidentally, many of these papers displayed numerous ads for slimming. Establishments once known as 'Beauty Institutes' were named 'Weight Loss Centers.' In media, photos advertising these city weight loss centres often showed the whole body of a female or male entering specialised machines from which it emerges transformed – slimmed – as an icon of the consumer consumed. To mind come the fruit and vegetable factories where food arrives green and emerges next morning red and ripe through accelerating chemical procedures.

These fat newspapers and magazines prominently featured also new anxieties over food consumption, particularly the infected foods by mad cow disease and genetic manipulation. This terror of transnational food consumption was in stark contrast to the apparent lack of anxiety concerning the overconsumption of globalised information which permeated public life. Collective cultural bloating on toxic information was effectively displaced onto the anxieties over the consumption of diseased and genetically altered substance from foreign parts.

Myth or reality, globalisation entered Greek consciousness as panic waves. These media scares, or 'thrillers' – a term accompanying all reportage – had a common motif: the degree to which social surfaces, taken-for-granted eco-systems like the ocean and the earth, and everyday items like the food we eat are contaminated both by poisonous substance and by foreignness. The dynamics of contamination entail delocalisation of regional ecology through pollution and the contamination of consumable substance causing us to question our own hold and ownership on local space.

People's gradual disenchantment with globalisation, originally seen as the promised era of unlimited flexibility, was intensified by the 'Greek crisis' and the subsequent austerity politics that threatened to wall-in all movement. Mediatic panics for toxic and altered transnational foodstuffs, which functioned as a powerful metaphor of the dangers and risks that accompany permeable national borders, were remediated by panics of starvation.

The mediatic pursuit of 'thrillers' shifted from toxic ingredients and contaminants arriving from outsiders to the hunger-stricken school kids and homeless insiders, Greek and Other. This mimetic replication of the American mediatic culture of catastrophilia is what Hal Foster (1996) would call 'traumatic realism.' The sensationalisation of the historical through the aesthetic of the 'thriller' would be enough to advance the attrition of Greek commensality, and to open the deterritorialisation and, consequently, the transnationalisation of the literal body, including the body-politic.

Transnationalised information transfers continue to grow, but the bodies of information consumers, previously instructed to shrink through dieting, are now forced to virtually cook and eat. Greek everyday experience, lately, has been bombarded by innumerable TV cooking programs coupled with massive advertising on Mediterranean dieting and delicious exotic recipes. Curiously, these instructive programs multiplied as austerity increased and its pains deepened.

They initially featured as information exchanges where women from various Greek locales exchanged recipes displaying their expertise in cooking. Traditionally, popular cookbooks were authored by women. When male chefs displaced female cooks, TV cooking was combined with in-land travelling to locate 'authentic' recipes. The cooking process was staged indoors and performed by women themselves. These mediatic male chefs functioned as directors or cultural managers that recovered and collectivised feminine domestic culture. To retain its allure, this performance was later transferred outdoors, in open spaces closer to nature with the kitchen acquiring the aura of an open-air museum. As cooking programs multiplied, they took the form of male touristic expeditions where chefs travelled the viewer out to foreign places for consuming the Other by cooking and eating the exotic ingredients of its food culture. Whether you watched one channel or switched channels, day or night, you could indulge in a continuous consumptive experience via successive virtual cooking programs. This consumptive euphoria was alternating with 'thrillers' of increasing starvation in Greece.

On the other side, the efforts and interests of local and national authorities to counterbalance this foreign invasion in food culture remained in the domain of economics and trade, mostly export-oriented, and aiming at protecting local and national business interests in the context of commercial internationalism. Thus, as trade networking with the outside world became more active, what and how was imported and consumed in the inside as both food and information were rather left to the unconscious. Indicative of this was the successful effort by media, businesses, and medical experts during the years of the economic bubble to socialise people to 'lighter oils' than their locally produced olive oil. Olive oil, 'the life and identity' of areas such as the Peloponnese, became guilty of high cholesterol levels causing death.

The subsequent crisis and austerity politics have probed some reflection on this deterritorialisation, as well as on its impact on the individual body and the body-politic. General efforts to reclaim territorialisation today are often metaphorised on the level of food and food culture. Olive oil has re-appeared in everyday life as a metaphor of an emergent *topophilia* (love of place) (Franco 1985).

Everyday people, particularly older women, seemed to have held a different perception of this invasion of their daily reality by virtual cooking programs and sites providing culinary tours and instructions on how to cook and consume food and information with foreign ingredients and mimic recipes and aesthetics of unknown origin. I asked old ladies if they like these instructive TV programs and if they cook the performed recipes. All responses revealed that the mediatic shift in performance from product to process is no neutral act: 'Never! what kind of cooking is this done anywhere and in no time?' some grumbled. The double meaning of their words pointed to the fact that these recipes have no history, they are copies with no original and their ingredients foreign and untranslatable, but more importantly, they cook in no time on TV. Cooking for them is a maturation journey, an experiential process that cannot be captured by television time. Nor is cooking a totally public affair. It is *poesis*, an act of tactile-affective creation that embodies self-reflexive 'secrets,' which cannot be sold out; they take place inside, long before they perform outside in a meal. Each dish acquires its meaning within an antiphonic combination of shared dishes that make up one's meal in this culture that favours tidbits over single, individually consumed dishes. Food eating is a shared taste.

Cooking and eating in Greece had been but one part of a complex set of practices and experiences linked together in a single experiential continuum by the ethics of care and tending of and for significant others, be those dead or alive people, trees, animals, or foreign immigrants. Olive trees, for instance, were treated like one's kids and olive groves had an identity, that of their owner and care-taker whose hand-prints were inscribed on the tree's body. Caring for them endowed them with affective value. Lack of care signified death, their abandonment to 'the outside.'

In the recent immigrant crisis, the icon of old ladies taking care and feeding washed offshore immigrant babies and kids of unknown origin became viral. What was simply 'natural' to these old ladies, the interiorisation of the other, became emblematic in the Western world in its quest for feminised refuge areas of stable social relations to counterbalance the present public violence. These island ladies were proposed for a Nobel prize, the name and purpose of which they ignored. For them, cooking and feeding others is the translation of multiple affectivities to a final product that carries and disseminates their mark, the mark of the one onto the others. This is their meaning of personal pleasure. And it is far from apolitical for it has been well documented that values of the inside can be expanded to social critique.

The invasion of their space is an appropriation of their concrete experiential values, their lifting from their local context and their abstraction into political and/or religious utopian norms, now manipulated rhetorically by state and media. At the same time, there is very little, if any, room left for them in zones of Western modernity where sensory experience is translated and stored by mediatic, scientific, or medicalised terms. The replacement of these local cooks by modern male chefs aiming at the re-education and modernisation of local cuisines to fit the new transnational context turned these women, along with local food varieties, into matter out of place. This sensory cosmetic re-education of the domestic masses often involves the handling of food with modern utensils and surgical gloves – a medicalised, one-use touch. As a 70-year-old lady, known for her good home cooking, confessed in a casual conversation: 'I feel left out.' These women poised at a historical cliff that they will not cross are destined to catharsis. The recuperation of their (food) culture is socially important and personally relevant.

On the other side, in academia, information and knowledge dissemination rested securely on its Platonic-Aristotelian split between body and mind, the sensible and the insensible, the inheritable and the uninheritable. In fact, in the world of formal education, *Paedeia*, the prevalence of this perspective, is not unrelated to the acute formalistic bias in the production and dissemination of knowledge, which renders the quotidian nonvisible and the present devoid of cultural meaning. Taste and touch, thus officially designated as the uninheritable, have been unconsciously and consciously suppressed.

In the era of globalisation, the witnessed inversion of the Aristotelian mind/body hierarchy aims at the re-biologisation and literalisation of the body – the body has come to signify with its physical functions alone, e.g. eating, sleeping. This facilitates a homogenisation, a discipline, and a functionalisation of food and eating cultures. The separation of the senses has also fostered the gradual identification of taste with food and its consumption, and with the social organisation of food and drink consumption. The result is the banality of individual pleasure – a pleasure whose constant pursuit, however, becomes in itself the experience, performance, and visualisation of mass identity (as in domestic cooking).

The project: taste and memory

My ethnographic explorations have been coupled with the designing and organisation of public educational events, whose main goal was to bridge commodified knowledge and everyday experience and to translate academic knowledge and ethnographic method to public knowledge: what ultimately matters is the dialogical interface between theory and practice. This is part of the dialogic nature of ethnography and its performance; performance is understood as the intervention in everyday structures that release hidden meaning from social and material relations. Performance, thus, as a practice of everyday life is *poesis* because it departs and embarks from the socially constructed sense of a naturalised, homogenised present.

The symposium *Taste and Memory* was designed as a response to globalisation's massive efforts to homogenise, discipline, cosmeticise, and functionalise Greek food and eating cultures through massive instructive cooking programs in the media. It was inspired by my abiding concern over bringing to consciousness the uninheritable, that is, any part of culture that is nonvisible in the sphere of what is officially designated as inheritance or heritage. Food and its consumption are one example. Here, taste is also presented as the gendered-specific subjugated knowledge of women, connected to practices of labour, care, and commensality.

Food consumption is one of the diverse sites of the everyday that can serve as registers of and for viewing globalisation. It has anticipated the effects of globalisation as developed and experienced today. In this transformative process, whole eating cultures could be silenced and turn to privatised memory. But memory is far from a mere property of an individual's mind and psychic. It is encapsulated and stored in artefacts, spaces, and temporalities of both 'making and imagining,' of sharing and exchange.

Cultural memory is embodied in performance practices, intentionally mediated by social actors, and can thus intervene in the meaning systems of the present. Philosophers of historiography maintain that event is not that which happens but that which is narrated; this alerts us to the fact that formulaic depictions can wall domains of experience out dehistoricised.

Taste and Memory thus was a field-based, participatory process and a public exhibition of the sensory memory of Messinia, a Southern Peloponnese region, before and during globalisation and more particularly the struggle for memory in wake of the defacement of Greek commensality.

This 'public anthropology' work was done in a milieu where the introduced core anthropological perspectives – e.g. culture as everyday life perspective, situational and positioned knowledge, and the dialogical foundation of social – were, and still are, themselves novel political interventions. It blended ethnographic knowledge and experience coming from both continents – Europe and America where I lived for 25 years – to a Greece where anthropology had no public face and a limited disciplinary status in an academic system that needs decentring itself. In this sense, this introduction of anthropological discourse into existing public arenas, where it had no formal or even rumoured existence, captures also many of the recent concerns in the discipline, such as its interface with the public.

Further, I propose, echoing Stanley Diamond's maxim, that 'anthropology is the study of people in crisis by people in crisis' (1999 [1972]: 401) and anthropologists are people who exist on the margins, who learn through and from other folks who dwell at the border of social and historical disaster. A central issue remains as to how memory is used in everyday life to respond to social crisis.

The symposio

In this context of cultural attrition, the public anthropology symposium *Taste and Memory* was organised with the aim to provide an alternative paradigm by adopting the dialogical performance of the Greek Tragic Poets on the premise that organisation is but the implementation of principles of cultural translation, which often involves historical translation.

To begin with, the mutation of the word 'symposium' from the original Greek meaning of *symposio* in academia has always troubled me. It is used interchangeably with 'conference' – a meeting of people who confer about a topic. A philological archaeology here would contribute in articulating the sensorial with affect.

Symposio means drinking and eating with others (the prefix *syn* means co- and *pino* means I drink) while exchanging ideas, (self)reflexive narrations of everyday life events and pains, poetry, games, and often music and dancing. These experiences are realised and granted as true and memorable when witnessed by others; they require a *syn-trofia* company, human or other. *Trefo* means I feed, cultivate, raise, as in raising a child, but also as in raising and cultivating hopes, dreams, desires.

Yet, today, symposium has ended up meaning a mechanism of exchange of ideas. What has been eliminated from the exchange is the sensorial dimension. The part took the name of the whole. Diotima's reminder to Socrates, though, echoes in the background:

> Any action which is the cause of something to emerge from nonexistence to existence is poesis, thus all craft works are kinds of poesis, and their creators are all poets.... Yet, ... they have different names.
>
> *(Plato 1976: 150)*

Complying with her definition have been modern women's practices and experiences. Like their embroidery series, which cohered into a visual, tactile story imprinted on cloth to ornament people and spaces, their elaborate preparation of olives and cooking of food were to nourish others within and beyond the household.

Rationale for taste and memory

My initial fieldwork included elementary and high schools of the area, with local cultural organisations and clubs following. (Re)sensitising young students and local citizens to the Greek meaning of *symposio* and the relation of commensality to memory was a first goal. A team of volunteers, two students, a young cameraman, and myself held repeated meetings with teachers and students to gradually acquaint them with the concept and theoretical presuppositions of the prospective event as they discussed their own experiences in exchange. The recipe, an uninheritable archive, became our medium of sensory communication with history. Anne Frank's diary with the description of meals cooked and consumed in her culture became an entry point to the texture of a modern memoir. My own personal archive of my late mother's handwritten recipes received by mail during my early years in the States triggered dozens of self-reflexive stories, of personal and familial re-collections. In the schools for the handicapped, discussion and exchange of experiences were generated via the story of the eighteenth-century French philosopher Diderot: his attention and interest to tactile sensibility were motivated by his encounter with a blind person. This opened a moving exchange of long-suppressed experiences in a social environment that the handicapped body has been perceived and treated as a visible display of disorder. The historical interaction of art and sign language, as in nineteenth-century France (Mirzoeff 1995), enhanced self-esteem and trust, and reinforced our dialogue, silent and vocal. The subjection of taste and touch – familiar means of exploration of the handicapped – to historical and theoretical contextualisation became the impetus for further reading on the subject for teachers and older students.

A prepared list of topics was left at the end with the teachers as a guide for motivating students to initiate dialogues with older generations.

On a later visit, while we videotaped their narrations and experiences on the process, participants expressed an enthusiasm that revealed also their need for educational programs that lift their everydayness off its formalistic routine. Ethnographic fieldwork was proven to all an invigorating journey and unparalleled educational tool.

As one female teacher stated:

> The magic of it was that both students and teachers discovered food differently, food as *history!* [her emphasis] At first, we thought this project would be another one of those TV cooking programs that we had no desire to join. But now, we talked with grandmothers, grandfathers, and through their narrations and those of their own ancestors, we ended up cooking ourselves recipes, some previously unknown to us; it was indeed a journey of embodying history.

'Shall I tell you my mother's favourite dish?' Another teacher jumped in the conversation. 'If you eat it, you'll flip, it's *a poem!*' While a third teacher interjected, 'we [teachers-students] did not just get recipes, we wrote poems, stories.' Diotima emerged smiling from the guts of history.

'It was some journey,' noted a music schoolteacher, 'to search for the note that best echoes each ingredient and its taste.' For sound tastes. Fruit and vegetables, for example, have their own speech sounds; watermelon, if tapped by hand, or touched by ear on the right spot, confirms 'I am ripe.' Its colour complements the dialogue. Sound, colour, and taste compose music or paintings.

'We also travelled out, to other areas,' a younger female teacher added, 'as we contacted oldies for recipes, we crossed Asia Minor, Mani, Messinia.' 'Oh, yes,' a young student rushed to confirm, 'I called my grandma in Northwestern Greece and gave me the recipe of her favourite dish. But she did not reveal the secret of its success [laughing].' There is always a secret hidden in each recipe; a touch of a herb can be one of these secrets.

Herbs, the invisible spices in food, signify specific emotions, some bring people closer by opening the bodies to communication, others close them up. Personal names in this context signify people's sensory identity. Female first names, once popular in the area and in local fairy tales re-emerged: Cinnamon, Lemon Tree, Sugaree.

In these journeys, herbs also triggered smells and reconfigured geography, turning it to a magical olfactory field in motion: roses grew in Delphi, oregano in Acropolis, sage in the Peloponnese. They were paralleled to the stars of the sky for 'the word gastronomy hides in it the word astronomy' as grandpa pointed in *Politiki Kouzina* (2003), and, by his fairytale lessons of the invisible worlds, the child learns to translate taste and touch into drawing: 'Pepper for Sun, it is hot and burns. Cinnamon for Aphrodite, the most beautiful woman, bitter and sweet.'

'Our talking and interacting with children,' a male teacher pointed emphatically, 'led us to the ingredients: where they come from, how they were cultivated and used. . . . It was difficult for children to comprehend that there were no sugar or candies in rural areas in the old days.'

Food preparation became the allegory for a musical composition *The Soup of Stone*, performed for the occasion by the 70-children choir of the Public Music School, and music calls for dancing. A short modern dance piece *In the Shadow of the Olive Tree* was choreographed to order and danced by a female adult dancer and a young girl of the Municipal Dance School. It was the first choreography on olive trees in Greece. Olive trees are always closely observed, for their appearance speaks of their sensorial and affective relation to both time and weather. They are old and young, dancing to the wind, lamenting in drought, washing in the rain, singing via countless cicadas in the summer, and displaying anger to negligence and deprivation of proper care by 'swivelling' the fruit. Urban directors and choreographers have yet to realise that 'taking' form and content from the local context results in inspirational ballet compositions that also are relevant to people's experiences. It takes two to tango in cultural translation.

These explorations made publicly visible a whole world previously thought of as strictly domestic. As a grannie exclaimed to her granddaughter when asked about her favourite recipe, 'does the university truly want to talk to me?!' Feeling 'left out' by higher-up chefs and others, she eagerly participated in a video taken by her granddaughter; allowed to proudly speak for herself, she narrated the history of her recipe.

A very young male student reminded us that today food is not always tied to family. It can be a medium of self-determination and personal enjoyment. Recalling Beauvoir (1960) and her favourite borscht, 'chief delight was in doing as I please.' Nor is food and its preparation an unvarying, natural, mundane practice of survival or bodily pleasure. He cooked secretly his own favourite dish one day, leaving his mother stunned, for, aside from his young age, it was far from natural for a boy to cook.

Twenty-five schools were gradually engaged in the project. Young students contributed with over 300 improvised paintings that 'dressed' the large exhibition Hall during the event. When asked what they felt while painting their dish, one of them responded shyly: 'I salivated!' Dali would confirm that all his own 'experiences are visceral.' These paintings were complemented by dozens of ceramic constructions featuring favourite dishes and fruits and vegetables of the area, all designed by students of the area's Ceramics Department. This tactile exploration of taste ran parallel with high-school students' explorations of ancient texts, popular music lyrics, and classic and modern poetry for gastronomic metaphors in language. Long-forgotten expressions emerged from their historical sleep: 'You ate me,' a common saying from mother to child expresses fatigue, torture; or 'the eye eats' which points to the 'translation and transcription of the senses,' as much as it extends taste and touch beyond the borders of the individual body (Seremetakis 1996). Some such examples brought laughter, while others raised serious questions pertaining to corporeal ideals and images, violence and *eros*, modernity and overconsumption. In this spirit, a theatrical play on eating, a humorous but poignant critique of overconsumption (of food and information) in our era, was prepared for the young and old by the Municipal Theater Company.

Different age groups have different modes and media of capturing everydayness. Older students contributed their own little video and photography productions based on their mini-field researches with 'the oldies.' Their videos on others and ours on them, featuring the event in process, were later discussed in terms of editing in and after the field: the tendency to privilege one's aesthetic perspective at the expense of documentation or the tendency to treat ritual performances as raw visual material.

The performance of food could not be missing from the polyphony of this event blending artistic expressions, popular films, and academic lectures. The preparation of a celebratory table allowed local social aesthetics to perform. Taste is both gustatory and aesthetic – despite Kant's disagreements. 'Show me how you eat to tell you who you are,' the saying goes. Appadurai (1986) would claim for Indian aesthetic theory what applies to Greek aesthetics of taste: a rather complicated concept, taste, engages 'aesthetic, emotional and sensuous appreciation.'

Reflections

Recalling Goffman (1959), the performance of food leads ultimately to the performance of self. A person behaving in a detestable way is 'an ugly person,' and was metaphorised as disgusting food. By naming 'inedible' the ugly person, students unconsciously brought home, down to context, abstract philosophical ideas and debates on disgust and the distinction between aesthetic and physical pleasure (as in Derrida 1987, Bourdieu 1990, Kant 1987). The example of the disgusting sauce with which Stoller was treated by the Songhai (Stoller 1989) to express anger was discussed next in order to reveal taste and smell as sensual, subjective, central ingredients in the recipe of social relations. This triggered the recalling of old Greek rituals, such as the sprinkling of salt behind the seat of an undesirable guest – a familiar extra-linguistic communication, for which though they lacked a translation.

The sensory performance of the *symposio* was complemented with the participation of the constituents in the exhibition of the materials they had created and collected themselves. Their concepts of ordering the exhibition space brought home 'the raw and the cooked,' 'the hot and cold,' 'wet and dry,' 'down and up,' and my own trajectory in the discipline's paths 'bitter and sweet.'

Taste is anything but static. It is often perceived as walled in a specific *topos* (Aristotle's definition of space) which renders both immutable. But isn't people's relation to food (re)formed by his or her biographically determined situation in space and time? The question arose, as my college students reflected on their experience of living themselves far from home and as they explored 'travelling tastes.' At this point of their lives, bus terminals have become their favourite spot in the city, as there arrive their favourite dishes cooked, packed, and sent by mothers or grandmothers from various cities. When received in a 'foreign' land, these affective tastes are never consumed in isolation. In fact, there one comes to like and share dishes never liked before. The recipient calls friends to join the table, to eat food in company, a *symposio* that seals the memory of past and future moments. Food sent afar travels as a patch to fulfil an unfinished dish with others.

Their moving narrations of such moments of sharing, of transforming feelings of isolation to a 'symbolic family,' awakened for me a taste I had unconsciously suppressed: the taste of those chocolate bars sent to me as letters from friends or family during my early years in the States.

Aftertastes

Conferences traditionally present ideas, and occasionally archive them in published volumes, with a beginning, middle, and conclusion. This symposium aimed at linking sensory fragments together, and theory to open-ended action. Given our present state of world affairs, multi-tools are needed that also translate knowledge to action. A dialogue among academic disciplines and arts and among so-called peripheries and centres, must lead to a questioning and reshaping of our public space itself, including our built and environmental space, for this reshapes our idea of citizenship by expanding it beyond the private. 'Aftertaste,' the last part in this *symposio*, proposed reflective actions on the university space itself. The eating hall of the university, resembling all other liminal, disciplinary, sensorially sanitised, public institutional spaces, was re-perceived. University spaces in Greece, as an old professor interviewed critically stated, are perceived and treated by their inhabitants 'as old bus- or train-stations for passing by,' or places you leave your consumables behind, or, places that, like the streets, are to be vandalised. In this context, students, inspired by the aesthetic transformation of the cultural hall that hosted the public event came up with imaginative ways of reshaping university's public-ness, of turning functional space into affective place by dressing and ornamenting it with colorful objects, artwork, and music, and every body and mind went to work to find ways to actualise them. It was an imaginary reinscription of a functional space into a place of **eros**. For food is **eros** actualised in place.

'Back-spaces' or 'junk spaces,' relegated to garbage or food-waste, emerged also as spaces of creative use. Recycling had been an unknown field of action. Raising awareness of what our culture chooses to exclude led to exploring possible installations of handmade compost bins on campus.

Conclusion

Local cultural organisations of the region comprised of several hundred members each embraced the idea of reenacting a *symposio*. These organisations have a long tradition in this culture. They connect everyday people with occasional feasting and various engagements in public local affairs. Women's preparation and organisation of buffets with ambrosia, the most delicious local varieties of homemade food and desserts, was a memorable gift expecting no return. Commensality is but an exchange of substances and objects incarnating remembrance and feeling. 'We remember well those days we worked together for that other memorable public event on [the] earthquake in the 90s,' several of them rushed to mention, pointing to our shared history. Our silent affective bond that released the old good local hospitality lay in those women's need to re-claim their culture. Its re-cognition by 'one of their own' and a representative of a higher institution, the university, which reached their doorstep first time, mattered.

The project, which mobilised a whole, regional *eros* – attended by hundreds of citizens including the local schools – attracted attention beyond the local. It was later invited and hosted by a national research centre in Athens (though of different theoretical orientation and organisational approach) and was presented as a model to a UNESCO state member meeting for the preservation of intangible heritage by the Greek Ministry of Culture. But in this rare dialogue of the local with the national and global, one part of the event was not transferable: the abundant inflow of homemade dishes by locals (as individuals or collectively as members of cultural organisations). That tactile donation, a gesture of pride, care, and affection, that had extended to the aesthetic preparation of the public cultural hall itself could only occur in places of shared memory. Abstract spaces, spaces of no history, could only be perceived and treated as museums. For objects, substances, and scents are in themselves histories of prior commensal events and emotional-sensory exchanges. These histories are exchanged at commensal events and spaces and qualify the latter as commensal to begin with. This sharply contrasted with the various TV cooking shows and other events restricted to the manual preparation of 'non-speaking' substances.

Whether definite or not, such ground-up initiatives that foster imagination and collective identity, and bridge institutional knowledge with everyday reality, indicate that history is not just materialised in the fossil as frozen trace or writing in suspension. Rather, it is living traces of alternative sensibilities long walled out of the sensorially numb public academicscape, which allows no transgression of its naturalised and outmoded spaces; spaces that, in the current austerity times, like cemeteries lying in ruins after an earthquake, have been left to their fate, to concretise the ruination of the historical (Seremetakis 2019).

References

Appadurai, A. ed. (1986). *The social life of things*. Cambridge: Cambridge University Press.
Bourdieu, P. (1990). *The logic of practice*, trans. R. Nice. Stanford, CA: Stanford University Press.
de Beauvoir, S. (1960). *The prime of life*. New York: Perseus Books.
Derrida, J. (1987). *The truth in painting*. Chicago: University of Chicago Press.
Diamond, S. (1999 [1972]). Anthropology in question. In: D. Hymes, ed. *Reinventing anthropology*. Ann Arbor, MI: University of Michigan Press, pp. 401–429.
Foster, H. (1996). *The return of the real*. Cambridge, MA: October Books/MIT Press.
Franco, J. (1985). Killing priests, nuns, women and children. In: M. Blonsky, ed. *On signs*. Baltimore: The Johns Hopkins University Press, pp. 414–420.
Goffman, E. (1959). *The presentation of self in everyday life*. Garden City, NY: Doubleday.
Kant, I. (1987). *Critique of judgement*. New York: Hachette Classics.
Mirzoeff, N. (1995). *Silent poetry: Deafness, sign, and visual culture in modern France*. Princeton, NJ: Princeton University Press.
Plato. (1976). *The symposium*, ed. I. Sikoutris. Athens: Estia.
Politiki Kouzina (A touch of spice), film. (2003). Tassos Boulmetis, director and scriptwriter. Capitol Films.
Seremetakis, C.N. (1991). *The last word: Women, death and divination in inner Mani*. Chicago: The University of Chicago Press.
Seremetakis, C.N. (2019). *Sensing the Everyday: Dialogues from Austerity Greece*. London/New York: Routledge.
Stoller, P. (1989). *The taste of ethnographic things: The senses in anthropology*. Philadelphia: Temple University Press.

31
FOODSHED AS MEMORYSCAPE
Legacies of innovation and ambivalence in New England's agricultural economy

Cathy Stanton

In many parts of the world, food reformers have been working for the past two or more decades toward what one early and influential set of authors (Kloppenburg et al. 1996) termed 'coming in to the foodshed' – that is, trying to create viable local and regional alternatives to the geographically extensive, highly consolidated, industrialised, and commoditised food supply chains that became the norm in most industrialised places over the course of the twentieth century. In its most general sense, 'foodshed' refers to the geographic area from which a particular population receives its food; given the scale of industrial food production and marketing, the foodshed for almost any given place now encompasses much of the globe. But the phrase 'coming in to the foodshed' – or what Wes Jackson has called 'becoming native to this place' (1994) – implies something much more specific and value-laden. This approach to 'rebuilding the foodshed' (Ackerman-Leist 2013) is grounded in ideals of emplacedness, along with practices that foster socioecological regenerativity and mutuality among human and non-human species and between living beings and the landscapes they inhabit – literally, their habitats. This approach draws on ideas from agrarianism, agroecology, ecospirituality and ecofeminism, post-colonialism, and environmentalism, among other strands of thought and activism. It takes form in projects and networks of exchange and association under terms like 'food justice' (Alkon & Agyeman 2011; Alkon & Guthman 2017), 'civic agriculture' (DeLind 2002; Lyson 2004), 'food sovereignty,' and 'solidary economics' (Holt-Giménez 2011). It is an important element of what is often referred to as 'the food movement,' which food writer Michael Pollan has characterised as a 'big lumpy tent' crowded with sometimes contradictory projects (Pollan 2010). But the task of coming in to the foodshed also frequently stands as a critique of the more mainstream elements of the food movement (Guthman 2007; Saraiya 2016; Genzlinger 2017). It overlaps with efforts to modify existing structures and practices, but in general envisions a much more radical rebuilding and reconceptualisation of society, economy, and ecology, with food-related changes as an entry point rather than an end in themselves.[1]

Attempts to come in to the foodshed are inherently in dialogue, and sometimes in tension, with the memoryscapes of particular places, understood as the layers of meaning that have accrued within a physical environment over time. Memoryscapes are shaped by and in turn shape the meanings that inhabitants and outsiders attach to a given place, creating both opportunities and limitations for the kind of deep re-envisioning that many neo-agrarians are working toward. Although food reformers are inherently focused on the future, they are also increasingly aware of the need to address legacies of past practices and inequities, including various forms of soil degradation and deep-rooted, often gendered, classed, and racialised patterns of displacement and exploitation. Some of those reconsiderations of the past take place on larger-than-local scales; for example, class action lawsuits in the United States over discrimination by federal agricultural officials against farmers of colour have been national in scope (Cowan & Feder n.d.). But given the inescapable linkage of food production with actual places and resources, even the broadest reform efforts inevitably come down to earth somewhere, creating inherent linkages with specific, emplaced histories and locations. As they do so, they intersect with existing commemorative practices and inherited understandings of particular memoryscapes. These encounters are enabled by what Anna Tsing (2005) has termed 'friction,' or the means by which widely circulating ideas and projects gain purchase in particular locales and allow both the global and the local to come into being in ways that can be complementary, confrontational, or both.

The six-state region of New England in the northeastern United States presents present-day food reformers with a particularly rich, layered, and complex ground on which to attempt the task of rebuilding a regionally scaled foodshed. New England is by no means alone in presenting a fairly coherent image of regional identity in which food plays a role; the construction and marketing of such imagery is very widespread, especially in areas that have had to reinvent themselves in the wake of economic and other changes. In the extensive scholarly study of 'invented traditions,' social memory and cultural geography in recent decades, commemorative and representational practices in such places have been explored very thoroughly. There is now a substantial body of literature that traces the construction of what have been called 'edible identities' (Brulotte & Di Giovine 2014) in regions like Tuscany (Gaggio 2011) and other rural parts of France (Trubek 2008), the Swiss Alps (Grasseni 2011), Britain's Lake District (Sims 2010), and the American South (Weiss 2016). Within these efforts, the crafting of New England's unique imaginary and the commemorative infrastructures that render it 'visitable' (Dicks 2003) and consumable stand out as particularly intentional and inventive; they have had remarkable longevity and reach; and they have a particularly complex and ambiguous relationship with both the natural resource base of the region and the industrial capitalist economy that has developed there and elsewhere over the past two centuries. But despite the presence of food-related issues within many of these processes of regional image-making, critical studies of place and memory have not yet deeply intersected the growing literature on the cultural politics of food reform in the present, in New England or elsewhere. In what follows, I trace the construction of some of the spatial, imaginative, and functional components of New England's pastoral memoryscape over the past two centuries and consider some of their implications for contemporary neo-agrarian projects of 'coming in to the foodshed.'

This chapter focuses on the agricultural and does not attempt to touch directly on New England's equally important and iconic maritime economy. Many of the issues of scale, nostalgia, and ideology faced by the region's farmers have been manifested in similar or parallel ways in fishing, and with the expansion of both fish farms and agroecological approaches to farming that situate the production of human food within larger biotic systems, the two may be moving closer together in the present and future. But in terms of inherited images and associations as encoded in New England's particular land and seascapes – the yeoman farmer and the hearty mariner, the 'wild' resource of fisheries and the cultivated landscapes of the farm – the histories, economies, and regulatory environments of farming and fishing remain quite distinct, making it difficult to cover both adequately in a survey of this length.

Limitations in an expansive economy

The two-century history of market-oriented agriculture in New England reflects the fundamental contradictions between farming at a small scale (whether through choice or necessity) and the logic of markets centrally organised around the search for profitability and the kinds of efficiencies and technologies that can produce it. Within that logic, the types of farming operations best suited to the region's resources of land and fertility have long been considered untenable or backward-looking by a range of actors, sometimes including farmers themselves. That association with backwardness and pastness has sometimes been pejorative, often nostalgic, and occasionally proudly embraced. In the aggregate, it has produced a profoundly ambivalent memoryscape that contains submerged or muffled memories of the illusory goal of both directing and containing capitalism's prodigious energies.

The earliest European settlers saw both an abundance of 'virgin' land and an antagonistic wilderness as they pushed beyond the fertile flatland farms of the coastal region into interior river valleys and then – as population pressures began to make themselves felt in the eighteenth century – farther into upland areas with thinner soils (Stilgoe 1982: 43–53; Judd 2014: 69–94). Throughout the colonial period and accelerating after the Revolution, and especially with the turn toward industrial production around the turn of the nineteenth century, well-capitalised and entrepreneurial elites seized on the new opportunities afforded by the expanding capitalist economy of the new nation (Clark 1992). More sedentary, 'slower' forms of wealth tied up in land quickly came into direct conflict with newer speculative ventures, a dynamic that has bedevilled farmers in New England and elsewhere ever since (Magdoff 2015; Holt-Giménez 2017: 23–56). A decade after the Revolution, in the agrarian uprising known as Shays' Rebellion, western Massachusetts farmers took an armed stand against the effects of what we now call 'financialisation' – the making of money *from* money rather than more directly from a natural resource base (Gross 1993; Richards 2002). The farmers were soundly defeated, but scholars have shown that the outcomes of the uprising were somewhat ambiguous (Pressman 1986; Gross 1993; Peet 1996; Goldscheider 2015). Most of the insurgents were pardoned rather than punished; there was a decisive shift toward stronger centralised government on both the state and national levels, and the event has always had a strangely muted presence within the subsequent commemorative landscape, suggesting that New Englanders in this period were already beginning to develop a collective ability to hold radical popular dissent in a continually unresolved though sometimes surprisingly productive tension with elite economic and political power. In subsequent centuries, the region's politics have often combined the famously liberal with a deep conservatism; one result has been a tendency to buffer rather than directly confront the problems caused by the power of capital.

Within both the memoryscape and the foodshed, this tension has historically manifested itself in a range of interconnected narratives, images, and experiences of New England's landscapes – especially its pastoral landscapes – over the nineteenth and twentieth centuries (Conforti 2001). In the visual arts (Truettner & Stein 1999), literature (Delbanco 2001; McWilliams 2004), and mass publishing, canonic works and consumable items of popular culture constructed and cemented what Julia Rosenbaum has called 'a distinct sense of home' (2006: 78) in the imaginations of many Americans in ways that have ranged from the quaint and nostalgic (Currier and Ives prints, *Little Women*) to the gothic (*Ethan Frome* and H.P. Lovecraft). These images often included or are even centred around the table, particularly around holidays like the Fourth of July (Waldstreicher 1997), Thanksgiving (Seelye 1998; Baker 2009), and Christmas (Nissenbaum 1997), when nationally circulating imagery and foods reinforced the notion of New England as an origin place for the nation.

This notion was carefully crafted and often directly linked with the projection of political and economic power within and beyond the region itself. As early as the first decade of the nineteenth century, influential white New England businessmen in other cities along the east coast and later in the Midwest established 'New England Societies' (Vartanian 1972; Seelye 1998) to 'cherish the memory and perpetuate the principles of the original settlers,' in the words of a historian of Cincinnati's chapter (Peet 1996: 23). Spatially, architecturally, and imaginatively, the iconic New England village formation – white-painted wooden houses, churches, and civic buildings surrounding a common or green – came to convey images of yesteryear to many in the region and far beyond (Wood 2001). These images were codified (or reinvented) in situ through the efforts of historic preservationists and exported widely throughout the nineteenth and twentieth centuries (Lindgren 1993; Holleran 1998). Nineteenth-century New England emigrés to upstate New York, Ohio, and points west reproduced this pattern of settlement along what some have termed the 'Yankee runway' (Conforti 2001: 150; Adams 2014) while colonial and craft revivalism movements of the twentieth century amplified the aura of 'pastness' associated with the region – especially its rural places – through the making and marketing of Shakers, gabled Cape-style suburban houses, popular domestic and holiday decorating traditions, the collecting of antiques (Greenfield 2009), and more. Assertions of the region as a a national and cultural origin place were also intended to bring people and wealth *into* the region through tourism, sentimental hometown reunions, and reinvestment in 'left-behind' upland and rural places (New England Magazine 1897; Sears 1989: 49–86; Brown 1995: 135–142; Albers 2000: 196–267).

Those places served as a kind of shadow of the westward-moving frontier and the scaling-up of agricultural production on ever-larger, more mechanised farms. Food production in the peri-urban belts around New England's industrial-era cities remained viable well into the twentieth century, but the mystique and nostalgia surrounding New England farming became attached to rural places that had been less able to meet the demands of industrialising agriculture from the outset. Indeed, as Stephen Nissenbaum has shown, 'the pastoral heart of New England' has been moving steadily northward for two centuries (Nissenbaum 1995: 39), always simultaneously one step ahead of and behind the edge of 'progress' and modernisation, with tourists (Harrison 2006), back-to-the-landers (Brown 2011), second-home buyers, and outdoor enthusiasts moving along with it and developers and businesses sometimes following.

As these images of regional identity were being crafted, promoted, and occasionally contested, a narrative about New England's inherent unsuitability for farming began to take shape, becoming more entrenched and taken for granted as time went on. According this narrative, competitive pressures beginning with the opening of the Erie Canal in the 1820s made it virtually impossible for most New England farms, with their limited soils and fertility, to remain economically viable, leading to widespread farm abandonment and a precipitous decline in agriculture as an economic driver in the region. Some observers blamed farmers themselves, pointing to poor practices of husbandry and soil management. Other versions of the tale, more aligned with the valorisation of the sturdy Anglo-Saxon yeoman farmer, celebrated settlers' initiative and resilience in carving homesteads out of the rocky hillsides in the first place, accepting the putative end of agriculture in New England as part of a more or less natural progression toward a more efficient and modern mode of living. Scholars have shown that, over time, variations on this decline-and-abandonment narrative have been enlisted in support of a wide range of projects including economic redevelopment, rural beautification, tourism promotion, and land and forest conservation (Barron 1984; Bell 1989; Donahue 2007). The narrative is reinforced by many features of the landscape itself, perhaps most famously the old stone walls running through tracts of reforested former farms, a kind of picturesque ruin that many observers have readily accepted as a sign of a lost, bygone era and way of life (Bell 1996; Lapping 2001; Ryden 2001; Wessels 2005).

This story of poor rocky farms and regional agricultural decline is both true and untrue, and farmers have had to maneuver carefully within and around it over time. As Dona Brown has noted, the whole notion of decline has been not so much mistaken as misunderstood (1995: 137–138). It has been very largely driven by regional chauvinisms and anxieties inflected in complex ways by a broad range of historical changes. In the early decades of the nineteenth century, New England's 'free soil' was enlisted within antebellum sectional debates; many Northerners felt that a demonstrable agricultural strength in the older American states was an essential part of the argument for northern ideological superiority, and feared that an actual or perceived decline weakened that argument. As the century went on, population pressures and demographic shifts linked with immigration patterns and the growth of cities continued to create new opportunities but also new stresses for New England's farmers. Growing cities expanded markets for food, but the excitement of innovation and the potential for high returns in the manufacturing economy set an unrealistic standard of growth for the more mature agricultural sector. At the same time, the industrialisation of farming itself meant that smaller, older farms trying to compete in expanding foodsheds had to invest more capital in order to keep up, to find new (often niche) markets, or both. The widely disseminated New England imaginary continued to work against farmers in a region that was both shaping and resisting change. If the agricultural landscape was central to the notion of New England as an origin place for the nation, signs of struggle or losses within that landscape threatened to undermine iconic images supporting high-profile narratives of national purpose and growth. By the early decades of the twentieth century, anxieties about agricultural decline had become full-blown pronouncements of crisis and moral decay in rural New England, intersecting in complex and often troubling ways with urban-led reform and even eugenicist projects (Brown 1995: 138, Gallagher 1999).

Landscapes of accommodation

In reality, the purported abandonment of agriculture and rural life was never as complete or straightforward as these jeremiads and elegies would suggest. Rather, variegated shifts in land use across the region reflected complex adjustments to changing markets, technologies, labour patterns, and transportation networks. New England's overall agricultural production actually continued to rise throughout the nineteenth century, the supposed century of decline, not peaking until 1910. The number of farms and farmers did decrease overall, but the steepest drop was not until after the Second World War when a more national-scaled, fossil-fuel-driven food system became dominant. Even today, acre for acre, New England farmland is as productive as almost any in the United States – just on a much smaller scale (Bell 1989: 49). The narrative of decline dramatically flattens the reality of continual readjustment and reassessment by those involved in food production in the region, as well as the persistence and inventiveness of New England's food sector over the more than two centuries since the logic of markets began to favor those with sufficient capital to pursue economies of scale and mechanised efficiencies of production.

The issue of scale, in terms of both the size of the regional land base and its potential fertility, has been central to the ways that farmers and many others in New England have negotiated the tension between their awareness – often a kind of moral consciousness – of ecological limitations and their accommodation or pursuit of the expansive promise at the heart of capitalism. Farmers have learned hard lessons about exceeding the often-limited carrying capacity of their land, and many of their most visible adaptations in the nineteenth and twentieth centuries – letting over-used upland pastures revert to woodlot, shifting fields into grass production to meet urban demand for hay in the age of horse-drawn transport, adding lumber to the mix of products on diversified small farms, specialising in protein and dairy rather than attempting to maintain more diversified crop production – were carefully considered course-adjustments rather than the wholesale retreat proclaimed by many politicians, economists, planners, and reformers (Barron 1984; Judd 2000; Donahue 2007).

New England's farmers have nimbly pursued niche opportunities and products, finding and creating markets for what we now call regional heritage foods (apples, cranberries, maple syrup) through a wide range of strategies including rural and farm tourism, various forms of value-added products and direct marketing, festivals, and cooperatives (Brown 1995: 135–168; Harrison 2006; Paxson 2012: 63–94; Lange 2017: 143–168). Renewed waves of farmers – southern and eastern European immigrants in the late nineteenth and early twentieth centuries, new homesteaders during the Depression and the 1960s and 1970s, college-educated neo-agrarians today – have brought skills and energy to older 'Yankee' farms, continually forging and re-forging relationships with still-expanding urban markets (Belasco 1989; Brown 1995; Kolinski 1995; Paxson 2010). Dairying has remained an essential component of New England's commercial farm economy and pastoral landscape for more than a century and a half, with 'cheese factories,' then liquid milk, then a turn toward specialty foods like ice cream and small-scale artisanal cheese production as strategies for countering the continuing trend toward consolidation and competition from ever-larger producers outside the region (Paxson 2012: 99–103; DuPuis 2002; Valenze 2011).

Over time, farmers have also entered into partnerships with a wide range of allies whose interests overlap with their own in many ways but just as often create new challenges. Tourists and tourism promoters have tended to value pastoral landscapes more in terms of their potential for recreation than for food production *per se* (Brown 1995; Albers 2000; Harrison 2006). Newer, more gastronomic enthusiasms like the farm-to-table movement may risk commoditising and fetishising the very types of small-scale production that have historically suffered most from commercial pressures in larger markets (Trubek 2008; Paxson 2010). The conservation movement that arose in New England in the late nineteenth century initially approached farms and other human elements as intrusions within natural landscapes (Judd 2000; Donahue 2007). That legacy continues to shape land conservation work in the present, although some conservationists are also beginning to recognise environmental values in small-scale working agricultural landscapes. A parallel shift is underway in the realm of historic preservation, although a similar legacy of earlier thinking – in this case, an often hagiographic focus on an idealised past – continues to reinforce rather than challenge the deep-seated declension narrative (Lindgren 2001; Stanton 2017). Meanwhile, state policies and projects have very often privileged conventional and expansive ventures at the expense of smaller, less 'legible' types of farming, while the newest wave of technological utopianism in the agricultural sector has been very largely focused on methods that eschew the challenges of soil and land altogether and look to the laboratory and the digital realm for solutions to the problems of viability that have long plagued farmers in New England and well beyond (DuPuis 1996; Gregg 2001; Strom 2016). Each of these alliances brings its own complex mix of accommodation and resistance to the logic of markets, further complicating the discursive and physical landscapes within which present-day food reformers are maneuvering.

New England currently boasts a very active contemporary network of food activism and food systems planning which has helped to aggregate and coordinate smaller-scale efforts into larger holistic initiatives, including at the state and regional scales (Carroll 2011; American Farmland Trust 2014; Donahue et al. 2014; Ruhf 2015).[2] Despite the vaunted parochialism of much of New England's civic life, such projects clearly reflect a widely shared conception of the region-as-foodshed (or at least potential foodshed). This conception rests in part on a sense that food reform involves rebuilding the kinds of local and regional systems that existed in the not-so-distant past and that have left many traces in New England's memoryscape – traces that reflect the essential ambivalence that people within this region have long felt about the productive capacity of their natural resource base as well as the capitalist expansion they themselves have done a good deal to advance. Contemporary food reform efforts must operate within this contradictory memoryscape, creating continued challenges for the project of uncoupling smaller-scale food production from the effects of larger economies that have emerged, in part, from the very kinds of innovation and adaptation at which New Englanders have long excelled.

Notes

1 The term 'foodshed' was coined by Walter P. Hedden in a 1929 book entitled *How Great Cities Are* but lay dormant until it was adopted by Arthur Getz in a 1991 article in *The Permaculture Activist*. Kloppenburg et al. took their inspiration from Getz's article, articulating an early and quite influential rationale for the rescaling and relocalisation of food systems. For some representative authors, groups, and publications in the more radical areas of the 'food movement,' see the works of neo-agrarianist Wendell Berry, ethnobotanist Gary Nabhan, seed sovereigntist Vandana Shiva, and ecospiritualist Gary Snyder, as well as land, farmworker, and peasants' rights organisations and movements like the Coalition of Immokalee Workers, Food First, the Land Institute, Movimento dos Trabalhadores Sem Teto, and La Via Campesina.
2 State-level initiatives that have been designed to integrate directly or indirectly with region-focused planning and policy include Vermont's Farm to Plate Strategic Plan (2011), the Massachusetts Food Plan (2015), the Maine Food Strategy Framework (2016), and "Relish Rhody," Rhode Island's Food Strategy (2017).

References

Ackerman-Leist, P. (2013). *Rebuilding the foodshed: How to create local, sustainable, and secure food systems*. White River Junction, VT: Chelsea Green.
Adams, B. (2014). *Old and new New Englanders: Immigration and regional identity in the Gilded Age*. Ann Arbor, MI: University of Michigan Press.
Albers, J. (2000). *Hands on the land: A history of the Vermont landscape*. Boston: MIT Press.

Alkon, A.H., and Agyeman, J. eds. (2011). *Cultivating food justice: Race, class, and sustainability.* Cambridge, MA: MIT Press.
Alkon, A.H., and Guthman, J. (2017). *The new food activism: Opposition, cooperation, and collective action.* Berkeley, CA: University of California Press.
American Farmland Trust, Conservation Law Foundation, and the Northeast Sustainable Agriculture Working Group. (2014). *New England food policy: Building a sustainable food system.* Accessed 30 January 2018, www.clf.org/wp-content/uploads/2016/03/1.New_England_Food_Policy_FULL.pdf
Baker, J.W. (2009). *Thanksgiving: The biography of an American holiday.* Lebanon, NH: University of New Hampshire Press.
Barron, H. (1984). *Those who stayed behind: Rural society in nineteenth-century New England.* Cambridge and New York: Cambridge University Press.
Belasco, W. (1989). *Appetite for change: How the counterculture took on the food industry.* Ithaca, NY and London: Cornell University Press.
Bell, M.M. (1989). Did New England go downhill? *Geographical Review,* 79 (4): 450–466.
Bell, M.M. (1996). Stone age New England: A geology of morals. *In:* E.M. DuPuis and P. Vandergeest, eds. *Creating the countryside: The politics of rural and environmental discourse.* Philadelphia: Temple University Press, pp. 29–64.
Brown, D. (1995). *Inventing New England: Regional tourism in the nineteenth century.* Washington, DC: Smithsonian Press.
Brown, D. (2011). *Back to the land: The enduring dream of self-sufficiency in modern America.* Madison and London: University of Wisconsin Press.
Brulotte, R.L., and Di Giovine, M. eds. (2014). *Edible identities: Food as cultural heritage.* London and New York: Routledge.
Carroll, J. (2011). *The real dirt: Toward food sufficiency and sustainability in New England.* Durham, NH: University of New Hampshire Press.
Clark, C. (1992). *The roots of rural capitalism: Western Massachusetts, 1780–1860.* Ithaca, NY: Cornell University Press.
Conforti, J.A. (2001). *Imagining New England: Explorations of regional identity from the Pilgrims to the mid-twentieth century.* Chapel Hill and London: University of North Carolina Press.
Cowan, T., and Feder, J. (n.d.). The Pigford cases: USDA settlement of discrimination suits by black farmers. Congressional Research Service Report for Congress 7–5700/RS20430. Accessed 30 January 2018, http://nationalaglawcenter.org/wp-content/uploads/assets/crs/RS20430.pdf
Delbanco, A. (2001). *Writing New England: An anthology from the Puritans to the present.* Cambridge, MA: Harvard University Press.
DeLind, L.B. (2002). Place, work, and civic agriculture: Common fields for cultivation. *Agriculture and Human Values,* 19 (3): 217–224.
Dicks, B. (2003). *Culture on display: The production of contemporary visitability.* Maidenhead, UK: Open University Press.
Donahue, B. (2007). Another look from Sanderson's Farm: A perspective on New England environmental history and conservation. *Environmental History,* 12 (1): 9–34.
Donahue, B. et al. (2014). *A New England food vision.* Durham, NH: Food Solutions New England.
DuPuis, E.M. (1996). In the name of nature: Ecology, marginality, and rural land use planning during the New Deal. *In:* E.M. DuPuis and P. Vandergeest, eds. *Creating the countryside: The politics of rural and environmental discourse.* Philadelphia: Temple University Press, pp. 99–134.
DuPuis, E.M. (2002). *Nature's perfect food: How milk became America's drink.* New York and London: New York University Press.
Gaggio, D. (2011). Selling beauty: Tuscany's rural landscape since 1945. *In:* N. Bandelj and F.F. Wherry, eds. *The cultural wealth of nations.* Stanford, CA: Stanford University Press, pp. 90–113.
Gallagher, N.L. (1999). *Breeding better Vermonters: The eugenics project in the Green Mountain State.* Hanover and London: University Press of New England.
Genzlinger, N. (2017). Michael Pollan and pangs of guilt, not hunger. *New York Times,* 16 February.
Goldscheider, T. (2015). Shays' rebellion: Reclaiming the revolution. *Historical Journal of Massachusetts,* 43 (1): 62–93.
Grasseni, C. (2011). Re-inventing food: Alpine cheese in the age of global heritage. *Anthropology of Food,* 8 (digital publication; no page range).
Greenfield, B. (2009). *Out of the attic: Inventing antiques in twentieth-century New England.* Amherst and Boston: University of Massachusetts Press.
Gregg, S.M. (2001). Can we 'Trust Uncle Sam'? Vermont and the submarginal lands project, 1934–1936. *Vermont History,* 69: 201–221.
Gross, R.A. ed. (1993). *In debt to Shays: The bicentennial of an agrarian rebellion.* Charlottesville and London: University of Virginia Press.
Guthman, J. (2007). Can't stomach it: How Michael Pollan et al. made me want to eat Cheetos. *Gastronomica. The Journal of Critical Food Studies,* 7 (3): 75–79.
Harrison, B. (2006). *The view from Vermont: Tourism and the making of an American rural landscape.* Lebanon, NH: University of Vermont Press.

Holleran, M. (1998). *Boston's 'changeful times': Origins of preservation and planning in America*. Baltimore: Johns Hopkins University Press.
Holt-Giménez, E. ed. (2011). *Food movements unite!* Oakland, CA: Food First Books.
Holt-Giménez, E. (2017). *A foodie's guide to capitalism: Understanding the political economy of what we eat*. Oakland, CA: Food First Books.
Jackson, W. (1994). *Becoming native to this place*. Berkeley, CA: Counterpoint.
Judd, R. (2000). *Common lands, common people: The origins of conservation in northern New England*. Cambridge, MA: Harvard University Press.
Judd, R. (2014). *Second nature: An environmental history of New England*. Boston and Amherst: University of Massachusetts Press.
Kloppenburg, J. Jr., Hendrickson, J., and Stevenson, G.W. (1996). Coming in to the foodshed. In: W. Vitek and W. Jackson, eds. *Rooted in the land: Essays on community and place*. New Haven, CT and London: Yale University Press, pp. 113–123.
Kolinski, D. (1995). Polish rural settlement in America. *Polish American Studies*, 52 (2): 21–55.
Lange, M.A. (2017). *Meanings of maple: An ethnography of sugaring*. Fayetteville, AR: University of Arkansas Press.
Lapping, M.B. (2001). Stone walls, woodlands, and farm buildings. In: B. Harrison and R. Judd, eds. *A landscape history of New England*. Boston: Massachusetts Institute of Technology Press, pp. 129–144.
Lindgren, J.M. (1993). *Preserving historic New England: Preservationism, progressivism, and the remaking of memory*. New York: Oxford University Press.
Lindgren, J.M. (2001). Preserving the illusion of being transported back into the past. In: B. Harrison and R. Judd, eds. *A landscape history of New England*. Boston: Massachusetts Institute of Technology Press, pp. 285–302.
Lyson, T.A. (2004). *Civic agriculture: Reconnecting farm, food, and community*. Medford, MA: Tufts University Press/ University Press of New England.
Magdoff, F. (2015). A rational agriculture is incompatible with capitalism. *Monthly Review*, 66 (10). Accessed 30 January 2018, https://monthlyreview.org/2015/03/01/a-rational-agriculture-is-incompatible-with-capitalism/
McWilliams, J. (2004). *New England's crises and cultural memory: Literature, politics, history, religion, 1620–1860*. Cambridge and New York: Cambridge University Press.
The New England Magazine. (1897). New Hampshire's opportunity, July, pp. 534–543.
Nissenbaum, S. (1995). New England as region and nation. In: E.L. Ayers, P. Nelson, S. Nissenbaum and P.S. Onuf, eds. *All over the map: Rethinking American regions*. Baltimore: Johns Hopkins University Press, pp. 38–61.
Nissenbaum, S. (1997). *The battle for Christmas: A social and cultural history of our most cherished holiday*. New York: Vintage Books.
Paxson, H. (2010). Locating value in artisan cheese: Reverse engineering terroir for new-world landscapes. *American Anthropologist*, 112 (3): 444–457.
Paxson, H. (2012). *The life of cheese: Crafting food and value in America*. Berkeley, CA: University of California Press.
Peet, R. (1996). A sign taken for history: Daniel Shays' memorial in Petersham, Massachusetts. *Annals of the Association of American Geographers*, 86 (1): 21–43.
Pollan, M. (2010). The food movement, rising. *New York Review of Books*. 20 May. Accessed 30 January 2018, https://michaelpollan.com/articles-archive/the-food-movement-rising/
Pressman, R.S. (1986). Class positioning and Shays' Rebellion: Resolving the contradictions of "The contrast". *Early American Literature*, 21 (2): 87–102.
Richards, L. (2002). *Shays's rebellion: The American revolution's final battle*. Philadelphia: University of Pennsylvania Press.
Rosenbaum, J.B. (2006). *Visions of belonging: New England art and the making of American identity*. Ithaca, NY and London: Cornell University Press.
Ruhf, K.Z. (2015). Regionalism: A New England recipe for a resilient food system. *Journal of Environmental Studies*, 5: 650–660.
Ryden, K.C. (2001). A walk in the woods: Art and artifact in a New England Forest. In: K.C. Ryden, ed. *Landscape with figures: Nature and culture in New England*. Iowa City: University of Iowa Press, pp. 135–157.
Saraiya, S. (2016). Michael Pollan has no answers. *Salon*, 22 February.
Sears, J.F. (1989). *Sacred ground: American tourist attractions in the nineteenth century*. Oxford and New York: Oxford University Press, p. 30.
Seelye, J. (1998). *Memory's nation: The place of Plymouth Rock*. Chapel Hill and London: University of North Carolina Press.
Sims, R. (2010). Putting place on the menu: The negotiation of locality in UK food tourism, from production to consumption. *Journal of Rural Studies*, 26: 105–115.
Stanton, C. (2017). Between pastness and presentism: Public history and local food activism. In: J. Gardner and P. Hamilton, eds. *Oxford handbook of public history*. New York: Oxford University Press, pp. 217–236.

Stilgoe, J.R. (1982). *Common landscapes of America, 1580–1845*. New Haven, CT and London: Yale University Press.
Strom, S. (2016). What's organic? A debate over dirt may boil down to turf. *New York Times*, 15 November. Accessed 30 January 2018, www.nytimes.com/2016/11/16/business/organic-certification-hydroponic-aquaponic-produce.html
Trubek, A.B. (2008.) *The taste of place: A cultural journey into terroir*. Berkeley, CA: University of California Press.
Truettner, W.H., and Stein, R.B. (1999). *Picturing old New England: Image and memory*. New Haven, CT and London: Yale University Press.
Tsing, A.L. (2005). *Friction: An ethnography of global connection*. Princeton, NJ: Princeton University Press.
Valenze, D. (2011). *Milk: A local and global history*. New Haven, CT and London: Yale University Press.
Vartanian, P. (1972). *The Puritan as a symbol in American thought: A study of the New England Societies, 1820–1920* (Unpublished PhD dissertation), University of Michigan.
Waldstreicher, D. (1997). *In the midst of perpetual fetes: The making of American nationalism, 1776–1820*. Chapel Hill and London: University of North Carolina Press.
Weiss, B. (2016). *Real pigs: Shifting values in the field of local pork*. Durham, NC and London: Duke University Press.
Wessels, T. (2005). *Reading the forested landscape: A natural history of New England*. Woodstock, VT: Countryman Press.
Wood, J.S. (2001). New England's legacy landscape. In: B. Harrison and R. Judd, eds. *A landscape history of New England*. Boston: Massachusetts Institute of Technology Press, pp. 251–267.

32
HISTORICISING HISTORICAL RE-ENACTMENT AND URBAN HERITAGESCAPES

Engaging with past and place through historical pageantry, c. 1900–1950s

Tanja Vahtikari

Introduction

Historical reenactment is a popular means of engaging with the past in the contemporary world: re-enactment is prevalent throughout popular culture, ranging from living history performances to museum exhibitions and television history programmes (de Groot 2009; Agnew 2004). In *Theatres of Memory* Raphael Samuel linked the popularity of historical re-enactment with the rise of local history 'do-it-yourself' projects and 'railway preservation mania' in the 1950s (1994: 175–176, 184). The close relationship between re-enactment and local history is important, however, the trajectory of the phenomenon can be traced further back than the 1950s. One medium through which to historicise re-enactment is historical pageantry. In historical pageants various communities and organisations perform episodes of their past that they understood as important. The high period of historical pageantry was the early decades of the twentieth century, but pageants remained a vibrant way for communities to engage with the past up to the 1950s, and even to the present. Historical pageantry became highly popular particularly in the Anglophone world, but was by no means limited to it. One of the largest interwar pageant spectacles was the Pageant of Empire, held at the Wembley Stadium in 1924 in London in association of the British Empire Exhibition.

This chapter has threefold aims. Firstly, as suggested previously, it argues that we should pay more attention to the historical trajectories of heritage and re-enactment. It thus concurs with David C. Harvey's (2001: 320) notion that heritage should not only be seen as a product of recent postmodern economic and societal changes. Secondly, and because of the distinct localness involved in historical pageantry, the pageants have a lot to contribute to discussions concerning memory and place, and the relationship between various, intertwined scales of heritage and senses of belonging. Historical pageants of the twentieth century were staged both in cities and in the countryside. In my own research I have explored historical pageantry in the context of post-war urbanisation and modernisation in Finland, in particular with regard to the cities of Helsinki, the capital of the country since 1812, and Tampere, the major industrial city, founded in 1779. Both cities organised large-scale and popular historical pageants in association with city jubilees, in 1950 and 1954, respectively (Vahtikari 2017, 2019 [forthcoming]). Thirdly, by focusing on urban historical pageantry, this chapter sheds light on the urban variable as part of memory/

heritagescapes. How did historical pageants – even though considered as passing events – work to anchor collective remembrance to the urban landscape?

Historical pageants: what, when, where and why?

Historical pageants were 'paratheatrical events in which performers impersonate(d) figures from the past' (Dean 2014: 1), and in which a community's or an organisation's history was presented in successive episodes. The ultimate high period of historical pageantry was the early twentieth century, a period that witnessed rapid societal change in the urbanising West. Recent scholarship has convincingly shown that historical pageants also remained popular and ambitious undertakings in the very different contexts of the inter-war period, pulling in 'huge crowds of both performers and spectators, all in the name of civic publicity' (Hulme 2017: 23) in both rural and urban locations. During the 1920s and 1930s, historical pageants also became an integral part of the commemorative culture of the First World War (Bartie et al. 2017). There was another high peak in historical pageantry during the immediate post-war period, which in Britain was associated with major events such as the Festival of Britain in 1951 and the coronation of Elisabeth II in 1953 (Freeman 2013).

The pageant form was used by a great variety of institutions and groups to commemorate a wide range of issues and events: to give a few examples, twentieth-century historical pageants were staged by city governments, historical societies, the co-operative movement, women's organisations, or the British Communist Party, which organised a Communist Manifesto Centenary Pageant in Kensington in 1948. Many historical pageants were organised in association with centenary celebrations of cities. The mid-1950s by no means marked the end of the phenomenon, but, as large-scale events that mobilised a large part of the community, historical pageants seem to have given way to other forms of both popular entertainment and historical re-enactment (Freeman 2013). The decision to stage a historical pageant for the opening ceremony of the 2012 London Olympics, watched by a global TV audience of 900 million people, is an example of the enduring significance of historical pageantry as a form of historical re-enactment.

There were two main types of historical pageants. In static, theatre-like pageants a community's or an organisation's history was performed to an audience that remained seated in one place. Important elements of the performance were drama, dance, and music with minimal dialogue. There also existed what may be called 'processional pageants,' in which the historical scenes from the life of a community or an organisation, often staged on platforms of trucks or on horse-drawn carriages, were presented to an audience gathered along a pre-determined route. Viewing a processional pageant, thus, to a certain extent at least, resembled the experience of visiting a museum exhibition: the main difference being that while at a museum, the audience is expected to tour a chronologically ordered route, in a historical pageant the members of the audience would stay in one place, and the historical narrative would unfold in front of their eyes. Because of their arrangement in the form of floating parades, the processional pageants were able to include more scenes than their theatre-like counterparts.

Both pageant types employed pageant masters who were theatre professionals and involved a great number of professional and volunteer actors, using public space as a stage. Many pageants relied more on visual spectacle than on the spoken word and dramatic narrative (Ryan 2007: 78). Historical pageants were represented and mediated in popular culture in numerous visual ways: they were reproduced in the media, film, photographs, and postcards. These visual forms of representation were supplemented by a broad variety of textual representations, such as written programmes, academic and popular history books, and newspaper articles.

In addition to the tradition of dramatic performance, the processional pageants were also linked to another long-term tradition, namely that of urban processions and parades of various kinds, whether or not their roots might be religious, royal, civic, or working-class. The examples given in this chapter draw from historical pageants staged by democratic governments; however, pageants were equally organised by totalitarian regimes. For example, National Socialist Germany commemorated the 700th anniversary of Berlin with a processional historical pageant as part of the wider jubilee programme. The focus of historical representation, instead of being on the more recent metropolitan past, was on medieval Berlin and Germany (Thijs 2008).

Chronology, relationship to modernity, and nationhood

Chronology was the main ordering principle of the episodes in both pageant types. The historical pageant organised in Helsinki in 1950, for instance, followed this principle rather meticulously. The pageant began by introducing two important opening units – one depicting the area of the city prior to the formation of an urban settlement, and the other, the founding of the city. These two scenes grounded the historical narrative presented in the pageant in the local place and, by producing continuity, they legitimised the position and power of the present municipal government. But they also made explicit the striking modernity of the present city when compared to its rural origins. These remained underlying themes in many scenes that followed, which depicted key moments in the urban development (e.g. 'Helsinki the fourth largest harbour in the country'), and a collection of 'firsts,' such as the unit presenting 'The first car' (1903). Several scenes made reference to former rulers, Swedish kings, and Russian emperors, as well as to the key figures of nineteenth-century Finnish nationalism. Equally many units presented the city's involvement in and survival from wars and other disasters (one pageant unit even referred to the most recent conflict, the Finnish involvement in the Second World War), tying the pageant narrative to a wider memory culture of post-war survival. The last step in the chronology was represented by a unit forward-looking called 'The Olympic youth.' The final scenes of Helsinki's pageant broke the chronology, as they presented different municipal institutions, such as the fire and police departments or social welfare institutions, in past-present comparisons. Nevertheless, these units co-produced the pageants' overall message of civic progress and an inclusive urban community. How near to the present the historical narratives unfolded varied significantly between different pageants. The post-war Finnish urban processional pageants usually ended with scenes depicting the contemporary and the future city, thus becoming a medium through which to imagine the future of the community (Vahtikari 2017, 2019 [forthcoming]).

One theme arising from pageants' commitment to chronology and to the narratives of progress has been their relationship with modernity. While historical pageants have been regarded as inherently anti-modern and antiquarian ventures (Glassberg 1990: 149–150), recent research has demonstrated the more nuanced relationship between modernity and anti-modernity involved in staging them. The modernity of early-twentieth-century pageants becomes clear when exploring their nature as modern mass-events, and their employment of modern visual techniques (Ryan 2007: 78). In addition, the research on historical pageantry of the twentieth century is necessarily a commentary on modernity's relationship to the past. As Paul Readman (2005: 199) has noted, history-consciousness mediated through the early-twentieth-century historical pageantry 'was not antithetical to contemporary engagement with "modernity," rather a counterpart to it.' There was a strong emphasis on continuity, which 'was bound up with emphasis on progress' (Readman 2005: 194). While post-war urban pageants were more inclined to make distinctions between the past, the present, and the future, the constructing of continuities, as well as a belief in the exemplary nature of the past, remained important features of them (Vahtikari 2019 [forthcoming]). The similar kind of active negotiation between continuity and discontinuity also characterised the nostalgic experiences that historical pageants facilitated, which involved more than just sentimental denial and longing for lost and idealised earlier times. There was also a utopian element in the pageant nostalgia (Vahtikari 2019 [forthcoming]; Hodge 2011: 120; Pickering & Keightley 2006: 921).

It has been debated with regard to historical pageants of the early twentieth century in England, to what extent their focus was on the narratives of the Empire versus a more locally grounded English nationhood (Readman 2005: 182–190; Ryan 1999). A link between historical pageantry and nationalism exists, and historical pageantry can be viewed as one medium that worked to localise national projects of identity and memory (in the Nordic context, see Aronsson et al. 2008). Historical pageants often included 'hot' signifiers of the nation, such as national flags, national anthems, or references to patriotic war narratives. Nationalism was also embedded in the programmes and practices of historical pageants in more subtle ways – e.g. in the language that was used – which in Michael Billig's (1995) terms could be called 'banal nationalism.' However, the manner in which people taking part in pageants reacted to these symbols and messages obviously varied. In Helsinki, in 1950, two pageant units in particular aimed to evoke patriotic sentiments in spectators: the 'Declaration of Independence in 1917,' where men dressed in white carried national flags, and the 'Year 1939,' a unit depicting the Winter War by showing a nameless group of military reservists. Building these two units into a seamless entity highlighted the patriotic version of Finnish history, as underscored in the conservative newspaper *Uusi Suomi* (12 June 1950):

> The most impressive unit of the pageant was the 'Declaration of Independence 1917,' in which vigorous men dressed in white carried fifty white-and-blue flags, the lion flag in the middle. On a fence sat a war-blinded soldier, to whom his comrade explained how the different units looked. When hearing the sea of flags passing by he stood up and took his hat off. He was still standing there when the next unit, Year 1939, passed by. That year was depicted, and how effectively, by a group of military reservists, some wearing fur hats, some having a package in their hands and some carrying a backpack.

For certain, many spectators shared the patriotic enthusiasm of the *Uusi Suomi* newspaper. But there also existed very different kinds of descriptions of the same event. For example, the Social Democratic newspaper, *Suomen sosiaalidemokraatti* (12 June 1950) reported on the 'Year 1939' scene by simply stating, in the voice of one spectator, 'that is how it was.' Historical pageantry, thus, is an example of, in Rhys Jones and Peter Merriman's (2009: 172) words, 'the multiplicity of nationalist discourses and practices affecting, and affected by, individuals and groups within particular places at specific times.' Meanings associated with national signifiers were sometimes contradictory (Jones & Merriman 2009; see also Jokela and Linkola 2013). Overall, it is important to note that while historical pageantry validated existing hierarchies (Woods 1999: 71), those who participated in pageants were much more than passive recipients in processes of heritage and memory (Vahtikari 2019 [forthcoming]; Freeman 2013; Ryan 2007, Readman 2005).

Beyond the nation and hegemonic narratives: historical pageant as experience

While it is essential to acknowledge the role of historical pageants in constructing the nation, and the empire, where it existed, it is equally important to acknowledge their parallel uses in creating senses of localism. On the one hand, this meant that the different conceptions of community and senses of belonging articulated in the context of historical pageants were not in any opposition to each other (Readman 2005: 178). On the other hand, the inter-war and post-war periods saw a more conscious turn to local community by means of historical pageants and the narrative themes of historical local self-government that they presented. This may be seen as a response to the contemporaneous tendencies towards national-level centralisation (Hulme 2017: 9; Freeman 2013: 427). The distinctive localness of historical pageants may also be seen in how they 'played an important role in supporting local community identities in the aftermath of collective trauma' of the First World War, thus becoming local 'sites of mourning' (Bartie et al. 2017). The same holds true of post–Second World War historical pageants, at least in the case of Finland, where urban pageants provided alternative arenas for the commemoration of the war, and for dealing with collective and individual traumas below the official state level of memory, which in the immediate war years was adapted to the realpolitik of Finnish-Soviet relations (Vahtikari 2017). The focus on the localism and voluntarism involved in historical pageants can also make visible social groups whose histories and experiences differed from nationally authorised heritage narratives.

Like research that deals with more recent heritage phenomena, studies on historical pageantry have often emphasised the representational side of pageants, and the messages they conveyed: which versions of the past were validated by official producers of heritage to the public. As with any memorial projects, historical pageants were equally about what was forgotten and deliberately disregarded as well as what was remembered. The pageant form as representation was necessarily selective: for example, the historical pageant organised in Helsinki in 1950 included 52 units, out of which some were ahistorical (Figure 32.1). In addition to these official, selective readings of the past and community that were put forward in the context of the pageants, there were co-existing private, vernacular, and class/gender-bound understandings that were not officially represented. Often, and not surprisingly, it was the difficult memories and the memories of minority groups that were marginalised as part of the public representation. The 1950 historical pageant in Helsinki, while otherwise keen on drawing survival lessons from previous wars and disasters, failed to mention the Civil War of 1918, which in the post-war period still represented a divisive experience within the Finnish society (Vahtikari 2017).

The level of spontaneity in the organisation of pageants varied in different settings. In Helsinki, the preparation of the pageant was fairly top-down. The city funded the whole event, except for the programme leaflet, which was paid for by local firms. The organisation was overseen by a city-government-appointed jubilee committee, including many prominent figures in culture and finance, as well as professional historians, and in particular by one of its 20 sub-committees. In addition, the city appointed two pageant masters,

Figure 32.1 Working-class history was represented by several units in the historical pageant staged as part of Helsinki's quadricentennial celebrations, 11 June 1950. The Civil War of 1918 between the Whites and the Reds, still a divisive experience within the post-war Finnish society, was not included amongst the pageant themes.

Source: Helsinki City Museum (Eino Heinonen).

who both were theatre professionals, to write the script and to direct the pageant. Despite this organisation from above, Helsinki's pageant also involved a fair amount of citizen participation and voluntarism. Around 800 people, including both professionals and amateurs, such as university students, gymnasts, and members of local societies, participated as actors. Some 100,000 people gathered on the city streets to follow the pageant, and an additional 40,000 participated in it at the Olympic Stadium, where the pageant became part of the city's wider 400th jubilee. This was quite a large audience in a city containing fewer than 400,000 inhabitants at the time, and while historical pageants were firmly part of middle-class urban culture, their audience was wider than that. In their capacity to bring together people and groups from different social backgrounds and possessing different goals, pageants may be seen as expressions of civil society (Vahtikari 2019 forthcoming). Historical pageants of the first part of the twentieth century were produced in the manner of huge theatre plays, but while adding elements of a very large and mostly amateur cast, interaction with public urban space, and drawing from competing local narratives of memory and history.

As with any study on heritage, the research on historical pageants involves the risk of privileging representation and pageants' symbolic functions in the construction of the nation, empire, or local community identity over interpreting pageants as bodily and emotional experiences, and how important these experiences were to an understanding of history. Framing historical pageantry in terms of heritage as practice and performance, and as something that is produced in 'the embodied and creative uses of heritage generated by people' (Haldrup & Bærenhold 2015; see also Crouch 2010), however, is important to our understanding of historical pageantry and its popularity. While history in pageants was represented, and subject to authorised heritage discourses, and the power/knowledge effects they created (Smith 2006), it was also felt, remembered, and lived 'in the dimensional complexities of the experiences' they created (Waterton & Watson 2015). Historical pageants were where the passage of time and the past-present-future continuities could be experienced (Vahtikari 2019 forthcoming).

The memorial landscape and pageant histories

Once constructed, memorials and urban memorial landscapes take on an apparent permanence, which masks the often conflicted conditions of their creation (Jordan 2006: 18; Dwyer & Alderman 2008: 168). Urban processional pageants were passing events, which traversed urban space, and lasted for only a few hours at a time. They did not produce any permanent memorials or sites of heritage. This notwithstanding, it may be argued that historical pageants became an integral part of urban memorial landscapes, which could be better seen, to borrow from Dwyer and Alderman (2008: 168), 'as open-ended, conditionally malleable symbolic systems, ones that are fashioned here-and-now in order to influence a near-at-hand tomorrow.' As they were staged to the very same city they represented, historical pageants provided physical linkage with the past and anchored collective remembrance to the urban landscape.

The historical pageant organised by the industrial city of Tampere in 1954 (Figure 32.2), serves as an example here. Not surprisingly, the industrial past of the city figured prominently in the pageant, and it included several units related to the history of the city's largest employer, the Finlayson cotton factory. The route of the pageant also passed by the Finlayson factory, placing the pageant's historical narrative in a relationship to the material city and its tangible heritage. As a result, a co-constructed experience of time and place, historical narrative and urban space was formed (Vahtikari 2019 [forthcoming]; Waterton & Watson 2015).

For those organising urban processional pageants, the route of the pageant in the public urban space, what was to be included and what was not, was a matter of as careful a consideration as that given to the building of the pageant programme. The route that was selected could reaffirm 'the symbolic centrality of certain streets, squares and buildings' (Gunn 2000: 181), and established sites of concentrated collective remembrance. In post-war Helsinki, this meant including key national and civic monuments on the pageant's route, but also including, in conformity with the strong social-democratic presence in the post-war

Figure 32.2 One of the displays depicting the Finlayson factory passing by the Finlayson factory gate at the 1954 historical pageant in Tampere.

Source: Museum Centre Vapriikki, Tampere.

city government, sites of memory associated with the history of the moderate left (Vahtikari 2019 forthcoming). In the process, some ordinary streets and urban spaces were also invested with a special meaning as part of the memorial landscape. The other side of the coin was the obscuring of undesirable urban areas from the public view, as was the case with the Ancoats area in the context of the Manchester Civic Week during the interwar period (Hulme 2013: 89). Much of the literature that discusses historical pageantry points out its nature as a form of civic education. That historical pageants took place in the public urban space and offered a way to associate these civic lessons with the material city.

As authenticity was one of the key concerns of historical pageantry, it was often the spatial design and the urban spaces in which pageants took place that served to authenticate the historical narratives included in the pageant programmes. Performing collective remembrance of the urban community through historical pageantry did not only shape and re-conceptualise time but also place and memoryscape (Dwyer & Alderman 2008: 173–174; Ryan 2007: 72). Professional historians often had an advisory role in deciding on which episodes should be included in historical pageants, and in this capacity they were able to control – at least to some extent – the overall narrative presented. Even though the degree of realism of pageants' programmes varied (Bartie et al. 2017), there was usually an endeavor to present historically authentic interpretations of the past. Historical pageantry provided professional historians with a means to spread up-to-date historical knowledge to wider audiences. As with many other elements of 'exhibitionary complex' (Bennett 1995), in historical pageants the two objectives of education and entertainment intertwined, and professional historians sometimes voiced concerns about authenticity (Vahtikari 2017). As Jeremy De Groot notes (2009: 105), when discussing contemporary re-enactment as a collectivised experience, it 'is defiantly outside mainstream professional ways of thinking about the past.' While professional historians in the post-war period were often involved in staging historical pageants, the realisation that control over historical interpretation was slipping out of their hands may have been part of their authenticity concerns. However, it is important to note that like many contemporary reenactors (Agnew 2004: 330), people taking part in historical pageants took their history seriously, and in fact shared the academic historians' concern regarding authenticity.

Conclusion: an early form of historical re-enactment

Even though there are some differences between contemporary historical re-enactments and historical pageants of the first half of the twentieth century, historical pageantry fits well within the framework of historical re-enactment, consisting of different history-themed genres 'linked by common methodologies, modes of representation, and choice of subject matter' (Agnew 2004: 327). Historical pageantry of the first half of the twentieth century and many contemporary historical re-enactments 'raise similar questions about the ways in which the past is represented, and "experienced"' (Freeman 2013: 425). While usually concerned with authenticity, they all involve imaginative and outside-the-mainstream ways of historical representation. An ambiguous relationship with academic history also characterises both contemporary historical re-enactments and the historical pageantry of the first half of the twentieth century.

Historical pageants, with their participatory nature, which allowed many groups coming from different societal backgrounds to take part and a large group of volunteers to be involved, served to democratise historical knowledge, even though the main narratives – which and whose pasts were to be included or excluded – often remained controlled from above. For those who participated, it was always possible to ignore the pageant messages, or at least to interpret them differently from how they were conceived by the organisers. Vanessa Agnew (2004: 327) distinguishes the 'winning combination of imaginative play, self-improvement, intellectual enrichment, and sociality' behind the contemporary popularity of reenactment. When adding a strong emphasis on local community to the equation, the same list of attributes could be easily adjusted to describe the earlier historical pageants. They were at the same time about learning history and entertainment, and about pursuing authenticity and possibilities for play, escapism, and fantasy. Some pageant performers even continued to wear their costumes weeks after the event had ended, 'producing a blurring of reality and fiction' (Ryan 2007: 75).

In Britain, historical pageantry was closely associated with the emerging urban preservationist moment, as often the same people contributed to both (Freeman 2013: 436). The connection between historical pageantry and tangible heritage would certainly warrant more research in other contexts as well. In post-war Finland, historical pageantry and urban conservation, albeit the latter mentioned still modest in the early-1950s, were both responses to urban transformation and modernity: both were concerned about the past and offered ways to negotiate the relationship to modernity. While place had a distinctive role in historical pageantry, pageant heritage was and is what scholars today would call intangible heritage. Historical pageants thus encourage us to ask, not only how they relate to historical re-enactment, but also how urban intangible heritage was understood before the more recent accelerating professional and academic interest in it arose.

References

Agnew, V. (2004). Introduction: What is reenactment? *Criticism*, 46 (3): 327–339. Accessed 8 September 2017, https://digitalcommons.wayne.edu/criticism/vol46/iss3/2

Aronsson, P., Fulsås, N., Haapala, P., and Jensen, B.E. (2008). Nordic national histories. *In:* S. Berger and C. Lorenz, eds. *The contested nation. ethnicity, class, religion and gender in national histories*. Basingstoke: Palgrave Macmillan, pp. 256–282.

Bartie, A., Fleming, L., Freeman, M., Hulme, T., Readman, P., and Tupman, C. (2017). 'And those who live, how shall I tell their fame?' Historical pageants, collective remembrance and the first world war, 1919–39. *Historical Research*, 90 (249): 636–661. Doi:10.1111/1468-2281.12189.
Bennett, T. (1995). *The birth of the museum: History, theory, politics*. London and New York: Routledge.
Billig, M. (1995). *Banal nationalism*. London: Sage Publications.
Crouch, D. (2010). The perpetual performance and emergence of heritage. In: E. Waterton and S. Watson, eds. *Culture, heritage and representation*. Aldershot: Ashgate, pp. 57–71.
Dean, J.F. (2014). *All dressed up: Modern Irish historical pageantry*. Syracuse, NY: Syracuse University Press.
De Groot, J. (2009). *Consuming history: Historians and heritage in contemporary popular culture*. London: Routledge.
Dwyer, O.J., and Alderman, D.H. (2008). Memorial landscapes: Analytical questions and metaphors. *Geoforum*, 73 (3): 165–178. Accessed 10 September 2017, https://doi.org/10.1007/s10708-008-9201-5.
Freeman, M. (2013). 'Splendid display; pompous spectacle': Historical pageants in twentieth-century Britain. *Social History*, 38 (4): 423–455.
Glassberg, D. (1990). *American historical pageantry: The uses of tradition in the early twentieth century*. Chapel Hill: University of North Carolina Press.
Gunn, S. (2000). *The public culture of the Victorian middle class: Ritual and authority and the English industrial city, 1840–1914*. Manchester and New York: Manchester University Press.
Haldrup, M., and Bærenholdt, J.O. (2015). Heritage as performance. In: E. Waterton and S. Watson, eds. *The Palgrave handbook of contemporary heritage research*. New York: Palgrave Macmillan, pp. 52–68.
Harvey, D.C. (2001). Heritage pasts and heritage presents: Temporality, meaning and the scope of heritage studies. *International Journal of Heritage Studies*, 7 (4): 319–338.
Hodge, C.J. (2011). A new model for the memory work: Nostalgic discourse at a historic home. *International Journal of Heritage Studies*, 17 (2): 116–135.
Hulme, T. (2013). *Civic culture and citizenship: The nature of urban governance in interwar Manchester and Chicago*. PhD. thesis, University of Leicester.
Hulme, T. (2017). 'A nation of town criers': Civic publicity and historical pageantry in inter-war Britain. *Urban History*, 44 (2): 270–292.
Jokela, S., and Linkola, H. (2013). 'State idea' in the photographs of geography and tourism in Finland in the 1920s. *National Identities*, 15 (3): 257–275. Doi:10.1080/14608944.2013.779644.
Jones, R., and Merriman, P. (2009). Hot, banal and everyday nationalism: Bilingual road signs in Wales. *Political Geography*, 28 (3): 164–173.
Jordan, J. (2006). *Structures of memory: Understanding urban change in Berlin and Beyond*. Stanford, CA: Stanford University Press.
Pickering, M., and Keightley, E. (2006). The modalities of nostalgia. *Current Sociology*, 54 (6): 919–941.
Readman, P. (2005). The place of the past in English culture c.1890–1914. *Past and Present*, 186: 147–199.
Ryan, D.S. (1999). Staging the imperial city: The pageant of London, 1911. In: F. Driver and D. Gilbert, eds. *Imperial cities: Landscape, display and identity*. Manchester: Manchester University Press, pp. 117–135.
Ryan, D.S. (2007). 'Pageantitis': Frank Lascelles' 1907 Oxford historical pageant, visual spectacle and popular memory, *Visual Culture in Britain*, 8 (2): 63–82.
Samuel, R. (1994). *Theatres of memory. Vol. 1, past and present in contemporary culture*. London: Verso.
Smith, L. (2006). *Uses of heritage*. Abington Oxon and New York: Routledge.
Suomen sosiaalidemokraatti. (1950). Menneiden helsinkiläispolvien elämä esiintyi nykyhetken helsinkiläisille suuressa historiallisessa kulkueessa, 12 June.
Thijs, K. (2008). *Drei Geschichten, eine Stadt: Die Berliner Stadtjubiläen von 1937 und 1987*. Cologne: Boehlau Verlag.
Uusi Suomi. (1950). Vaikuttava historiallinen kulkue, 12 June.
Vahtikari, T. (2017). 'Washing away the dirt of the war years': History, politics and the reconstruction of urban communities in post-world war II Helsinki. In: H. Kaal and S. Couperus, eds. *(Re)constructing communities in Europe, 1918–1968: Senses of belonging below, beyond and within the nation-state*. Abingdon, Oxon and New York: Routledge, pp. 65–84.
Vahtikari, T. (2019). Post-war urban pageants in Finland: Representation, participation and power. In: S. Gunn and T. Hulme, eds. *New approaches to governance and rule in urban Europe Since 1500*. Abingdon, Oxon and New York: Routledge.
Waterton, E., and Watson, S. (2015). A war long forgotten: Feeling the past in an English country village. *Angelaki*, 20 (3): 89–103.
Woods, M. (1999). Performing power: Local politics and the Taunton pageant of 1928. *Journal of Historical Geography*, 25 (1): 57–74.

PART VII

Ritual

Eerika Koskinen-Koivisto

Introduction

The concept of 'ritual landscape' emerged in archaeology in the 1980s to conceptualise sacred archaeological complexes of the distant past with a variety of constructions and ceremonial artefacts, which were used as evidence for ritual activity. In the following decades, its meaning became more ambiguous, since many saw it as imposing modern spatial constructs on prehistoric perceptions (Robb 1998). In ethnographic research, the study of ritual and landscape has focused on spatialised ritual practices and constructions of mythic landscapes (e.g. Hastrup 2008). Ethnographic studies of ritual landscapes have emphasised the interactions between human and non-human actors in the making of ritual landscapes exploring how landscapes are produced, and how they, in their turn, produce certain types of being (Virtanen et al. 2017).

According to social anthropologist Paul Connerton (1989), social memory is embodied and transmitted through social interaction and performance such as community rituals. The body and its ritualised and habitual movement is also essential for the capacity to remember landscapes. Pilgrimage, for example, constitutes a liminal space that connects the pilgrim not only to the past but also to future. Many pilgrims wish to leave some kind of physical trace of their journey, for example by building a pile of stones as a route marker by the road or by adding a stone to an already existing marker (Peelen & Jansen 2007: 83; Österlund-Pötzsch 2011).

In this section, contributors analyse mnemonic rituals embedded in particular places and the materialised and embodied dimensions of these rituals. The landscapes vary from historical heritage sites to urban environment. Some of them have appeared to commemorate contemporary legends and heroes, whereas other sites have continued to act as scenes of rituals and beliefs that have transformed over time.

In this first chapter, Hilary Orange and Paul Graves-Brown, study the creation of tribute sites for singer David Bowie by his fans and contextualise them in relation to other sites of celebrity pilgrimage. These shrines are subject to ongoing processes of development, elaboration, and destruction, as land and property owners and managers, authorities, and heritage professionals answer to the spontaneous appearance of flowers, photographs, drawings and paintings, greeting cards, and LP covers. Hayley Saul, on her part, discusses another form of contemporary pilgrimage where the memory work builds on spontaneous and momentary experience: Tibetan Buddhist pilgrimage to Himalaya's hidden valleys. The starting point for her reflection is that the concept of memory yields Tibetan Buddhist philosophy (*Terma*), which considers existence as a constant state of flux. What follows is that the experience of pilgrimage is not directed towards the past, but to enlivening enduring visualisations of an immediate present.

In their study of English and Welsh churchyards, Howard Williams and Elizabeth Williams scrutinise how the practice of cremation has transformed the churchyard. They argue that as ashes can be staged to symbolise the absence, the cremation facilitates the creation of varied and distinctive landscapes of death and memory and a more nuanced commemoration of the dead. Ana Mayorgas addresses the evolution of ritual commemoration and memory landscapes in ancient Rome, showing how different variables such as political context and religious conceptions shape social recollection. Her foray into the present memory of a remote past brings a fascinating nuance to the collection of chapters not only in this section, but in the wider volume, by challenging the notion that memory is something that is remembered by eyewitnessing practices or by individuals directly experiencing events that engender recollection.

Ceri Houlbrook introduces the concept of ritual recycling, illuminating it through the case of ritual landscape of Isle Maree from the Scottish Highlands. Tracing the pagan and Christian layers of the natural space in literary and narrative sources as well as archaeological research at the site, she offers a dynamic view to the rituals taking place in the Island. Reuse of materials serves as a powerful metaphor for the past being incorporated in the present. The last chapter of the section by Jeanmarie Rouhier-Willoughby addresses the contested memory of the Holy Springs of Lozhok in Western Siberia, Russia, by drawing on scholarship on legend. Exploring the different versions of the legend, she demonstrates the extensity of this tradition; the spring is a locus for members of different communities to reconcile the conflicting narratives about the past in the present. This contribution challenges the accepted memory paradigms of Soviet Russia by incorporating a sense of the sacred into the perception of the troubled past of the region.

The texts in this section illuminate the processes of emergence, adaptation, and recycling of spatialised ritual practices in the hands of varying collectives as agents of memory of different times: citizens, religious groups, villagers, or fans. Ritual landscapes have characters of both repetition and ephemeral, continuity, and transformation. Hopefully, readers will be able to fully immerse themselves in these dynamic and fascinating realms of memory and tradition, experiencing place through the sacred dimension of the pervasive present-past.

References

Connerton, P. (1989). *How societies remember*. Cambridge: Cambridge University Press.
Hastrup, K. (2008). Icelandic topography and the sense of identity. *In:* M. Jones and K.R. Olwig, ed. *Nordic landscapes: Region and belonging on the northern edge of Europe*. Minneapolis: University of Minnesota Press, pp. 53–76.
Österlund-Pötzsch, S. (2011). The ephemeral act of walking: Random reflections on moving in landscapes of memory (loss). *Ethnologia Scandinavica*, 41: 110–127.
Peelen, J., and Jansen, W. (2007). Emotive movement on the road to Santiago de Compostela. *Etnofoor*, 20 (1): 75–96.
Robb, J.G. (1998). The 'ritual landscape' concept in archaeology: A heritage construction. *Landscape Research*, 23 (2): 159–174.
Virtanen, P.K., Lundell, E., and Honkasalo, M-L. (2017). Introduction: Enquiries into contemporary ritual landscapes. *Journal of Ethnology and Folkloristics*, 11 (1): 5–17.

33
'MY DEATH WAITS THERE AMONG THE FLOWERS'

Popular music shrines in London as memory and remembrance

Hilary Orange and Paul Graves-Brown

Introduction

The singer, musician, actor, and producer David Robert Jones (David Bowie) was born in Brixton, London on 8 January 1947 and died in New York on 10 January 2016 at the age of 69. Reportedly, he has no grave, but was cremated in New Jersey and his ashes were scattered in Bali. The problem for David Bowie fans was and is that, except for his apartment on Lafayette Street, New York, no obvious locus exists for mourning in the aftermath of his death. Thus, unofficial memorials have been created by fans in London, New York, and Berlin. These, in turn, have become focal points for the remembrance of Bowie's life and music: places of musical pilgrimage, where memory is given materiality in the form of tributes and graffiti.

For us, the absence of a grave reflects the ambiguity that was a hallmark of Bowie's chameleon identity. Indeed, 'David Bowie' was essentially a theatrical role adopted by David Jones, upon which he superimposed and discarded other roles, such as the flame-haired, sexually ambiguous Ziggy Stardust, or the Thin White Duke. By the time of his 1983 Serious Moonlight tour, Bowie had achieved megastardom and at that time 'What David wanted was to have the past erased' (Zanetta & Edwards 1986: xiii). Yet in later life Bowie accumulated an archive of over 70,000 items of memorabilia that formed the basis of the *David Bowie is* exhibition, which opened in March 2013 at the Victoria and Albert Museum in London. Even his death seems to have been carefully stage managed, as the radio presenter Paul Gambiccini commented he was the 'Only artist to make art out of his death' (London Live 2016), referring to Bowie's last album *Blackstar*.

Like Freddie Mercury before him (see later), Bowie seems to have abhorred the concept of a pilgrimage associated with a last resting place, and perhaps wished to retain a degree of control of his identity and image even after death, reflecting an awareness of a 'post-self' that 'continues to change and grow after death' (Naylor 2010: 251). The rise in popularity of cremation has also created a degree of uncertainty about where the famous dead should be commemorated. In some cases, the location where ashes were scattered is accessible. Where death has been traumatic and unexpected, the place of death itself can serve as a focus; roadside shrines have been a common phenomenon for many years (MacConville 2010) and when Diana, Princess of Wales, died in Paris in 1997, flowers were left at the entrance to the Pont de l'Alma tunnel, the scene of the road traffic accident that caused her death. Such shrines or tribute sites can also be compared to public, high profile graves which draw in pilgrims and tourists; for instance, Elvis Presley's grave at Graceland, Memphis has been visited by more than 20 million people (Graceland website 2016).

In his account of Graceland as a 'locus sanctus,' Vikan (1994: 155) says 'Unlike the tourist, who goes to places mostly to see, the pilgrim has a distinctly tactile notion of travel.' While tourists come to gaze (Urry 1990), pilgrimage is both tactile and transcendental, it is about seeking a deeper physical, emotional, or spiritual connection with the object of veneration (Margry 2008: 17; Vikan 1994). Pilgrimage is also a form of self-expression that raises 'fan cultural capital' (Hills 2002), and has also been positioned as a form of healing (Recuber 2012). Such characteristics are then connected to and amplified by the process and performance of tribute giving and graffiti making. From the selection of the tribute object or graffiti, to the carrying of materials and eventual placement, observation of the scene, thoughts or prayers directed to the deceased, and physical interaction with the tribute site and other pilgrims/fans. The same is true of other memorials to the dead; 'Touching war memorials, and in particular, touching the names of those who died, is an important part of the rituals of separation that surround them' (Winter 1995: 113).

Therefore, in creating shrines, fans do transitional, emotional, and material work and are, in Alderman's terms, both 'author' and 'agent' of the celebrity's memory (see Alderman's 2002 discussion of Graceland). In the literature such tribute places are often referred to as spontaneous shrines, though, as we will discuss, most of the David Bowie shrines are sited due to one or other criteria including historical/biographical association, symbolic value, and access.

However, it is not only fans who make choices. There are other agents, such as land and property owners/managers, authorities, and heritage professionals, who are behind the creation, maintenance, dismantling, and destruction of memorials and tribute sites. In this chapter, we discuss the initial findings of our longitudinal study of Bowie tribute sites in London (Graves-Brown & Orange 2017) and seek to contextualise them in relation to other sites of celebrity pilgrimage. The shrines are subject to ongoing processes of development, elaboration, and destruction whose outcome is uncertain and we end the chapter by discussing indications that some are moving from being unofficial to becoming official.

The Bowie tribute sites

There are three principal shrine sites to David Bowie in London, as well as sites in New York and Berlin. After his death, tributes were left outside his apartment in New York (Kelly Brit pers. comm.) and in Berlin, where Bowie lived in the late 1970s, tributes were left at his former home Hauptstraße 155, Schöneberg, and at the Hansa TonStudio, a recording studio near Potsdamer Platz. The tributes that we have seen in London are often floral, but there have also been a plethora of other items including photographs, drawings and paintings, greeting cards, and LP covers. Stranger items that we have seen include an old vinyl record deck and containers of milk, seemingly reflecting the era when the cocaine-addicted Bowie lived on little else (Yentob 1975; Jones 2017). Most intriguing, perhaps, are the sealed envelopes addressed to the star (sealed envelopes have also been observed at other memorial sites, such as the 2010–2011 Canterbury earthquakes in New Zealand [Sarah Murray, pers. comm.]).

In London, the 'oldest' place of Bowie pilgrimage is Heddon Street, adjacent to Regent Street (Graves-Brown 2012). We know that Bowie-related graffiti has been present in the phone box on the site since at least 1985 and that fans have visited the site since the 1970s (Jones 2017). Heddon Street was the locus of the cover photo shoot for *The Rise and Fall of Ziggy Stardust and the Spiders from Mars* album of 1972. The choice of venue was largely coincidental as photographer Brian Ward had his studio in the street. Bowie was photographed in a series of poses and the images taken outside No. 23 Heddon Street and inside the K2 phone box were used on the cover. The original phone box was replaced by a more modern one in the early 1980s but then re-replaced by a red K6 box around 1996 and continues to attract graffiti and tributes. There was a long campaign to erect some form of plaque to Bowie/Ziggy, originally proposed as part of the Westminster City Council's Green Plaque scheme (see Graves-Brown 2012). However, it eventually fell to the Crown Estate, the principal property owner in the street, to erect one of its black plaques to 'Ziggy Stardust' in March 2012, on the 40th anniversary of the photo shoot

(Tallulah Morris, pers comm). As the *Daily Telegraph* (27 March 2012) noted, the plaque is unusual in commemorating a fictitious character, Ziggy rather than David, other examples of such sites around the world are those dedicated to Ianto Jones (Beattie 2014), Shakespeare's Juliet, Goethe's Werther, Sherlock Holmes, and Lara Croft. On Bowie's death, the site attracted many tributes both outside No. 23 and in the phone box, but on the subsequent anniversary of his death in 2017, there were in fact few tributes left at the site.

The second and by far the most popular location is the mural of Aladdin Sane/Bowie painted by Australian street artist Jimmy C in 2013 on the wall of Morley's Department Store, Tunstall Road, Brixton (Figure 33.1). Jimmy C is the alias of Australian artist James Cochran, whose work can be seen in several cities around the world. Since Bowie's death, the site has been the focus for offerings of flowers, other tributes, and graffiti. As at Elvis' Graceland and other sites, such as Abbey Road Studios, Bowie shrines have attracted extensive amounts of graffiti. Rapidly the wall around the Aladdin Sane mural, and parts of the mural itself, were covered with graffiti, extending as far as the corner of Bernay's Grove (some 30 metres). Often the graffiti takes the form of 'RIP Starman' or 'Thank you for the music,' but some evince a more spiritual or even political sentiment; at Tunstall Road Brixton, graffiti includes 'You were my best friend during the hours of darkness,' 'thank you for giving my [dad] the confidence,' 'Thanks for saving me when I needed you most,' and also 'Thank you for freeing society.' At Heddon Street, one pilgrim wrote 'When I first saw you photographed in this spot, the shock to my system was great enough to give me the courage I needed to pursue what I wanted from life rather than what I was told I should want.' The department store has chosen to leave most of the graffiti in place, although that around the window display adjacent to the entrance of the store was removed and painted over in the summer of 2016. Moreover, a sign was

Figure 33.1 Mural of Aladdin Sane/Bowie in Tunstall Road, Broxton. Mural painted by Jimmy C. in 2013.
Source: Photo by Hilary Orange.

placed on the mural asking visitors not to graffiti the mural itself and, in January 2017, immediately before the anniversary of Bowie's death, a polycarbonate screen was installed to protect the mural.

The bandstand in Croydon Road Recreation Ground, Beckenham, is the third major site of pilgrimage. Built between 1880 and 1921 and made by Glasgow firm of McCallum & Hope Ltd, the bandstand was the focus of the Beckenham Free Festival in August 1969. The festival was organised by Bowie and other members of the Beckenham Arts Lab, which frequently met at the nearby Three Tuns pub (now an Italian restaurant). The Beckenham Arts Lab was inspired by the Arts Lab founded in 1967 by Jim Haynes at 182 Drury Lane, London. Since 2015, an annual festival has been organised at the site, both to celebrate David Bowie's life and to raise funds for the restoration of the bandstand. After Bowie's death, and on the first anniversary, a large number of tributes were left on the bandstand, partly facilitated by the organisers of the festival who, in January 2017, provided a visitors' book for people to sign. Tributes were also left outside the former Three Tuns and a small number of floral bouquets and other objects (such as album covers and candles) were left at Bowie's childhood homes, No. 4 Plaistow Grove, Bromley, and 40 Stansfield Road, Brixton.

Why Bowie tribute sites are where they are

Regarding Bowie tribute sites (and the comparative sites that we have studied, see the next section) we have determined that there are three categories of tribute site: 1) sites pertaining to the artists' history or biography; 2) symbolic sites, often associated with secondary artworks; and 3) sites associated with song lyrics.

The homes of the famous are often the locus for memorials either formally (e.g. the English Heritage blue plaque scheme) or informally, as in the 'sea of flowers' left outside Kensington Palace in 1997 in tribute to Princess Diana (Monger & Chandler 1998). The tribute sites in London are largely associated with the boroughs of Bromley and Lambeth, where Bowie lived as a child and young adult and thus there is a biographical link to the artist. It is clear though that some obvious biographical sites have not become significant tribute attractors. Only a small number of tributes were left at Bowie's birthplace, 40 Stansfield Road, despite only being about 0.4 miles from Tunstall Road. We have found no evidence for tributes at the site of Haddon Hall, a large gothic villa on Southend Road, Beckenham, where Bowie lived with his wife Angela Bowie (nee Barnett) from 1969 to 1972, and where his career was, to all intents and purposes, launched. This site has presumably avoided tribute activity as the hall was demolished in the early 1980s to make way for a block of flats.

Places also become strongly associated with musical artists through song lyrics, for example, the Beatles wrote songs about two places in Liverpool: 'Strawberry Field,' a children's home, and 'Penny Lane' ('Penny Lane' and 'Strawberry Fields Forever' both released in 1967). In the case of Bowie, the Croydon Road Recreation Ground (immortalised by Bowie in the track 'Memory of a Free Festival,' recorded in 1970) is the only tribute site associated with his song lyrics as well as being a biographical site.

Tunstall Road, however, has no direct biographical link to the artist beyond being a street in the Borough of Lambeth. This tribute site, we suggest, is located where it is due to its existing symbolic value, wherein the artwork of the mural has become an attractor to further tribute activity and fan/media attention. Its exact location appears to have been accidental in that Jimmy C sought a Brixton location, the wall provided a suitable canvas, and Tunstall Road is placed in a central high street location. This third category of site type – a symbolic site – thereby provides a far looser association with the artist's life and career and rather represents a more random, or fluid process whereby place can gain value, or indeed where place-makers can manufacture value through the deliberate creation of monuments to the living and to the deceased.

Heddon Street shares two of these characteristics (biographical and symbolic), connected as it is to the night of the photo shoot and the cover artwork. A fourth site category could include private shrines and tributes (e.g. those erected by family members, close friends, or fans in private contexts). However, so far within this research, we have not found evidence of these.

The memorial locations that we outline here rely on one unifying theme, that to be created and reinvested with meaning over time they must be publicly accessible and usable as an agora or space of mourning. In this we can compare Elvis' grave at Graceland, a National Historic Landmark and a museum and visitor attraction, to the inaccessible grave of Princess Diana which is located on an island within an ornamental lake within Althorp Estate's pleasure gardens, Northamptonshire. As well as being publicly accessible there also have to be usable features (pavements, walls, trees, and street furniture) on to which tributes can be placed, hung, written, or otherwise fixed. For instance, tributes have been placed on grass and pavement areas, tied to or pushed between railings or placed within structures while graffiti has been written on walls, feeder pillars, and structures. The Heddon Street phone box is a three-dimensional tribute object and perhaps, given the prevalence of mobile phone usage, is more often used as a tribute space than a place to make phone calls. Another characteristic of the Bowie sites (and many sites to other musicians as we will go on to discuss later) is that they appear on pedestrian streets (Heddon Street and Tunstall Road) or in parks/gardens. Therefore, the sites present no hindrance in terms of overlapping with highway boundaries and becoming a distraction to motorists (see Sanders 2010 for a discussion on local authority responses to roadside shrines).

Other tribute sites to musicians and their characteristics

The veneration of the famous, both real and imaginary, is not a new phenomenon. In the eighteenth century, 'Werther mania' swept Europe in the aftermath of the publication of Goethe's *The Sorrows of Young Werther* in 1774. 'At Garbenheim, an innkeeper heaped up a mound of earth and would solemnly tell visitors that it was Werther's grave' (Hulse 1989: 12). In 1817, the death of Princess Charlotte, the only daughter of George IV, created a public furore that prefigured the Diana phenomenon of 1997 (Behrendt 2005). The supposed former home of Juliet Capulet in Verona, Italy, has attracted pilgrims and tributes in the form of love notes since at least the 1930s, prefiguring shrines to more recent fictional personages (The Juliet Club, n.d.). But such activity seems to have intensified in the last 40 years.

Our comparative research has thus far focused on other tribute sites to music business celebrities in London, namely Marc Bolan, Freddie Mercury, Amy Winehouse, and George Michael, which constitute the principal examples in the city. Again, these tributes sites mostly relate to places of former residence, or they have another biographical link to the artist.

Marc Bolan (born Marc Feld) died on 19 September 1977 in a road traffic accident in Queens Ride, Barnes Common, South London. The site of the accident has been turned into a memorial garden, known as 'Marc Bolan's Rock Shrine' (Figure 33.2). In 1997, a memorial stone (donated by the Performing Right Society) and a bronze bust of artist (commissioned by the T-Rex Action Group) were installed. The early history of the site is vague, but it appears that pilgrims/fans were visiting the site not long after he died (Hardy 2013) and on visits to the site in 2016 and 2017, there were clear signs of recent tribute activity. In addition, there are several plaques to Marc Feld/Bolan located elsewhere. At 25 Stoke Newington Common, his birthplace and residence from 1947–1962, a plaque was put up by the Borough of Hackney in the 1990s, but it is unclear exactly when. Bolan was cremated at Golders Green Crematorium where there are two plaques in the West Cloister and a marker in the crematorium gardens that locates the place where his ashes were spread (West Statue Rosebed, Section 5, Plot E 35979), the latter is the site of annual pilgrimage on anniversaries of his death.

Freddie Mercury (born Farrokh Bulsara), lead singer of the rock band 'Queen,' died at home on 24 November 1991 at the age of 45 as a result of complications from the HIV virus. Freddie's family moved from Zanzibar to Feltham, a town in west London, when he was 17. In 2009, to much ceremony, a star-shaped Hollywood style memorial was set into the pavement of Feltham's shopping centre. It featured a gold star in a black pentagon in a red star in a yellow pentagon in a white circle in a black circle with the inscription 'FREDDIE MERCURY, MUSICIAN, SINGER & SONGWRITER' and was decorated

Figure 33.2 Bronze bust of Marc Bolan, Barnes Common.
Source: Photograph by Paul Graves-Brown.

with two clef symbols. Just two years later it had fallen into such a state of disrepair through weathering and vandalism (probably by souvenir hunters) that Hounslow Council replaced it with a smaller memorial stone on Feltham High Street that is still extant. In 2016, an English Heritage blue plaque was unveiled at 22 Gladstone Avenue, Feltham, the family home (Kennedy 2016).

The wall of Mercury's home 'Garden Lodge' in Earl's Court was covered in fan graffiti (and paper tributes that were protected by a plexiglass screen), with additional graffiti written on the surface of the screen (Figure 33.3). All material was removed in 2017 by Mary Austin, Mercury's close friend and subsequent owner of the home, apparently due to pressure from neighbours (Burke 2017). Following his cremation

Figure 33.3 The tribute wall to Freddie Mercury, Earl's Court. The tributes were removed in November 2017.
Source: Photograph by Paul Graves-Brown.

at Kensal Green Cemetery, there has been some speculation as to the location of his ashes. A plaque was erected in the cemetery in 2013 reading 'In Loving Memory of Farrokh Bulsara. Pour Etre Toujours Pres De Toi Avec Tout Mon Amour' and signed 'M' but was removed by the end of the year (Eisen 2013; Evening Standard 2013). A bronze statue of Freddie Mercury was erected close to the waterfront in Montreux, Switzerland, in 1996, and is another site of fan pilgrimage. Mercury had a home there and some have claimed that his ashes were scattered on the lake shore while others believe that his ashes are in a dedicated Parsee plot in Brookwood Cemetery, Surrey (Jones 2011: 333).

There are two tribute locales for the singer/songwriter Amy Winehouse in Camden Town, an area of north London with which she was strongly associated. Winehouse died of alcohol poisoning at the age of 27 on 23 July 2011 and a life-size statue of the artist, with her trademark beehive hairdo, is in the Stables Market, Camden. The statue, created by sculptor Scott Eaton, was unveiled in September 2014, the date that would have been Winehouse's 31st birthday. On visits to the Market, we have noted that the statue has been decorated with friendship bracelets and visitors pose next to the statue to have their photographs taken (Figure 33.4). The second site is a spontaneous shrine located opposite Winehouse's former residence in Camden Square Gardens. The tributes originally covered the residence's wall, pavement, and, according to press photographs, a parked car, but now the shrine site has relocated to the Gardens directly opposite the house. Here, graffiti and tributes focus on a row of wooden posts and tree-shrines. Handwritten notes, flowers, crushed beer cans, cigarette packets, business cards, and other material have been pushed into spaces between the tree trunks and protective matting, presumably placed there by Camden Council to protect the trees. Following a funeral at Edgwarebury Cemetery, Winehouse was cremated at Golders Green cemetery. Almost a year later, a gravestone dedicated to Amy and

Figure 33.4 The statue of Amy Winehouse by sculptor Scott Eaton, Stables Market, Camden.
Source: Photograph by Hilary Orange.

her grandmother Cynthia was placed in Edgwarebury Cemetery, attracting a collection of small stones, laid by visitors according to Jewish custom.

Tribute materials also amassed at the former homes of George Michael, former member of the 1980s pop group 'Wham!' and later a megastar as a solo artist. The singer died at his home in Goring-on-Thames, Oxfordshire on Christmas Day 2016 of natural causes at the age of 53 and tributes amassed outside the frontage of the house after his death, and on subsequent anniversaries of his death. Michael was buried next to his mother in Highgate West Cemetery. The cemetery is only accessible through guided tours and, in consequence, a large and complex shrine has also established itself in 'The Grove' opposite his former residence in Highgate, north London, having been relocated from the frontage of the house (Figure 33.5). We believe the plot belongs to the house and that the shrine may well have had the semi-official sanction of his family, although more recently they have asked for material to be removed.

As noted earlier, where sites are inaccessible or difficult to access (as in the case of the inaccessibility of George Michael's grave), fans will create their own tribute spaces. In the cases described, these are largely connected to places and areas of former residence, with official plaques or statues acting as memorials and attractors, or public gardens and trees providing the necessary space. Although some grave sites/markers are accessible it seems that tribute activity often focuses more on grassroot memorials established by fan groups in prominent city locations. Tribute materials, as with David Bowie, are a mixture of generic tributes (such as flowers and soft toys), and items that are personal to the giver (a handwritten note with the fan's name and country of residence) or to the receiver (for instance, cigarette packets given to Amy Winehouse, who was a smoker).

Figure 33.5 The shrine to George Michael, Highgate.
Source: Photograph by Hilary Orange.

Unofficial becoming official and official becoming unofficial/lost

The fact that the tribute sites and memorials to Marc Bolan and Freddie Mercury demonstrate a longevity of activity, over 40 years in the case of Bolan, suggests that the Bowie sites may continue to be visited by fans for many years to come. The memorial garden to Marc Bolan has undergone phases of elaboration over time. But how do tribute sites change and, in some cases, develop into forms of 'official' heritage? Conversely, as in the case of the 'lost' shrines (whether through deliberate removal or natural erosion) to Freddie Mercury, how do shrines go from being official to disappearing from public view? An issue which has parallels in the fate of First World War memorials, many of which have been removed since their erection after 1918 (Winter 1995).

The extent to which a spontaneous tribute site remains (although the tribute objects themselves change) depends on the site's visibility to passing pedestrians and traffic and its 'promotion' on social and other media. Of course, it also depends on continuing fandom, on an individual level in terms of the discovery of the music by members of subsequent generations and on the continuing membership of fan groups. Then there is the agency of property and landowners and managers over time, where properties are resold and where local authorities deal with the material and graffiti left at the sites.

In the case of Bowie, building owners (such as Morley's Department Store), highways and transport, and local authorities have control over how long tributes are left in situ, and in turn attract further tribute activity. At the Bowie sites, perishable materials, such as flowers, have been removed by site owners/managers or street cleansing teams once material has started to decay. Speaking to the team from the Crown Estate regarding Heddon Street, they were keen that tributes were not ruined by weather or present a hazard for office workers using nearby entrance doors and in consequence materials were cleaned up on a regular basis (pers. comm. Martin Brazier, the Crown Estate). At Tunstall Road, we have seen a street cleansing team member carefully sorting material (decaying flowers from more recent deposits), while Ian Baker (Head of Service, Highways & Transport) has been in regular contact with the store manager of Morley's Department Store to coordinate the tidy-up operations, including the periodic removal of dead flowers and litter from the area (Ian Baker, pers comm).

We are also in contact with several archives that have received donations of tribute material or have been proactively collecting material. Tributes from Tunstall Road have been collected by Lambeth Archives (Jon Newman, pers comm) and material from Heddon Street has been collected by the Crown Estate and donated to the Victoria and Albert Museum (Ramona Riedzewski, pers comm.). Similarly, tributes from outside the former Three Tuns on Beckenham High Street and the bandstand in Croydon Road Recreation Ground are now archived with the Bromley Historic Collections and a small number of tributes are now part of their new permanent exhibition, within a section on the theme of creativity, at Bromley Central Library (Jane Cameron, pers. comm.).

The development of archives and exhibitions leads us to also consider the preservation of graffiti at such sites. Morley's Department Store has, to date, chosen to leave most of the graffiti in place, now behind a protective screen. However, as we have seen in the case of Freddie Mercury, such screens protect graffiti, but can be later removed.

We are also following conversations regarding plans for more 'permanent' memorials to David Bowie. Some have already appeared outside of London. In August 2016, a plaque commemorating Bowie's residence in Schöneberg was unveiled in Berlin (Oltermann 2016). The plaque is one of a series of Berlin *Gedenktafeln* that have been erected around the city since 1985, resembling English Heritage's familiar Blue Plaque scheme. Back in the United Kingdom, in June 2017, a Blue Plaque dedicated to Bowie was unveiled at the Royal Star Arcade, Maidstone, Kent. Bowie played there (as Davie Jones) with one of his bands, The Manish Boys, between 1964 and 1965. The plaque was one of 47 created for BBC Music Day 2017 and voted for by listeners to 40 BBC local radio stations and the Asian Network. A statue of Bowie, entitled 'Earthly Messenger,' was unveiled in the Market Square, Aylesbury, Buckinghamshire in 2018, and almost immediately vandalised. Bowie played several times at the nearby Friars Club and launched his Ziggy Stardust incarnation there in 1971.

At the unveiling of the plaque in Berlin, the Mayor of the City, Michael Müller, remarked that 'David Bowie belongs to Berlin, David Bowie belongs to us' (Oltermann 2016). In the period after Bowie's death, London also claimed Bowie. For instance, the marquee of the Ritzy Cinema in Brixton carried the words 'David Bowie Our Brixton Boy RIP.' In February 2017, a crowdfunded campaign to raise a nine-metre monument, the 'ZiggyZag,' adjacent to the Tunstall Road mural was launched. The monument was designed to resemble the lightning bolt makeup featured on the 1973 album *Aladdin Sane*. Subsequently, the organisers terminated their campaign when it became apparent that they would not raise the estimated £900,000 needed to erect the monument. Meanwhile, the bandstand in Croydon Recreation Ground is in the process of becoming a Bowie memorial through the activities of the Bandstand Restoration Group, which focuses largely on Bowie's performance at the 1969 Free Festival.

Discussion and conclusion

The David Bowie shrines form part of a growing array of tribute sites to both real and fictional celebrities. Whilst such phenomena have a long history, they seem to have become far more prevalent in recent years, perhaps reflecting changing social values and the growing predominance of a celebrity culture. In many ways this reflects what Riesman (1950) called 'other directedness'; that the decisions of fans as to how to commemorate their idols are formed by a kind of social 'radar' which has been magnified by the advent of social media. Such popular or collective action contrasts with the traditional or at least formal channels through which individuals are memorialised, be it in cemetery markers, blue, green, brown, and black plaques or the honour of inclusion in Poets Corner in Westminster Abbey. It is also interesting that such collective memorialisation increasingly includes fictional characters, be it Benedict Cumberbatch's Sherlock (at the St Bartholomew's Hospital in Giltspur Street, London), or the character of Ianto Jones from the BBC's *Torchwood*, who is memorialised in Cardiff Bay (Beattie 2014). The form, location, and developmental trajectory of such sites is shaped by a range of factors including the manner of the person's demise, the places they lived and worked and, on occasion as at Tunstall Road, more seemingly random factors. Site development usually requires a physical focus, be it a building, mural, or a phone box. Moreover, such sites obviously need to be publicly accessible, but also need to provide a space or agora within which tribute activity can safely and sustainably take place. Equally there are pressures for the removal of shrines. Stars such as Freddie Mercury and George Michael ended up owning homes in the more 'upmarket' areas of London, where neighbours, probably conscious of the value and situation of their property, are not entirely sympathetic to the memorialisation of former residents, except in the more 'official' form of plaques. For example, opposite the now controversial George Michael memorial garden, there is a (brown) English Heritage plaque on 3 The Grove dedicated to J.B. Priestley. Other sites, such as the Marc Bolan shrine, now owned and managed by fans, do not experience this negativity, whist the graffiti in the Heddon Street phone box experiences a grudging tolerance – it is periodically removed, but always reappears. In some loci, this iterative process has a formality to it; at Abbey Road Studios, in St Johns Wood, London, and on the boundary wall of Elvis' Graceland estate, graffiti are periodically removed with the tacit understanding that they will be replaced.

The Bowie shrines are subject to ongoing processes of development and elaboration whose outcome we do not yet know. At present, shrines to David Bowie and other music stars are still in their first, spontaneous stage of development, with indications that they are entering a second more formal stage. Thinking of the First World War, we know that spontaneous shrines to those who had gone to war appeared as early as 1914, and the process of formal memorialisation began during the war and continued into the 1920s, but as Winter (1995) points out, there is a third stage in which memorials are either forgotten or melt into obscurity – who now takes note of the Crimean War memorial in Lammas Street in Carmarthen, or has any sense of why numerous streets in Britain are called Alma Road? Whether the Bowie sites eventually sink into obscurity remains to be seen and will be, in part, dependent on whether fans continue or cease to visit as well as official management of such sites of pilgrimage.

Acknowledgements

We extend thanks to Rodney Harrison (UCL Institute of Archaeology) for reporting on developments at Tunstall Road, Brixton, and to the following individuals and organisations for supporting our research: Ian Baker (London Borough of Lambeth), Martin Brazier and Tallulah Morris (Crown Estate), Kelly M. Brit (Brooklyn College, New York), Jane Cameron (Bromley Historic Collections), Sarah Murray (Canterbury Museum, New Zealand), Jon Newman (Lambeth Archives), and Ramona Riedzewski, V&A Museum.

References

Alderman, D.H. (2002). Writing on the Graceland wall: On the importance of authorship in pilgrimage landscapes. *Tourism Recreation Research*, 27 (2): 27–33.
Beattie, M. (2014). A most peculiar memorial: Cultural heritage and fiction. In: J. Schofield, ed. *Who needs experts? Counter-mapping cultural heritage*. London: Taylor & Francis, pp. 215–224.
Behrendt, S.C. (2005). *Mourning, myth and merchandising: The public death of Princess Charlotte*. In: C. Riegel, ed. *Responses to death: The literary work of mourning*. Edmonton: University of Alberta Press, pp. 75–96.
Burke, D. (2017). Freddie Mercury's ex causes outrage after removing shrine to rock legend outside the £20million home where he died. *The Mirror*, 19 November. Accessed 20 March 2018, www.mirror.co.uk/news/uk-news/freddie-mercurys-ex-causes-outrage-11547879

Daily Telegraph. (2012). David Bowie's Ziggy Stardust commemorated on plaque. *Daily Telegraph*, 27 March. Accessed 15 April 2018, www.telegraph.co.uk/culture/music/9168943/David-Bowies-Ziggy-Stardust-commemorated-on-plaque.html

Eisen, J. (2013). Freddie Mercury's plaque vanishes from Kensal Green cemetery. *Evening Standard*, 3 March. Accessed 22 April 2018, www.standard.co.uk/showbiz/celebrity-news/freddie-mercurys-plaque-vanishes-from-kensal-green-cemetery-8520387.html

Evening Standard. (2013). Is Freddie Mercury laid to rest in Kensal Green Cemetery? *Evening Standard*, 25 February. Accessed 15 April 2018, www.standard.co.uk/showbiz/celebrity-news/is-freddie-mercury-laid-to-rest-in-kensal-green-cemetery-8509585.html

Graceland. (2016). *Graceland celebrates 20 million visitors*. 3 May. Accessed 6 December 2018, www.graceland.com/news/details/graceland-celebrates-20-million-visitors/8044/

Graves-Brown, P. (2012). Where the streets have no name: A guided tour of pop heritage sites in London's west end. *In:* S. May, H. Orange and S. Penrose, eds. *The good, the bad and the unbuilt: Handling the heritage of the recent past.* Studies in Contemporary and Historical Archaeology 7. Oxford: Archaeopress, pp. 63–76.

Graves-Brown, P., and Orange, H. (2017). "The stars look very different today": Celebrity veneration, grassroot memorials and the apotheosis of David Bowie. *Material Religion*, 13 (1): 121–123.

Hardy, F. (2013). *Resting places of the rich and famous*. London: RW Press Ltd.

Hills, M. (2002). *Fan cultures*. London: Routledge, Taylor & Francis.

Hulse, M. (1989). Introduction. *In:* J.W. von Goethe, ed., M. Hulse, trans. *The sorrow of young werther*. London: Penguin Classics, pp. 5–22.

Jones, D. (2017). *David Bowie: A life*. London: Preface.

Jones, L. (2011). *Freddie Mercury: The definitive biography*. London: Hodder & Stoughton.

Juliet Club. (n.d.). *About us*. Accessed 22 April 2018, www.julietclub.com/en/

Kennedy, M. (2016). Freddie Mercury's modest London home gets blue plaque. *The Guardian*, 1 September. Accessed 15 March, www.theguardian.com/music/2016/sep/01/freddie-mercury-feltham-london-home-blue-plaque

Lambeth Council. (2016). Bowie mural to be listed. Accessed 15 November 2018, https://lambethnews.wordpress.com/2016/03/21/bowie-mural-to-be-listed/

London Live. (2016). The man who stole the world. *First Broadcast*, 5 April.

MacConville, U. (2010). Roadside memorials. *Bereavement Care*, 29 (3): 34–36.

Margry, P.J. (2008). *Shrines and pilgrimage in the modern world. new itineraries into the sacred*. Amsterdam: Amsterdam University Press.

Monger, G., and Chandler, J. (1998). Pilgrimage to Kensington Palace. *Folklore*, 109 (1–2): 104–108.

Naylor, A.K. (2010). Michael Jackson's post-self. *Celebrity Studies*, 1 (2): 251–253.

Oltermann, P. (2016). David Bowie: Berlin plaque commemorates time in city of 'Heroes.' *The Guardian*, 22 September. Accessed 17 January 2018, www.theguardian.com/music/2016/aug/22/david-bowie-berlin-plaque-commemorates-singers-time-in-city

Recuber, T. (2012). The prosumption of commemoration: Disasters, digital memory banks and online collective memory. *American Behavioral Scientist*, 56 (4): 531–549.

Riesman, D. (1950). *The lonely crowd*. London: Yale University Press.

Sanders, J. (2010). Roadside memorials. *Bereavement Care*, 29 (3): 41–43.

Urry, J. (1990). *The tourist gaze: Leisure and travel in contemporary societies*. London: Sage Publications.

Vikan, G. (1994). Graceland as locus sanctus. *In:* G. DePaoli, ed. *Elvis + Marilyn. 2X Immortal*. New York: Rizzoli, pp. 150–170.

Winter, J. (1995). *Sites of memory, sites of mourning: The great war in European cultural history*. Cambridge: Cambridge University Press.

Yentob, A. (1975). *Cracked actor*. Omnibus. BBC documentary.

Zanetta, T., and Edwards, H. (1986). *Stardust: The David Bowie story*. London: Michael Joseph.

34

AN ETHNOGRAPHY OF MEMORY IN THE SECRET VALLEYS OF THE HIMALAYAS

Sacred topographies of mind in two *Beyul* pilgrimages

Hayley Saul

It is the middle of April in 2013 and I am stumbling along a steep trail about halfway up the Langtang Valley in Nepal. After every ten geologically slow steps, I pause bemusedly to survey my weary muscles once again, panting at the exertion of just placing another foot on the uneven stones. I mentally chastise myself for persisting with a dinner of 'Sherpa Dal Bhat' the night before; the rice soaked in warm yak butter that, in hindsight, was a theme park for listeria. Barely an hour after eating it I was dragging myself to a sketchy toilet bowl in my guesthouse to vomit (and would continue to spend the next two solid days doing so every half hour). And so, it was in this not-so-plucky state, barely able to lift my head, that I entered the sacred Beyul of Langtang (Tib. *sbas yul*) for the first time. I had been interested in these mythical landscapes since reading Ian Baker's (2004) *The Heart of the World*; one of the most comprehensive physical and literary explorations of the Beyul tradition I have come across, and a beautiful rendering of the multi-dimensional geography of the Tsangpo Gorge (Beyul Pemako). Beyul, as Baker (2004) notes, are valleys hidden by the powerful intention of Guru Rinpoche (*Sanskrit: Padmasambhava*) in the eighth century AD. It was in this early historic period that Guru Rinpoche journeyed throughout the Himalayas on foot, hiding secret wisdoms (as scrolls, objects, or simply thoughts) in rocks, caves, trees, rivers, and even the sky. This tradition of transmitting wisdom via text or by imbuing an object directly with the power of revelation, is called the Terma Tradition, so-called because the knowledge 'treasures' are named *Terma* (Tib. *gter ma*). It is within this tradition that Guru Rinpoche is also acknowledged to have hidden entire valleys that would one day be revealed to a few to act as a refuge in times of crisis for humanity. Those Terma that describe how to access these Beyul valleys are called *Neyig*.

Not only are they often physically remote, but these landscapes are disguised 'by barriers formed out of our habitual ways of perceiving our surroundings' (Baker 2004: 11). In this sense then, they occupy geographies that transcend to the esoteric; to alternate realities brought about by variations in the 'ordinary' perception of space. As a result, discovery of the innermost, secret regions of Beyul is only possible by spiritual adepts – those that have cultivated a pure mind: 'naked, immaculate, transparent, empty, timeless, uncreated, unimpeded; not realizable as a separate entity, but as the unity of all things' (Baker 2004: 21). These adepts, or Tertons, as they are known in the Terma Tradition, are often evoked as reincarnations of Guru Rinpoche's original disciples, reborn multiple times until such a moment that humanity needs the Beyul and the esoteric wisdoms that inhabit such places.

What struck me upon further investigation of this remarkable legacy was that revealing those arcane spaces of the Beyul and the hidden Terma contained within is occasionally likened, in the tradition's commentaries, to a process of *remembering*. Both the landscape itself, but also 'symbolic scripts on scrolls of paper are used as the key to awaken the recollection of the teaching that has been concealed in the essential nature of minds' (Thondup Rinpoche 1997: 61). Only Tertons were bestowed with the revelations of a Beyul's hidden geographies originating from the period when the tradition was first established, but anyone willing to invest in his or her own spiritual attainment could benefit from pilgrimages to Beyul. Indeed, the '*Essential Inventory of Yolmo*' text of the Terma tradition suggests that even the mere intention to travel to a Beyul is meritorious:

> This is a most auspicious sanctuary,
> Where longevity, merits and resources all multiply,
> Those living in the degenerate age, who would practise what I, (Padmasambhava) have taught,
> Go find that sanctuary!
> Those who think of it and long for it,
> Have accumulated merit over limitless aeons.
>
> (cited in Dondrup 2009: 6)

And so, acknowledging that the phenomenal experience of a pilgrim whilst travelling through/to Beyul starts from an 'anamnestic attitude,' in this chapter I want to explore these hidden valleys as a tradition of memory work. I pick up on the ways that memories disrupt the perception of place in the Terma tradition and, conversely, are evoked and transmuted by those disruptions. Further, I seek to engage with the Terma tradition's peculiar range of what Olick (2008) refers to as 'retrospective products, practices, and processes' that could otherwise be masked beneath the generalising notion of 'tradition' or 'collective memory.' And to do this, I draw on auto-ethnographic experiences of pilgrimage routes predominantly to the Langtang Valley (recognised as the location of Beyul Dagam Namgo [Tib. *zla gam gnam sgo*]) and to a lesser degree the valley of the Melamchi Ghyang (revered as the location of the Yolmo Beyul). These are contextualised by the few published *Neyig* and commentaries.

Despite the distraction of my ailing health, it was impossible not to be stunned by the beauty of my surroundings. It was the season of rhododendron blooms, and their wispy florets peppered the forested hillsides in every shade of crimson and magenta up to at least 3,000 metres. Grey langur monkeys peered inquisitively from their leafy branches as I picked my way over fallen logs. Sub-tropical forest plants had started to give way to the occasional temperate hemlock and larch, their furry foliage thick enough to conceal a Himalayan black bear, or so it seemed to my dreamily exhausted mind. Memory is a concept that is surprisingly absent from Tibetan Buddhist philosophy, at least partly because it presents problems for explanations of the ego. The recognition of mundane objects through everyday, fleeting acts of recollection – perceiving something 'to be' in the present because of its similarity to the past – is a leading culprit. For in Buddhism, which considers existence as a constant state of flux, such a recollection is ontologically unfounded, and therefore vulnerable to falsity (Gyatso 1992). What is more, failing to truly accept this leads to habitual ways of thinking, what Sarah Ahmed (2004) would call the 'stickiness' of things, an intensity of attachment that underpins a wholly substantialist theory of self-hood. Sticky objects accumulate a patina of thoughtful and emotional experience which lends them the guise of an 'I.' It is the 'I,' the ego, that troubles the soteriological pursuits of Tibetan Buddhism, matting the fabric of an otherwise flowing 'essential nature' (Tib. *Kham rig*) and resulting in a mistaken conflation of habitual perception with unquestionable fact.

Griffith's (1992) observation that a systematic phenomenology of remembering is missing from Buddhist philosophy holds true because of these unhelpful consequences for the project of salvation. Conversely, descriptions of striking *acts* of memory as evidence of a realised state are not uncommon in Buddhist verses; the Terton master Mañjuśrīghosa, for example, could recite the full teachings of Guru Rinpoche simply by concentrating on the words so that they arose in his mind (Third Dondrup Chen Rinpoche 1997, cited in Thondup Rinpoche 1997). The earliest mention of memory of the past from an analytical perspective comes from a first-century treatise called the Mahāvibhāsā, an explanation of the Buddha's teachings originating in India mere centuries after his awakening (Walser 2009). The seemingly striking absence of memory theory in Tibetan Buddhism could, alternatively, result more from a skewed expectation of what such a body of thought should look like though. Though the landmark framing of the experience of memory by Bergson (1990) did much to dispel static renderings of 'memory as a storage device' and replace them with appreciations of its active, fluid qualities, those understandings still rested upon the position of memory as a product of the past. But, this dominant relegation of memory to a backwards-looking gaze, oriented solely to the past, is to conceive the very nature of memory in narrow, restrictive terms (Gyatso 1992). Furthermore, as this chapter will explore, such a narrow definition serves to illuminate more about an Anglophone scholarly appreciation of memory than a Tibetan Buddhist one. For a truly comparative philosophy, we must include an adjustment to our very notion of what it means to remember, and the social-spiritual affordances of doing so in the Terma Tradition. It was into this thoughtful initiation that I stepped when I entered the Dagam Namgo Beyul in 2013, conscious only that if a reciprocal thread exists between perception and memory then the true perception of the Beyul's hidden aspects might somehow require me practice a different type of memory work.

Extra-human actants in memory webs

At the summit of each hillock was a vista of arresting beauty. Such is the tremendous incline of the Dagam Namgo Valley that the scene back down to the trailhead at Syabru Besi cascades in a tumble of forest blooms, crumbled yak herder huts and a vein of glacial waters. This almost technicolour beauty, I was told, is a clue about the presence of a Beyul. When walking the valley, the first two days are spent in a steep-sided gorge, ascending quickly before the valley opens out on the higher pastures and ends in a horseshoe of crisp white peaks (Figure 34.1). Those moments certainly startled me from the inconsequential mutterings of my internal dialogue that seemed hellbent on encouraging me to have a nice sit-down. As I descended from these knolls, out of the sunshine and under the rhododendron canopy, the grand panoramas gave way to more immediate vistas: gashes made from rockfall, the precarity of the path that wound beside the monsoon-engorged river, the claustrophobia of the cliffs. This was the broken and weathered underbelly of the Beyul, unstable and dangerous. Simply the scale of my witnessing provoked its varying aspects. I simultaneously experienced both uneasiness at the unpredictable fluidity of the rocky terrain, and its Romantic beauty, the latter buffered by tales I'd read of plant hunting explorers (Lancaster 1981), and a rich botanical material culture that I'd seen in the United Kingdom's Royal Botanical Gardens, accumulated since the naturalists of 1802 visited these mountains. These 'social frameworks in which we are called on to recall are inevitably tied up with what and how we recall,' observes Olick (2008: 115), and I was entrapped by their habitual presence in my mind. Such mediated framings mesmerised every viewpoint, inspiring me to sink into familiar fantasies of reminiscence.

It is this totalising tendency of collective memory that is resisted in the Terma Tradition. To do this, known Terma manuscripts are often written in a poetic and indirect script, described by Baker (2004: 10) as a 'twilight' language because of its obscurity. They demand deciphering for the very purpose of disrupting any *a priori* conflation of collective memory and knowledge. But this poses a peculiar problem for the transfer of knowledge. The antithesis of totalising narratives – where shared experience makes up the connecting principle between individuals – is something akin to a reductionist reading of memory forwarded by Bergson (1990) who emphasises the variability of individual memory, and therefore the limits of its reproduction collectively. It is this latter perspective that is championed by the Terma Tradition. In the absence of any residue of connectivity from shared memory, the Terma Tradition of knowledge is arguably established on epistemological foundations that design inherent instability into its practice. The tradition allows for

Figure 34.1 Looking up the Dagam Namgo valley to the horseshoe of white peaks, the physical centre of the Beyul.

collective practice by anyone but excludes the development of awareness to those individuals unwilling or unable to realise the impermanence of mind, and the inherent fallibility of memory. Recall memory has no authority in the Terma Tradition (Wayman 1992), and consequently knowledge based on recall doesn't either. Habitual patterns of knowledge, acquired through remembered teachings, are navigated around by charting routes between individual and collective experience.

How can a tradition that guides a mode of awareness do so without recourse to collective representational means? An embodied encounter with the esoteric aspects of the Beyul landscape is one important way that this happens. For there to be a transfer of knowledge or awareness – a teaching, in other words – there needs to be a connecting principle linking agents into webs of interaction. Whilst collective memory is censured in the tradition, it is replaced by the concept of *Tantra*. *Tantra* is a connecting principle that is not oriented to the past, but to the present. A nebulous term, it encompasses all forms of practice that challenge a sense of separation that arises because of an erroneous belief in an autonomous 'self' (Baker 2004). It is through *Tantra*, its own form of memory work directed at the present rather than the past, that practitioners gain access to and harvest enlightened consciousness that courses fluidly through all form and matter and can be harnessed to achieve certain creative outcomes that include emancipation (Gordon White 2000) and expanded awareness. How is *Tantra* a form of memory work? I would argue there are two main ways, both of which the Beyul landscape provokes: enduring visualisation and the *Vajra* body.

Enduring visualisation: presencing memory

If memory is not exclusively that which is directed to the past (and its recollection) in Himalayan cultures, then other candidates for the title of 'remembering' must be considered. An important part of *Tantric* practice involves powerful visualisation; in other words, it is a memory praxis that involves creating an enduring projection (be it image, concept, sensation) that is retained, and its richness enlivened. When undertaken as a visualisation of the Buddha or Bodhissatva, it is akin to commemoration (Gyatso 1992). Simultaneously though, the practice of visualisation requires that recollective intrusions into that singular concentration *not* be attended to, that they be observed to drift into consciousness and, according to their fleeting nature, exit unimpeded. The objective of Tantric visualisation is to invest the focus of concentration so singularly on the thought-form that its qualities are actualised. Those qualities might be conceptualised as 'energies' or 'intensities.' Such terms, as Gordon White (2000) observes, are numerous; for orthodox practitioners 'teachings' may be more appropriate, whereas for many of the inhabitants of Dagam Namgo 'beings' may make more sense. Capturing their essence lexicologically is problematic because these energetic 'teachings' are affective and sensuous.

As such, visualisation as memory is not strongly mind-oriented; mind only serves to translate the sublimity of the macrocosm into mundane embodied affect. It was not to the grand romance of Dagam Namgo's panoramic views that I therefore looked for these teachings. The sensations that spilled over me from those vistas were fleeting and too easily indulged my childhood venturer's imaginary. It was to the underbelly of Dagam Namgo that I was invited to dwell on the disquieting sensations of vulnerability that the landscape and its 'beings' elicited at the immediate scale. As I approached the horseshoe of white mountains on the third day, still weary from food poisoning, and now with heart palpitations from some second-hand antibiotics I'd been gifted from a dusty old tin the day before, I felt dubious about how welcome I was in the Beyul. On the outskirts of the village of Langtang, my eye was persistently drawn to a cliff face, upon which black lichen clung, dripping from an elevated moraine bed above, which sat at the foot of Langtang Lirung mountain. The stain looked like a handprint. A local man, seeing my repeated glances, gestured towards the cliff and told me it was the hand of Guru Rinpoche, etched onto the stone to hold back avalanches. I was reminded that most Beyul Terma describe 'protector deities' as beings that inhabit Beyul. These protectors could be either peaceful or wrathful, and both literal forces in the outer world as well as constitutive of resonances within a sensuous, inner world. In Terma, everything external has corresponding affects for the body. Protector deities arrived into the Buddhist canon from the Bonpö 'old gods,' pre-Buddhist divine beings that were subdued by Guru Rinpoche and convinced to apply their wrathful propensities in the defence of the Dharma and its Beyul. The handprint was massive and quagmired my thoughts at the level of observation: the search for evidence of danger. For Massumi (2010: 53), 'Threat is from the future. It is what might come next,' and whilst true, the experience of threat is vulnerability, which inflects with the immanence of isolation and helplessness. I was present, attuned to a pre-cognitive unease at the scale of the handprint and what that meant for the scale of the avalanche that a wrathful deity could send down.

Other indications of Guru Rinpoche in the form of a series of footprints moulded into rocks are encountered sequentially as one walks through the valley (Saul & Waterton 2017), each serving to presence the sublimity of the Guru. At each of these, the local Langtangpa had marked their location with upright prayer flags on poles, and in one case a footprint lay on top of a large boulder (Figure 34.2), the sides of which were densely engraved with colourful 'om mani padme hum' prayers. Such presencing of divine power is common to other Beyul too. In the *Guide to the Hidden Land of Pemako* neyig, the location of the sacred valley is indicated by, amongst other waymarkers, 'Guru's caves of rumination, within which lay the impression of his feet in solid rock and mantric seed syllables growing from the walls themselves' (Khamtrul Jamyang Dondrup Rinpoche 1959: np). Some days later, as I descended the Gosaikund Pass into the adjacent Beyul of Yolmo I would once again witness these miraculous bodily impressions of Guru Rinpoche at Yangdak Chok meditation cave. In the *Variegated Jewel Garland*, a neyig about Yolmo by Shamar Chökyi Wangchuk (cited in Dondrup 2009) this cave is said to be at the centre of the Beyul and, 'above the cave, on a broad and even boulder, footprints of the Dakinis[1] are quite clear, and there is a Parikrama[2] path of the Dakinis. Inside the cave is the Guru's headprint.' Here, Mixter and Henry's (2017) observation, that analysing interactions with more-than-human agents in memory processes allows for multiple viewpoints

Figure 34.2 One of the footprints of Guru Rinpoche that occupies the Dagam Namgo Beyul, sitting atop a boulder that is intricately carved with 'om mani padme hum' prayers.

to be recognised, becomes important not just for their intended work of unpacking the establishment and legitimation of power relations, but as a method of generating awareness of the phenomenal world of the more-than-human. In this case, the webs of interactions I was tentatively establishing were with extra-human agents,[3] beings on the fringes of ordinary phenomenal experience, responsible for awakening patterns of visualisation and reflection as a form of transcendent collective memory work specific to the Terma tradition. What is particularly interesting about those extra-human webs of interaction with Dagam Namgo's 'protector deities' is that they are an affective provocation; subtle intensities acting upon me that, despite superficially seeming to be benign 'stories' or 'places,' exerted malevolence of a sort with effective agency. At least in part this was because the extra-human interactions expanded and contracted the scale of the Beyul; it became a landscape that pulsed with divinities. Individually, the impressed footsteps of Guru Rinpoche that paced through Dagam Namgo were only twice the size of my own, but the stride between each reached several kilometres. My own phenomenal world was confronted by the swellings of these powerful beings which left me feeling dwarfed and exposed. But through those 'fun house' visualisations my presence was also invited to magnify and distort with them, and to experience in some small way the divine alterity of those extra-human others.

The Vajra body: authenticating shared experience

In *Tantric* visualisation the memory process is not just about generating a visualisation with durability, but it is about doing so in order that the visualisation can prompt access to perfect wisdom (Tib. *Pha rol tu phyin pa*), which is collective at a transcendent level beyond shared reminiscences. The Vajra body is the instrument for accessing this latent, mutual knowledge. In other words, rather than embodied experience proceeding from a generic, undifferentiated 'body,' the Vajra is a bodily form *disposed towards* engaging with a collective transcendent wisdom that resembles shared memory. A Vajra is a diamond, hard and resistant, so the Vajra body evokes the powerful possibilities of specific embodied experience that can cut through illusive perceptions of mind and arrive at authentic wisdom-knowledge. Cultivating the Vajra body is a reductive process involving sensuous engagement with increasingly subtle 'energies' or 'teachings' that lead, ultimately, to a stable core from which to appreciate authenticity.

The process of pursuing a Vajra body, it seemed to me as I perambulated towards the valley-head, led not just to knowing something from multiple perspectives – an idea that hinges on the Derridean notion of the inherent dispersion of the 'object' world amidst mutable meanings – but knowing the gross and subtle aspects of those meanings too. It is this latter, attentively introspective praxis that has affiliations with memory work. Numerous *Neyig* establish a map-like resonance between external Beyul landscape and internal excitations (Rawson 2012). In the Pemako Beyul, for example, the Terton Stag-sham-pa received visions that the valley was a physical manifestation of the goddess Vajravārāhi (Sardar-Afkhami 1996). Traversing through the valley replicated the movement of this goddess' 'subtle wind' (Tib. *rlung*) – the energies that animate and sustain life and have tremendous power to transform it if mastered (Baker 2004) – that course through her body's chakras[4] and energy channels. Attuned to the intensities of the goddess' *rlung* by effectively moving through her body, the pilgrim metaphysically becomes her *rlung* and can resonate with the subtleties of divine reality. The sensation of subtle *rlung* energies is an indication that a practitioner has cultivated his or her Vajra body.

Back in Dagam Namgo, The *Neyig* of Rigdzin Gödem (Tib. *Rig-'dzin rGod-lden 'phru-can*) (1337–1408) mentions 'four large gates' to the Beyul, oriented on the cardinal points (Ehrhard 2007: 343). It is commonly agreed by such a reference that Dagam Namgo, the 'Heavenly Gate of the Half Moon Form,' is a mandala, and it is therefore through this device that the Vajra body is achieved here. A mandala is a complex graphic, usually composed of mathematically significant arrangements of circles, squares, and triangles, depicting the ideal realm of existence of a *Tantric* deity (Baker 2004). Though a two-dimensional construction, mandalas actually exist as a sort of 'blueprint' for a three- (or more) dimensional space (Saul & Waterton 2017). Walcott locates them as powerful 'dwellings wherein mobile practitioners and spirits could metaphysically cohabitate' (2006: 81), in essence, esoteric designs for 'an ideal environment for the pilgrimage' (ibid.: 75) that exists in three spheres: outer, inner, and secret.

The closer I came to the valley head and its horseshoe of white mountains the more densely laden with stories and special places the landscape became. At around 3,800 metres a boulder caught my eye in the distance. It was covered in 'stone men'; little piles of rocks towered into precarious-looking cairns. Upon it was an impression of a snake, or so it seemed, though I could see no evidence of engraving and nor did it present itself as some geological phenomenon. Its origins confounded me, and little was mentioned about it by my local companions except that it was called 'snake rock' (Figure 34.3). Could it be that I was (physically, at least) moving from the outer level of the mandala towards the deity in the central palace? Snakes (called *nagas*) are common depictions on 'palace mandala' designs, located in each of the four quadrants. According to Walcott (2006: 77) this class of deity symbolises 'rainbow bridges' facilitating the traverse from gross to increasingly subtle teachings.

I make no claims to have experienced even a portion of the subtle *rlung* of an otherwise spiritual adept. Though I opened myself up to sensations that challenged the reminiscences of my habitual mind and, indeed, the exhaustion of food poisoning even helped me to detach from investing in these mundane memory patterns, I remained a novice at the type of subtle memory work a truly Vajra body could perceive. What did occur to me as I stood, after several days, in the mandala's physical centre, surrounded on three sides by magnificent crystal peaks, was a small meditation on the relationship between memory, authenticity, and intuition. It is with this culmination that I would like to conclude. Throughout my journey along this Beyul pilgrimage route, my discomforting encounters with the valley's 'protector deities' in their variously mighty and modest aspects had accumulated a sensorial intensity with which I had begun to resonate.

Figure 34.3 'Snake rock', the engraved impression of a *naga* being, often known as 'rainbow bridges' between realms in the Beyul.

Perhaps this was the process that Rawson (2012: 68) describes when he says, 'During meditation or ritual the mandala figures invoked are kept in states of radiant presence. A figure does not just disappear when it gives place to another. It may either be absorbed into the successor or continue to emanate from it.' But to what degree had I resonated with the Beyul's sublime powers (if at all), and how could I know if this Terma-specific memory praxis had aroused any latent wisdoms? The answer, I would argue, is a final piece in the puzzle of memory praxis in the Terma tradition.

At the beginning of this chapter, I argued that recollective memory holds no authority in Tibetan Buddhism, and that this poses problems in the Terma tradition for the manner in which its teachings – conventionally something held as collective memory, whether oral or written – can be transmitted. The Terma-specific practices that resolve this paradox chart a novel route between individual experience and collective 'shaping' of that experience to guide, but not impose, subtle awareness of the 'essential nature' of reality. Those practices harness a special type of memory work that is not directed towards the past, with its habitual patterns of mind, but to enlivening enduring visualisations of an immediate present. To do so requires that a practitioner simultaneously hold a visualisation in mind whilst allowing intrusive thoughts to brush past and dissolve into their changing nature. This form of memory is secondarily supported by the cultivation of the Vajra body, a mode of Being and experiencing the world that is unconventionally sensitive to subtle intensities of energy that flow through existence. Such a notion finds common traction with expressions of affect: the 'name we give to those forces – visceral forces beneath, alongside, or generally other than conscious knowing, vital forces insisting beyond emotion – that can serve to drive us towards movement, toward thought and extension' (Seigworth & Gregg 2010: 1). How could I know whether I had cultivated Vajra qualities, then? Since the Vajra body is the vehicle for cutting through illusive perceptions manifest in habitual recollective-patterns, one might rephrase this question as how do I know if I had an authentic experience of the Beyul? The answer I was led to is an area related to affect theory that is only modestly explored outside the discipline of psychology: that of intuition.

Though characterisations of intuition share similarities with definitions of affect, the two are not synonymous with each other just as, as Sara (2007: 51) notes, 'intuitive seemings remain distinctive conscious states in their own right, without collapsing into beliefs.' Common to both is their 'forcefulness' (Sara 2007, Seigworth & Gregg 2010). However, one might always affect and be affected in encounters (Massumi 2015), but one does not always necessarily intuit. Whilst affective forces build from intensities with an encounter (Seigworth & Gregg 2010), intuition is probative (Sara 2007), its relationship to encounter is not always clear. Intuition hones a diffusion of possibilities into a nagging spur of familiarity and definitude. But insofar as intuition is a felt attraction or aversion to a certain proposition, it inherently arrives with doubt; the 'pull of the rejected proposition,' notes Sara (2007: 51), 'is not removed but overcome.' In part, what seems to distinguish the two classes of impulse stems from the degree of consciousness that goes into them. Following a reading of Deleuze, affect is non- or pre-cognitive (Pile 2009), in common with the assertion about intuition that we simply do not inhabit enough of ourselves to be aware of all the deep associations that are being made by our body-mind (Weintraub 2012). Importantly, in intuition the associations that are being made through the attractions and aversions of its unconscious forces are multiple and complex, but rather than persist with their polyvalency and ultimately be characterised by such a diffusion of intensities, as in affect, intuition congeals those pre-conscious recognitions and patterns them into a conscious sensation of knowing. Intuition, therefore, is commonly conceived as a type of recognition, which places it within the realm of memory. And yet, it is a process that circumvents *rational* knowing, with all of the fallacies of habitual patterns of thinking that can gain no authoritative traction in Tibetan Buddhism. It is pre-cognitive memory, felt memory, a sensible amalgamation of innumerable experiences, scenarios, and possibilities that interact in non-linear ways. It therefore sidesteps memory *narratives* that take their power as knowledge from the consensus of the collective. Intuition is a felt recollection of authentic knowledge that surfaces only when one gives in to the dispersion and flow that Tibetan Buddhism advances as the nature of reality.

Notes

1 A Dakini, or 'Sky Goer' from the Tibetan translation, is a female deity, often wrathful, that embodies wisdom (His Holiness the Dalai Lama 1995). Their affiliation with the sky means they are associated with the energy of movement through space, and, with that, they embody the potentiality of transformation for a practitioner (Monaghan 2010).
2 A Parikrama is the route of a circumambulation around a site of religious significance. In other words, the path a pilgrim would take as he or she circles a spiritually important place in a clockwise direction. Again, this reference to Parikrama refers to the energy of movement associated with Dakinis.
3 See Saul (2019) for a fulsome explanation of more-than-human beings that exist on the fringes of what can be materially experienced but are still substantial because of their involvement in cause-and-effect entanglements.
4 'Pools' or wheels of *rlung* energy that are potent centres of transformative potential (Johari 2000).

References

Ahmed, S. (2004). *The cultural politics of emotion*. Edinburgh: Edinburgh University Press.
Baker, I. (2004). *The heart of the world: A journey to Tibet's lost paradise*. London: Penguin.
Bergson, H. (1990). *Matter and memory*. New York: Zone.
Dondrup, K.N. (2009). *Guide to the hidden land of the Yolmo snow enclosure and its history*. Helmu, Nepal: Vajra Publications.
Ehrhard, F-K. (2007). A 'hidden land' in the Tibetan-Nepalese borderlands. In: A.W. Macdonald, ed. *Mandala and landscape*. New Delhi: Printworld, pp. 335–364.

Gordon White, D. (2000). Tantra in practice: Mapping a tradition. *In:* D. Gordon White, ed. *Tantra in practice.* New Delhi: Motilal Banarsidass, pp. 3–38.

Griffiths, P.J. (1992). Memory in classical Indian Yogācāra. *In:* J. Gyatso, ed. *In the mirror of memory: Reflections on mindfulness and remembrance in Indian and Tibetan Buddhism.* Albany, NY: State University of New York Press, pp. 109–132.

Gyatso, J. (1992). Introduction. *In:* J. Gyatso, ed. *In the mirror of memory: Reflections on mindfulness and remembrance in Indian and Tibetan Buddhism.* Albany, NY: State University of New York Press, pp. 1–20.

His Holiness the Dalaia Lama. (1995). *The path to enlightenment.* New Delhi: Montilal Banarsidass.

Johari, H. (2000). *Chakras: Energy centers of transformation.* New York: Simon & Schuster.

Khamtrul Jamyang Dondrup Rinpoche. (1959). *A concise guidebook to the hidden land of Pemako.* Accessed 20 September 2018, www.scribd.com/document/357451830/Orgyen-Norla-Guidebook-to-the-Hidden-Land-of-Pemako-Dudjom-Beyul

Lancaster, R. (1981). *Plant hunting in Nepal.* London: Croom Helm.

Massumi, B. (2010.) The future birth of the affective fact: The political ontology of threat. *In:* M. Gregg and G.J. Seigworth, eds. *The affect theory reader.* Durham, NC and London: Duke University Press, pp. 52–70.

Massumi, B. (2015). *Politics of affect.* Cambridge: Polity Press.

Mixter, D.W., and Henry, E.R. (2017). Introduction to webs of memory, frames of power: Collective remembering in the archaeological record. *Journal of Archaeological Method and Theory*, 24 (1): 1–9.

Monaghan, P. (2010). *Goddesses in world culture.* Santa Barbara, CA: Praeger.

Olick, J. (2008). From collective memory to the sociology of mnemonic practices and products. *In:* A. Erll and A. Nünning, eds. *Cultural memory studies: An international and interdisciplinary handbook.* Berlin: Walter de Gruyter, pp. 151–161.

Pile, S. (2009). Emotions and affect in recent human geography. *Transactions of the Institute of British Geographers*, 35 (1): 5–20.

Rawson, P. (2012). *In dwelling places for the divine, the mysterious arts of sacred Tibet interweave Shamanism, Buddhism, magic and myth.* London: Thames & Hudson.

Sardar-Afkhami, H. (1996). An account of Padma-Bkod: A hidden land in southeastern Tibet. *Kailash*, 18 (3–4): 1–22.

Sara, E. (2007). Intuitions: Their nature and epistemic efficacy. *Grazer Philosophische Studien*, 74: 51–67.

Sardar-Afkhami, H. (1996). An account of Padma-Bkod: A hidden land in southeastern Tibet. *Kailash*, 18 (3–4): 1–22.

Saul, H. (2019). "Challenging demoniacal beings": Extinction, materialities and the mortal frontiers in Alexandra David- Neel's journeys in the Himalayan Highlands. *In:* H. Silverman, E. Waterton and S. Watson, eds. *Affective geographies of transformation, exploration and adventure: Rethinking frontiers.* London and New York: Routledge, pp. 102–116.

Saul, H., and Waterton, E. (2017). Restoring a Nyingma Buddhist monastery, Nepal. *In:* H. Silverman, E. Waterton and S. Watson, eds. *Heritage in action: Making the past in the present.* Cham: Springer, pp. 33–46.

Seigworth, G.J., and Gregg, M. (2010). An inventory of shimmers. *In:* M. Gregg and G.J. Seigworth, eds. *The affect theory reader.* Durham, NC and London: Duke University Press, pp. 1–28.

Thondup Rinpoche, T. (1997). *Hidden teachings of Tibet: An explanation of the Terma tradition of Tibetan Buddhism.* Boston: Wisdom Publications.

Walcott, S. (2006). Mapping from a different direction: Mandala as sacred spatial visualisation. *Journal of Cultural Geography*, 23 (2): 71–88.

Walser, J. (2009). The origin of the term 'Mahāyāna' (the great vehicle) and its relationship to the Āgamas. *Journal of the International Association of Buddhist Studies*, 30 (1–2): 219–252.

Wayman, A. (1992). Buddhist terms for recollection and other types of memory. *In:* J. Gyatso, ed. *In the mirror of memory: Reflections on mindfulness and remembrance in Indian and Tibetan Buddhism.* Albany, NY: State University of New York Press, pp. 133–148.

Weintraub, S. (2012). *The hidden intelligence: Innovation through intuition.* London: Routledge.

35
CREMATION AND CONTEMPORARY CHURCHYARDS

Howard Williams and Elizabeth Williams

Introduction

Cremation is a complex and variable fiery technology. Across the human past and present, fire has been variously deployed to transform the dead in a range of spatial and social contexts. Often operating together with other disposal methods, cremation has risen and fallen in popularity in association with many shifts in mortuary practice since the Stone Age (Cerezo-Román & Williams 2014; Williams et al. 2017). Yet 'cremation' is far more than just the fiery dissolution of the human cadaver: in the human past and present it is often part of a multi-staged mortuary process that can afford a range of distinctive spatial and material possibilities for the translation and curation of the 'cremains' or 'ashes' together with a range of other material cultures and substances. By rendering cadavers fragmented, shrunken, and distorted, burning bodies not only denies decomposition and speeds corpse transformation, it renders the dead portable and partible. In a range of subsequent post-cremation practices and beliefs, 'ashes' from pyres can be considered a versatile mnemonic and numinous substance which might be consigned to graves and tombs, but also readily strewn over land and water or integrated into above-ground architectures and portable material cultures. Hence, not only does cremation involve fiery transformation, it facilitates the creation of varied and distinctive landscapes of death and memory through the deposition and commemoration of the dead in which ashes facilitate remembering and forgetting through their presence and their staged absence.

This characterisation prompts the archaeological investigation of cremation in the present as well as the past. For this chapter we address examples from the English county of Cheshire, and the Welsh county borough of Wrexham to reflect on broader tends in the Global West: how Christian churchyards have been adapted in response to the rise of cremation practices over the last century. We refer to these environments as 'deathscapes' to encapsulate the social, emotional, material, and spatial dimensions of cremation burials and memorials in churchyards (Maddrell & Sidaway 2010). By deathscapes, we refer not only to the individual cremation plaques and gravestones, and the burial plots in which they are situated, but also the networks of relationships (both planned and perhaps incidental) between cremation burials with other churchyard features, including gates, walls, paths, borders, inhumation graves, war memorials, and church buildings.

We contend that through these spatial and material intersections, as much as the individual depositions and memorials themselves, cremation has increasingly configured social memory for mourners and churchyard visitors of the last half century. The creation, use, and reuse of specific arrangements of cremation memorials stage the presence and absence of the dead body (cf. Sørensen 2009), and thus show the tensions between diocesan rules and regulations, and choices negotiated by parochial councils, church wardens, and local people regarding how their churchyard is managed and used (see also Rugg 2013a and b). As such, the patterns and the variability we encounter show the late-twentieth- and early-twenty-first-century development of the churchyard as a site of memory.

An archaeological approach to this theme can shed light on trends and variations hitherto unexplored by other disciplines who, as Julie Rugg (2013a) has rightly criticised, have tended to overlook the detailed and nuanced resistance to regulations through individual and group agency in churchyard use. Furthermore, Rugg challenges the tendencies of researchers to equate the rise of cremation with a decline in the desire for monumental memorialisation (Rugg 2013a: 223). Tackling the material evidence from churchyards counters such characterisations, and paves the way for a new 'contemporary archaeology of cremation' in the churchyard that responds to recent arguments for foregrounding the material and spatial dimensions of death in modernity. In particular, it shows how a fine-grained archaeological perspective affords a detailed and contextual approach to death and memory in contemporary landscapes, which responds to sociological critiques with a single grand narrative about mortality in the modern world (e.g. Woodthorpe 2010a). Simultaneously, we provide a case study in how contemporary archaeologists can address global themes regarding memory and landscape through relatively small-scale local and regional investigations of mortuary environments (cf. Harrison & Schofield 2011).

Background: cremation in the modern world

Cremation re-emerged in Europe and North America in the closing decades of the nineteenth century and has risen inexorably in popularity over the last 150 years (Davies with Mates 2005: 432–473). Cremation today is often crudely caricatured as part of the 'modernist' project, and in particular owes much to developments in disposing of the dead in the post–Second World War era (Mytum 2004a: 164–165; Grainger 2005; Parsons 2005). For example, almost four in five people in the United Kingdom are now disposed of by cremation (Rugg 2013b: 340). Likewise, across much of the Global West as well as other parts of the world, under the technological, urban, demographic, economic, environmental, medical, and consumer trends and pressures of the twentieth and early twenty-first centuries, cremation has emerged as the dominant mode of disposal and commemoration practised by many different cultural and religious groups.

The relationship between cremation and religion is complex and changing. Not only is there diversity within as well as between religious traditions in attitudes towards cremation, but the choice and manner of cremation is also influenced by a host of other social, cultural, economic, and political factors.

Yet in general terms, while some major world religious traditions have been staunchly opposed to cremating the dead – notably Islam (Turner 2005: 271–273) and Judaism (Pursell 2005: 286–287) – many religious traditions have long preferred cremation (Hinduism, Sikhism, and Buddhism). In Europe and elsewhere, Catholic and Orthodox Christian churches have staunchly support inhumation and discourage cremation (Davies with Mates 2005: xx; Newton 2005: 108–109), yet cremation has continued to rise in popularity in nations influenced by these traditions. Meanwhile Protestant Christian and secular nations have become the most cremation dominated (Davies with Mates 2005: xxi). For Catholics in the United Kingdom, cremation has been permitted since 1963, while Anglican attitudes have shifted and become increasingly positive towards cremation, mirroring the widespread popularisation of cremation from the 1950s onwards (Jupp 2005).

Hence, cremation cross-cuts many religious traditions and is the choice disposal method for many who have no fixed or firm cultural or religious views on the afterlife. Reflecting both intense regulation over both the dead body and cemetery space, but also individual agency regarding the choice of disposal and memorial form, location, and elaboration (see Dawdy 2013; Anthony 2016; Williams & Wessman 2017), cremation is thus a global phenomenon of modernity. As such, it has participated in the emergence of new, complex, and distributed mortuary environments (e.g. Williams 2011a). Furthermore, cremation's material impact on the landscape has augmented and transformed traditional cemeteries and churchyards, originally designed exclusively or primarily for the inhumed (unburned) disposal of corpses.

Studying contemporary cremation

As the dominant disposal method, cremation has been subject to interdisciplinary research from a range of historical, anthropological, and sociological perspectives. These include exploring crematoria architecture and their gardens of remembrance (e.g. Davies 1996; Grainger 2005, 2010), cemeteries (Woodthorpe 2010a 2010 b, 2011), and the relationship between cremation burials and the 'traditional' grave (Kellaher et al. 2005). Discussions have also focused on ash-scattering and memorialisation in the 'new' designed landscape and the countryside (e.g. Prendergast et al. 2006; Hockey et al. 2007; Kellaher & Worpole 2010; Kellaher et al. 2010).

What can archaeology bring to such studies? Contemporary archaeology – exploring the material cultures and landscapes of modernity – is a burgeoning field of research. It deploys specifically archaeological methods and/or focuses attention on the testimony of material culture and landscapes to reveal processes and patterns in human societies (Harrison & Schofield 2011). While contemporary archaeologists have tackled a wide range of present-day environments and themes, from parks to council houses, most studies of the 'contemporary past' have largely eschewed mortuary environments because of their attendant ethical challenges (Harrison 2011; Harrison & Schofield 2011; Graves-Brown et al. 2013).

While not denying there are multiple ethical issues with recording extant and still-used mortuary environments during fieldwork and writing, there is considerable potential for archaeological explorations of both below- and above-ground contemporary deathscapes (Williams 2011a and b; see also Anthony 2016). Indeed, it might be argued that archaeologists need to actively challenge and contextualise the perceived 'taboo' nature of contemporary mortuary environments for academic discussion (see also Woodthorpe 2010a), especially as the study of 'past life' to the exclusion of death is integral to the tropes of archaeology as a modernist project (cf. Harrison 2011). Hence, the absence of mortuary and commemorative traces from theoretical debates regarding the nature of archaeological investigations of the contemporary world reveals the struggles we continue to tackle in conceptualising the archaeologist's theoretical and methodological toolkit for investigating today's societies. For instance, churches and their yards can be considered a set of 'surfaces' constituted by generations of intercutting graves and memorials. Alternatively, they might be viewed as an accumulating and fluctuating 'assemblage' of both living people (worshippers, mourners, and others) and material traces of past lives.

A small range of archaeological researchers have begun to buck the life-focused trend of contemporary archaeology to explore the material and spatial dimensions, above and below ground, of later-twentieth- and early-twenty-first-century death ways (notably Mytum 2004a and b; Sørensen & Bille 2008; Sørensen 2009; Corkill & Moore 2012; Rebay-Salisbury 2012; Dawdy 2013; Parker & McVeigh 2013; Anthony 2016). Within this research, cremation has been both directly and indirectly tackled, focusing on how crematoria operate in technological and ritual terms (Back Danielsson 2009; Oestigaard 2013), as well as how cremation has augmented and transformed Europe's suburban cemeteries (Parker Pearson 1982; Anthony 2016; Williams & Wessman 2017). There have been discussions of relationships between ash-disposal and memorialisation in new environments, notably zoos and animal sanctuaries (Williams 2011a).[1] Furthermore, cremation's materialities and spatialities can be understood in relation to a wide range of cenotaphic memorial landscapes (e.g. Williams 2014). Indeed, the relationship between present-day cremation and the way we display cremation from the human past in modern museums sheds light on the role of archaeological and museological practices in engagements with mortality past and present (Williams 2016).[2]

Yet how might we apply these archaeological perspectives on cremation towards churchyards specifically?

Despite the relatively greater distances of rural churchyards from crematoria (in contrast to municipal cemeteries) cremation has become an integral part of churchyard transformations in the later twentieth century. However, cremation's impact on churchyards has been ignored in detail (see Rugg 2013a and b). To date, this specific dimension of cremation in late modernity has received limited material and spatial attention beyond a recognition that PCCs (parish county councils) often chose different memorial options and regulation for cremation interments and scattering (e.g. Rugg 2013b: 346–348). This is where an archaeological perspective can help: focusing on the material and spatial arrangements of cremation practices and offering new perspectives on this phenomenon.

The approach adopted here is therefore broadly framed in relation to recent trends in the archaeology of the post-medieval and contemporary past (see also Mytum 2004a; Williams 2011b; Anthony 2016) rather than the use of the present to provide a specific set of analogies for the interpretation of the human past (see Parker Pearson 1982; Downes 1999). Equally, the approach proposed is as much an extension of recent approaches to emotion, personhood, and social memory in mortuary archaeological research. Specifically, it can be considered in relation to the subfields of the archaeology of cremation (Cerezo-Román & Williams 2014; Williams et al. 2017) and the public archaeology and heritage of death and memory (Holtorf & Williams 2006; Giles & Williams 2016), as well as a contribution to the contemporary archaeology of death (see Anthony 2016).

Archaeologists are, therefore, well placed to identify the complexity and variability of how the technology of cremation both facilitates and offers new opportunities for memorialisation in the most traditional of spaces: the churchyard. Previous research has explored the materialities and spatialities of Danish churchyards, including their 'lawn cemeteries' (Sørensen 2009) and of Swedish rural churchyards and their 'minneslundar' (gardens of remembrance) (Williams 2011c, 2012). For the United Kingdom, we have tackled how cremation plots constitute distinctive elements of churchyards to be studied and explored within public archaeology projects (Williams & Williams 2007). We now return to this phenomenon following years of conducting archaeological surveys of churchyards and field visits in Cheshire and north-east Wales.[3]

Archaeologists can readily identify themes, but also local variations, in considering the interplay of cremation-related spaces and practices in churchyards, thus challenging simplistic narratives about the rise of cremation in modernity. Rather than the rise of cremation and its memorial indices being seen as anonymous and amorphous: a weakening of the connection between body and landscape with ritual and mourners (Curl 2002; Worpole 2003: 183–187; Mytum 2004a: 164–165), this chapter looks at the intersectional agencies of diocesan regulations, parish councils and individual mourners to create a diverse range of new memorial environments. Likewise, the consistency and modesty of cremation memorials are also regarded as evidence of the strict regulatory control over how the dead are to be commemorated in twentieth-century societies (e.g. Mytum 2004b), and yet, as Rugg (2013a) clearly argues, this approach denies the agency of mourners to select and adapt the memorial choices available to them within capitalist consumerist societies in which death is often a money-making industry as well as a regulated civil service. In this regard, cremation is not simply associated with restricted disposal choices, on the contrary, cremation facilitates a wide range of new commemorative strategies, fuelled by new technologies and business initiatives. The memorial modesty and/or neglect associated with cremation is therefore not a symptom of those opting for cremation failing to memorialise per se or the triumph of regulation over individuality, but a shifting relationship between memory and material culture and the ongoing tension between regulation and individual/family strategies of commemorative expression and material consumption during death rituals.

Within the diversity of ash disposal strategies associated with modern cremation, many may indeed choose to do nothing with the ashes or else have them disposed of in ephemeral ways outside the cemetery. For those choosing to deploy traditional churchyards, cremation promotes the extension and reuse of the environment, facilitating ongoing bonds with the dead and the places connecting the living and the dead. Churchyards in this sense have a series of distinctive and evolving roles.

Churchyard features and the church: cremation switchback

Mytum (2004b) notes how historic churchyards often exhibit horizontal stratigraphy. Later gravestones are situated increasingly farther from the church over time in a regulated fashion, with many often located in one or more churchyard extensions.

This applies to cases where extensions append the historic churchyard, and widespread examples where space is not available for such extensions and instead separate parochial cemeteries are established disconnected from the original yard. Cremation memorials support, but can also subvert, this spatial trend: a phenomenon I refer to as 'cremation switchback.'

This refers to the opportunity afforded by cremation burial to reuse older – usually nineteenth-century – parts of the churchyard Such plots often afford close proximities to the church building itself and other prominent 'landmark' features of the historic churchyard, such as sundials and First and Second World War memorials (see also Walls 2011). Cremation interments are shallow, and therefore, reoccupying such spaces need not disturb earlier inhumation graves located far deeper in the ground.

Therefore, cremation memorials constitute a *surface* reuse: overlaying but not disrupting earlier graves.

A good, modest-sized example of this trend is the cremation plot at St Peter's, Delamere (Cheshire). Here, there is a hedged cremation burial plot that contains a patio of small, square cremation plaques of consistent grey colour dating to between the 1980s and 2000s (Figure 35.1). The plot is situated beside the path immediately south of the church. As well as proximity to the church and its principal south door, the memorials wrap around two nineteenth-century grave-slabs, so the respect for, and integration with, the existing, historic space is articulated.

A memorial bench provides a place to reflect and look out over the cremation plot and the wider churchyard. The space is therefore between church and churchyard, situated in its own space.

In further instances, the cremation switchback is complemented by further cremation plots established in the most recent churchyard extensions or parochial cemeteries. Such instances sometimes appear when

Figure 35.1 Cremation memorials beside the medieval church of St Peter's, Delamere (Cheshire)
Source: Photograph by Howard Williams (2009).

the original cremation plot close to the church becomes full, but in other occasions the cremated dead are afforded contemporaneous options for location either beside the church or in the churchyard extension.

The most elaborate example encountered is situated at the east end of St Bartholomew, Great Barrow church (Cheshire), where there is an ornate and enclosed rockery garden with a diversity of styles of memorials to the cremated dead dating to the 1980s, 1990s, and 2000s.

These include headstones, horizontal plaques of different stone colour and shape, a bird bath, memorial trees, and rose bushes as well as memorials built into the front wall of the rockery.

These are adjacent to the path running around the east end of the church and opposite a series of surviving nineteenth-century crosses, headstones, and chest tombs. Amidst the rockery, and operating as the focus of the cremation memorial, is a late nineteenth-century diminutive memorial to the daughter of the Rector of the parish. This space is accessed from the rest of the churchyard through a portal through a high hedge. This private, concealed space for the cremated dead, reusing the site of a child's grave, therefore sits in close proximity to the church, while a larger garden of remembrance also operates in the churchyard extension.

The same relationship was observed at St Mary's, Chirk (Wrexham) where cremation memorials are situated in an open lawned plot at the east end of the church and dating from the 1960s to the 2000s.

This is augmented by a larger dedicated cremation burial plot in the churchyard extension from the mid-1970s to the 2000s, but clearly both locations have operated and been augmented in parallel (Figures 35.1, 35.2 and 35.3). In these instances, the church and its yard are revitalised with the condensed clustering of cremation memorials in multiple locations.

Figure 35.2 Cremation memorials at the east side of the medieval church of St Mary's, Chirk, Wrexham.

Figure 35.3 Cremation memorials on the south side of the churchyard extension of St Mary's, Chirk, Wrexham.

Populating paths and walls

Cremation memorials are often small, but through their location, arranged in rows or small plots, they might be prominent within churchyard spaces. Memorials from the 1980s to the present can be seen prominently interpolated within areas of surviving Victorian graves in many churchyards, as at All Saints, Gresford (Wrexham) and St James', Christleton (Cheshire): in the latter case the plot was succeeded by more formalised cremation plots in the churchyard extension from the mid-2000s (Figure 35.4).

Figure 35.4 Cremation memorials in defined plots at St James' Christleton (Cheshire).
Source: Photograph by Howard Williams (2009).

There are many further cases where cremation memorials interact with the layout of churchyards in a distinctive set of ways, filling in gaps too narrow for inhumation graves, and often situated alongside paths and/or beside walls. The cremated dead are thus connected to key routes of movement through the churchyard and to the church. At St Dunawd's, Bangor on Dee (Wrexham), the main line of cremation memorials – largely consistent black headstones of standard height comparable to inhumation graves and dated from the late 1980s through to the early 2000s–flank the principal path on its west side from the lychgate towards the south door of the church (Figure 35.5a). Further cremation plaques are situated east of the church dating to the 2000s.

Figure 35.5a and b Cremation memorials beside paths at (a) St Dunawd's, Bangor on Dee (Wrexham), (b) St Chad's, Farndon (Cheshire)

Source: Photographs by Howard Williams (a: 2018, b 2009).

Figure 35.6 Cremation graves outside the original churchyard wall in a graveyard extension, St Marcella's, Marchwiel (Wrexham).

A fuller-still integration of cremation memorials can be found at St Chad's, Farndon (Cheshire). Here, all the main pathways around the south-side of the churchyard are framed by the cremated dead of the late 1960s to 2000s, interpolated between Victorian gravestones and framing approaches to the sundial and the war memorial (Figure 35.5b). They are thus situated in contrasting orientations as they face the path itself rather than a conventional W–E alignment of the nineteenth-century headstones and chest tombs. At St Deiniol and St Marcella's, Marchwiel (Wrexham), cremation memorials wrap around the east end of the church dating from the 1980s, but also they occur in the border between the churchyard wall and the path in the extension during the 1990s and 2000s (Figure 35.6).

Complementing these examples where paths are the foci, there are instances where historic churchyard walls are deployed to frame new cremation memorials. At St Boniface, Bunbury (Cheshire), a regular distinctive yellow-grey colour of miniature headstone has been adopted for most memorials since the late 1990s. They are positioned inside the churchyard boundary looking inward towards the church and over

Figure 35.7a and b Cremation memorials beside churchyard walls at (a) St Boniface, Bunbury (Cheshire) and (b) St Mary's, Eccleston (Cheshire).

Source: Photographs by Howard Williams (2009).

the Victorian memorials (Figure 35.7a). Looking outwards, cremation burials of different forms are situated along the low, original churchyard boundary for the now-abandoned ruined historic church at St Mary's, Eccleston (Cheshire) (Figure 35.7b). Whether paths, walls, or both, these examples show how cremation plots, and individual memorials, can be integrated into the deathscape of the historic churchyard, juxtaposed next to far older memorials and thus revitalising the space for those traversing it en route to the church.

Elaborating extensions

We have already alluded to cremation plots in churchyard extensions and it is to these we will now turn. Mytum (2004b) offers a regulatory view of cremation plots in discussing the longue durée of the traditional churchyard at Kellington, North Yorkshire, suggesting they reflect an 'industrial management' affecting inhumations and cremations alike. However, in management and appearance, churchyard extensions offer a range of options for cremation plots that challenge Mytum's characterisation.

At St Helen's, Tarporley (Cheshire), the cremation memorials are clustered in a comparable plot to the inhumation graves, but their small size and close proximity affords them a distinctive appearance. Another fine example of this phenomenon is at St Peter's, Waverton (Cheshire), where the extension to the churchyard took place in two stages to the south of the historic churchyard (Figure 35.8). The principal WSW–ENE orientation of the inhumation graves in ordered rows towards a path that runs NNW–SSE close to the

Figure 35.8 Cremation memorials in relation to churchyard extensions at St Peter's, Waverton (Cheshire).
Source: Photograph by Howard Williams (2009).

Figure 35.9a and b Cremation memorials at St Andrew's, Tarvin (Cheshire), (a) a garden of remembrance as a focal point of inhumation graves, and (b) cremation graves against the outside of the old churchyard wall.

Source: Photographs by Howard Williams (2018).

eastern border of the plot. While the inhumation memorials face ENE, in contrast, the cremation memorials – small headstones mostly dating from the 1980s onwards – are situated in two rows on the narrow space between the path and the fence, facing WSW towards the path and opposing the inhumation memorials.

The forms of the memorials vary, but they are generally reduced-size versions of standard headstones.

At St Andrew's, Tarvin, the second, most recent, churchyard extension contains inhumation graves aligned upon a centrally situated 'garden of remembrance,' with a bench, a dwarf wall around which memorial flower holders are situated, and an inner area for floral offerings (Figure 35.9a). Further memorials dating from the very late 1990s onwards are fixed to the outside of the old churchyard wall, augmented by a parallel dwarf wall (Figure 35.9b).

These examples suffice to show how churchyard extensions divide the dead between those inhumed and those that cremate, but the extension also facilitates a dialogue or interplay between the memorial spaces. The cremated dead are afforded a distinctive set of alternative deathscapes upon both horizontal and vertical surfaces, set apart from, but in close proximity to, the inhumed dead.

Conclusions

Julie Rugg (2013a) has robustly criticised how the twentieth-century (and we would add, also early-twenty-first-century) history of death is largely under-researched, with attempts to mischaracterise modernity and its attitudes to death in crude and general terms in contrast to a caricatured Victorian way of dead. She identifies increasing regulation as resulting in more modest commemorative expressions rather than 'disengagement' from mortality. However, researchers have yet to explore in detail the new and emergent deathscapes for the cremated dead of the latter part of the twentieth century and the first decades of the twenty-first century (Rugg 2013b: 316, 346–348).

Contemporary archaeology, focusing on the emergent surfaces and spaces, materials, and material cultures of death, offers one avenue to counter grand narratives regarding death in modernity, but also to see modernity as an unfinished project (see Harrison 2011: 153). Indeed, exploring within the gates of the cemetery and churchyard challenges the bracketing of mortuary spaces as ethically 'out of bounds.' It also challenges the romanticisation of these spaces as ruins and relics of earlier epochs, thus making us look afresh at spaces traditionally investigated by medieval and early modern archaeologists. It allows us to explore the fine-grained detail of how mourner's personal/family agency, and community agency more broadly, interacts with regulations controlling cemeteries. Archaeological surveys and observations reveal how cremation has facilitated new, modest, miniature, and condensed memorials. The superficial 'banality' or 'monotony' of these memorials and motifs (Rugg 2013a: 229–230) conceals considerable personal choice and variability, sometimes planned, often improvised (contra Mytum 2004b). A dismissive view of modern churchyard memorials thus fails to recognise the distinctive ways cremation memorials interact with churchyard space and architecture. Whether by design or happenstance, these planned yet cumulative and condensed, almost intimate yet collective, arrangements of cremation memorials have created new forms of emotive and mnemonic deathscape, perpetuating and diversifying engagements between the living and the dead within the churchyard space for mourners and visitors alike.

In the examples discussed in this short chapter, we can see how even a brief survey of trends reveals the fluctuating and innovative uses of churchyard space, including the revitalisation of areas around church buildings, paths, and walls, and churchyard extensions with small and miniature, by intense and vibrant memorial practices. Rather than complementing existing narratives of death in England in the later twentieth and early twenty-first centuries, the evidence reveals varied avenues by which local communities adapt and adopt churchyards for the memorialisation of the dead, often deliberately interpolating the cremated dead with far older features to integrate them with the perceived antiquity and sacrality of holy ground.

This highlights the enduring mnemonic and numinous significance of ashes' relationship to place and memory. Also, it serves as a key example of what Harrison (2011) constitutes as a key shift in archaeology's focus to the present and the future, yet challenges the still near-exclusive focus of archaeology on modern 'life' and 'living' without adequate dialogue with other discipline's explorations of contemporary death ways. Yet how we engage with mortality and memorialisation is central to our concerns, and challenging the life focus of archaeology is to critique one of archaeology's most distinctive traits as a modernist project. This has implications on individual and community levels, but also for our present-day landscapes, including holy sites. By exploring contemporary archaeologies of the churchyard as a deathscape, we reveal the complex interplay between inhumation and cremation are fostering emergent dialogues between the living and the dead.

Acknowledgements

This chapter began its life as a conference presentation at the 2010 *A Good Send Off* conference organised by the Centre for Death and Society at the University of Bath. We thank the organiser and audience of this event, and subsequently, after years in gestation, to the editors for their invitation to contribute and their constructive input.

Notes

1 Ash-scattering, for example, is a regular feature of heritage sites including castle ruins – https://howardwilliamsblog.wordpress.com/2017/01/04/ashes-at-the-castle/; https://howardwilliamsblog.wordpress.com/2016/11/09/ashes-around-castell-dinas-bran/; and https://howardwilliamsblog.wordpress.com/2016/08/30/the-secret-dead/. For a cremation memorial on country estates in the care of the National Trust: https://howardwilliamsblog.wordpress.com/2017/11/05/whose-ashes-rest-here-cremation-and-memorial-archaism-at-attingham-park/. A further example is this private burial ground beside a café and motel: https://howardwilliamsblog.wordpress.com/2017/12/03/an-island-of-the-cremated-dead/.

2 For other discussions of cremation in museums, see: https://howardwilliamsblog.wordpress.com/2017/05/14/cremation-on-display-at-the-museum-of-london/

For memorial environments, see: https://howardwilliamsblog.wordpress.com/2017/06/10/the-alrewas-beaker-the-use-of-the-bronze-age-in-the-21st-century-commemoration/

3 See also the Archaeodeath blog for entries relating to cremation in the contemporary churchyard for: Pennant Melangell: https://howardwilliamsblog.wordpress.com/2014/07/19/horizontal-stratigraphy-and-cremation-switchback-in-the-welsh-churchyard/; Bangor-is-y-Coed: https://howardwilliamsblog.wordpress.com/2015/09/28/cremation-switchback-at-bangor-is-y-coed/; Meifod: https://howardwilliamsblog.wordpress.com/2015/10/15/sun-time-and-cremation-meifod-memorials/; Wrexham: https://howardwilliamsblog.wordpress.com/2017/10/23/eliugh-yales-tomb/

References

Anthony, S. (2016). *Materialising modern cemeteries: Archaeological narratives of Assistens Cemetery, Copenhagen*. Doctoral Thesis, University of Lund, Lund.

Back Danielsson, I-M. (2009). A rare analogy: Contemporary cremation practices. *In:* I-M. Back Danielsson, I. Gustin, A. Larsson, N. Myrberg and S. Thedéen, eds. *On the threshold: Burial archaeology in the twenty-first century*. Stockholm: Department of Archaeology, University of Stockholm. Stockholm Studies in Archaeology 47, pp. 57–80.

Cerezo-Román, J.I., and Williams, H. (2014). Future directions for the archaeology of cremation. *In:* I. Kuijt, C.P. Quinn and G. Cooney, eds. *Transformation by fire: The archaeology of cremation in cultural context*. Tucson: University of Arizona Press, pp. 240–255.

Corkill, C., and Moore, R. (2012). 'The Island of blood': Death and commemoration at the Isle of man TT races. *World Archaeology*, 44 (2): 248–262.

Curl, J.S. (2002). *Death and architecture*. Third Edition. Stroud: Sutton.

Davies, D.J. (1996). The sacred crematorium. *Mortality*, 1 (1): 83–94.

Davies, D.J., and Mates, L.H. eds. (2005). *Encyclopedia of cremation*. Aldershot: Ashgate.

Dawdy, S.L. (2013). Archaeology of modern American death: Grave goods and blithe mementoes. *In:* P. Graves-Brown, R. Harrison and A. Piccini, eds. *The archaeology of the contemporary world*. Oxford: Oxford University Press, pp. 451–465.

Downes, J. (1999). Cremation: A spectacle and a journey. *In:* J. Downes and T. Pollard, eds. *The loved body's corruption: Archaeological contributions to the study of human mortality.* Glasgow: Cruithne Press, pp. 19–29.

Giles, M., and Williams, H. (2016). Introduction: Mortuary archaeology in contemporary society. *In:* H. Williams and M. Giles, eds. *Archaeologists and the dead.* Oxford: Oxford University Press, pp. 1–20.

Grainger, C. (2005). *Death redesigned: British crematoria: History, architecture and landscape.* Reading, MA: Spire.

Grainger, C. (2010). Maxwell Fry and the 'anatomy of mourning': Coychurch Crematorium, Bridgend, Glamorgan, South Wales. *In:* A. Maddrell and J.D. Sidaway, eds. *Deathscapes.* Farnham: Ashgate, pp. 243–262.

Graves-Brown, P., Harrison, R., and Piccini, A. eds. (2013). *The archaeology of the contemporary world.* Oxford: Oxford University Press.

Harrison, R. (2011). Surface assemblages. Towards an archaeology *in* and *of* the present. *Archaeological Dialogues,* 18 (2): 141–161.

Harrison, R., and Schofield, J. (2011). *After modernity: Archaeological approaches to the contemporary past.* Oxford: Oxford University Press.

Hockey, J., Kellaher, L., and Prendergast, D. (2007). Sustaining kinship: Ritualization and the disposal of human ashes in the United Kingdom. *In:* M. Mitchell, ed. *Remember me: Constructing immortality – beliefs on immortality, life, and death.* London: Routledge, pp. 35–50.

Holtorf, C., and Williams, H. (2006). Landscapes & memories. *In:* D. Hicks and M. Beaudray, eds. *Cambridge companion to historical archaeology.* Cambridge: Cambridge University Press, pp. 235–254.

Jupp, P. (2005). The UK: Catholic and Anglican churches. *In:* D. Davies and L.H. Mates, eds. *Encyclopedia of cremation.* Aldershot: Ashgate, pp. 113–116.

Kellaher, L., Hockey, J., and Prendergast, D. (2010). Wandering lines and cul-de-sacs: Trajectories of ashes in the United Kingdom. *In:* J. Hockey, C. Komaromy and K. Woodthorpe, eds. *The matter of death: Space, place and materiality.* London: Palgrave Macmillan, pp. 133–147.

Kellaher, L., Prendergast, D., and Hockey, J. (2005). In the shadow of the traditional grave. *Mortality,* 10 (4), 237–250.

Kellaher, L., and Worpole, K. (2010). Bringing the dead back home: Urban public spaces as sites for new patterns of mourning and memorialisation. *In:* A. Maddreell and J.D. Sidaway, eds. *Deathscapes.* Farnham: Ashgate, pp. 161–180.

Maddrell, A., and Sidaway, J.D. eds. (2010). *Deathscapes: Spaces for death, dying, mourning and remembrance.* Farnham: Ashgate.

Mytum, H. (2004a). *Mortuary monuments and burial grounds of the historic period.* New York: Kluwer/Plenum.

Mytum, H. (2004b). Rural burial and remembrance: Changing landscapes of commemoration. *In:* D. Barker and D. Cranstone, eds. *The archaeology of industrialization.* Leeds: Maney, pp. 223–240.

Newton, J. (2005). Catholic church. *In:* D. Davies and L.H. Mates, eds. *Encyclopedia of cremation.* Aldershot: Ashgate, pp. 107–109.

Oestigaard, T. (2013). Cremations in culture and cosmology. *In:* S. Tarlow and L. Nilsson Stutz, eds. *The Oxford handbook of the archaeology of death and burial.* Oxford: Oxford University Press, pp. 497–509.

Parker, G., and McVeigh, C. (2013). Do not cut the grass: Expressions of British Gypsy-Traveller identity on cemetery memorials. *Mortality,* 18 (3): 290–312.

Parker Pearson, M. (1982). Mortuary practices, society and ideology: An ethnoarchaeological study. *In:* I. Hodder, ed. *Symbolic and structural archaeology.* Cambridge: Cambridge University Press, pp. 99–113.

Parsons, B. (2005). *Committed to the cleansing flame: The development of cremation in nineteenth-century England.* Reading, MA: Spire.

Prendergast, D., Hockey, J., and Kellaher, L. (2006). Blowing in the wind? Identity, materiality, and the destinations of human ashes. *Journal of the Royal Anthropological Institute,* 12: 881–898.

Pursell, T. (2005). Judaism. *In:* D. Davies and L.H. Mates, eds. *Encyclopedia of cremation.* Aldershot: Ashgate, pp. 284–286.

Rebay-Salisbury, K. (2012). Inhumation and cremation: How burial practices are linked to beliefs. *In:* M.L.S. Sørensen and K. Rebay-Salisbury, eds. *Embodied knowledge: Historical perspectives on technology and belief.* Oxford: Oxbow, pp. 15–26.

Rugg, J. (2013a). Choice and constraint in the burial landscape: Re-evaluating twentieth-century commemoration in the English churchyard. *Mortality,* 18 (3): 215–234.

Rugg, J. (2013b). *Churchyard and cemetery: Tradition and modernity in rural north Yorkshire.* Manchester: Manchester University Press.

Sørensen, T.F. (2009). The presence of the dead: Cemeteries, cremation and the staging of non-places. *Journal of Social Archaeology,* 9: 110–135.

Sørensen, T.F., and Bille, M. (2008). Flames of transformation: The role of fire in cremation practices. *World Archaeology,* 40 (2): 271–273.

Turner, C. (2005). Islam. *In:* D. Davies and L.H. Mates, ed. *Encyclopedia of cremation.* Aldershot: Ashgate, pp. 284–286.

Walls, S. (2011). 'Lest we forget': The spatial dynamics of the church and churchyard as commemorative spaces for the war dead in the twentieth century. *Mortality,* 16 (2): 131–144.

Williams, H. (2011a). Ashes to asses: An archaeological perspective on death and donkeys. *Journal of Material Culture*, 16 (3): 219–239.

Williams, H. (2011b). Archaeologists on contemporary death. *Mortality*, 16 (2): 91–97.

Williams, H. (2011c). Cremation and present pasts: A contemporary archaeology of Swedish memory groves. *Mortality*, 16 (2): 113–130.

Williams, H. (2012). Ash and antiquity: Archaeology and cremation in contemporary Sweden. *In:* A.M. Jones, J. Pollard, M.J. Allen and J. Gardiner, eds. *Image, memory and monumentality: Archaeological engagements with the material world.* Oxford: Oxford University Press, pp. 207–217.

Williams, H. (2014). Antiquity at the national memorial arboretum. *International Journal of Heritage Studies*, 20 (4): 393–414.

Williams, H. (2016). Firing the imagination: Cremation in the modern museum. *In:* H. Williams and M. Giles, eds. *Archaeologists and the dead: Mortuary archaeology in contemporary society.* Oxford: Oxford University Press, pp. 293–332.

Williams, H., Cerezo-Román, J.I., and Wessman, A. eds. (2017). Introduction: Archaeologies of cremation. *In:* J.I. Cerezo-Román, A. Wessman and H. Williams, eds. *Cremation and the archaeology of death.* Oxford: Oxford University Press, pp. 1–24.

Williams, H., and Wessman, A. (2017). The contemporary archaeology of urban cremation. *In:* J.I. Cerezo-Román, A. Wessman and H. Williams, eds. *Cremation and the archaeology of death.* Oxford: Oxford University Press, pp. 266–296.

Williams, H., and Williams, E.J.L. (2007). Digging for the dead: Archaeological practice as mortuary commemoration. *Public Archaeology*, 6 (1): 45–61.

Woodthorpe, K. (2010a). Buried bodies in an East London cemetery: Re-visiting taboo. *In:* A. Maddrell and J.D. Sidaway, eds. *Deathscapes.* Farnham: Ashgate, pp. 57–74.

Woodthorpe, K. (2010b). Private grief in public spaces: Interpreting memorialisation in the contemporary cemetery. *In:* J. Hockey, C. Komaromy and K. Woodthorpe, eds. *The matter of death: Space, place and materiality.* London: Palgrave Macmillan, pp. 117–132.

Woodthorpe, K. (2011). Sustaining the contemporary cemetery. Implementing policy alongside conflicting perspectives and purpose. *Mortality*, 16 (3): 259–276.

Worpole, K. (2003). *Last landscapes: The architecture of the cemetery in the West.* London: Reaktion.

36
RITUAL, PLACE, AND MEMORY IN ANCIENT ROME

Ana Mayorgas

The first and most comprehensive approach to the relationship between memory, ritual, and place in Antiquity was produced by Egyptologist Jan Assmann in his book, *Das kulturelle Gedächtnis. Schrift, Erinnerung und politische Identität in frühen Hochkulturen* (1992), translated into English in 2011 (Assmann 2011; see also Assmann 2010). Drawing on the work of sociologist Maurice Halbwachs, historian Pierre Nora, and anthropologists Lévi-Strass and Jan Vansina, he argued that ancient societies remembered their origins through ceremonies and festivals under the supervision of specialists. The remote past was ritually re-enacted in those events in which material culture and memory landscapes – *mnemotopes* – played also a significant role. For this collective recollection, Assmann coined the term 'cultural memory.' Assmann considered such memory exceptionally relevant, as it provided the group with a common identity and the means to understand their present situation. In contrast, what he called 'communicative memory' involved the non-institutional transmission of events from the recent past, which did not reach back more than three generations. Communicative memory, conveyed by informal means, was meant to be forgotten unless it became a meaningful past for the present group. That was the case with the genealogies of elite families, whose remembrance of the lineage tried to overcome the memory vacuum – the 'floating gap' in Vansina's words – dividing the historical events of the last 100 years from the mythical origins with which they sought to connect.

Assmann's conceptual framework had an initial cold reception among classicists. Only recently there seems to be an awareness of its potential, in parallel with a growing interest in exploring the interplay between memory and religion in antiquity (Bommas 2012; Cusumano 2013). But few attempts have actually been made to rethink ancient evidence through the concepts of 'cultural memory' and 'communicative memory,' although significantly all of them focused on ancient Rome (Hölkeskamp 2004: 137–168, 2006; Rodríguez Mayorgas 2007, 2010; Eckert 2012, 2014). Some voices have warned against a too-rigid implementation of Assmann's categories (Galinsky 2016: 11–15). Certainly as with any other theoretical framework, the right methodological approach entails using Assmann's reflections to ask new questions to the ancient evidence, and not simply to look for the confirmation of a pre-established scheme (Hölkeskamp 2014). In fact, such a scheme could hardly be of any use in the case of Rome, which moved from being an Italic city to the largest and longest lived empire in antiquity. Forcibly, Roman memory and identity changed over time, while the spread of literacy and the emergency of historiography introduced a new means of relating to the past. The aim of this contribution is to show how ritual, place, and memory interplayed in ancient Rome. The three elements interacted in manifold and complex ways throughout the history of Rome, and as will be evident, this interaction evolved from the Republic to the Empire.

Roman rituals of memory and power

During the Republic (509–31 BC), there was a twofold memory. On the one hand, the origin of the city was associated with the myth of the twin brothers, Romulus and Remus, and celebrated in festivals. On the other, the ruling class was commemorated in pageants and rituals in which the individual and collective merit of the oligarchy was exhibited. No connection was established between both memories. Contrary to the Hellenic genealogical tradition, the Roman leading families did not claim descent from gods or semi-divine figures until the Late Republic. Their legitimacy lay elsewhere, in office holding and military victories. Certainly Romulus, the first ruler of Rome, set an example as conqueror. From the beginning of Roman historiography in the late third century BC, the narration of his reign was full of military encounters and victories. He was the first to celebrate a triumph and to dedicate the spoils of war to Jupiter. Nevertheless, these aspects were not highlighted in the two festivals that commemorated the origin of the city: Lupercalia (15 February) and Parilia (21 April).

Originally both feast days were devoted to the ritual purification and protection of the herd, but at some point after the emergence of Rome as a city-state, they were associated with the twin brothers, in all likelihood because Romulus and Remus were thought to have grown up anonymously as shepherds until their royal ancestry was exposed. The pastoral atmosphere of both festivals reminded the Romans of the twins' youth. Therefore, the ritual performances carried out in those days were reinterpreted to fit the founding story of Rome (Beard 1987). The Parilia were celebrated in the countryside as well as in the city of Rome where a priest conducted a public ceremony (Scullard 1981: 103–105). The festival ended with the crowd jumping through bonfire. This purification rite was interpreted as the symbolic act carried out by the Roman population of the times of Romulus after abandoning and setting fire to their humble huts in order to found a new city. Accordingly, 21 April was considered the birthday of Rome, i.e. the day the city had been founded by Romulus.

The Lupercalia was also a festival originally devoted to the protection of cattle. A group of priests, called Luperci, performed a ritual sacrifice at the Lupercal, a sacred cave of the Palatine hill, where according to tradition a she-wolf had suckled the twins under a fig tree, *ficus Ruminalis*. Afterwards they ran around the hill almost naked striking with thongs the bystanders (Scullard 1981: 76–78). The ritual race was the most remarkable part of the festival. It started at a very symbolic place, the Lupercal. In Republican times, when the hill was highly urbanised, the cave and the fig tree were a reminder of the natural of the natural environment of the origins. Furthermore, from the early third century BC a statue of the twins fed by the she-wolf was set up nearby. Therefore it is not surprising that the performing priests were associated with Romulus and Remus. The race supposedly re-enacted the wild lifestyle of the twins when they were herdsmen. The relevance of the place, the Palatine hill, as a memoryscape or *mnemotope* of the origins is unquestionable. Up the hill, close to the Lupercal, the Romans preserved an old hut identified as Romulus' dwelling. Its sacred character is shown by the fact that public priests took care of its maintenance, avoiding the use of new materials. Therefore, place and ritual converged to commemorate the origin of Rome and to reproduce a memory of the humble and rustic Roman life of the foundation (Edwards 1996: 30–43).

These festivals remained popular throughout the Roman Empire. The Lupercalia is actually one of the pagan ceremonies known to have survived beyond the late fifth century AD despite the opposition of the Christian popes (McLynn 2008: 174–175). The Parilia were renamed as Romaia in the early second century AC by the emperor Hadrian, who also ordered the construction of a temple of Venus and Rome in the Forum, the civic centre of the city, to become the main stage of celebration. In contrast with the remembrance of the origins, the Roman elite developed its own means of memorialisation through ritual with the Forum as the main scenario. Its prestige relied on military success and office-holding, and was based on individual merit as well as on the collective glory of the family. The theatrical way of conveying this fame is very particular of Roman culture. It appeared in the late fourth century BC when magistrates started to capitalise on the organisation of public events in the city and were collectively called *nobiles*, that is, 'well known' or 'conspicuous,' stressing above all the relevance of their publicity and public profile (Flower 2004: 322–325). Two rituals organised by these *nobiles* were especially remarkable: triumphs and funerals.

Romans relied considerably on the gods in warfare and developed a wide range of war rituals to proceed according to the gods' will in order to win their favour. The triumphal parade was one of those rituals (Beard 2007). It was the more sought-after honour for any Roman aristocrat. It was granted by the Senate to those generals who achieved an outstanding military victory that ended a war or at least inflicted heavy damage to the enemy. The aim of this public celebration was to thank the main Roman god, Jupiter Optimus Maximus, for protecting the city and citizenry, so in essence the parade was a religious one whose last stop was the temple of this divinity at the Capitol. There, the victorious commander made a public sacrifice on behalf of the whole community. However, the main figure of the festive event was the Roman general. He was entitled for a day to stand out from his peer aristocrats who as senators had agreed on the concession in recognition of his military merits. Dressed as Jupiter, and wearing a crown of laurel, the commander rode on a special chariot in the middle of the procession (Beard 2007: 225–238). In front of him the magistrates paraded, along with booty, prisoners of war, models of captured cities, and representations of battles. The soldiers marched after him singing in joy. The celebration took up the city centre for a day, or even several days when the amount of spoils and captives increased significantly from the wars of conquest of the second century BC onwards. The processional route could be subjected to changes but invariably started from the Campus Martius, went through the Circus Flaminius, Forum Boarium, Circus Maximus, and Forum Romanum, and finally ascended to the Capitol (Beard 2007: 92–105). The celebration involved the whole population of Rome who could accompany the procession or watch it pass from the Circus seats or from the wooden stands installed in the Forum.

The triumphal parade left many traces in the city contributing to keep alive the memory not only of individual Roman generals but also of Rome's military success. The name of the victorious commander entered public record. The procession was reproduced in different manners. It was re-enacted at theatre and circus events, and especially represented in domestic objects, paintings, reliefs, coins, and friezes like those that decorated triumphal arches transmitting the solemn and ritual nature of the celebration (Favro 2014: 93–99). The triumphal arch was the monument that most vividly reminded people of the festive event. It appeared in the early second century BC to commemorate the triumphal parades most likely in imitation of the gate through which the procession entered the city. But other military monuments were also set up in the route of the triumph: statues, columns, and temples commissioned and dedicated mostly – although not exclusively – by triumphal commanders. The temples are especially relevant. Vowed on the battlefield by generals seeking to gain the divine favour, they were built from the profits of war and had a festival day associated for the anniversary. Finally war spoils as well as models and representations from the triumphal parade were deposited in temples. All these monuments crowded particularly the Capitoline hill and the Forum, turning the city centre into a real memoryscape that materialised the prestige of the ruling aristocracy (Hölkeskamp 2006: 483–491). These memorialising dynamics were reoriented in imperial times when Roman aristocracy was deprived of military glory by the emperor, who from that moment was the only one, along with his male relatives, entitled to celebrate military victories. The number of triumphs decreased but the emperor assumed its paraphernalia for certain ceremonial occasions enjoying an almost permanent triumphal status (Beard 2007: 272–277).

Roman funerals were a different kind of ritual, although equally theatrical and memory oriented (Flower 1996: 91–127). Every Roman family could celebrate a private funeral in the cemetery, but only those men who have held a high office were allowed to have a public funerary parade from his house to the Forum where the main stage of the procession took place. There, in the platform for political and judicial addresses (the *Rostra*) the oldest son of the deceased delivered a eulogy in front of the crowd stressing his father's merits. Afterwards the funeral cortège headed for the necropolis, where the proper ritual of internment took place privately. The speaker praised especially his father's military victories and the magistracies he held but the funeral procession was above all a show of power of the whole family. The coffin was accompanied by actors representing the deceased's ancestors who have also been elected for high offices. They wore realistic wax masks of the ancestors' faces and the appropriate garb of their highest magistracy. Those impersonating triumphant generals walked surrounded by paintings, models, and booty from the original parade. Escorted by attendants, a symbol of power in Rome, these ancestors solemnly sat down on chairs over the platform where the speaker recalled also their military deeds and political achievements. There could not be a more impressive display of family memory and prestige, re-enacted every time a funeral was celebrated. After the burial the recollection of the family fame continued at home where the ancestors' masks along with war booty were displayed in the public space of the house for all the visitors to admire (Flower 1996: 185–222). As in the triumph, the whole populace of Rome was summoned to the aristocratic funeral, which could be complemented with gladiatorial shows, theatrical performances, and a public banquet at the family's expense. By contrast, the individual merit that was so bombastically exhibited in the triumphal parade became in this occasion counterbalanced by the weight of the kinship group. On both events space was a fundamental factor. The Capitoline hill and the Forum were the heart of political and religious activity in Rome and consequently the natural stage for demonstration of political power. For this reason, triumphal parades went across the Forum by the so-called Via Sacra up to the Capitol, and the main stage of aristocratic funerals was the *Rostra* instead of the necropolis.

Rome as memoryscape of the empire

The new political regime established by Augustus (27 BC–AD 14) curtailed aristocratic competition. From then on, the only manner for the elite to excel socially was the loyal service to the emperor. This altered considerably the Republican dynamics of ritual, place, and memory. As was mentioned, triumphal parades were restricted to the imperial family while aristocratic funeral rites were progressively confined to the domestic space of the house and the necropolis. The display of ancestors' masks in the cortège fell into disuse in the first century AD. Besides, only prominent senators were granted the privilege of a funerary public eulogy at the *Rostra*, although not pronounced by the eldest son or a relative but delivered by another senator, hindering any possibility of boosting elite family glory (Bodel 1999: 265–272). Solely the imperial family was to enjoy in its own right a public procession before the burial, which starting from the noble tradition of the Republic turned into a dazzling spectacle of imperial power.

From Augustus' funeral, whose protocol he bothered himself to write down, the main elements of the rite remained, although with a new meaning. The cortège did not consist of the deceased's kin but senators and knights; the masks did not represent the family's ancestors but in a broad sense the forebears and leaders of Rome including Aeneas and Romulus; and an open display of grief was expected from the populace as a whole during the parade, followed by an official mourning period. The son or heir of the deceased still delivered a speech of praise; however that was not the end of the public spectacle as previously but the beginning. From the Forum cortège and spectators proceeded to the Campus Martius where the body – and effigy – was cremated in a pyre, which became increasingly more elaborate, and finally buried (Price 1987: 62–70). Therefore from early first century AD onwards aristocratic exequies gave way to state funerals, in which the memory of the glorious past of Rome was exhibited endorsing the emperors.

Augustus was also conscious of the evocative force of the urban landscape. By the time of his accession to power the civic centre of Rome was crowded with monuments recalling the military merits of the Republican aristocracy which could endanger the one-man rule he intended to impose. After 31 years of government, his vast programme of construction and reconstruction had reshaped dramatically Roman memory in the city consigning to oblivion a significant amount of Republican commissioners and creating a new forum as the political centre. Temples were one of his main objectives. Augustus himself claimed to have restored 82 temples in the city. There is no evidence for all of these interventions but the procedure is well documented. The emperor not only funded the reconstruction but also rededicated the shrine on a new date, most often his birthday, 23 September, and secondarily the day of a war victory obtained by someone from his entourage. Thus the memory of the original commissioner and the motive for the construction vanished. The new *dies natalis* of the temple, a celebration, which implied a ritual procession through the city, took place. It did not commemorate any longer the political or military achievement of an aristocrat from the Republic but the glory of Augustus as saviour of Rome and restorer of ancestral traditions (Orlin 2016: 122–124).

Augustus built also a new forum which superseded the Republican one as the centre of political activity in Rome. It was dominated by a temple of Mars that the emperor, as was the custom of Roman generals, had vowed to the god, on this occasion before the battle of Philippi (42 BC) against Caesar's assassins. Inside, stood statues of Mars, Venus – the divine forbear of the Julian family – and probably Julius Caesar, Augustus' adoptive father. By order of the Senate a monument of Augustus on a triumphal four-horse chariot was set up in the middle of the square. An inscription recalled all his military successes and the granting of the honorific title of "Father of the Fatherland" (Pater Patriae). But the new forum was not only a place to extol the figure of the new ruler of Rome. The exedrae and colonnades that flanked it were decorated with images of the mythical ancestors of the Romans – the Trojan Aeneas and the kings of Alba Longa and Romulus – along with about 100 statues of Rome's great men which transmitted the idea of a glorious and unproblematic past without a trace of the aristocratic competition which had ended in several civil wars (Zanker 1990 [1987]: 239–264). This decorative programme was in line with the ritual and ceremonial functions of the new forum. There, young men assumed the toga of manhood entering into the citizen body; those belonging to the elite class of the knights (*equites*) celebrated an annual festival to exhibit their prowess as cavalry; senators were summoned to debate on peace and war matters; generals and provincial governors met the emperor before leaving Rome; and the populace attended the rituals in honour of Mars on the *dies natalis* of the temple bringing to memory Augustus' victory over the assassins of this adoptive father. No doubt, the setting of those public ceremonies inculcated a harmonious and triumphalist memory of Rome in the new generations (Woolf 2015: 220–223).

In contrast with the dynamism of the old civic centre, whose monuments were always being renovated, the Forum of Augustus, presented a still image of the past and left no room for the future with the first emperor as the culmination of the history of the Roman Empire. His successors did not modify notably the memoryscape of Rome until the arrival of a new dynasty, the Flavians (69–96 AD). Following the example of Augustus Vespasian and his sons sought to leave again their mark on the city, a policy which recurred every time the imperial succession could be questioned.

Deprived of any means of achieving military glory or claiming family power, the new aristocracy of the Empire became spectators instead of actors of the Roman past. Traditional rituals and urban monumentalisation were monopolised by emperors to transmit exclusively an imperial memory in the city and beyond. In effect, contrary to Republican aristocratic tradition, remembrance of emperors overstepped Rome's walls and spread out the Empire in the shape of a ruler cult. In the East, emperor worship was easily adopted due to the previous cult of Hellenistic kings, who strikingly continued to be venerated along with the new foreign rulers (Noreña 2015: 88–93). Indeed, posing no threat to Roman authority, the Greeks were allowed to keep festivals, sanctuaries and statues in honour of Alexander the Great and his successors while taking the lead to pay the same homage to Roman emperors. Shortly afterwards and mostly based also on local initiative, emperor worship emerged in the western provinces of the Empire. The result was a mosaic of autonomous imperial cults located in main cities which never became a unitary phenomenon (Woolf 2008: 241–242). This does not mean that emperors were unconcerned thereof. Augustus himself orchestrated the divinisation of his adoptive father Julius Caesar after death and the construction of his temple in the Forum. Besides, he promoted worship to the *genius Augusti* ('life spirit of Augustus') in the city of Rome (Gradel 2002: 109–139) and established a significant number of entries in the traditional Roman calendar commemorating his own achievements, which were to be remembered annually through prayers and sacrifices the same way as the Lupercalia and Parilia recalled the figure of Romulus (Wallace-Hadrill 1987: 225–227). Successors followed the example. As a result, calendrical commemoration, divinisation of dead predecessors, and temple-building throughout the Empire ended up generating an imperial memory of glorious and worthy rulers whose merits sustained the Empire. Contrary to the well-studied cases of the eastern provinces, there is much work to do so as to understand how evolved the interplay of the previous local memory, place, and ritual in the western empire (Kamash 2016) and how this remembrance coexisted with the Roman memory of imperial cult.

Finally, the triumph of Christian religion in the fourth century AD brought major changes to Roman memory. Imperial cult started to decline from the reign of Constantine the Great (AD 306–337): royal funerals abandoned pagan ritual practices, no more were emperors divinised, temple building in their honour ceased, and their commemorative feasts plunged into oblivion. To top it all Rome lost its status as capital of the empire. In the East, Constantinople assumed this role in AD 330 and imperial interest and benefaction moved to the new seat of power. The management of the urban landscape of Rome remained in the hands of local aristocracy who showed interest not only on keeping imperial memory alive but also on showcasing Roman past in a broader sense (Machado 2006: 169–185). Despite the effort, at the same time as the metropolis lost political relevance, its pagan memoryscape finally decayed in the fifth century AD due to several sacks of the city and civil conflict. Meanwhile a new Christian memory map emerged throughout the Empire, in which the old capital did not hold such a relevant position as it did before. Instead, the Christian past, Christ's life and death, as well as the martyrdom and sanctification of his followers, dictated new places and times of ritual memorialisation. In this new memory scheme, the Holy Land played a prominent role when pilgrims started to look for the location of significant episodes of Christ's life (Halbwachs 2008 [1941]). Pagan festivals were Christianised (Salzman 1999) and pagan temples progressively lost their clientele. To construct Christian churches, previous religious sites were mostly avoided and when they were reused – mainly in the East due to urban constrictions – the process of resignification caused the complete loss of the pre-Christian memory (Ward-Perkins 2003).

Concluding remarks

This brief summary of the evolution of ritual commemoration and memory landscapes in ancient Rome has shown how complex and changing the phenomenon was due to different variables. A determinant one was the political context. In effect, who held the power and how this power was conceived were major factors in the articulation of place, ritual and memory, and it accounts for the choice of certain locations as well. Thus, the strong civic dimension of the ruling elite in the Roman Republic determined that the only scenario for memorialisation was the political centre of the city, while other spaces such as the tomb were suitable to remember the emperors, who exercised a dynastic power. In any case, the more conspicuous policies of remembrance never left the city, not even when the emperors ruled a large empire, because no other location was ever conceived as a proper seat of power on an equal foot with Rome. Combined with the huge investment that Republican aristocrats and subsequent emperors made on their self-aggrandisement, this fact easily explains why Rome turned into a packed and multi-layered memoryscape which was unparalleled in the ancient world. Its dependency on political power became apparent when emperors did not reside in the city anymore in the fourth century AD and the court moved to the East. As a result, the memory landscape of Rome lost its vitality and turned more than ever into a museum.

Religious conceptions are also relevant to understand the shape of social recollection. Contrary to many other cultures, Romans did not believe originally that any man or woman could achieve a superhuman or divine status. A significant break divided mortals and gods, so even the greatest citizens were considered human beings and remembered as such, including Romulus, the founder of the city. He was just a man, although conceived by the god Mars and a royal princess. Nevertheless this fact did not prevent victorious generals from acquiring a certain kind of divine aura in rituals like the triumphal parade, which ultimately explains the emergence of emperor worship. The classical conception of divinity enabled this process to happen, since it implied the existence of different and graded manifestations of the godlike nature. Thus, by the end of the Republic – first century BC – the distinction between mortals and gods became blurred. Romulus was assimilated to the god Quirinus and some leaders approached a divine status. Consequently the means of commemoration changed in the imperial period, giving way to the cult of the emperor, a kind of worship similar to, and to some extend heir to, the Hellenistic ruler cult (Chaniotis 2003). It is no coincidence that emperor worship emerged when Rome became a monarchy in the same way that ruler cult appeared in Greece with the formation of the Hellenistic kingdoms. In both cases, Greeks and Romans placed monarchs among the gods in recognition of the immense power they held in life; and therefore granted them the same kind of honours and celebrated them in similar rituals.

Finally, the content and means of memory transmission have to be taken into account to understand its interplay with place and ritual. Romans never thought that they had a glorious past nor did they develop a tradition of oral epic to remember the deeds of their ancestors. Instead, Romans remembered their humble origins in certain festivals in which they re-enacted the rural atmosphere of the times of Romulus and Remus; and they preserved as sacred the locations related to the twins in the Palatine hill. As Florence Dupont suggested, it was precisely the lack of a significant mythological poetry that made Roman memory take root in the city landscape (1992 [1989]: 73–74). In any case, that humble past was clearly at variance with the imperialist nature of the Roman Republic. Perhaps for this reason Romans felt a strong need to memorialise their recent imperial past through ritual ceremonies and large monumentalisation projects.

References

Assmann, J. (2010). Communicative and cultural memory. In: A. Erll and A. Nünning, eds. *A companion to cultural memory studies*. Berlin and New York: De Gruyter, pp. 109–118.
Assmann, J. (2011 [1992]). *Cultural memory and early civilization. Writing, remembrance and political imagination*. Cambridge: Cambridge University Press.
Beard, M. (1987). A complex of times: No more sheep on Romulus' birthday. *Proceedings of the Cambridge Philological Society*, 33: 1–15.
Beard, M. (2007). *The Roman triumph*. Cambridge, MA: Harvard University Press.
Bodel, J. (1999). Death on display: Looking at Roman funeral. In: B. Bergmann and C.H. Kondoleon, eds. *The art of ancient spectacle*. Washington, DC: National Gallery of Art, pp. 259–281.
Bommas, M. (2012). Introduction: Sites of memory and the emergence of urban religion. In: M. Bommas, J. Harrisson and P.H. Roy, eds. *Memory and urban religion in the ancient world*. London: Bloomsbury, pp. xxvi–xxxviii.
Chaniotis, A. (2003). The divinity of Hellenistic rulers. In: A. Erskine, ed. *A companion to the Hellenistic world*. Malden, MA: Blackwell, pp. 431–445.
Cusumano, N. (2013). Memory und religion in the Greek world. In: N. Cusumano, V. Gasparini, A. Mastrocinque and J. Rüpke, eds. *Memory and religious experience in the Graeco-Roman world*. Stuttgart: Franz Steiner Verlag, pp. 17–19.
Dupont, F. (1992 [1989]). *Daily life in ancient Rome*. Oxford and Cambridge, MA: Blackwell.
Eckert, A. (2012). *Lucius Cornelius Sulla in der antiken erinnerung. Jener Mörder, der sich Felix nannte*. Berlin-Boston: De Gruyter.
Eckert, A. (2014). Remembering cultural trauma. Sulla's proscriptions, Roman reponses and Christian perspectives. In: E-M. Becker, J. Dochhorn and E.K. Holt, eds. *Trauma and traumatization in individual and collective dimensions. insights from Biblical studies and beyond*. Goettingen: Vandenhoeck and Ruprecht, pp. 262–274.
Edwards, C. (1996). *Writing Rome. textual approaches to the city*. Cambridge: Cambridge University Press.
Favro, D. (2014). Moving events: Curating the memory of the Roman triumph. In: K. Galinsky, ed. *Memoria Romana: Memory in Rome and Rome in memory*. Ann Arbor, MI: University of Michigan Press, pp. 85–101.
Flower, H.I. (1996). *Ancestor masks and aristocratic power in Roman culture*. Oxford: Clarendon Press.
Flower, H.I. (2004). Spectacle and political culture in the Roman Republic. In: H.I. Flower, ed. *The Cambridge companion to the Roman Republic*. Cambridge: Cambridge University Press, pp. 322–343.
Galinsky, K. (2016). Introduction. In: K. Galinsky, ed. *Memory in ancient Rome and early Christianity*. Oxford: Oxford University Press, pp. 1–39.
Gradel, I. (2002). *Emperor worship and Roman religion*. Oxford: Clarendon Press.

Halbwachs, M. (2008 [1941]). *La topographie légendaire des Evangiles en Terre sainte: Étude de mémoire collective*. Paris: Presses Universitaires de France.
Hölkeskamp, K-J. (2004). *Senatus populusque Romanus. Die politische kultur der Republik—dimensionen und deutungen*. Stuttgart: Franz Steiner Verlag.
Hölkeskamp, K-J. (2006). History and collective memory in the middle Republic. *In:* N. Rosenstein and R. Morstein-Marx, eds. *A companion of the Roman Republic*. Malden, MA: Blackwell, pp. 478–495.
Hölkeskamp, K-J. (2014). In defense of concepts, categories, and other abstractions: Remarks on a theory of memory (in the making). *In:* K. Galinsky, ed. *Memoria Romana: Memory in Rome and Rome in memory*. Ann Arbor, MI: The University of Michigan Press, pp. 63–70.
Kamash, Z. (2016). Memories of the past in Roman Britain. *In:* M. Millett, L. Revell and A. Moore, eds. *The Oxford handbook of Roman Britain*. Oxford: Oxford University Press, pp. 681–696.
Machado, C. (2006). Building the past: Monuments and memory in the *Forum Romanum*. *In:* W. Bowden, A. Gutteridge and C. Machado, eds. *Social and political life in late antiquity*. Leiden and Boston: Brill, pp. 157–192.
McLynn, N. (2008). Crying wolf: The Pope and the Lupercalia. *The Journal of Roman Studies*, 98: 161–175.
Noreña, C. (2015). Ritual and memory: Hellenistic ruler cults in the Roman empire. *In:* K. Galinsky and K. Lapatin, eds. *Cultural memories in the Roman empire*. Los Angeles: J. Paul Getty Museum, pp. 86–100.
Orlin, E. (2016). Augustan reconstruction and Roman memory. *In:* K. Galinsky, ed. *Memory in ancient Rome and early Christianity*. Oxford: Oxford University Press, pp. 115–144.
Price, S. (1987). From noble funerals to divine cult: The consecration of Roman emperors. *In:* D. Cannadine and S. Price, eds. *Rituals of royalty. power and ceremonial in traditional societies*. Cambridge: Cambridge University Press, pp. 56–105.
Rodríguez Mayorgas, A. (2007). *La memoria de Roma: Oralidad, escritura e historia en la República Romana*. Oxford: John and Erica Hedges Ltd.
Rodríguez Mayorgas, A. (2010). Romulus, Aeneas, and the cultural memory of the Roman Republic. *Athenaeum*, 98 (1): 89–109.
Salzman, M. (1999). The Christianization of sacred time and sacred space. *In.* W.V. Harris, ed. *The Transformation of Urbs Roma in Late Antiquity. Journal of Roman Archaeology Supplementary Series*, 33: 123–134.
Scullard, H.H. (1981). *Festivals and ceremonies of the Roman Republic*. London: Thames & Hudson.
Wallace-Hadrill, A. (1987). Time for Augustus: Ovid, Augustus and the Fasti. *In:* M. Whitby, Ph. Hadie and M. Whitby, eds. *Homo viator: Classical essays for John Bramble*. Bristol: Bristol Classical Press and Bolchazy-Carducci, pp. 221–230.
Ward-Perkins, B. (2003). Reconfiguring sacred space: From pagan shrines to Christian churches. *In:* G. Brands and H-G. Severin, eds. *Die spätantike stadt und ihre Christianisierung*. Wiesbaden: Reichert Verlag, pp. 285–290.
Woolf, G. (2008). Divinity and power in ancient Rome. *In:* N. Brisch, eds. *Religion and power: Divine kingship in the ancient world and beyond*. Chicago, IL: The Oriental Institute of the University of Chicago, pp. 235–251.
Woolf, G. (2015). Mars and memory. *In:* K. Galinsky and K. Lapatin, eds. *Cultural memories in the Roman empire*. Los Angeles: J. Paul Getty Museum, pp. 206–224.
Zanker, P. (1990 [1987]). *The power of images in the age of Augustus*. Ann Arbor, MI: University of Michigan Press.

37
RITUALLY RECYCLING THE LANDSCAPE

Ceri Houlbrook

Introduction

British landscapes have long been the stage for ritual. From the enigmatic stone circles of prehistory to the coins we toss into fountains today, a sense of the ceremonial has permeated our environments for millennia. But, as demonstrated throughout this volume, these environments are not static. The landscape is subject to the same mercuriality as the societies inhabiting it. Ritual spaces shift and alter, especially natural sites which, subject to organic processes, undergo constant changes of growth, decay, and destruction, often in defiance of humanity's best efforts to retain continuity.

Just as landscapes are mutable and malleable, so too are rituals. 'Old rituals die hard,' writes archaeologist Ralph Merrifield. 'If something has always been done, it may be safer to continue to do it, though it may on occasion be necessary to find a new explanation to reconcile it with the beliefs that are currently acceptable' (1987: 107). People can observe the essential elements of a ritual, but for it to be relevant to them it must prove malleable enough for the participants to shape and colour it to their liking (Bascom 1965: 29). Indeed, malleability is integral to a ritual's perpetuity. Like all things, rituals must adapt if they are to survive, and in a society of rapidly shifting popular beliefs, rituals must, by necessity, adapt quickly or risk extinction.

This chapter demonstrates how even seemingly isolated landscapes can be beset by cultural and environmental changes. Drawing on a particularly remote case study from the Scottish Highlands, it explores how such landscapes are adapted to accommodate changing rituals and, vice versa, how rituals are adjusted to fit within altered environments. These are the processes of ritual recycling, which I have previously applied to objects, in exploring how a mundane item can become sacred by reaching the end of one stage in its biography and entering the next (Houlbrook 2013). However, ritual recycling is a term that can also apply to a landscape, which is reformed into something new by the ritual activities of those who use it.

A brief history of excavating the palimpsest

Ritually recycling is only one of the many metaphors used to express these processes. It has long been recognised that landscapes, and our cultural perceptions of them, are mutable. Their biographies are complex, non-linear, and inalienable to the biography of society. In scholarship, this characteristic has come to be viewed as more than an incidental aspect of the landscape, but as the crux of our understanding of it. A prime example is Keith Thomas's *Man and the Natural World* (1983), a diachronic exploration of the shifting perspectives of British society towards their trees and woodland. In his chapter on 'The Worship of

Trees' (1983: 212–223), Thomas describes how woodlands were increasingly imbued with symbolic value in the early modern period. From the eighteenth century onwards, they became emblematic of a community's continuity, of the nation's strength, and of a family's ancestry. Such meanings were not inherent to these landscapes; they were endowed by society.

Since archaeologist Osbert Crawford first applied the palimpsest metaphor to the 'surface of England' as a 'document that has been written on and erased over and over again,' historians, archaeologists, and geographers alike have been employing a variety of imagery to illustrate the complex nature of our relationship with the natural environment (1953: 51). In the 1950s, William Hoskins was advocating the study of landscape 'as though it were a piece of music, or a series of compositions of varying magnitude, in order that we may understand the logic that lies behind the beautiful whole' (1955: 20). In the 1970s, geographer Donald Meinig was describing the landscape as an 'accumulation ... an enormously rich store of data about the peoples and societies which have created it' (1979: 44). And in the 1980s, Oliver Rackham was expressing the landscape as a library in a detailed analogy that deserves to be repeated in full:

> The landscape is like a historic library of 50,000 books. Many were written in remote antiquity in languages which have only lately been deciphered; some of the languages are still unknown. Every year fifty volumes are unavoidably eaten by bookworms. Every year a thousand volumes are taken at random by people who cannot read them, and sold for the value of the parchment. A thousand more are restored by amateur bookbinders who discard the ancient bindings, trim off the margins, and throw away leaves that they consider damaged or indecent. The gaps in the shelves are filled either with bad paperback novels or with handsomely-printed pamphlets containing meaningless jumbles of letters. The library trustees, reproached with neglecting their heritage, reply that Conservation doesn't mean Preservation, that they wrote the books in the first place, and that none of them are older than the eighteenth century; concluding with a plea for more funds to buy two thousand novels next year.
>
> *(1986: 29–30)*

Here the landscape is portrayed as a perpetually changing environment, with aspects being lost and others introduced in an inevitable narrative of revisions and transitions.

Another analogy drawn on is Simon Schama's description of how the landscape is built up from strata of myth and memory lying beneath the surface – strata to be excavated – in his almost poetic ode to *Landscape and Memory* (1996: 7). While Alexandra Walsham, in her seminal biographical treatment of *The Reformation of the Landscape*, traces the changing perceptions of the religious landscape throughout the early modern period by presenting the natural environment as 'a porous surface upon which each generation inscribes its own values and preoccupations without ever being able to erase entirely those of the preceding one' (2011: 6).

The term 'ritual recycling' encapsulates all these observations. It describes a site, a monument, a whole landscape that is being adapted to fulfil a certain ceremonial purpose. Physically the recycled landscape might be the same, or the changes do not quite conceal past uses – like an under-layer of Crawford's palimpsest – but it has been restructured to fulfil a different purpose. Landscapes are old and new at the same time. The base material remains largely constant but its uses can be in constant flux. To illustrate this process, I have drawn on the complex ritual narrative of a small and remote landscape in the Scottish Highlands: Isle Maree.

Introducing Isle Maree

Stretching for 12 miles in a northwesterly direction, Loch Maree in Wester Ross is the fourth largest freshwater loch in Scotland, and accommodates more than 60 islands. One of these islands shares its name with the loch. Situated 250 metres from the northern shore, Isle Maree is of triangular shape and covers six acres, making it one of the smaller islands. However, despite its modest size, it is seen as Loch Maree's 'principal island' (Reeves 1857–1860: 268), and it is a landscape that boasts a long narrative of ritual recycling.

Ecologically, it is one of the more interesting islands on the loch, with its densely forested centre. In the eighteenth century, traveller and antiquarian Thomas Pennant described it as 'the most beautiful of the isles; the others have only a few trees sprinkled over their surface,' whilst Isle Maree was richly adorned with oak, ash, birch, willow, hazel, and holly (1775: 330). Anthropologist Gertrude Godden, visiting the island over a century later, describes it with equal admiration: 'so covered with luxuriant foliage that a fragment of green forest seems to have been carved out and placed in the loch, set in a border of golden sand' (1893: 498).

Both the island and the loch are named after the area's patron saint, St Maelrubha (also known as Maree). Born in County Down, Ireland, in 642, St Maelrubha came to Scotland and founded the monastic community of Applecross, Ross, in 673 (Reeves 1857–60: 259; Mitchell 1863: 254–255). He is often credited with the introduction of Christianity into this region of Scotland and, following his death in 722, he became the patron saint of the district (Mitchell 1863: 254–255).

Isle Maree was purportedly the saint's 'favoured isle' and, according to popular belief, it became the site of his chapel (Pennant 1775: 330). Why he chose this particular island over the many others remains a mystery, but it was possibly a case of ritual recycling. There is a theory that the saint was supplanting an earlier pagan centre, a well-known method of Christianisation. Pennant describes the remains of a structure at the island's centre, consisting of a circular dike of stones with a narrow entrance, which Pennant believed 'to have been originally *Druidical*, and that the ancient superstition of *Paganism* had been taken up by the saint, as the readiest method of making a conquest over the minds of the inhabitants' (1775: 330, emphases in original). However, evidence of a structure is not evidence of paganism. Whatever was there could have post-dated St Maelrubha.

However, the island's profusion of oak trees (traditionally associated with Druids) has also been cited as possible evidence that the island was a pagan cultic centre before St Maelrubha's arrival, and it is popularly claimed that the saint planted holly on the island, a tree with Christian connotations, to counterbalance the more pagan oak. There is no way to prove or disprove this, but if it was the case then this is a prime example of ritually recycling the landscape: subtle alterations to the natural environment reflecting changes in how and by whom the site is used.

The holy well of Isle Maree

The same process of ritual recycling was applied to the island's well, which sits in the southwest corner of the island. Believed to have been consecrated by St Maelrubha, this water source became a 'holy well,' widely held in the surrounding districts to cure lunacy. Many pilgrims were said to have visited the island (sometimes of their own volition but often in custody) and drank from the well in order to affect a cure for mental illness (Pennant 1775: 330; Reeves 1857–1860: 288–289; Mitchell 1863: 251–262; Queen Victoria (Duff 1968: 332); Dixon 1886: 151; Godden 1893: 500–501; Muddock 1898: 437–438; Barnett 1930: 113). This well may be further equivocal evidence that the site possessed some pre-Christian significance. Many of the wells that dot the British landscape are believed to have originally been employed as part of pagan hydrolatry, but were later adopted by Christianity, the wells transferring to the guardianship of Christian saints (Daly 1961; Rattue 1995).

Under St Maelrubha's custodianship, the well's perceived powers appear to have lasted centuries, surviving the Scottish Reformation and stretching into the Victorian period. It was probably last resorted to for the cure of insanity in the 1850s. Godden dates the last appeal to the holy well to 1857 (1893: 500), while John Dixon cites an example from 1858 of a woman from Easter Ross having been taken to the island for a cure (1886: 151). Certainly by this time, however, belief in the well had waned – possibly following the desecrating act of a farmer lowering his afflicted dog into the well earlier that century. And by the time

Arthur Mitchell, physician and historian, visited Isle Maree in 1863, the well was dry 'and full of last year's leaves, and the flat stone which serves for a cover we found lying on the bank' (1863: 262).

By the 1950s, when travel writer Brenda Macrow visited the island, she remarked on how difficult it was to determine the site of this well (1953: 88). It is unknown whether the well was filled up deliberately, having purportedly lost its efficacy, or whether this was simply the result of many years of disuse. Either way, no visible trace remains of either the well or its stone slab covering today. The well was not, however, the last feature in this landscape to have been ritually employed.

In Pennant's 1775 description of the island, he writes of how pilgrims would drink from the holy well and then leave an offering to the saint on a nearby tree stump, which he describes as 'an altar, probably the memorial of one of stone' (1775: 330). The landscape has been ritually recycled once more: a tree has been transformed into a votive altar, possibly replacing a stone predecessor. The tree does not seem to have been significant in itself: more for its convenient location. Trees that stood beside holy wells were often used as receptacles for offerings. John Campbell writes of a similar arrangement on the Hebridean island of Islay, where votive objects, such as copper caps, pins, and buttons, were 'placed in chinks in rocks and trees at the edge of the 'Witches' Well' (1860: 134).

However, in many cases, a tree connected with a holy well is not incidental, but central to the ritual itself. Trees can become sacred via osmotic transference; their association with the holy wells bestows sanctity upon them also (McPherson 2003 [1929]: 74; Lucas 1963: 40). This bestowal may be literal as well as symbolic, with the holy water from the well travelling into the tree itself (Shephard 1994: 2). At St Patrick's Well, Enfield, Co. Westmeath, for example, the water which emerges from between the roots of an ash tree is believed to cure eye disease, while at Easter Rarichie, Ross and Cromarty, the healing spring known as Sul na Ba flowed through a tree trunk, endowing that tree with curative properties itself (Bord & Bord 1985: 59; Milner 1992: 139). It is not uncommon for a tree to replace desecrated or polluted holy wells as the objects of people's veneration, thus becoming 'holy wells' themselves.[1]

This transference of sanctity not only imbues the tree with power, but allows it to establish itself as a ritual structure independent from the holy well, so that it may subsequently outlive it. For example, the holy well on the River Sullane, Ireland, had run dry by the early twentieth century, but the surrounding briar bushes were still heavily affixed with votive offerings (Hull 1928: 108). The same process occurred on Isle Maree: the tree outlived the well. Visiting the site in the 1920s, Lieutenant Colonel George Edington in fact wrote that the tree had 'been fixed into the filled-up holy well' (cited in McPherson 2003 [1929]: 75), revealing that the tree had replaced the well physically as well as ritually. While this tree may have originally been ritually recycled simply because of its association with the holy well, it went on to supplant that well.

The tree of Isle Maree

When Pennant described the tree in 1775, he did not specify what form of objects the pilgrims deposited. Later sources, however, refer to its use as a rag-tree. Hartland describes how pilgrims, seeking a cure from the holy well of St Maelrubha, attached pieces of clothing to the nearby tree (1893: 453), and Barnett reports that the patients brought to the island would tie rags or ribbons to its branches (1930: 114). On Mitchell's visit to Isle Maree in 1863, the tree was apparently studded with nails: 'To each of these was originally attached a piece of the clothing of some patient who had visited the spot' (1863: 253). This use of rags corresponds with the connection of Isle Maree to the cure of insanity. In the British Isles, rag-trees were most commonly employed to affect cures. It was widely believed that rags contained whatever ailment the depositor wished to be rid, and, by affixing them to a tree, the ailment is transferred from person to tree via 'contagious transfer' (Hartland 1893: 460; Frazer 1900: 39; Foley 2010, 2011).

However, at some point during its ritual career, the tree on Isle Maree shed its rags and ribbons and became predominantly a nail-tree. Mitchell describes how the tree was 'studded with nails' (1863: 253), whilst Hartland observes how 'Many of the nails are believed to be covered with the bark, which appears to be growing over them' (1893: 453–454). This reincarnation as a nail-tree is a logical next evolutionary step from the rag-tree; pins and nails were particularly popular devices for contagious transfer. Knocking nails into an oak tree, for example, was a well-known remedy for toothache in Cornwall. The toothache was believed to transfer into the tree, from the sufferer, through the nail (Walhouse 1880: 99; Porteous 1928: 188). Pins were also employed as cures for warts; inserted into each wart, then into the bark of an ash tree, this was believed to transfer the affliction to the tree (Wilks 1972: 121). It may not only have been the physical properties of nails, permitting easy insertion, that gave these objects such perceived powers, but also their association with the crucifixion and thus their history of ritual significance within Christianity (cf. Hutton 1995: 108).

The transformation of the rag-tree into a nail-tree, however, was probably wholly incidental. As has been observed, the rags and ribbons were attached to the tree using nails (Mitchell 1863: 253; Hartland 1893: 453). The nails were therefore convenient tools for securing offerings, rather than offerings themselves. However, cloth decomposes much faster than iron; the nails would therefore survive long after the rags and ribbons had decayed. This natural process, leaving the tree studded with nails rather than adorned with rags, likely led to its gradual transformation into a nail-tree. Pilgrims to the island began inserting nails as votive offerings or vehicles of contagious transfer in and of themselves.

The coin tree

The tree on Isle Maree did not remain a nail-tree for long. In 1877 the island's reputation warranted it a visit from Queen Victoria, whose tour of Scotland had led her to the loch. Writing in her diary, Queen Victoria described the tree, and how it had become the custom 'for everyone who goes there to insert with a hammer a copper coin, as a sort of offering to the saint' (Duff 1968: 332). The tree had become a coin-tree. Indeed, by the 1890s, it was being referred to as 'the money tree' (Muddock 1898: 437), and by Colonel Edington's visit in 1927, no pins or nails were visible in the bark of the tree, only coins (McPherson 1929: 75) – so many coins, in fact, that Edington describes the tree as 'covered with metallic scales . . . something like what is depicted on a dragon' (cited in McPherson 1929: 75).

The hundreds of coins inserted into clefts and cracks have no doubt taken their toll on this tree, which is now dead (Figure 37.1). McPherson writes that this 'holy tree shared the fate of the holy well – the devotion of pilgrims has proven its undoing. The coins, hammered in and destroying the bark, have killed the object of their veneration' (1929: 75). The death of the tree, however, has not led to the death of the tradition. Indeed, it appears to have proliferated. As the original tree became too densely coined, the custom appears to have spread to surrounding trees.

By the 1950s, people had begun inserting coins into the stake used to prop up the original tree as well as into the barks of surrounding trees (Macrow 1953: 88–89). In 2002, when the North of Scotland Archaeological Society conducted a survey of the site, they observed that the original coin-tree (referred to in their report as a votive tree, and catalogued as VT1) was leaning on six spars, also embedded with coins, and counted two subsidiary parts of VT1, scattered some distance away, VT2 and VT3 (North of Scotland Archaeology Society 2002: 22–23). In total, they catalogued nine boles and spars embedded with coins on Isle Maree.

In the intervening decade between the 2002 survey and my own fieldwork, on 14 April 2012, this number had further increased. Clustered around what is believed to have been the original tree are 12 further spars, logs, stumps, and living trees that have been embedded with coins. I counted approximately 2,000 coins, although this probably only represents the tip of the iceberg: many coins had fallen from the decaying trees and become buried over time. The majority of the coins were decimal one pennies, half pennies, and shillings, and while most of the years of issue had long since become illegible, I was able to date one to 1875. Surprisingly, however, there were also many coins that were post-decimalisation, the most recent one dating to 2010. There was also some foreign coinage, including a Dutch five-guilder coin from 1985, ten

Ritually recycling the landscape

Figure 37.1 The remains of the Isle Maree tree: coin-encrusted spars, propped up against each other.
Source: Photograph by author, 2012.

Dutch cents, a one Euro cent, and a one US cent. The custom of leaving an offering in this landscape has far from fallen out of popularity, even in the twenty-first century (Houlbrook 2014, 2015a, 2015b, 2018).

The wishing tree

The custom is, however, no longer observed for affecting cures. By the time the tree had become a cointree in the late nineteenth century, its purpose had changed. It had become a 'wishing-tree.'[2] It was now believed that, as described by McPherson, a 'wish silently formed when any metal article was attached to the tree, or coin driven in, would certainly be realised' (2003 [1929]: 76). No longer associated with healing, the tree became imbued with the power to grant wishes or to ensure good luck (MacLeish 1968: 420), the only two traditions which participants seem to observe today. Local residents in Gairloch, for example, associate the tree with only two things: wish-making and good luck. The tree has therefore shed its (Christianised) curative properties and become a wishing-tree instead, a custom much more inclusive – albeit perhaps less earnestly observed.

Whether or not all of the stories told about Isle Maree are grounded in historical truth, the landscape of this island has a long narrative of ritual recycling within the popular imagination. With the rise of Christianity, Isle Maree's possible pagan associations were supplanted by St Maelrubha, who was believed to have sanctified the island and consecrated the well. Isle Maree thus became Christianised, and for centuries it was a destination for pilgrims hoping to affect cures. However, believed to be polluted by the farmer's dog, the well fell out of use and was filled in – to be replaced, both symbolically and literally, by the tree, which was used as a receptacle for offerings and devices of contagious transfer: rags, nails, and then coins. However, with the declining faith in the power of saints and folk cures, the tradition needed to adapt to retain its popularity. And so, ritually recycled once more, Isle Maree became the site of a modern-day wishing-tree, demonstrating that even an island is not an island in the metaphorical sense of detached seclusion. It can sustain a sense of the sacred but can never be isolated from external changes and events.

Future directions

In 1955, Hoskins lamented that 'especially since the year 1914, every single change in the English landscape has either uglified it or destroyed its meaning, or both' (1955: 238). Rackham was less direct in his condemnation of contemporary changes, but his disapproval is still clear: the older volumes in the metaphorical landscape have been lost or taken, and the 'gaps in the shelves are filled either with bad paperback novels or with handsomely-printed pamphlets containing meaningless jumbles of letters' (1986: 30). Thankfully, perceptions have moved on, and the contemporary changes in our landscapes are no longer being dismissed as irrelevant to their narratives. Instead, they are recognised as the latest set of words inscribed on the palimpsest: as the topmost layer in the strata of myth and memory or as the most recent recycled state.

Where might this topic go from here? As with the ritually recycled landscape, while it is impossible to predict with any certainty what its future state will be, it will no doubt continue to adapt. Because our environments and our beliefs will never stop changing, the subject is unlikely to dry up. Natural spaces will continue to be used as sacred places, even if what constitutes 'sacred' is likely to change – and even if what constitutes 'natural' is also likely to change. For this topic to be adequately explored, a wide chronological scope is required. Historians and archaeologists will need to engage with ethnographic material to understand how the sites they study have been ritually recycled in the present. Likewise, anthropologists exploring contemporary sacred landscapes will need to trace the histories of their sites in order to appreciate their ritual narratives. Only by adopting a broad and inclusive scope, as well as a variety of methodologies, will we be able to fully comprehend the extent to which our landscapes have been, and will continue to be, ritually recycled.

Notes

1 In his catalogue of the sacred trees of Ireland, Lucas lists numerous examples: Lady's Well, Skirk, Co. Laois; The Tree of Castlebellew, Cloonoran, Co. Galway; the Pin Well, Tartaraghan, Co. Armagh; Mary's Well, Rockspring, Co. Cork; and St Margaret's Well, Cooraclare, Co. Clare (1963: 41).
2 'Wishing tree' is a term employed by Dixon (1886: 150), Godden (1893: 499), McPherson (2003 [1929]: 76), Barnett (1930: 114), and Macrow (1953: 88–89) to describe the Isle Maree tree.

References

Barnett, T.R. (1930). *Autumns in Skye, Ross & Sutherland*. Edinburgh: John Grant Booksellers Ltd.
Bascom, W.R. (1965). Four functions of folklore. In: A. Dundes, ed. *The study of folklore*. Englewood Cliffs: Prentice-Hall, Inc., pp. 279–298.
Bord, J., and Bord, C. (1985). *Sacred wells: Holy wells and water lore in Britain and Ireland*. London: Paladin Books.
Campbell, J.F. (1860). *Popular tales of the West Highlands*, Vol. 2. Edinburgh: Edmonston and Douglas.
Crawford, O.G.S. (1953). *Archaeology in the field*. London: Phoenix House Ltd.
Daly, J.F. (1961). Curative wells in old Dublin. *Dublin Historical Record*, 17 (1): 13–24.
Dixon, J.H. (1886). *Gairloch in North-West Ross-shire: Its records, traditions, inhabitants, and natural history: With a guide to Gairloch and Loch Maree and a map of illustrations*. (4th reprint). Fort William: Nevisprint Ltd.
Duff, D. ed. (1968). *Victoria in the highlands: The personal journal of Her Majesty Queen Victoria: With notes and introductions, and a description of the acquisition and rebuilding of Balmoral Castle*. London: Frederick Muller Ltd.
Foley, R. (2010). *Healing waters: Therapeutic landscapes in historic and contemporary Ireland*. Farnham: Ashgate Publishing.
Foley, R. (2011). Performing health in place: The holy well as a therapeutic assemblage. *Health and Place*, 17: 470–479.
Frazer, J.G. (1900). *The golden bough*, Vol. 1. London: Macmillan & Co. Ltd.
Godden, G.M. (1893). The sanctuary of Mourie. *Folklore*, 4 (4): 498–508.
Hartland, E.S. (1893). Pin-wells and rag-bushes. *Folklore*, 4 (4): 451–470.
Hoskins, W.G. (1955). *The making of the English landscape*. London: Hodder and Stoughton.
Houlbrook, C. (2013). Ritual, recycling, and recontextualisation: Putting the concealed shoe into context. *Cambridge Archaeological Journal*, 23 (1): 99–112.
Houlbrook, C. (2014). The mutability of meaning: Contextualising the Cumbrian coin-tree. *Folklore*, 125 (1): 40–59.
Houlbrook, C. (2015a). 'Because other people have done it': Coin-trees and the aesthetics of imitation. *Journal of Contemporary Archaeology*, 2 (2): 309–327.
Houlbrook, C. (2015b). Small change: Economics and the British coin-tree. *Post-Medieval Archaeology*, 4 (1): 114–130.
Houlbrook, C. (2018). *The magic of coin-trees from religion to recreation: The roots of a ritual*. Basingstoke: Palgrave Macmillan.
Hull, E. (1928). *Folklore of the British Isles*. London: Methuen & Co. Ltd.
Hutton, R. (1995). The English reformation and the evidence of folklore. *Past & Present*, 148: 89–116.
Lucas, A.T. (1963). Sacred trees of Ireland. *Journal of the Cork Historical and Archaeological Society*, 68: 16–54.
MacLeish, K. (1968). The highlands of Scotland. *National Geographic*, 133 (3): 420.
Macrow, B. (1953). *Torridon highlands*. London: Robert Hale & Company.
McPherson, J.M. (2003 [1929]). *Primitive beliefs in the Northeast of Scotland*. Whitefish, MT: Kessinger Publishing Co.
Meinig, D.W. (1979). The beholding eye: Ten versions of the same scene. In: D.W. Meinig and J. Brinckerhoff Jackson, eds. *The interpretation of ordinary landscapes: Geographical essays*. New York: Oxford University Press, pp. 33–48.
Merrifield, R. (1987). *The archaeology of ritual and magic*. London: B.T. Batsford Ltd.
Milner, J.E. (1992). *The tree book: The indispensable guide to tree facts, crafts and lore*. London: Collins & Brown Ltd.
Mitchell, A. (1863). On various superstitions in the north-west highlands and islands of Scotland, especially in relation to lunacy. *Proceedings of the Society of Antiquaries of Scotland*, 3: 251–288.
Muddock, J.E. (1898). The land of Santa Maree. *The Celtic Magazine*, 3: 435–461.
North of Scotland Archaeology Society. (2002). *Isle Maree, Ross and Cromarty: An Archaeological Survey*. (s.l.): (s.n.).
Pennant, T. (1775). *A tour in Scotland and voyage to the Hebrides, MDCCLXXII*, Vol. 2. Dublin: A. Leathley.
Porteous, A. (1928). *Forest folklore, mythology, and romance*. London: G. Allen & Unwin.
Rackham, O. (1986). *The history of the countryside*. London: Weidenfeld & Nicolson.
Rattue, J. (1995). *The living stream: Holy wells in historical context*. Woodbridge: Boydell Press.
Reeves, W. (1857–1860). Saint Maelrubha: His history and churches. *Proceedings of the Society of Antiquaries of Scotland*, 3: 258–296.
Schama, S. (1996). *Landscape and memory*. London: Fontana Press.
Shephard, V. (1994). *Historic wells in and around Bradford*. Loughborough: Heart of Albion Press.
Thomas, K. (1983). *Man and the natural world: Changing attitudes in England 1500–1800*. Harmondsworth: Penguin Books.
Walhouse, J. (1880). Rag-bushes and kindred observances. *The Journal of Anthropological Institute of Great Britain and Ireland*, 9: 97–106.
Walsham, A. (2011). *The reformation of the landscape: Religion, identity, and memory in early modern Britain and Ireland*. Oxford: Oxford University Press.
Wilks, J.H. (1972). *Trees in the British Isles in history and legend*. London: Frederick Muller Ltd.

38
CONTESTED MEMORY IN THE HOLY SPRINGS OF WESTERN SIBERIA

Jeanmarie Rouhier-Willoughby

Defining legend

Legends about a place often play a central role in defining a location and creating memoryscapes. This chapter will examine the role of legend in the negotiation of memory regarding a holy spring in Western Siberia located on the site of a former Stalin-era prison camp (GULAG). The folk genre of legend resists simple definition, but today generally conveys a disparaging attitude toward the content in the story, 'it's just an urban legend.' However, legends should not be summarily dismissed, given their important role in constructing understandings of reality, of core socio-cultural beliefs, and indeed of the places associated with these narratives. Before turning to the legends of the holy spring and their role in creating memoryscapes, it behooves us to get a better understanding of this elusive genre.

Stories referred to by the term 'legend' feature a plethora of characters, including humans, gods, saints, and animals. Plots feature encounters with the supernatural, heroic battles, and religious miracles, as well as much more mundane events, such as criminal activities, sexual escapades, and university pranks. They may be told from the first- or third-person perspective and are often presented as true, even when the teller may not believe the legend being narrated. There are no limitations on when they may be told and they are not associated with any particular ritual. The legend is the most prolific and persistent form of folk narrative in today's world, much more productive than tale or myth.

Linda Dégh, perhaps the foremost legend scholar in the twentieth century, states (2001: 97) that 'a legend is a legend once it entertains debate about belief . . . the sounding of contrary opinions is what makes a legend a legend.' Gillian Bennett (2005: xi) expands upon Dégh's definition, arguing that legends

> are situated somewhere on a continuum between myth and folktale at one pole and news and history at the other, moving along this continuum depending on the individual story and the whims or objectives of the individual storyteller who relates it. . . . however fantastical the protagonists and events, the scene is always set in the world as we know it, and the stories seem to be giving information about that world. For this reason, readers or hearers are sooner or later faced with a challenge: where on the continuum between folktale and news, myth and history, should these stories be placed? Are they true, like news and history, or true in some symbolic way, like myth, or fictional, like folktales?

Bill Ellis (2003: 11–12) describes the process of legend telling as

> *a communal exploration of social boundaries*. By offering examples of the extremes of experience – unusual, bizarre, inexplicable, unexpected, or threatening incidents – members of the legend-telling circle attempt to reach some consensus on the proper response to what is arguably 'real.'... Legend telling... is a means of expressing anxieties about a group's cultural worldview, as well as a way of redefining it in light of individual experiences. It also provides a safe way of questioning what important institutions define as 'real' and 'proper.'

Bennett (2005: xii), like Ellis, focuses on the content of a story and the process of evaluating the narrative as legend in her definition,

> The question of whether or not a story is a legend is critical to recognising it as a legend... because the information (or misinformation) contained in the legend is challenging in some way, it creates... 'cognitive dissonance'... by juxtaposing the world as we know it with something every different or by melding two cultural categories we think of as quite separate.

The first situation, according to Bennett, produces traditional legends about the supernatural or religious experiences, while the latter results in contemporary legend (often called urban legend) about vicious killers, infanticide, and dangerous commercial products, for example.

The fact that legend offers debate about belief and the nature of reality results in ostensive action, according to Dégh and Vásonyi (1983: 18–19). In their view, legends not only tell about reality, but have the potential to create it, to turn words into facts. For example, they cite legends about Halloween candy poisonings (first documented in the 1940s), which have resulted in a drastic shift in the trick-or-treating tradition in the United States to 'safer' alternatives (Dégh & Vásonyi 1983: 12) on the basis of little actual historical evidence that the practice was dangerous to children. Legend thus holds the power to play a fundamental role in the way people interpret events and develop beliefs about reality and about places associated with them. These beliefs can hold sway despite significant scientific evidence to the contrary. In fact, they (and the legends) are often built on as many historical facts as the contrary opinions held by experts, despite the widespread opinion that legends connote untruth or represent flawed reasoning. Legend then is a genre that represents an exploration of social and cultural norms, particularly relating to the uncanny or threats in the contemporary world.

Legend and place

Because legends describe the real world, they are often tied to a particular place or historical event. In the case of supernatural or religious legends, localisation results in a particular space being associated with the legendary events that purportedly took place there. Legend contributes to the memoryscapes about a physical location and results in a particular type of ostensive action, the so-called legend trip (see Ellis 1983, 1996; Tucker 2007, 2015). In a legend trip, a group of people who have shared or heard a legend, particularly of a haunting or violent past event, travel to the location and interact with the site; then they narrate the experience at the site to others, thereby continuing the cycle of transmission of belief about the location. As Ellis (2015: 196) argues,

> The supernatural is not 'out there,' but rather it is 'here among us' in the shared milieu that allows legends to be told and actively discussed. But this context also includes ways in which we socially construct the space around us, both the landscape and the ways in which humans alter it for their

use. Seen that way, 'place' is like the supernatural in that it is an experience filtered into discourse with the help of previous stories of its history, use, and social significance. We experience a place not just in terms of sensory impulses, but also as the most recent unfolding of the spot's ongoing history.

From the folklorist's point of view, legend thus constructs not only belief, but historical memory of a place. The physical space becomes numinous, associated with the mystical or supernatural events described in the stories about the location. The experiences at the site are interpreted through the narratives and ascribed meaning in large part because of them.

Typically, legend trippers are adolescents in search of a thrill, but as Carl Lindahl (2005) describes in his story of legend trips to the San Antonio Ghost Tracks, a legend trip to a haunted location may be akin to religious pilgrimage. Legends about the site tell of a bus of school children killed by a train. Lindahl writes (2005: 166) that

> At the core of the train tracks legend is a 'gravity hill' phenomenon: people visit the scene by car drive slowly (and at least seemingly) upward along a gradual incline toward the tracks. Stopping short of the rails, the driver shifts the car neutral, and the car seems to roll uphill and over the tracks in defiance of gravity.

Legend trippers dust the trunk with powder and, after the car has been pushed forward, they inspect it for traces of children's handprints. Lindahl (2005: 174) documents one such legend trip by a family to illustrate the spiritual relevance of the site for the local Hispanic community:

> Nevertheless, most experienced more immediately the thrill of a good scare, and only later a deeply spiritual sense well-being. For Lydia and her boyfriend's family, the journey to the train tracks [was] an exercise in both positive and negative ostension. The family went to the train to play the roles of the victims, the children who died there. The sense of connection the family felt with the children was deep indeed. . . . According to Lydia and her mother, one visits the tracks not only to step into the roles of the dead children, but also to perform a kind of ostensive healing of their ghosts. A trip to the tracks may fill the need of the children's souls to save others, an act through which the pilgrims in turn help save themselves.

This sense of the legend trip as a destination for religious pilgrimage (rather than for ghost hunting) is key to an understanding of the beliefs surrounding the holy springs in Western Siberia. The spring at Lozhok, for example, is given sacred status precisely because of the deaths that occurred at the prison camp. The legends describe how the religious dead, viewed as martyrs for the faith, have sanctified this site associated with past atrocities. As a result of the stories, it has now become the site of pilgrimage and serves as the centre for the life of the local congregation.

Legends alone, however, do not encapsulate the holy spring's resonance among the local populace or believers. Pilgrimage to a sacred site must be understood through the prism of vernacular religion. Leonard Primiano (1995: 44) writes that

> Vernacular religion is, by definition, religion as it is lived; as human beings encounter, understand, interpret and practice it. . . .Vernacular religious theory involves an interdisciplinary approach to the study of religious lives of individuals with special attention to the process of religious belief, the verbal, behavioral, and material expression of religious belief, and the ultimate objects of religious belief.

Primiano (2012: 384) emphasises that vernacular religion must not be interpreted an alternative to official religion or religious institutions; rather, he argues, official institutions have 'vernacular religion as their foundation.' This approach to religious belief forms the basis of the analysis of the role of place and memory through the lens of legends about the holy spring of Lozhok. The narratives reveal how people view the site and negotiate the debates presented within the legends themselves and how they form part of their vernacular religious beliefs. The contrary views of the past revealed by the legends allow for the creation of a memoryscape that accommodates a variety of positions about that past and about the former GULAG where the spring lies.

History and legends of the Lozhok holy spring

As Anne Applebaum documents (2003: xxx), the history of Siberia as a place of exile and imprisonment was not invented by the Bolsheviks. The tsars had long relied on Siberia as the location for disruptive elements in society, be they criminals or those calling for political reform. However, the systematised incarceration of political prisoners and criminals known as the GULAG was perfected by Stalin beginning in 1929 (Applebaum 2003: 46–47). The camps provided an affordable system of labor during the shift from an agrarian to an industrial country in the Soviet Union. The GULAG system was also a means for Stalin to consolidate power, by denouncing his enemies using Article 58 of the Soviet constitution that made 'counter-revolutionary' activity a crime. The GULAG was ultimately denounced after Stalin's death in Khrushchev's 1956 'secret speech' to the Communist Party Congress (Applebaum 2003: 508), which resulted in the rehabilitation of many current and former prisoners, including those who had died in the camps.

The legends of three Siberian holy springs demonstrate how memory and place interact in the contemporary Russian context. Lozhok, in the Iskitim *oblast'* is located approximately 50 kilometres from the largest city in this region, Novosibirsk, and was the site of a GULAG camp (a quarry) from 1929 to 1956 (Zatolokin 2007: 17, 25; Applebaum 2003: 247). There are four common legends told about the site: 1) 40 priests and monks had been executed by prison guards there; 2) 12 priests were buried alive on the spot; 3) a group of priests interned in the camp asked the guards for water; the guards instead gave them salted fish, whereupon the priests prayed for water, and the spring bubbled up in answer to their prayers; 4) two priests sought haven at the spring (either during the Civil War after the Bolshevik Revolution or while imprisoned at the camp). These legends have prompted ostensive action on the part of the church and the local populace. People, even those who are not ardent believers, visit the site to bring the water home, since it is said to prevent or cure disease. They also bathe in the water; while bathing occurs year-round, this practice is especially popular on 19 January, the holy day of Christ's baptism by John the Baptist. Since 2007, on the Day of All-Russian Saints, believers have participated in a pilgrimage from Berdsk (the pilgrimage shifted to departure from Iskitim in 2014) to the site. In addition, Father Igor Zatolokin and his congregation have built a memorial park on the site (discussed in more detail later).

Therefore arguably the most shameful event of the Soviet Union's past has produced a sacred site in this region. Based on my fieldwork in these areas over the last ten years, I have uncovered a set of interconnected factors that have influenced this conception among the local populace: 1) nostalgia for the Soviet Union and Stalin's rehabilitation, 2) resurgence of the Russian Orthodox church and condemnation of the Soviet past, and 3) regional identity issues.

The Siberian region is known for being conservative politically. It has consistently supported the Communist party, voting against Putin and Medvedev in the last three national elections. While Soviet ideology is perceived favorably in many ways, it has also been perceived as a threat to native traditions, particularly religious practices. When faced with the historical record (and local knowledge) about the GULAG in their region, people are torn between their desire to restore the stability represented by the Soviet Union and the fact that it was bought at such a horrific price.

Novosibirsk and its satellites were quintessentially Soviet, founded as academic and industrial centres designed to foster the national technological infrastructure, home to premiere research institutions in the sciences and space industry. Those aspects of the Soviet legacy are still honored. The city has not been 'desovietised' through renaming, and a statue of Lenin still stands in the city's main square. It is located just across from a chapel built to honor the 300th year anniversary of the House of Romanovs that was destroyed in 1930 and replaced by a statue of Stalin. Stalin's statue was removed during Khrushchev's reign, and the chapel (dedicated to Saint Nicholas) was rebuilt in 1991. On the Second World War Victory Day 2016, the Communist Party, who, as of 2014, controls the city government, hung billboards across the city featuring an image of Stalin. The Novosibirsk Regional Governor Vladimir Gorodetsky, of the United Russia party, expressed dismay with this decision at a press conference on 11 May 2016 'I have always been convinced that Victory Day is an extremely bright holiday, which plays a unifying, consolidating role. I am convinced that celebrations of Victory Day with the portrait of Stalin do not in fact foster this unity' (Novosib Room 2016). The city and region are characterised by both pro- and anti-Soviet sentiments, often expressed by the same person. The Lozhok holy spring represents a site of negotiation of significant strands in local history about the Soviet experience and belief and is a locus for the reframing of memory about the GULAG system in this region.

Negotiation of memory at the Lozhok holy spring

The construction of the Cathedral of the Russian New Martyrs and Confessors near the holy spring in Lozhok began in 2006. Also located there are a kiosk (selling bottles, religious items and souvenirs), a 'chapel' with the icon of the Madonna of the Ever-full Chalice, a baptistery, a memorial stone, an open-air bathing area, a gender-segregated enclosed bathing area, and a wedding chapel dedicated to Peter and Fevronia. The congregation intends to engrave the memorial stone with the names of the 40 religious executed at the spring. Research in local archives for documents relating to this mass execution have produced no concrete information. The archive held by the State Administration of Internal Affairs has been inaccessible to the local priest Igor Zatolokin, who spearheaded the spring complex from its founding in 2002 until 2014, when he was assigned to another parish in the region.

The Cathedral of Russian New Martyrs and Confessors was completed in 2015. On 4 October 2015, it was consecrated by Metropolitan Pavel of Minsk and Zaslavsky, the Patriarchal Exarch of Belarus. Also participating in the service were Metropolitan Tikhon of Novosibirsk and Berdsk, three local bishops, and an American of Russian descent, Nick Buick, Cathedral Warden of the Holy Virgin Joy of All Sorrows Cathedral (San Francisco) of the Russian Orthodox church Abroad. In addition, the Communist mayor Anatoly Lokot' of Novosibirsk and the Lieutenant Governor Viktor Shevchenko (a representative of Putin's United Russia party) gave remarks during the consecration. The service illustrated the ongoing negotiation of the conflicted and conflicting conceptions of local and national Russian identity and memory of the past that the spring represents.

In 1996 the church canonised Nicholas II and his family, naming them (and others) to be Russian New Martyrs and Confessors in the church role of saints. The Order of Russian New Martyrs and Confessors grapples with the socialist legacy and the independence of its flock from the institution of the church for seventy years. The church has generally taken the position that anyone who died at the hands of the Soviet authorities, regardless of his or her personal beliefs, were exemplars of these 'new martyrs.' In fact, in his history of the GULAG and the spring, Father Igor Zatolokin (2007: 22) makes specific reference to this issue:

> Despite the unlikelihood of mass shootings of priests, for many years believers prayed at the spring for the repose of the dead and those who suffered during the years of the persecution of Orthodox Christians, saying the prayer of the New Russian Martyrs. And on the basis of prayers to the

New Martyrs, the waters of the spring, which existed even before the many horrible events that transpired here, became blessed, people who came here began to be healed of various illnesses. In the neighborhood of Lozhok the Holy Spring flows – like an non-corporeal memorial to all those who suffered without guilt. This water is blessed by the torments of the people who suffered in this terrible punishment camp.

Metropolitan Paul expressed similar sentiments about the GULAG victims as martyrs in his sermon at the consecration:

> it is rare that we can remember the persecutors . . . of Christ's church. Their names are rarely preserved, and if they are . . . then they sound a note of warning for all of us. Fear those names in history: they are bloodthirsty individuals, who did not spare the upper echelon of our civilisation. After all, in the twentieth century they hunted down and destroyed the advanced and wise people, the scholars, and they, as a rule, were all believers.
>
> *(Orthedu 2015)*

The ambiguous phrase 'they were, as a rule, all believers' illustrates the dilemma the church faces in discussing GULAG victims. Many were committed atheists or believers in faiths other than Orthodoxy. The congregation is well aware of these facts, but the Metropolitan emphasises the association between Orthodoxy and the persecution of the prisoners to conform to the church's understanding of the camps.

Similarly, Orthodox authorities are faced with the dilemma of Stalin, whose name was never mentioned by any official, religious or secular, at this event. His legacy as both victor in the Second World War and as author of the repression complicates discourse about GULAG history, in particular in the face of his ongoing rehabilitation and his 're-Christianisation' through legends relating how Saint Matrona of Moscow blessed him during the Second World War (Rouhier-Willoughby & Filosofova 2015; Antropologicheskii forum-Online 2010). To accuse him by name risks alienating the laity who do honor him. If church authorities hope to keep the flock's trust, they too may (genuinely or not) advocate for a more tolerant view of the Soviet period, or at least not be too overt in condemnation of Stalin himself. Paul only refers to the 'bloodthirsty individuals who did not spare the upper echelon of our civilization' and killed 'most progressive and wise people, the scholars.' Yet he did not go so far as to utter their names at an event designed to commemorate their victims.

He likewise neglected to mention that many of the progressive people who were victimised by these nameless tyrants were also Bolsheviks or activists against the church. He held up only select few of the guilty and presented the victim's belief systems to fit the prevailing view of holy site. One reason for this emphasis may be that he is targeting an audience derived from the intellectual centre of Western Siberia. But another might be that he is attempting to fight against the ongoing rehabilitation of the Soviet era by highlighting how it destroyed the best and brightest and brought about a 'dark ages' in (Soviet) Russian cultural heritage and in society.

The laity have a similarly restricted view of the past. Every variant of the legend of the holy spring at Lozhok focuses on the religious dead alone, not on the laypersons who died there. While it is clear that untold numbers of people perished in the camp, in the local imagination, the site is sacred because of the religious victims. This persistent focus on the religious dead alone is the key to how the memory of the GULAG in this region is being reframed in the post-socialist era. Because narratives about the spring foreground only the religious dead, narrators and the audience focus their attention on the single most negative trait of the USSR they agree on: that religious belief was discouraged. Stories about religious martyrs serve to redeem the violent past, and the spring may be seen as a sign from God that they have been forgiven for any complicity they may have had in the camp system. In addition, this view of the past is tied to a conception of Russians as a people that had never abandoned faith in Orthodoxy, despite its acceptance of core Soviet values.

The political guests at the consecration were also circumspect in their comments. The communist mayor, Lokot', did not mention the camp history at all, but rather offered remarks reminiscent of a tourism brochure: 'Here, to this holy spring, for years, decades, not only residents of Novosibirsk and the region have visited, but also from many corners of Russia' (Orthedu 2015). His interpretation of the spring as a local tourism site, rather than as a sacred spot deflects attention from the originator of the camp system as well and plays into local conceptions that the spring makes this region special to the outside world. He made no mention of the church's role in commemoration of Stalin's victims either. Recall that this mayor arranged for billboards featuring Stalin's face to celebrate Victory Day, so that it is not surprising that he chose to deflect all attention away from the darker aspects of his legacy. The United Russia representative, Lieutenant Governor Shevchenko, spoke more overtly about the camp. He uses the official church language for GULAG victims (*bezvinno ubiennym* 'innocent victims'), indicating his support of their mission and an awareness of official doctrine about camp victims as martyrs (Orthedu 2015). However, he did not mention the instigator of these atrocities in his remarks either. Like the religious authorities, he is also aware that it is politically risky be too overt in one's criticism of the socialist past, in particular of Stalin, in the current climate.

The legends about the spring allow this community to negotiate the complex intersection between memory, belief, and identity. The legends promote ostensive action: visits to the site, consecration and construction of a memorial complex. The spring is the locus of local pride of place that reflects Soviet ideology and faith, seemingly radically conflicting attitudes, and yet attitudes that both official institutions and the general populace convey. The Lozhok spring is a locus for community members of all stripes to reconcile these conflicting narratives about the past to remake memory through the lens of the present.

The former socialist spaces across the globe are ripe for folkloric investigation of the intersection between ostensive action prompted by legends about a place, particularly a place of violence, and the memory of the past. The socio-political shifts brought by the collapse of socialist nations, the return of the religious institutions, and the difficulty of dealing with a violent history have already resulted in similar studies from anthropologists, geographers, and historians, but much work remains to be done by those in the field of folklore. Folklorists in Europe and the United States have not often focused on the post-socialist transition, but disciplinary interest in belief and legend is growing significantly. The result will likely be more such studies on vernacular religion, including questions of pilgrimage and the legends of holy spaces, in many cultures. They are a much-needed contribution to the literature that folklorists are well placed to provide with their understanding of the interplay among text, practice, and belief.

References

Antropologicheskii forum-Online. (2010). J. Kormina, Politicheskie personazhi v sovremmennoi agiografii: Kak Matrona Stalina blagoslovila. 12. Accessed 25 February 2018, http://anthropologie.kunstkamera.ru/files/pdf/012online/12_online_kormina.pdf
Applebaum, A. (2003). *GULAG: A history*. New York: Anchor Books.
Bennett, G. (2005). *Bodies: Sex, violence, disease, and death in contemporary legend*. Jackson, MS: University of Mississippi Press.
Dégh, L. (2001). *Legend and belief*. Bloomington: Indiana University Press.
Dégh, L., and Vásonyi, A. (1983). Does the word 'dog' bite? Ostensive action: Means of legend telling. *Journal of Folklore Research*, 20 (1): 5–34.
Ellis, B. (1983). Legend-tripping in Ohio. A behavior survey. *Papers in Comparative Studies*, 2: 61–73.
Ellis, B. (1996). Legend trips and Satanism: Adolescents' ostensive traditions as 'cult' activity. *In*: G. Bennett and P. Smith, eds. *Contemporary legend: A reader*. New York: Garland, pp. 167–186.
Ellis, B. (2003). *Aliens, ghosts, and cults: Legends we live*. Jackson, MS: University Press of Mississippi.

Ellis, B. (2015). The haunted Asian landscapes of Lafcadio Hearn: Old Japan. *In:* T.J. Banks Thomas, ed. *Putting the supernatural in its place: Folklore, the hypermodern, and the ethereal.* Salt Lake City: University of Utah Press, pp. 192–220.

Lindahl, C. (2005). Ostensive healing: Pilgrimage to the San Antonio ghost tracks. *The Journal of American Folklore*, 118 (468): 164–185.

Novosib Room. (2016). Gorodetskii osudil plakaty so Stalinym. [Gorodetsky condemned the posters with Stalin], 11 May. Accessed 25 February 2018, http://novosib-room.ru/gorodetskij-osudil-plakaty-so-stalinym-57881

Orthedu. (2015). Na sviatom ischochnike v Lozhke osviashchen kram v chest' Novomuchenikov i ispovednikov tserkvi Rossii. [Cathedral in honour of the New Martyrs and Confessors of Russian consecrated at the holy spring in Lozhok], 4 October. Accessed 25 February 2018, www.orthedu.ru/eparh/13879-na-svyatom-istochnike-v-lozhke-osvyaschen-hram-v-chest-novomuchenikov-i-ispovednikov-cerkvi-russkoy.html

Primiano, L.N. (1995).Vernacular religion and the search for meaning in religious folklife. *Western Folklore*, 54 (1): 37–56.

Primiano, L.N. (2012). Afterword: Manifestations of the religious vernacular: Ambiguity, power, and creativity. *In:* M. Bowman and U.Valk, eds. *Vernacular religion in everyday life: Expressions of belief.* Sheffield: Equinox, pp. 382–394.

Rouhier-Willoughby, J., and Filosofova, T.V. (2015). Back to the future: Popular belief in Russia today. *In:* S. Brunn, ed. *The changing world religion map*, Vol. III, Dordrecht: Springer, pp. 1531–1554.

Tucker, E. (2007). *Haunted halls: Ghostlore of American college campuses.* Jackson, MS: University Press of Mississippi.

Tucker, E. (2015). Messages from the dead: Lily Dale, New York. *In:* T.J. Banks, ed. *Putting the supernatural in its place: Folklore, the hypermodern, and the ethereal.* Salt Lake City: University of Utah Press, pp. 170–191.

Zatolokin, I. (2007). *Lozhok: Iz istorii Iskitimskogo katorzhnogo lageria.* Iskitim: Novisibirsk Diocese.

INDEX

Note: page numbers in *italic* indicate a figure and page numbers in **bold** indicate a table on the corresponding page. Page numbers followed by 'n' refer to notes.

9/11 239, 244

A435 Alcester-Evesham Bypass *256–257*
Abaeian, Nasim 59, *60*
abandonment 24
Abercrombie, Patrick 113
absences 215–216, 218
adoption 143
adventure playgrounds 111–112
affect 238–239, 241–242, 251
affective atmospheres 244
Africa 10; and storytelling 148–149; and United Kingdom 42–51
Aghyaran 298–302
Aldridge, Jeff 209
Aldridge, Terry 188–190, 191
Aldwych station 223, 228–229, *230*
Algodres 308, 311–312, 313
alienation 185, 192n1
alterity 280
Amicale Austerlitz-Lévitan-Bassano 71–72
And While London Burns 166
Antigonas Tribunal de Mujeres 277
Anzac Day 1, 250
Apache 139
apartheid 31, 33
aplianisis 258
applied theatre 278
apps 164
archaeology 150–151, 155, 215, 220, 254–255, 258–259; *see also* industrial archaeology
Argentina 68
Army Corps of Engineers 218–219
art 114; *see also* murals

Asfaw, Menen 47
ashes 344, 367, 370, 380, 381n1; *see also* cremation memorials
Asia Minor 13–21
Asia Minor catastrophe 15–16
Assmann, Jan 384
Associação Transumância e Natureza (ATN) 306–307, 311–312, 313
Asta Trail: Planes, Trains and Graffiti Walls 160, *161*
ateliers nomades 263
Athabasca 139
Atlanta 121
Atlascine 54, *54*
atmospheres 242–244, 245
ATN *see* Associação Transumância e Natureza (ATN)
attachment 188, 189–190, 191
Augustus 387–388
Auschwitz 71, 74
Austerlitz 67, 71
Austin, Mary 350
Australia 1
Australian Remembrance Trail 103
authenticity: and experiences 18; and historical pageants 340–341
autoethnography 42

BAE Systems 196–197
Bakarian, Meghri 61–62, *62–64*
Baker, Khadija 62, *62–64*
Bandholtz, H.H. *133*
Barthes, Roland 258, 271
Bath 45, 46
Batthyány, Lajos *133*
Battle of Bogside 293–295

Battleship Row 240, 243
Beckenham Arts Lab 348
Beckford, William 45
Beckford's Tower 45, 46
Belene 67, *82–83*, *85*, *86*, 87n2
Belene Island 77–87, *79*
Benjamin, Walter 163, 223, 258
Bergen-Belsen 71, 74
Bergen County 124–125
Berlin 346, 354
Berlin, Mike 161
Beyul 357–365
Bigwood, Jonathan 163
Bitar, Lilia 59–61, *60*
Blitz, the 111, 113
Bloody Sunday 293
'blue guides' 271
Blue Plaques 354
Boas, Franz 140
bodies 5
Boga, Cindy 167
Bogdanovdol 83
Bolan, Marc 349, *350*, 353, 355
bombs 110
bombsites 109–114, *112*
Bowfin Submarine Museum and Park 237, 241
Bowie, David 345, 346–349, *347*, 354, 355
Bowl around Bedlam with the Brainless, Brazen Brothers, A 166
breaching experiments 185
breweries 188–189
bright objects 237, 239, 244, 246
Bristol 45, 46
Brittany 272
Brixton Munch 164, *167*
Bruce, John 208
Buchan, Norman 196
Budapest *133*
Buddhism 343, 358–359
buildings 172
Bulgaria 67, 77–87
bunkers 225
Burgos 103–104
buried memories 225–226
Bus: a linear history project 163
buses 169n5
bus routes 163
Butler, Judith 233

C, Jimmy 347
Cade, Rosana 262
Camden Town 351
Campus Martius 387
Canada 10, 178–179; and Ktunaxa First Nation 145; and Stó:lō 139
Cape Breton 180
Cape Town 10, 31–40, *32*
Capilano, Joe 143–144

capitalism 187
Capitoline hill 387
Capulet, Juliet 349
caring 16
Carrickanaltar *300*
catharsis 20
Catholics 298–301, *300*
cattle 385
celebrity shrines 5
cemeteries 92–93
Centre for Oral History and Digital Storytelling (COHDS) 53
Chambers, Ajit 229
Channel Heritage Way *165*
Channel Heritage Way 164
Charlie, Jimmie 142
Charlie, Kelsey 142
Chehalis, George 140
chess sets 13, *15*
Chi, Tshei *50*, 51n3
childhood 39
children 111, *112*
Chipperfield, David 109
Chisikana Spring 153
chorography 292
Christ Church Greyfriars 113
Christianity 389, 398
Chthonic Reverb 262
churchyards 367–368, 380; and cremation memorials 369–378; extensions of 378–380
Cidadelhe 308–310, 313
cities 224
C.K. 57, *58*, 59
class politics 199–200
Clemens, Samuel Langhorne 268
Clyde Dock Preservation Initiative 200
Côa Valley 305–307, *306*, 308
Côa Valley Grand Route 308
Cochran, James *see* C, Jimmy
coin trees 396–397, *397*
Cold War 134
Cole, Thomas 221
collective memories 25, 69–70
collective milieus 72–73
Colombia 277–284
colonialism 5, 9; and literacy 141–142; and maps 269, 272; and memoryscapes 139; and Zimbabwe 151
'Comfort Women' 117, 119–127, *120*, *125*
Commonwealth War Grave Commission (CWGC) 104, 106
communicative memories 384
communism 81
communities 9
community projects 158–161
concentration camps 71–74; *see also* forced-labour camps
conflict 277–284; *see also* wars
Constantine the Great 389

Constantinople 389
contact zones 9
Convolutes method 163
cooking 316–318
Cortezi, Paolo 84
counter-mapping 271, 273
Crafted with Pride 181
cranes 172, 193, *194*, 196–199
cremation memorials 367–368, 370–381, *371–373, 374, 375–377, 377, 378*
cremations 344, 367–370
crosses *301*
Croydon Road Recreation Ground 348
cultural landscapes 148–149, 155, 306–307
cultural memories 384
cultural traumas 16
CWGC *see* Commonwealth War Grave Commission (CWGC)
Cyprus 105

Dagam Namgo 359, *360*, 361, *362*
Daily Record, The 50, 51n3
dairying 329
Dakinis 361, 365n1
Daleman's Journey, A 163
Daly, Edward 295
dams 216–219
dance 62
Danube 77, 79
Dawson, Henry 189–190, 191
Deakin, Roger 262
death 91
deathscapes 367
'Declaration of Independence in 1917' 337
Delaware River 172, 214–221
Delaware Water Gap 215
dérive 261
Derry 288, 293–298, *294, 296–297, 298, 302,* 303
Detroit 177, 181–182
Detroit 1967 Oral and Written History Project 177–178
Devlin, Bernadette 293, *294*
Diana, Princess of Wales 345
diffraction 4
Dimitrovas, Stefanie 56, *57*
displacement 2, 5
dissonant heritage 27
District Six 31–40, *34*
dockyards 190–191
"Dolphin, The" 15
Dominion 139
double-consciousness 44
Downriver, or, The Vessels of Wrath: a narrative in twelve tales 262
Drancy 74
Dresden 114
drifts 261
Drumawark cross *301*, 302

Duddy, Jackie 295
Dunn, Billy 198
Durban 33

Eadie, Bruce 166
Easter Rising 298
edible identities 326
education 158–161
Edward VII, King of England 143
Emma Bridgewater Factory 264
emotions 239
encounters 214, 238
Epano Skala 13
escapes 214
ESRI Story Maps 56–57, *58*
ethics 16
Ethiopia 47
ethnic cleansing 67; *see also* Holocaust
Eveleigh Railway Workshops 172, 203–211
exhumation 100–101
exiles 65n1; and maps 53–56, 64–65; and story maps 56–59
experiences 2–3
experiential maps 274n3

factories 172
Faia Brava 306, 308, 309, 310, 313
Fairfield House 43, 47
Fairfield shipyard 193, *194*
fairy forts 145
families 10
Farmer, Stephen 197–198
farming 326, 328–330
farm-to-table movement 329–330
feelings 241, 241
Feld, Marc *see* Bolan, Marc
femininity 203
festivals 385
Finland 337, *338*
Finlayson factory *340*
First World War: and historical pageants 335, 337; and Istria 22; and memorials 130, 355
Fisher, Archie 196
fishing 24
food cultures 288, 316–323
foodsheds 325–330
footprints 361, 362
forced-labour camps 79–80, 131; *see also* concentration camps
Ford, Laura Oldfield 262
forensic archaeology 99–100
forensis 100
forgetting 91
Forum 387
France 67, 272
Frauenkirche 114
Free Territory of Trieste (FTT) 22, 23

friction 326
Fromelles 104–105
FTT *see* Free Territory of Trieste (FTT)
funerals 386–387; *see also* mortuary practices

Gallen, Danny 299, *300*
Garrano horses 309
gender 203–211, 317
gender displays 208
gentrification 178, 179
geographic information system (GIS) *see* GIS
geopoetics 263–264
George, Rosaleen 141
Germany 1
ghosts 187
Gilbert, Dan 178
GIS 52
Gisimba 55
glaciers 139
Glasgow 172, 199–200
Glasgow Miles Better 199–200
Glendale 126
Glen Livet 151
Govan 193–199, 200–201
Govan Graving Docks 200
graffiti 347, 354
graves 80, 90; *see also* cremation memorials; mass graves
Great Britain *see* United Kingdom
Great Enclosure *150*
Great Zimbabwe 118, 148–156, *150*, *151*
Greece 10, 13, 316, 390
Greek tragedy 315–316
greening 181
Griffintown 179
Guinness 188
Gulags 5, 403, 405
Guru Rinpoche 357, 361, *362*

Habimana, Emmanuel 55
Hackney Service Center 166
Halbwachs, Maurice 69–75, 129, 223
Hall, Stuart 43, 44, 45
Hamilton, Mina 219
Hands-Off campaign 33
hanging trees 144
Harry, Prince, Duke of Sussex *50*
hauntings 39, 263, 264
Hawaii 237
Heddon, Deirdre 263
Heddon Street 348, 349
Helsinki 334, 336–337, *338*, 338–339
herbs 321
heritage 237, 245, 247, 305; and re-enactments 334, 339; and rewilding 313
heritagisation 27
heterotopias 38
Hidden London 229

historical pageants 334–341
Holocaust 1, 67; and Hungary 132; and Paris 71–74
home 38, 43–44, 327
hooks, bell 43
Horozov, Krum 85–87
Horthy, Miklós *133*, 134
Hubbard, Michelle 'Mother' 46
Hungarian Uprising *131*
Hungary 118, 130–136

imagination 2, 6, 9
India 1
Indigenous people 138–139; *see also* Athabasca; Ktunaxa First Nation; Stó:lō; Tlingit
inductive visualisation 53, 65n2
industrial archaeology 194–195
infrastructure 224–225
intangible heritage 341
International Union for Conservation of Nature (IUCN) 307
interpretative mapping 270
intuition 365
Inuit Land Use and Occupancy Project (ILUOP) 64
invented traditions 326
Inxilio: el Sendero de Lágrimas 277, 278, 282–284
IRA *see* Irish Republican Army (IRA)
Irish Republican Army (IRA) 295, 298, *298*, 299
Isle Maree 393–398, *397*
Istria 22–28
Italy 1, 22–28
IUCN *see* International Union for Conservation of Nature (IUCN)
Izmir 18

Japan 119–121, 123–124
Jenness, Diamond 142
Jews 71–74, 132, 135; *see also* Holocaust
Johannesburg 33
Johnson, Keith 209
Jones, David Robert *see* Bowie, David
Julius Caesar 388
Jupiter Optimus Maximus 386

Kaiser Wilhelm Memorial Church 113
Kanom, Halima 166
Kashmir 1
Kayiganwa, Emmanuelle 55
Keele Chapel 265
keimelia 16–19
knowledge 254–255
Korea 119–121, *120*, 123–124
Ktunaxa First Nation 145

La Candelaria 90, 91
Lake Kivu *54*
landscapes 259, 274, 308; of death 90–91, 95; and narratives 292; and palimpsests 393; and rituals 392

Index

Langtang Valley 358
language 26–27
La Salida Más Conveniente 277
Lee, John 209
legends 344, 400–402
legend trips 401
Lemmington Terrace *32*, 34–35, *36*
Lenni Lenape 217–218, *219*
Lesvos 10, 13
lieux de mémoire 223, 234, 251
Linked 263
lions 155
listening 234, 279–283
literacy 141–142
Loch Maree 393–394
London 68, 110, *112*, 113, 118, 173, 346
London Consolidation Crew (LCC) 229
London Memorandum 22
looting 74
Los Angeles Riots 124–125
loss 38
Lough Derg *301*
Lough House 299
Loutra 15
Love Canal 181
Lozhok 403–406, *404*
Lupercalia 385
Lutz, Carl *133*
lynching 144
Lyons, Antony 305–306, 310

Macdonald, Anna 266
MacDonald, James 199
Macfarlane, Robert 262
MacHalsie, Sonny 143
Maelrubha, St 394
Maison d'Haïti 54
Making Community History Trails 158–159
Malaysia 145, 271
Manchester Civic Week 340
mandalas 363
Mandela, Nelson 39
Manifest Destiny 139
mapping 52–53, 234
Mapping the Life Stories of Exiles 53
maps 3, 10, 268–273; and exiles 53–56, 64–65; and prisoners 85; and refugees 20
Markle, Meghan *50*
Markov, Georgi 77–78
Mars 388
Marsh Farm 256
Marx, Leo 221
Marxism 185, 192n1
masculinity 203, 205, 206–209, 211
masks 387
mass graves 99–100; at Belene 80; and Spanish Civil War 52, 100–106

Matthews, Bob 208
Mauch, Karl 152
Mayfest 200
Meakin, Charles 264
Mees, Sarah 164
Meier, Silvio 227–228, *228*
Melvin, Jean 199
memorials *131*, *133*, 225; and bombsites 113; in Budapest *131*; and Meier, Silvio *228*; and Pearl Harbor Memorial Complex *238*, *245*; and public memory 129–136; and tribute sites 346; and USS *Arizona* Memorial *240*
'Memories of Mining' 265–266
memory constructs 25
memory gaps 25
memory holes 74
memory work 279
mental maps 274n3
Mercury, Freddie 345, 349–351, *351*, 353, 355
Michael, George 352, *353*, 355
migrations 234; and Istria 22–24; and multiculturalism 122–123
mills 176
mirroring 37
Mitchell, Bill 163
Mitchison, Laura 164–165
Mladic, Ratko 101
mnemonic communities 187, 224
mnemonic rituals 343
mnemotopes 384
mnemonic mapping 268–273
mobile workshops 263
mobility 9
modernity 336
Molano, Ramírez 277
Montana 181
Montreal Life Stories Project 53
moral integration 123
morality 124
mortuary practices 90–94; *see also* funerals
mothers 35–37, 175
mourning 94
Mt. Haggin 181
Muir, Olive *50*, 51n3
Mujejeje 156
Munitions Annexe 207–208
murals 20, 293–295, *294*, *296–297*
Museum of La Plata 95
Museum of Refugee Memory, The 14–21
Muslims 19
Myers, Misha 263
Mytilene 13

Nagy, Imre *133*, 134
nail-trees 395–396
narratives 6, 148–149, 155–156, 205, 252; of decline and abandonment 328; and families 10; national 250; and

resistance 9; and restorative justice 277; and shared knowledge 287; *see also* legends; storytelling
nationalism 336–337
networks 226
Neues Museum 109
New England 288, 326–329
New England Societies 327
New Jersey *125*, 214
'new ruins' 264
New Scotland thesis 180
newspapers 316
Newstead Abbey 42, 44–45
New York 181, 346
New Zealand 1
Nolli, Giambattista 270
Northern Ireland 292–298
nostalgia 5, 26, 171; and attachment 191; and deindustrialisation 1193; and the future 200; and loss of home 38; and work 186–187
nostophobia 180
Nova Scotia 181
Novosibirsk 404

oaks 394
objects 2, 16–19; *see also* sticky objects
Occupy LSX 168, *168*
Occupy LSX Audio Trail *168*
Old London Underground Company (OLUC) 229
olive oil 317
O'Mongan, Cornelius 299
oral histories 4
Order of Russian New Martyrs and Confessors 404
O'Rourke, Donny 199
Ottomans 13

Pacific Aviation Museum 237
Page Rwanda 54
Paisley Thread Mills Memoryscape 180
Pakistan 1
Pákozd *131*, 132
palimpsests 392–393
Parcel 301 118, *131*, 131–132
Parikramas 362, 365n2
Parilia 385
Paris 71–74
Paris Peace Treaty 22
paths 261, 376
Pearl Harbor 234, 237–238, 239–247
Pearl Harbor Memorial Complex *238*, *245*
Pearl Harbor Virtual Reality Center 248n1
performances 278, 318, 322
performativity 233
Peters, Bertha 141–142
Pfuko yaKuvanji 153
Philoctète, Alexandra 53, 55, 56–57, *57*
Phokaia 15

photographs 256–259
photography 255
photology 255
Pierre, Peter 142
pilgrimages 51, 343, 346, 357–365, 394–395, 402
Piran 24
place attachment 5
place-making 291
place-worlds 291
Point, Steven 145
Poland 1
Pollan, Michael 325
Pollitt, Stuart 190–191
Ports of Call 159, 160–161, *162*
Portugal 288, 305, *306*
post-conflict scenarios 278
postmemory 10
Presley, Elvis 345–346
prisons 77–87
Pritchard, Bill 198–199
prosthetic memories 244
Protestants 298–301
public memory 129–136
public transportation 173; *see also* railroads; underground railways

Quigley, Colin 198
quilting 122, 125

racism: and Detroit 178; and South Africa 31–40; and United States 43
Rafaela Ishton Community 94–95
rag-trees 395–396
railroads 216; *see also* underground railways
rainbow bridges *364*
Rákoskeresztúr Cemetery 118, *131*, 131–132
Rakowitz, Michael 111
Rastafari 43
Reagan, Ronald *133*, 134
recipes 320–321
Recsk prison camp *131*, 134
recycling 344, 392–393
re-enactments 288, 334–341
refugees 9–10, *14*, 14–21
remediation 181–182
Remus 385, 390
restorative museums 14
Restrepo, Alvaro 277, 278, 282
reterritorialisation 122, 124, 126
rewilding 305–312
Rewilding Europe 307
ritual landscapes 343
rituals 392
River Clyde 196
Rix Centre 164
RIX Wiki 164
Roma 136

Rome 384–390
Romulus 385, 389–390
Ross, Joyce 177–178
rubble 68, 111
Rubble Films 114
Rudkino cemetery 135
ruination 2, 171, 176
ruins 264
Rwanda *54*

sacred places 5
Saint Andrew's, Tarvin *378–379*, 380
Saint Bartholomew, Great Barrow 372
Saint Boniface, Bunbury 376, *377*
Saint Chad's, Farndon *375*, 376
Saint Deiniol 376
Saint Dunawd's, Bangor on Dee 374, *375*
Saint Helen's, Tarporley 378
Saint James' Christleton *374*
Saint Marcella's, Marchwiel 376, *376*
Saint Mary's, Chirk 372, *372–373*
Saint Mary's, Eccleston *377*
Saint Paul's Cathedral 110
Saint Peter's, Delamere 371, *371*
Saint Peter's, Waverton 378, *379*
Sam, Louie 144
Samariterstrasse station 223, 227–228, *227*
Sands, Bobby 295, *298*
Sane, Aladdin *347*
San Francisco 121
Santos, Juan Manuel 278, 282–283
Sarawak 271
Satizabal, Carlos 277
Save the Delaware Coalition 219
Schaffer, Lucille 177
Scotland 180, 392, 393–394
Seams 265
Second World War 67, 100; and 'Comfort Women' 117, 119–120; and Hungary 135; and industry 203–204; and Istria 22; and mass graves 105; *see also* Holocaust; Pearl Harbor
Selassie, Haile 43, 47, 49, *50*, 51n3
Self, Will 262
Selk'nams 68, 89–96
senses 250–252; *see also* tastes
Seoul *120*
Serbia 105, 135
shared memories 16
Shaw, Emory 57, *58*
Shays' Rebellion 327
'Shipyard Apprentice, The' 196
'Shoes on the Danube Bank' 132
Shona 149
shrines *see* tribute sites
Shukaitis, Nancy 218
Siberia 403
Sinclair, Iain 262

sites 150
Skala Loutron 13
Skochev, Borislav 81
slavery 10, 42, 45, 70
Slave Trade Legacies 46, 51n2
Slovenia 10
Smith, Sam 167
snake rocks *364*
social justice 2
social morphology 70
Solnit, Rebecca 262
songs 310
Sorrows of Young Werther, The 349
Sotelo, Luis C. 277
souls 16, 18–19
South Africa 10
souvenirs 9
Soviet Union 403–406
Spanish Civil War 52, 68
Spinoza, Baruch 238–239
Spode Works 264
springs 402, 403–406
Srebrenica massacre 102
Sri Lanka 278
Stalin, Joseph 403, 404, 405
Stardust, Ziggy 346–347; *see also* Bowie, David
'Statue of Peace' *120*, 121
sticky objects 358
Stoke-on-Trent 264
Stó:lō 118, 138–146
stories *see* narratives
story maps 56–59
storytelling 292; *see also* legends; narratives
structured feeling 291
Sturgeon Falls 176
Sydney 172
Sydney Tar Ponds 181
symposio 319
Syria 19

tactics 9
Tafari, Judah 42–43
Tampere 334, 339, *340*
Tantra 360–361, *362*
Tanzania 272
'Tapping Ware' 264
taskscapes 194
tastes 318, 320, 322
temples 386, 387–388
Terma Tradition 359–361, 364
Tertons 357–358
testaments 18–19
testimonies 18, 280–281
Thomas, Jo 160
threats 361
'thrillers' 316
Tierra Del Fuego 68, 89

Tlingit 139
Tocks Island 218
toponymy 27
topos 15
tourism 329
tourist brochures 270–271
Tourli, Efthalia 14, 16, 17, 18, 19
trauma 315
travel 269
tree maps *60–61*, 61
trees 392–393, 395, *397*, 398n1
tributes 352, 353–354
tribute sites 343, 345–355
triumphal arches 386
truth-telling 280
Tunstall Road 348
Turkey 13, 19
Turner, Richard 163
Twain, Mark 268

U-Bahn 223, 226
Újvidék 'raid' 135
uncanny 31–32, 34
Underground 173, 223, 226
underground railways 223–229; *see also* railroads
UNESCO World Heritage Sites 307
United Kingdom: and Africa 42–51; and Battle of Bogside 293; and historical pageants 341; and South Africa 33
United States: and 'Comfort Women' 121; and racism 43
urban heritage 23–28
urban regeneration 178
Uribe Vélez, Alvaro 281
USS *Arizona* Memorial 237, 240, *240*, 246
USS Bowfin 241, 246
USS *Missouri* 241
USS *Oklahoma* 243

Vajra bodies 362–363, 364–365
vernacular religions 402–403
Vespasian 388
Victor, Aggie 142
Victoria, Queen of England 396
Victory Day 404
Vilbr, Rosa 166

village tourism 309
virtual reality 249
visualisation 3
Vivificar 277
voicing 234, 280–283

walking 234, 261–266
Walking: Holding 262
Walking Library, The 263
Wander East Through East 166
Ward, Brian 46
wars 5, 67, 237–247; *see also* conflict
warts 396
Warwickshire 255–256
Warwickshire Museum 255
Washoe Smelter 181
water 153–154; *see also* springs; wells
Way Home: a walk around Hackney's housing history, The 166
Wednesday Demonstrations 120, *120*
wells 394–395, 398
Wemyss Bay 49
Wildman, Thomas 44–45
Wilson, Allan *151*
Winehouse, Amy 351–352, *352*
Winter War 337
wishing-trees 397–398, 398n2
witnesses 18
women 203
work 197–199
working classes 2, 171–172, 175–182
workplaces 185–192, 203–211
World War I *see* First World War
World War II *see* Second World War
worth 141
wounds 264, 265

Xe:Xá:ls 140

Yangdak Chok 361
'Year 1939' 337
Yugoslavia 10, 22–28, 105

Zanzibar 272
Zatolokin, Igor 403, 404
Zimbabwe 149; and colonialism 151; *see also* Great Zimbabwe

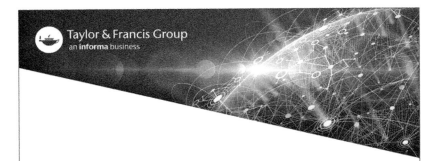